Asymmetric Synthesis in Organophosphorus Chemistry

Asymmetric Synthesis in Organophosphorus Chemistry

Synthetic Methods, Catalysis, and Applications

Oleg I. Kolodiazhnyi

WILEY-VCH
Verlag GmbH & Co. KGaA

Author

Professor Oleg I. Kolodiazhnyi
National Academy of Sciences
Physiologically active phosphorus comp.
1, Murmanskaya Street
02154 Kiev
Ukraine

Library of Congress Card No.: applied for

British Library Cataloguing-in-Publication Data
A catalogue record for this book is available from the British Library.

Bibliographic information published by the Deutsche Nationalbibliothek
The Deutsche Nationalbibliothek lists this publication in the Deutsche Nationalbibliografie; detailed bibliographic data are available on the Internet at http://dnb.d-nb.de.

Print ISBN: 978-3-527-34150-4
ePDF ISBN: 978-3-527-34151-1
ePub ISBN: 978-3-527-34153-5
Mobi ISBN: 978-3-527-34152-8
oBook ISBN: 978-3-527-34154-2

Typesetting SPi Global, Chennai, India

Printing and Binding Markono Print Media Pte Ltd, Singapore

Printed on acid-free paper

Dedicated to my best friend and darling woman, Galina.

Contents

Preface

Chiral phosphorus compounds play an important role in many areas of science, including biologically active pharmaceuticals, agrochemicals, and ligands for transition metal complexes. In the last few years, enormous success has been achieved in the asymmetric synthesis of organophosphorus compounds and many new developments, finding considerable use in industry, have taken place,. Asymmetric synthesis and asymmetric catalysis have been, and remain, one of the most important research directions in chemistry, attracting the interest of many scientists and chemical groups. As a consequence, the asymmetric synthesis of organophosphorus compounds is studied extensively in many scientific centers, including academic and industrial research laboratories. Many methods for the preparation of enantiomerically pure organophosphorus compounds including classical resolution via diastereoisomers, chemical kinetic resolution, enzymatic resolution, asymmetric metallocomplex catalysis, and organocatalysis have been developed. Complexes with transition metals containing PAMP, DIPAMP, DIOP, CHIRAPHOS ligands are widely used for the asymmetric formation of C–H and C–C bonds. Over the last few years, great success has been achieved in the asymmetric synthesis of organophosphorus compounds, primarily with phosphine ligands for catalyzed asymmetric hydrogenation reactions, and many articles devoted to the synthesis of chiral organophosphorus compounds have been published. In the last 10–15 years, many excellent reviews and multivolume monographs dedicated to the stereochemistry of organophosphorus compounds have been published. Several journals dedicated to asymmetric synthesis and chirality, *Tetrahedron: Asymmetry* and *Chirality* being among the foremost, have also gained in popularity.

The importance of stereochemistry in drug action and differences in the physiologic action of enantiomeric antipodes is well-known and intensively studied now. The requirements stipulated for new drugs by the Food and Drug Administration in the United States and by similar regulating agencies in other countries has made this more obvious. Some amino- and hydroxyphosphonic acids, as well as synthetic phosphonic acids, have been found to have effective medicinal properties; they have been applied in pharmacology and medicine. Detailed information about these functionalized phosphonates and phosphonic acids can be found in the chapters of this monograph.

This book emphasize the importance of chiral organophosphorus compounds and their asymmetric synthesis. There has been no monograph devoted to the asymmetric synthesis of organophosphorus in the chemical literature although such a study would be of great interest. This is what encouraged and inspired us to prepare this book. Our book is intended to be used by chemistry experimenters, professors of universities, as well as research students, as a source of basic knowledge and convenient reference. The literature covered is up to 2016.

Chapter 1 of this monograph is dedicated to the fundamentals of stereochemistry of organophosphorus compounds, including general theoretical concepts, common nomenclature, and analytical methods related to the stereochemistry of organophosphorus compounds. The other chapters of the book survey the various types of asymmetric reactions and chiral organophosphorus compounds.

Chapter 2 discusses methods for the synthesis of compounds with chiral phosphorus atoms, including compounds with dicoordinated phosphorus atom, three-coordinated trivalent compounds, compounds of tetracoordinated phosphorus, and also compounds of penta and hexacoordinated phosphorus, and so on.

Chapter 3 describes the methods for asymmetric synthesis of phosphorus compounds bearing chiral centers in side chains. These reactions are particularly important for the production of pharmaceutical products and intermediates.

Chapter 4 presents asymmetric catalysis with complexes of transition metals, that is, asymmetric catalytic hydrogenation and stoichiometric reduction of various unsaturated compounds. Asymmetric hydrogenation is the simplest way to create new chiral centers and the technology is a flagship for chiral synthesis. Because asymmetric synthesis is a highly application-oriented science, examples of industrial applications of the relevant technologies are appropriately illustrated throughout the text.

Chapter 5 is devoted to organocatalysis, which is especially intensively investigated. The most important principles of organocatalysis and examples of preparative and practical applications are discussed. In particular, the use of alkaloids of quinine and its derivatives, sparteine, proline, and amino acid and their derivatives as catalysts is described.

Chapter 6 considers the use of enzymes and others biological methods in asymmetric synthesis. Methods of kinetic resolution of racemic organophosphorus compounds, biocatalytic transesterification, dynamic kinetic resolution of α-hydroxyphosphonates, enzymatic resolution of aminophosphonates, and the biosynthesis of compounds with C–P bonds are discussed. Microbiological synthesis of chiral phosphorus compounds with yeast, bacteria, fungi are considered.

The book discusses methods for the asymmetric synthesis of chiral organophosphorus compounds with many applications in stereoselective synthesis and asymmetric catalysis with reference to updated literature findings as well as the author's original researches performed over the last 15–20 years.

Kolodiazhnyi O.

Abbreviations

Ac	acetyl group
AC	absolute configuration
AD mix-α	reagent for asymmetric dihydroxylation
AD mix-β	reagent for asymmetric dihydroxylation
ALB	Al-Li-bis(binaphthoxide)
Ar	Aryl
BCL	*Burkholderia cepacia* lipase
BINOL	2,20-dihydroxyl-1,10-binaphthyl
BINAP	2,20-bis(diphenylphosphino)-1,10-binaphthyl
Bn	benzyl group
BOC	*tert*-butoxycarbonyl group
Bz	benzoyl group
CALB	*Candida antarctica* lipase B
CBS	chiral oxazaborolidine compound developed by Corey, Bakshi, and Shibata
CCL	*Candida cyclindracea* lipase
CD	circular dichroism
CDA	chiral derivatizing agents
COD	1,5-cyclooctadiene
CIP	Cahn–Ingold–Prelog
Cp	cyclopentadienyl group
CPL	circularly polarized light
CRL	*Candida rugosa* lipase
CSR	chemical shift reagent
DBU	1,8-diazobicyclo[5.4.0]undec-7-ene
de	diastereomeric excess
DEAD	diethyl azodicarboxylate
DET	diethyl tartrate
DHQ	dihydroquinine
DHQD	dihydroquinidine
DIBAL-H	diisobutylaluminum hydride
DIPT	diisobutyl tartrate
DKR	dynamic kinetic resolution
DMAP	4-*N*,*N*-dimethylaminopyridine
DME	1,2-dimethoxyethane
DMF	*N*,*N*-dimethylformamide
DMSO	dimethyl sulfoxide

DMT	dimethyl tartrate
l-DOPA	3-(3,4-dihydroxyphenyl)-L-alanin
DYKAT	dynamic kinetic asymmetric transformation
ee	enantiomeric excess
GC	gas chromatography
HMPA	hexamethylphosphoramide
HOMO	highest occupied molecular orbital
HPLC	high performance liquid chromatography
Ipc	isocamphenyl
IR	infrared spectroscopy
L*	chiral ligand
LDA	lithium diisopropylamide
LDBB	lithium 4,4′-di-*tert*-butyldiphenylide
LLB	Ln-Li-is(binaphthoxide)
LHMDS	$LiN(SiMe_3)_2$
MEM	methoxyethoxymethyl group
Mnt	menthyl
MOM	methoxymethyl group
MPA	methoxyphenylacetic acid
Ms	methanesulfonyl, mesyl group
MTPA	α-methoxyltrifluoromethyl-phenylacetic acid
NAD(P)H	nicotinamide adenine dinucleotide (phosphate)
NHMDS	$NaN(SiMe_3)_2$
NME	*N*-methylephedrine
NMMP	*N*-methylmorpholine
PFL	*Pseudomonas fluorescences* lipase
PLE	pig liver esterase
PTC	phase transfer catalyst
R*	chiral group
RAMP	(*R*)-1-amino-2-(methoxymethyl) pyrrolidine
Salen	*N,N′*-disalicylidene-ethylenediaminato
TBAF	tetrabutylammonium fluoride
Tf	trifluoromethanesulfonyl group
THF	tetrahydrofuran
TMS	trimethylsilyl group
TMSCN	cyanotrimethylsilane
Ts	tosyl group
TS	transition state

1

Fundamentals of the Stereochemistry of Organophosphorus Compounds

1.1 Historical Background

Natural processes are subordinate to geometrodynamics – the theory describing physical objects, geometrical spacetime, and associated phenomena completely in terms of geometry, and her elder sister – symmetry. Symmetry/asymmetry is one of the basic concepts in modern natural science [1]. Research into this field began in the Middle Ages, when the birefringent properties of calcite were discovered. In 1669, Bartholinus observed the double refractive properties of the calcite Iceland spar. Later, in 1801, the mineralogist Haui found that quartz crystals are enantiomorphic, representing mirror images of one another. In 1815, another French naturalist J.-B. Biot discovered that certain chemical compounds rotate the plane of a beam of polarized light [2]. Biot constructed the first polarimeter and he also discovered that many natural compounds exhibit optical activity, that is, they rotate the plane of circularly polarized light. Studying crystals under a microscope, Biot discovered two types of crystals. The sample consisting of crystals of one type turned polarized light clockwise and that from another type in the opposite direction. A mixture of the two types of crystals had a neutral effect on polarized light. The nature of this property remained a mystery until 1848, when Louis Pasteur proposed that it had a molecular basis originating from some form of dissymmetry [3]. Pasteur separated the left and right hemihedral crystals of the sodium-ammonium salt of D,L-tartaric acid under a microscope, and connected the opposite optical activity to the mirror image of these crystals. Pasteur termed the mixture creating polarization as dissymetric and the phenomenon as dissymmetry (asymmetry). The term *chirality* was proposed by Lord Kelvin in 1894 and introduced into chemistry by Mislow in 1962. Dissimmetry, as discovered by Pasteur, is found in nature, whereas compounds obtained from living organisms are chiral or nonracemic. In 1852, Pasteur discovered that resolution could also be achieved by using a chiral base (quinine and brucine) and by using microorganisms. He discovered that paratartaric acid could be separated under the influence of optically active natural bases such as quinine or brucine. Pasteur developed a method for the separation of paratartaric acid with the help of *Penicillium glaucum*, leading to the formation of levorotatory tartaric acid, thus creating the basis for microbiological separation of racemates. J. Wislicenus came to the conclusion that the right- and non-superimposable levorotatory lactic acids have an identical structure, and he noticed that the only difference between the isomers is the order in which the radicals are distributed in space [4]. The origin of chirality itself was finally discovered in 1874, when van't Hoff and Le Bel independently proposed that this phenomenon of optical activity can be explained by the assumption that the four saturated chemical

Asymmetric Synthesis in Organophosphorus Chemistry: Synthetic Methods, Catalysis and Applications, First Edition. Oleg I. Kolodiazhnyi.

bonds between carbon atoms and their neighbors are directed toward the corners of a regular tetrahedron [5]. This concept led to the explanation for the observed optical activity by recognizing that a carbon atom with four different substituents exists in two mirror images: that is, it is chiral. The study of enantioselective reactions began with Emil Fisher [6], who studied the addition of hydrogen cyanide to sugars. In 1912, Bredig and Fiske [7] described the first catalytic enantioselective reaction. They studied the addition of hydrogen cyanide to benzaldehyde catalyzed by cinchona alkaloids. Although the mandelic acid that they obtained after hydrolysis of the initially formed benzcyanohydrin was of low optical purity (3–8%), Bredig and Fiske showed that it was possible to synthesize optically active compounds out of achiral precursors by using a chiral catalyst. Unlike Fischer, Marckwald performed an enantioselective reaction upon an achiral, unnatural starting material, although with a chiral organocatalyst [8]. In a paper titled "Ueber asymmetrische Syntheses," Marckwald gave the following definition of asymmetric synthesis: "Asymmetric syntheses are those reactions which produce optically active substances from symmetrically constituted compounds with the intermediate use of optically active materials but with the exclusion of all analytical processes." Fifty years later, Horst Pracejus reported the asymmetric organocatalytic reaction of methyl(phenyl)ketenes with alcohols catalyzed by alkaloids, leading to the formation of enantiomerically enriched esters of α-phenyl-propionic acid [9].

Louis Pasteur (1822–1895)

Hermann Emil Fischer (1852–1919)

The first work devoted to the asymmetric synthesis of aminophosphonates by catalytic hydrogenation of unsaturated phosphonates was published approximately 30 years ago. The development of enantioselective synthesis was initially slow, largely owing to the limited range of techniques available for their separation and analysis. It was not until the 1950s that real progress began with the development of new techniques. The first of these was X-ray crystallography, which was used to determine the absolute configuration (AC) of an organic compound by Bijvoet *et al.* [10]. During the same period, methods were developed to allow the analysis of chiral compounds by NMR, either using chiral derivatizing agents (CDAs), such as Mosher's acid [11], or europium-based shift reagents, of which $Eu(DPM)_3$ was the earliest [12]. Chiral auxiliaries were introduced by Corey and Ensley in 1975 [13] and featured prominently in the work of D. Enders. Around the same time, enantioselective organocatalysis was developed and enzyme-catalyzed enantioselective reactions became more and more common during the 1980s, particularly in industry, with their applications including asymmetric ester hydrolysis with pig-liver esterase. The emerging technology of genetic engineering has allowed the tailoring of enzymes to specific processes, permitting an increased range of selective transformations.

Today, the asymmetric synthesis of organophosphorus compounds is an extremely dynamic research domain in modern chemistry. Contributions to the development of asymmetric synthesis was made by many outstanding chemists.

Thus, L. Horner studied the electrochemical cleavage of quaternary phosphonium salts leading to the discovery that tertiary phosphines with three different substituents are chiral [14, 15]. This knowledge formed the basis of the pioneering work of Horner on enantioselective catalysis, especially enantioselective homogeneous hydrogenation [15], which was published independently in the same year as the work of W. S. Knowles [16] – work that was honored by the Nobel Prize and which was based on the chiral phosphines discovered by Horner [15]. Knowles developed one of the first asymmetric hydrogenation catalysts by replacing the achiral triphenylphosphine ligands in Wilkinson's catalyst with chiral phosphine ligands. He developed an enantioselective hydrogenation step for the production of L-DOPA (3-(3,4-dihydroxyphenyl)-L-alanine), utilizing the DIPAMP ligand. L-DOPA later became a mainstay for treating Parkinson's disease. Noyori Ryōji won the Nobel Prize in Chemistry together with W. S. Knowles for the development of the atropoisomeric ligand BINAP (2,2′-bis(diphenylphosphino)-1,1′-binaphthyl) and study of chirally catalyzed hydrogenation [17]. In 1985, Schöllkopf *et al.* [18] reported asymmetric hydrogenation of *N*-[1-(dimethoxyphosphoryl)-ethenyl] formamide, using a rhodium catalyst with (+)-DIOP chiral ligand to afford the L-(1-aminoethyl) phosphonate in good yields and 76% ee enantioselectivity. The initially formed formamide was hydrolyzed with concentrated hydrochloric acid to give the aminophosphonic acid. Crystallization from water/methanol increased the enantiomeric purity of the product up to 93% ee.

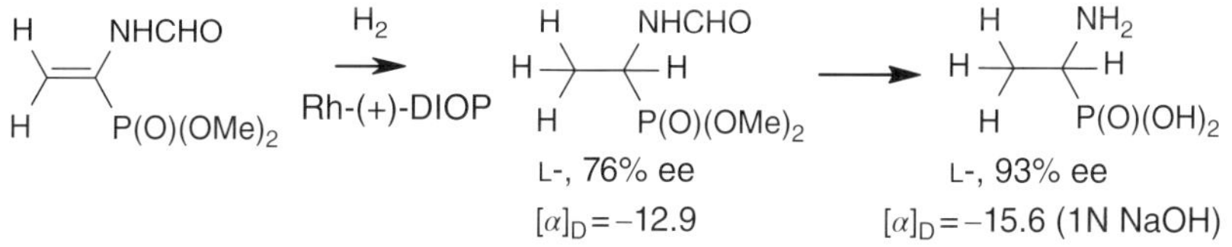

Significant contribution to the development of asymmetric synthesis of organophosphorus compounds was made by Henry Kagan, a member of the French Academy of Sciences. He developed C_2-symmetric phosphinic ligands, including DIOP, for asymmetric catalysis. These ligands have wide practical applications in the chemical industry [19].

The Japanese chemist Imamoto developed many types of phosphine ligands, which found practical applications [20]. The French chemist Juge created the accessible "ephedrine" method for the preparation of chiral phosphines named "the Juge-Stephan method." Together with Imamoto, he developed phosphine-boranes [21]. The American chemist William McEwen developed the fundamentals of the stereochemistry of organophosphorus compounds [22]. The Polish chemists Kafarsky [23] and Mikolajchyk [24] conducted important research studies in the application of phospha and sulfur reactants for the preparation of bioactive and natural compounds. Pietrusiewicz *et al.* [25], Kielbasisky and Drabowich [24, 26] are now continuing these studies. Methods for asymmetric synthesis and the synthesis of chiral organophosphorus compounds are of great interest to a number of powerful industrial firms and scientific research institutes, notable among them being the Leibniz Institute for Catalysis at the University of Rostock (LIKAT), the largest publicly funded research institute in

Europe. Professor A. Börner of the Institute has been working on the development of new phosphinic chiral ligands and their practical applications [27]. In addition to those mentioned above, hundreds of highly professional chemists in many scientific centers are working in the domain of asymmetric synthesis of organophosphorus compounds. Their names and achievements can be found in the chapters of this monograph.

1.2 Some Common Definitions in Stereochemistry

Some common terms in the field of stereochemistry are explained in this section. These terms appear repeatedly throughout this book. Therefore, it is essential that we establish common definitions for these frequently used terms [28].

Absolute configuration. The spatial arrangement of the atoms of a physically identified chiral molecular entity (or group) and its stereochemical description (e.g., (*R*) or (*S*), (P) or (M), (D) or (L)).

Absolute configuration. A chemist's term that refers to chiral molecules. Note particularly that this refers to both the entity under consideration, namely, the crystal structure versus molecule, as well as the symmetry restrictions.

Asymmetric compounds. Absence of all elements of symmetry. An asymmetric molecule is optically active. It has an additional molecule which is its non-superimposable mirror image. Together they are termed a *pair of enantiomers.* Some asymmetric molecules may exist not only as enantiomers but also exist as diastereomers.

Assigning the absolute configuration - the R–S sequence rules. In order to assign the stereochemistry of a stereocenter, the priority of the groups attached to the stereocenter must be determined.

The CIP (**C**ahn–**I**ngold–**P**relog) priority rules are a standard process to name the stereoisomer of a molecule. *R*/*S* descriptors are assigned by using a system for ranking priority of the groups attached to each stereocenter. The atomic numbers (*Z*) of the atoms directly attached to the stereocenter are compared. The group having the atom with the higher atomic number receives higher priority. Priority increases as the atomic number increases: I > Br > Cl > S > P > O > N > C > H > electron pair.

After the substituents of a stereocenter have been assigned their priorities, the molecule is oriented in space so that the group with the lowest priority is pointed away from the observer. A center with a clockwise sense of rotation is an (*R*) or rectus center and a center with a counterclockwise sense of rotation is an (*S*) or sinister center. The order of substituent priority in tetrahedral phosphorus compounds differs from that in carbon compounds with a true C=O multiple bond (Alk < R–O–C < C=O). The P=O bond in phosphates, phosphonates, and related compounds is traditionally represented as a double bond, although it is more correct to treat it as a single bond with two electron pairs localized on the oxygen atom. This is the reason that substituents at tetrahedral phosphorus have the following priority order: Alk < P=O < R–O–P [29, 30]. In tricoordinate phosphorus compounds, the group with the lowest priority is the electron pair.

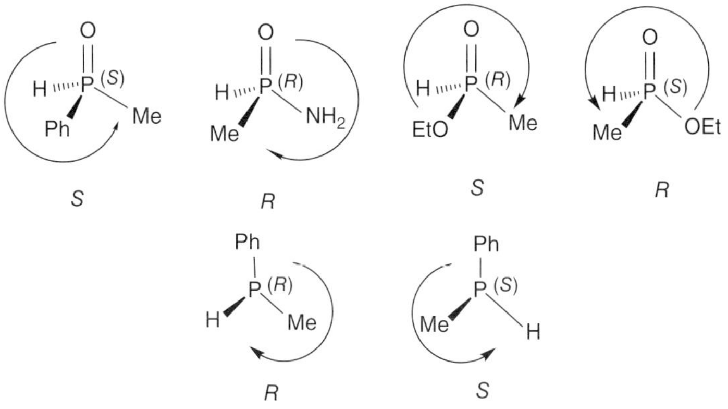

Biocatalysis. Biocatalysis is the chemical process through which enzymes or other biological catalysts perform reactions between organic components. Biocatalysis makes use of biological compounds ranging from isolated enzymes to living cells to perform chemical transformations. The advantages of these reagents include very high ee and reagent specificity, as well as mild operating conditions and low environmental impact.

Chirality. The geometric property of a rigid object (or spatial arrangement of points or atoms) of being nonsuperposable on its mirror image. Such an object does not have symmetry operations of the second kind. If the object is superposable on its mirror image, the object is described as being achiral, and is modified for H-M symbols. Hermann–Mauguin notation is used to represent the symmetry elements in point groups, plane groups and space groups [28].

Chiral auxiliaries. A chiral auxiliary is an organic compound that couples to the starting material to form a new compound which can then undergo enantioselective reactions via intramolecular asymmetric induction. At the end of the reaction, the auxiliary is removed under conditions that will not cause racemization of the final product. It is typically then recovered for future use.

Dissymmetric compounds. Compounds lacking an alternating axis of symmetry and usually existing as enantiomers. Dissymmetry is the property of non-superimposability of a molecule on its mirror image. A dissymmetric molecule may have a simple axis of symmetry, yet it will be optically active and exist as a pair of enantiomers. Both asymmetric and dissymmetric molecules are optically active.

Prefixes d or l. Dextrorotatory or levorotatory according to the experimentally determined rotation of the plane of monochromatic plane-polarized light to the right or left.

Prefixes D *or* L. Absolute configurations assigned to a molecule through experimental chemical correlation with the configuration of D- or L-glyceraldehyde; often applied to amino acids and sugars, although (*R*) and (*S*) are preferred.

Diastereoisomer. Stereoisomers with two or more chiral centers, where the molecules are not mirror images of one another, for example, derythrose and *d*-threose. The term diastereoisomer is often contracted as diastereomer.

Enantiomerically pure/enantiopure. A sample in which all molecules have (within the limits of detection) the same chirality sense. The use of homochiral as a synonym is strongly discouraged (Moss [28]).

Enantioselective synthesis, also called *chiral synthesis* or *asymmetric synthesis.* This is defined by IUPAC as "a chemical reaction (or reaction sequence) in which one or more new elements of chirality are formed in a substrate molecule and which produces the stereoisomeric (enantiomeric or diastereoisomeric) products in unequal amounts."

Enantioselective organocatalysis. Organocatalysis refers to a form of catalysis where the rate of a chemical reaction is increased by an organic compound consisting of carbon, hydrogen, sulfur, and other nonmetal elements. When the organocatalyst is chiral, enantioselective synthesis can be achieved; for example, a number of carbon–carbon bond-forming reactions become enantioselective in the presence of proline with the aldol reaction being a prime example. Organocatalysis often employs natural compounds as chiral catalysts.

Enantiomer. Two stereoisomers that are non-superimposable mirror images of each other.

Enantiomer excess (ee). Enantiomeric excess (ee) is a measurement of purity used for chiral substances. It reflects the percentage by which one enantiomer is in excess over the other in a mixture of the two. A racemic mixture has an ee of 0%, while a single completely pure enantiomer has an ee of 100%: $ee = [(E_1 - E_2)/[(E_1 + E_2)] \times 100\%$.

Enantiotopic. The stereochemical term enantiotopic refers to the relationship between two groups in a molecule which, if one or the other were replaced, would generate a chiral compound. The two possible compounds resulting from that replacement would be enantiomers.

Erythro/threo. Terms derived from carbohydrate nomenclature used to describe the relative configuration at adjacent stereocenters. Erythro refers to a configuration with identical or similar substituents on the same side of the vertical chain in Fischer projection. Conversely, a *threo*-isomer has these substituents on opposite sides. These terms came from the nomenclature of two carbohydrate compounds, threose and erythrose.

Flack parameter. The parameter x in the structure-amplitude equation G:

$$I(hkl) = (1 - x)\left[F(hkl)\right]^2 + x\left[F(-h - k - l)\right]^2$$

Homotopic groups. Groups that can be exchanged by a symmetry axis. It follows that any achiral or chiral molecule which has an axis of symmetry contains at least one set (usually a pair) of homotopic groups.

Meso compounds. Compounds whose molecules not only have two or more centers of dissymmetry but also have plane(s) of symmetry. They do not exist as enantiomers, for example, *meso*-tartaric acid.

Optical activity. Experimentally observed rotation of the plane of monochromatic plane-polarized light to the observer's right or left. Optical activity can be observed with a polarimeter.

Optical isomer. Synonym for enantiomer, now disfavored, because most enantiomers lack optical activity at some wavelengths of light.

Optical purity. The optical purity of a sample is expressed as the magnitude of its optical rotation as a percentage of that of its pure enantiomer (which has maximum rotation).

Optical rotation. Enantiomers that rotate the plane-polarized light clockwise (to the right) are said to be dextrorotatory and are indicated with a lowercase "*d*" or a positive sign (+). Those that rotate the plane counterclockwise are called *levorotatory* and are indicated with a lowercase "*l*" or a negative sign (−).

P-Chirogenic. In the literature, a phosphorus atom bonded to three different substituents is called *P-stereogenic, P-chirogenic,* or *P-chiral.* It should noted that "P-chiral" is not strictly correct because chirality is a property of a molecule as a whole.

Prochirality. Refers to the existence of stereoheterotopic ligands or faces in a molecule that, upon appropriate replacement of one such ligand or addition to one such face in an achiral precursor, gives rise to chiral products.

Pro-R and Pro-S. Refer to heterotopic ligands present in the system. It is arbitrarily assumed that the ligand to be introduced has the highest priority, and replacement of a given ligand by this newly introduced ligand creates a new chiral center. If the newly created chiral center has the (*R*)-configuration, that ligand is referred to as *pro-R*,; while pro-S refers to the ligand replacement that creates an (*S*)-configuration.

Racemate. An equimolar mixture of a pair of enantiomers. It does not exhibit optical activity. The chemical name or formula of a racemate is distinguished from those of the enantiomers by the prefix rac- (or racem-) or by the symbols *RS* and *SR*.

Racemization: The process of converting one enantiomer to a 50 : 50 mixture of the two.

Re and Si. Labels used in stereochemical descriptions of heterotopic faces. If the CIP priority of the three ligands a, b, and c is assigned as a > b > c, the face that is oriented clockwise toward the viewer is called *Re*, while the face with a counterclockwise orientation of a < b < c is called *Si*.

Scalemic. Compounds existing as a mixture of two enantiomers in which one is in excess. The term was coined in recognition of the fact that most syntheses or resolutions do not yield 100% of one enantiomer.

Stereoisomer. Molecules consisting of the same types and same number of atoms with the same connections but different configurations.

1.3 Determination of Enantiomer Composition

Stereochemistry and chirality are of great importance in many different fields as the molecular properties and biological effects of the stereoisomers are often significantly different. Determination of ee's of the drug samples may allow for individualization and tracking of drug distribution routes. Aside from the classical methods of polarimetry and chemical resolution, some of the most popular current methods for ee. determination include chromatography (i.e., gas chromatography (GC), high performance liquid chromatography (HPLC)), and other techniques that may be considered related variants, such as capillary zone electrophoresis, micellar electrokinetic chromatography, and supercritical fluid chromatography (SFC). These techniques can be applied directly to the samples, or some achiral reagent may be used for sample modification, for instance, the acylation of an amine for improved chromatographic separation. To determine how much one isomer is in excess over the other, analytical methods based on HPLC or GC on a chiral column have proved to be most reliable. Chiral chemical shift reagents and chiral solvating agents for NMR analysis are also useful, and so are optical methods [31–34].

The enantiomer composition of a chemical compound may be described by the ee, which describes the excess of one enantiomer over the other. Correspondingly, the diastereomer composition of a sample can be described by the diastereomer excess (de), which refers to the excess of one diastereomer

$$\text{enantiomeric excess } (\%\text{ee}) = \frac{[R]-[S]}{[S]+[R]} \times 100\%$$

$$\text{diastereomeric excess } (\%\text{de}) = \frac{[S*S]-[S*R]}{[S*S]+[S*R]} \times 100\%$$

where (*R*) and (*S*) are the composition of *R* and *S* enantiomers, respectively, (*S*,*S*) and (*S*,*R*) are the composition of the diastereomers.

A variety of methods are also available wherein the compound under investigation can be converted with a chiral reagent to diastereomeric products, which have readily detectable differences in physical properties. If a derivatizing agent is employed, it must be ensured that the reaction with the subject molecule is quantitative and that the derivatization reaction is carried out to completion [31].

1.3.1 Method of Nuclear Magnetic Resonance

Spectroscopic techniques, primarily NMR, are highly useful for determination of ee's by the observation of ^{1}H, ^{13}C, ^{19}F, or other nuclei. NMR methods have employed direct methods, using chiral lanthanide shift reagents or chiral solvating agents, but also can use indirect methods [32–39]. One typical indirect NMR method is the use of a chiral reagent to transform substrate enantiomers into stable diastereomeric derivatives. Any NMR approach hinges on observing separate absorptions (different chemical shifts) for corresponding nuclei in the substrate enantiomers.

1.3.1.1 Chiral Solvating Agents

In organophosphorus chemistry, the chiral solvating agents (CSA), quinine, cinchonine, derivatives of amino acids, chiral phosphonic acids, and Kagan's amides are most often applied (Table 1.1) [35–55]. Use of cinchona alkaloids (quinine and cinchonidine) as chiral solvating agents is a convenient method for determination of the enantiomeric composition of hydroxyphosphonates [32–34]. Determinations are carried out by the addition of an alkaloid solvent in $CDCl_3$ to a hydroxyphosphonate placed in the NMR tube and subsequent recording of NMR ^{31}P-[^{1}H] spectra. The signals of diastereomers in the spectrum are well resolved, thus allowing the integration. The optimal magnitude of $\Delta\delta_P$ signals was attained at a 1 : 4 molar ratio of hydroxyphosphonate/alkaloid (Figure 1.1) [40].

It was found that the determination can also be achieved in achiral solvents in the presence of certain chiral compounds, namely, chiral solvating agents. In these cases, the determination is achieved on the basis of diastereomeric interaction between the substrate and the chiral solvating agent. It is possible to use such deuterated solvents as C_6D_6 or $CDCl_3$ which do not interfere with the solvating action of the alkaloid; however, the use solvents such as deuteromethanol leads to negative results that play a key role in the formation of hydrogen bridges between the alkaloid and the hydroxyphosphonate, leading to discrimination of the enantiomers in the NMR spectra. (*S*)-(1)-*N*-(3,5-dinitrodibenzoyl)-1-phenylethylamine and the corresponding (*S*)-(1)-1-naphthyl derivative (Kagan's amide) are effective CSAs for tertiary phosphine oxides and phospholene oxides. Association with 2-phospholene-1-oxide derivatives causes characteristic perturbations of the ^{31}P resonance that correlate with the AC [41–43].

Table 1.1 Some CSA used for determinations of enantiomeric excesses.

Entry	Reagent	Types of phosphorus compounds	References
1	*N*-Fmoc-*N*′-Boc-L-tryptophan	Phosphonates, phosphinates, phosphates, phosphine oxides, and amidophosphonates	[35]
2	Quinine	Hydroxyphosphonates	[36–39]
3	Cinchonidine	Hydroxyphosphonates	[40]
4	Kagan's amides 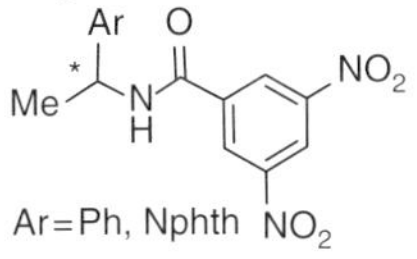 Ar=Ph, Nphth	Tertiary phosphine oxides, phosphine oxides	[41–43]
5	Macrocycles 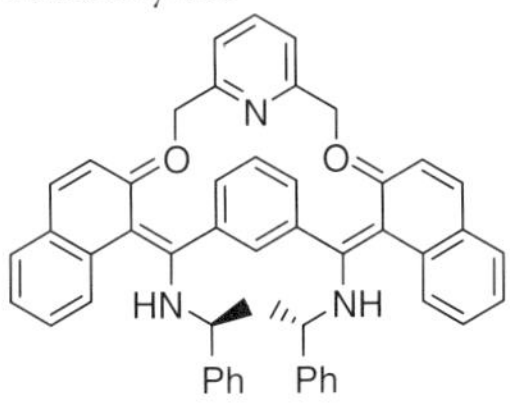	Phosphinic, phosphonic, and phosphorus acids	[44]
6	*t*-Bu(Ph)P(S)OH	P-chiral phosphonates and α-substituted phosphonates, tertiary phosphine oxides, phosphinamides, phosphinates, phosphinthioates, phosphinites	[45–47]
7	Cyclodextrins	Hydroxyphosphonates, aminophosphonates	[48–54]

The enantiomeric discrimination is usually larger with a greater shielding naphthyl derivative. (*R*)-(1) or (*S*)-(2)-*tert*-butylphenylphosphinothioic acid is also an effective chiral NMR-solvating agent that associates with compounds containing a phosphine oxide unit. A similar association occurs with phosphinate esters and phosphorus thioacids and thionates. In each case, shielding from the phenyl group causes consistent trends in the ^{1}H and ^{31}P NMR spectra that correlate with the AC. The same occurs with (*R*)-(1) and (*S*)-(2)-(*N*-phenyl)methylphenyl phosphinic amide, which associates with other phosphinic amides through hydrogen bonds [47]. Kafarski *et al.* used α- and β-cyclodextrins as chiral selectors for the ^{31}P NMR determination of the ee of aminoalkanephosphonic and aminoalkanephosphinic acids. Most of these acids form inclusion complexes with α- and β-cyclodextrins and upon increasing the cyclodextrin to aminophosphonic acid molar ratio, the ^{31}P NMR signals for (*R*)- and (*S*)-enantiomers separate. When a racemic mixture of an aminophosphonic acid was dissolved in a solution containing cyclodextrin, two diastereomeric complexes were formed and in most cases two signals were observed in the ^{31}P NMR spectra [48–54].

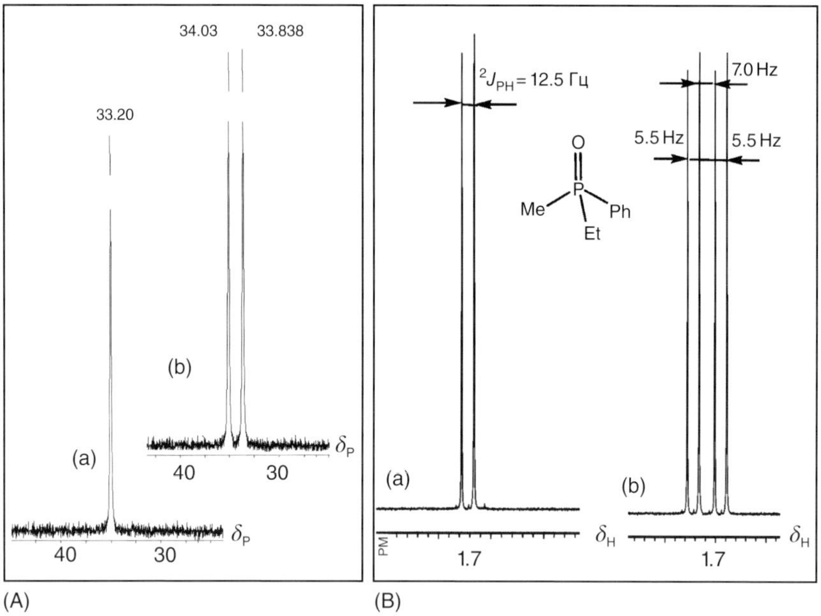

Figure 1.1 ^{31}P-[^{1}H] NMR spectra of racemic $(EtO)_2P(O)CH(Cl)Bu$-*t* without solvating reagent and with cinchonidine (A). NMR spectra of racemic methyl-ethyl-phenylphosphine oxide without a solvating reagent and with Kagan's amide (B).

1.3.1.2 Complexes of Metals (Shift Reagents)

One of the most useful applications of lanthanide shift reagents is in the determination of optical purity by the use of chiral ligands on the lanthanide. Two of the more effective reagents developed are Eu(facam)$_3$ (tris(3-trifluoroacetyl-D-camphorato)europium(III)) and Eu(hfbc)$_3$ (tris(3-heptafluorobutyryl-D-camphorato)europium(III)). Chiral chemical shift reagents (CSR) may be used to enhance the anisochrony of diastereomeric mixtures to facilitate their quantitative analysis. CSR are paramagnetic complexes of certain lanthanides, such as europium and ytterbium, with ligands designed to make them soluble in organic solvents. They are also chemically inert and, in some cases, improve the solubility of compounds in nonaqueous solvents. When added to NMR samples, they coordinate weakly to polar functional groups, such as amines, esters, ketones, and alcohols, and create a strong local magnetic field that produces large chemical shift changes. Examples of chiral CSRs (which are commercially available) are shown in Scheme 1.1. If a CSR binds a substrate possessing a stereocenter, two diastereomeric complexes can be formed from its enantiomers, which, in principle, will exhibit different chemical shifts. This leads to distinct resonance peaks (^{1}H or ^{13}C) of the two enantiomeric forms of the sample. Chiral shift reagents usually form adducts with enantiomeric substrates containing diastereomeric protons showing well-resolved NMR signals; for example, complexes of lanthanides such as camphor derivatives, in particular tris-(3-3-(+)-camphorato) lanthanum (III) (R=C(CH$_3$)$_3$, C$_3$F$_7$, and others) or Eu(hfc)$_3$ (europium tris[3-(heptafluoropropyl-hydroxymethylene)-(+)-camphorate]) **1**. The resolution in NMR spectra of an (*R*)/(*S*)-isomeric mixture in the presence of chiral shift reagent of europium tris[3(heptafluoropropylhydroxymethylene)-(−)-camphorate] is

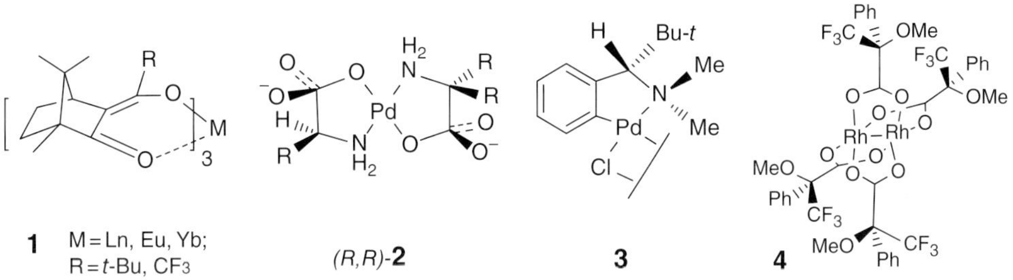

Scheme 1.1 The metal complexes increasing anisochronous signals of tested materials.

sufficient for determination of their enantiomeric ratio [56–59]. Lanthanide complexes can serve as weak Lewis acids. In nonpolar solvents (e.g., $CDCl_3$, CCl_4, or CS_2), these paramagnetic salts are able to bind Lewis bases, such as amides, amines, esters, ketones, and sulfoxides. As a result, protons, carbons, and other nuclei are usually deshielded relative to their positions in the uncomplexed substrates, and the chemical shifts of those nuclei are altered. The extent of this alteration depends on the strength of the complex and the distance of the nuclei from the paramagnetic metal ion. Therefore, the NMR signals of different types of nuclei are shifted to different extents, and this leads to spectral simplification. The spectral nonequivalence observed in the presence of chiral CSR can be explained by the difference in geometry of the diastereomeric CSR-chiral substrate complexes, as well as the different magnetic environment of the coordinated enantiomers that causes the anisochrony. For example, it was found by NMR that the interaction of Pd(II) ions with 1-aminophosphonate ligands (L) yields, in a non-diastereoselective manner, diastereoisomeric chelate pairs (PdL_2) observable in alkaline D_2O solutions [56]. The chelate complexes **2** give two peaks in the ^{31}P NMR spectra; one corresponds to the chiral species (both ligands are (*R*) or (*S*) enantiomers) and the second to the *meso*-forms (*R*) and (*S*)-ligands). Under the experimental conditions used, values in the range 0.03–0.18 ppm were observed. Enantiomers of the monodentate phosphine *t*-BuP(Ph)C_6H_4Br-4 were resolved by chromatographic separation of their diastereomeric adducts with a homochiral *ortho*-palladated resolving agent **3** derived from α-*tert*-Bu-substituted tertiary benzylamine. The conformation of the palladacycle and the AC of the phosphine were determined using ^{1}H NMR spectroscopy and confirmed by an X-ray diffraction study of both diastereomeric complexes [60]. The enantiomerically pure dirhodium complex (*R*)$Rh_2(MTPA)_4$ **4** is a good auxiliary for chiral recognition of a variety of organic compounds by using ^{1}H or ^{13}C NMR spectroscopy. The dirhodium method works particularly well with functional groups which are soft bases. Therefore, it is a good supplement to the methods using CSR, which are known to complex rather hard bases. The diastereomeric complexes of **4** with the P=S and P=Se derivatives exhibit significant differences in the chemical shifts which allows their determination by ^{1}H, ^{13}C, and ^{31}P NMR spectroscopy [59].

1.3.1.3 Chiral Derivatizing Agents for NMR

CDAs are optically active reactants that react with the enantiomers that are to be analyzed. CDAs are chiral auxiliaries used to convert a mixture of enantiomers into diastereomers in order to analyze the quantities of each enantiomer present within the mixture. Analysis can be conducted by spectroscopy or by chromatographic methods.

Typically, the reaction involves the formation of a covalent bond, although in some cases it may involve the formation of a soluble salt. In many cases, the diastereomeric complexes exhibit chemical shift patterns that correlate with the AC. For the CDA to be useful in determining ee, it is essential that no kinetic resolution occur during the derivatization reaction.

Dale and Mosher [11] proposed α-methoxy-α-phenyl-α-trifluoromethyl acetic acid (MTPA) in both the (*R*)- and (*S*)-forms. MTPA is now known as *Mosher's acid*. The chloride of MTPA reacts with chiral alcohols (mostly secondary alcohols) to form diastereomeric mixtures called *MTPA esters* or *Mosher's esters*. There are two advantages in using MTPA: The epimerization of the chiral α-C is avoided because of the absence of the α-proton; and the introduction of a CF_3 group makes it possible to analyze the derivatives by means of ^{19}F NMR, which simplifies the analysis process. The presence of a second NMR active nucleus (^{19}F) provides another way to determine ee and possibly the AC. Peak overlapping is generally not observed, and the ^{19}F NMR signals are far better separated than are the ^{1}H NMR peak configurations [61]. The derivatization of amino- and hydroxyphosphonates with MTPA (Mosher acid) [61–64], camphoric acid [63], mandelate acid [64], phosphono-didepsipeptides [65, 66], diazaphospholidine chloride [67], and others has been described. (*S*)-Naproxene® chlorides and (*S*)-ibuprofen® chlorides are convenient chiral derivatizing reagents for determination of the enantiomeric purity of α- and β-hydroxyalkylphosphonates by ^{31}P NMR spectroscopy [68]. ^{1}H NMR spectroscopy of chiral 1-(1-naphthyl)ethylamine salts of hydroxyphosphonic acids [55], NMR in chiral medium, GLC with chiral stationary phase [69], and other methods were also used (Figure 1.2).

N-Substituted (L)-amino acids were used for the determination of the enantiomeric composition of chiral 1-hydroxyalkylphosphonic acids by means of ^{31}P NMR spectroscopy (Scheme 1.2) [11, 61–65]. The ^{1}H NMR spectroscopy of chiral 1-(1-naphthyl)ethylammonium salts of hydroxy phosphonic acids, NMR in chiral media, and other methods were also used [51, 70–77]. The diastereomeric esters formed can be analyzed by ^{1}H, ^{19}F, and ^{31}P NMR spectroscopy. The signals of the P atoms in ^{31}P NMR spectra of (*S*)-hydroxyphosphonate esters are usually located in a lower field than those of (*R*)-hydroxyphosphonates.

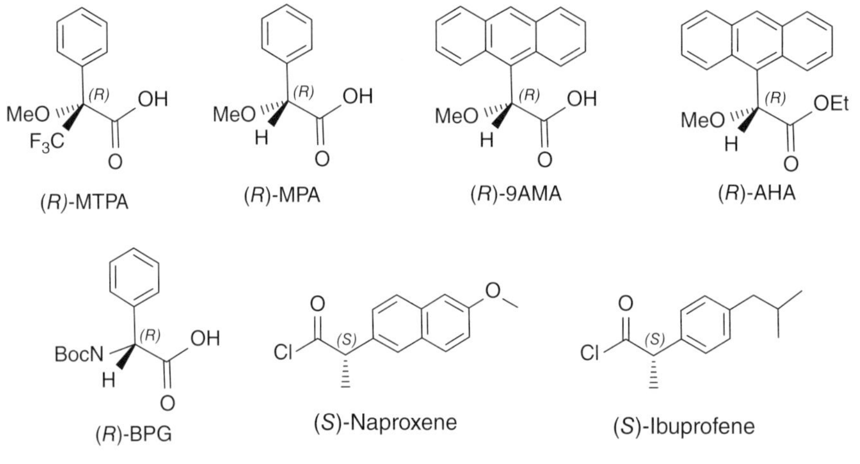

Figure 1.2 Some chiral derivatizing agents.

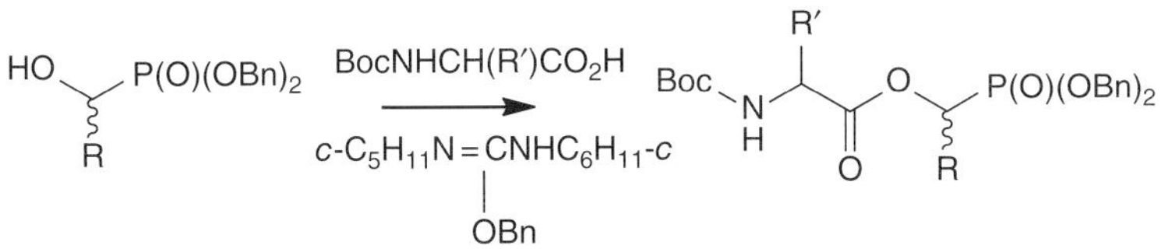

Scheme 1.2 Determination of optical purity by means of N-substituted L-amino acids.

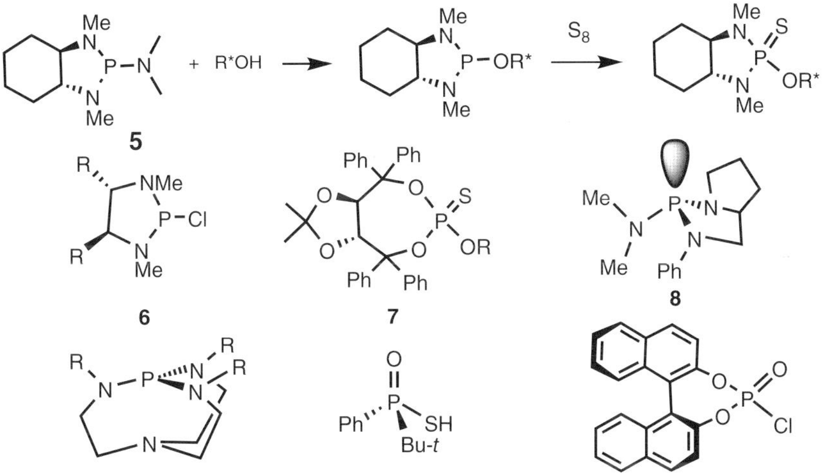

Scheme 1.3 Phosphorus-containing CDA.

A number of chiral phosphorus derivatizing agents **5–11** were described and used for the determination of enantiomeric purity of organophosphorus compounds by NMR [70–76] (Scheme 1.3). For example, the phosphorus derivative of TADDOL 7 was used for the enantiomeric discrimination of alcohols and carboxylic acids by ^{1}H and ^{31}P NMR spectra. P(III) and P(V)-phosphorus derivatives of C_2-symmetric ligands (1,2-diphenyl-1,2-bis(*N*-methylamino)ethane or 1,2-bis(*N*-methylamino)cyclohexane) are suitable for the determination of the optical purity of carboxylic acid. Cyclic phosphorochloridites prepared by reaction of PCl_3 with chiral butane-2,3-diol or hydrobenzoin were used to measure the ee of chiral alcohols. A phosphorus derivative of (*S*)-2-anilinomethylpyrrolidine was useful for the enantiomeric discrimination of halohydrins (Scheme 1.3).

Derivatized diastereomers of various alcohols and amines allow to exactly define the diastereomeric ratio and optical purity of samples by ^{31}P NMR spectra, even in reaction mixtures. For example, the derivatizing reagent obtained from tartrates or from 1,2-diaminocyclohexane allow to attain the $\Delta\delta$ of derivatizing compounds in several ppm.

Dimenthylchlorophosphite is a convenient chemical derivatizing agent for the determination of the enantiomeric purity of hydroxyphosphonates, aminophosphonates, amino acids, and alcohols by ^{31}P NMR. The diastereomeric hydroxy phosphonate derivatives formed upon the reaction with this reagent differ appreciably in the

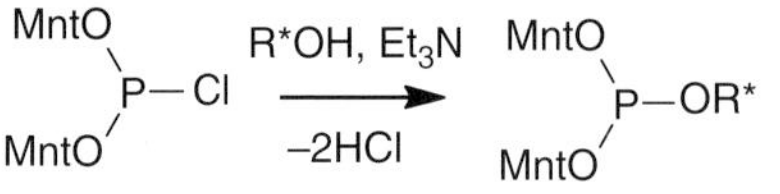

Scheme 1.4 Dimenthylchlorophosphite as derivatizing reagent.

chemical shifts δ_P; therefore, the enantiomeric ratio can be easily determined by integrating the ^{31}P NMR signal intensities [77] (Scheme 1.4).

1.3.2 Chromatographic Methods of Analysis

Both GC and HPLC provide fast and accurate methods for enantiomeric separation of chirogenic organophosphorus compounds and allow quantitation of both mass and even optical rotation for HPLC, if appropriate detection devices are used. Chromatographic methods are among the most useful for chiral separation. There are two approaches: indirect, which utilizes derivatizing agents and direct, which uses chiral stationary phases or chiral mobile phase additives. In the indirect method, a racemic mixture is made to react with a chiral reagent to form a pair of diastereomers and then chromatographed using an achiral column. Because diastereomers possess different physiochemical properties, they can be separated in an achiral environment. The following are the advantages of the indirect approach: (i) it is less expensive, that is, conventional chromatographic columns can be used, (ii) it is more flexible because various achiral columns and mobile phase conditions, as in HPLC, can be used, and (iii) numerous types of derivatization chemistry are available. On the other hand, the following are the disadvantages of this method: (i) it involves a long analysis time that includes sample preparation and verification of the derivatization chemistry. (ii) There is inconvenience, specifically in preparative chromatography, when reversal of derivatization is needed to recover the pure enantiomers. (iii) The need arises to synthesize noncommercially available pure derivatizing reagent. (iv) Biased results are obtained for enantiomeric composition due to partial racemization of the derivatizing agent or unequal reaction rate.

Direct separation of enantiomers on an achiral column using a chiral mobile phase additive is applied only in HPLC. In GC, the mobile phase is an inert carrier gas, where the possibility of selective interactions with the analyte or the stationary phase is minimal. However, in HPLC, the mobile phase is a dynamic part of the system that influences both analyte and stationary phase interaction. In this method, enantiomeric separation is accomplished by the formation of a pair of transient diastereomeric complexes between the racemic analyte and the chiral mobile phase additive.

Many racemic mixtures can be separated on conventional achiral LC columns by using an appropriate chiral mobile phase additive. Additives such as β- and γ-cyclodextrins have been successful. The advantages of this technique are as follows: (i) it is less expensive conventional LC columns can be used. (ii) A wide variety of possible additives are available. (iii) Different selectivities can be obtained from the chiral phases. However, the problems with this technique include the following: (i) many chiral additives are expensive and sometimes have to be synthesized. (ii) The mode of operation is complex and inconvenient for preparative applications because the chiral additive must be removed from the enantiomeric solutes.

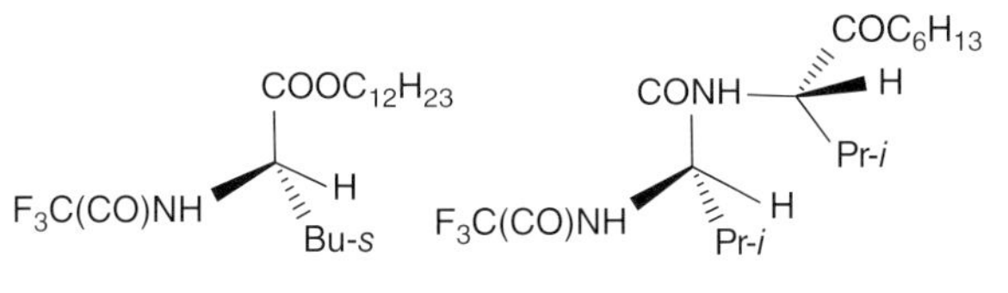

Scheme 1.5 Chiral stationary phases for gas chromatography.

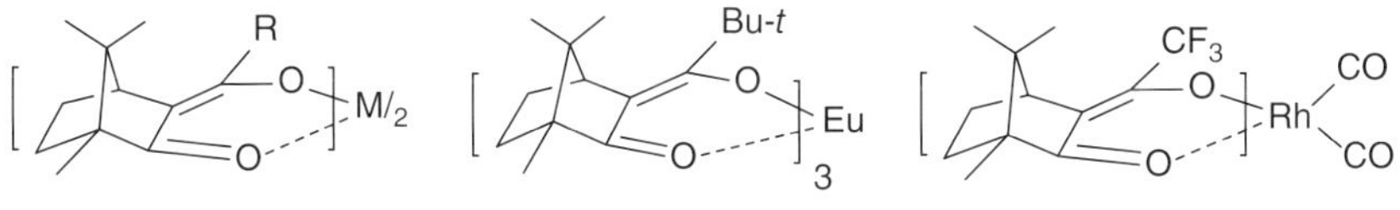

M = Ni, Mn; R = CF_3, C_3F_7

Scheme 1.6 Chiral metallochelates for gas chromatography.

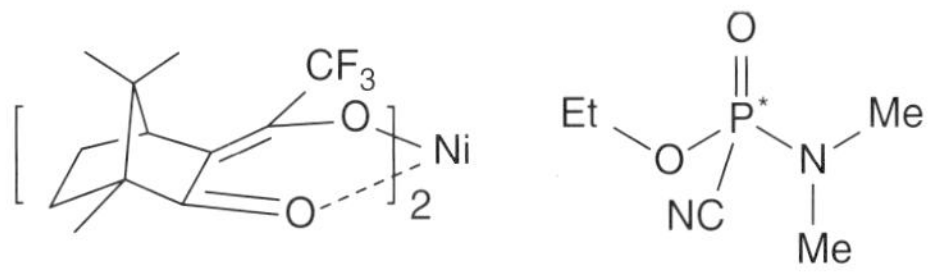

Scheme 1.7 Structure of the chiral metal chelate Ni(II)Bis[(IR)-3-(heptafluorobutyryl) camphorate].

1.3.2.1 Gas Chromatography

Chiral GC is a very commonly used method for the analysis of mixtures of enantiomers. The method is based on the principle that molecular association between the chiral stationary phase and the sample may lead to some chiral recognition and sufficient resolution of the enantiomers. The chiral stationary phase contains an auxiliary resolving agent of high enantiomeric purity (Schemes 1.5–1.7). The enantiomers to be analyzed undergo rapid and reversible diastereomeric interactions with the stationary phase and hence may be eluted at different rates. Separation of enantiomeric or diastereomeric mixtures by GC is a good method for determining enantiomer compositions. However, this method is limited to samples that are both volatile and thermally stable. Normally, if the compound to be separated has a low boiling point (e.g., <260 °C), or it can be converted to a low boiling substance, and no racemization occurs during the analysis, it is possible to analyze it by GC. If the compound has a high boiling point, or the compound tends to decompose or racemize at high temperature, HPLC would be the choice of separation [78, 79].

Benschop [80] reported the gas chromatographic separation of the stereoisomers of several chiral organophosphorus compounds using the glass capillary columns coated with the nonchiral phases SE-30 and Carbowax 20 M, or with the chiral in OV-phases Chirasil-Val and Ni(II)bis[(1*R*)-3-(heptafluorobutyryl) camphorate]. The Chirasil-Val column was extended with a Carbowax 20M column, thus giving complete separation of the four Soman stereoisomers. Besides Soman, the enantiomers of isopropyl methylphosphonofluoridate (Sarin) and cyclohexyl methylphosphono-fluoridate (cyclohexylsarin) were resolved. The enantiomers of *O*-ethyl *N*,*N*-dimethylphosphoramidocyanidate (Tabun) were separated by GC on a capillary column coated with Ni(II)-bis[(1*R*)-3-(heptafluorobutyryl)camphorate] in OV-101 (L = 14 m, i.d. = 0.44 mm, T. oven = 120 °C). Keglevich *et al.* resolved the 1-*n*-butyl-3-methyl-3-phospholene 1-oxide with TADDOL derivatives by chiral GC on a 30 m capilar column Supelco BETA DEX 120 [81].

1.3.2.2 Liquid Chromatography

Chiral HPLC is one of the most powerful tools to prepare enantiopure standards of chiral compounds. HPLC procedures were developed and successfully employed for the separation of a high number of organic and organophosphorus compounds. Because of their specific adsorption character, cyclodextrins and cyclodextrin derivatives immobilized on silica gel have been frequently applied for the improvement of the separation parameters of various chromatographic methods [82–89]. Aryl(hydroxymethyl) phosphonates were separated by HPLC on chiral stationary phase **12**. It was noted that (*R*)-(+)-hydroxyphosphonates are retained by the chiral stationary phase (3*R*,4*S*)-Whelk-0-1 **13** more strongly than (*S*)-(−)-hydroxyphosphonates [90]. This is due to the formation of a stronger hydrogen bond between the oxygen atom of the P=O group and the hydrogen atom of the amide NH group in the case of (*R*)-isomers than between the hydroxyl and NH-groups in the case of (*S*)-hydroxyphosphonates. For example, the enantiomers of diethyl α-hydroxybenzylphosphonate containing para, or ortho substituents or other aromatic rings (1-naphthyl, 2-naphthyl, and 2-thienyl) were separated by HPLC on a Whelk-O-1 chiral stationary phase **13**, which is superior to other CSPs (Scheme 1.8) [84].

In order to study the retention and chiral recognition mechanism, the method of quantitative structure-enantioselectivity retention relationships (QSERRs) has been investigated from the quantitative equations established between the chromatographic retention of enantiomers and their molecular descriptors of physicochemical properties [88, 89]. The separation of enantiomers of a series of 18 novel nitrogen mustard-linked phosphoryl diamide derivatives was investigated on the phenyl carbamate derivative β-cyclodextrin bonded phase in normal-phase HPLC. Some of the enantiomers were separated in baseline. The retention and separation mechanism involves the external association and inclusion between the substituent R^2 and the hydrophobic pocket [89].

The stationary Pirkle's phase for the super- and subcritical fluid chromatography separations of enantiomeric pairs of phosphine oxides was successfully used [90].

Kobayashi [91] reported the chromatographic resolution of racemic compounds containing phosphorus as chiral center. The compounds were resolved by HPLC on optically active (+)-poly(triphenylmethyl methacrylate). The resolved compounds

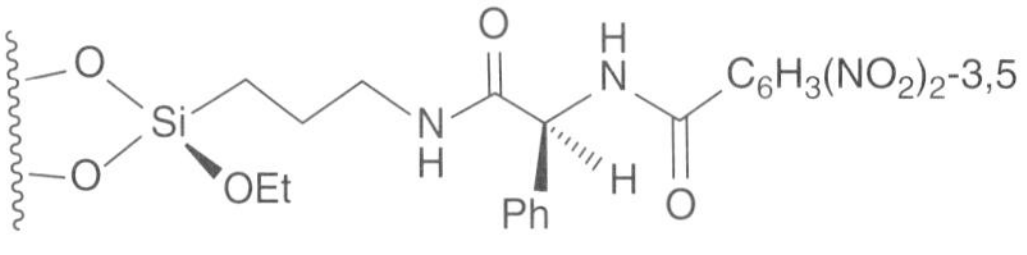

12

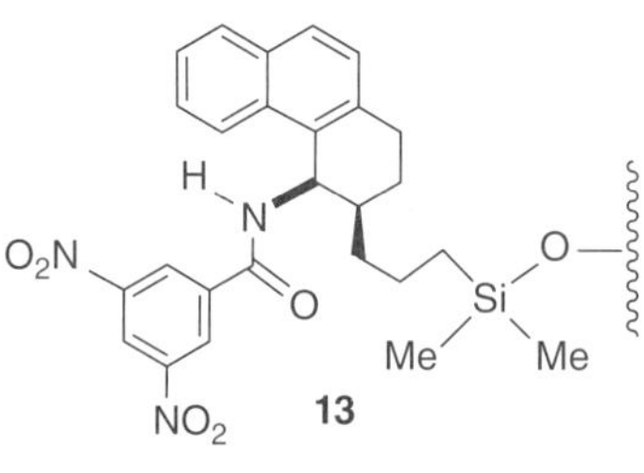

13

(3*R*,4*S*)-Whelk-0-1-CSR

Scheme 1.8 Effective chiral phase for resolution of amino acids [86, 90].

include insecticides such as *O*-ethyl *O*-(4-nitrophenyl) phenylphosphonothionate (EPN), *O*-(4-cyanophenyl)-*O*-ethyl phenylphosphono-thionate (cyanofenfos), and 2-methoxy-4*H*-l,3,2-benzo-dioxaphosphorin 2-sulfide (salithion).

The separation of a number of P-chiral racemates and C-chiral organophosphorus pairs of enantiomers was achieved on a commercial cellulosic tris-(3,5-dimethylphenyl-carbamate) stationary phase (Lux Cellulose-1, Phenomenex) in SFC [92].

1.4 Determination of the Absolute Configuration

In the area of asymmetric synthesis, one of the most important parameters is the configuration of the major product of an asymmetric reaction [93]. The methodologies to determine the ACs of the chiral compounds are classified into nonempirical methods for determining ACs and the relative methods using an internal reference with known AC. The main nonempirical methods are the Bijvoet method [10] by X-ray crystallography and the circular dichroism (CD) exciton chirality method. In X-ray crystallography, the anomalous dispersion effect of heavy atoms can be measured very accurately under proper conditions and the absolute stereostructure obtained is clear and unambiguous. However, the X-ray method needs crystals of suitable quality for good X-ray diffraction, and there is the problem how to obtain such single crystals.

The CD method for determination of chirality is also useful because the AC can be determined in a nonempirical manner, which does not require crystallization. Moreover, chirality of some biological reactions can be monitored by CD, and even the ACs of unstable compounds can be determined by this method [94].

The ACs of secondary alcohols and hydroxyphoshontes can be determined by the Mosher's method using MTPA. This method is very convenient as it does not require crystallization of compounds. Although this method was first applied to the secondary alcohols, it can be used for other kinds of compounds. The AC of new compounds can also be determined by the method of chemical correlation and comparison of optical rotation, $[\alpha]_D$, or CD spectra with reference compounds.

1.4.1 X-ray Crystal Analysis

There are several ways to determine the absolute structure by X-ray crystallography. For example, a comparison of the intensities of Bijvoet pairs [10] or of the R factors for the two possible structures can suggest the correct absolute configuration (AC). One of the most effective approaches is application of the Flack parameter, because this parameter unambiguously indicates the AC of molecule. In this case, the AC can be easily defined using the chirality of the auxiliary introduced as an internal reference. Consequently, the samples do not need to contain heavy atoms for an anomalous dispersion effect. The result obtained is one-valued, even when the final *R*-value is not small enough owing to poor quality of the single crystal. The AC can be determined, even if only the relative configuration is obtained. A number of methods to link an internal reference to the target molecule have been developed and described, for example, ionic acidic or basic salts, covalent esters or amides, or various inclusion complexes.

Normal X-ray crystallography cannot distinguish between enantiomers. If the sample includes only light nuclei, the interference pattern is determined only by internuclear separation, and the phase coincidence is independent of the spatial orientation of these nuclei. Thus, from the diffraction pattern, it is possible to calculate various internuclear distances and constitutions in the molecule and to deduce the relative positions of these nuclei in space. One can build the relative configuration of a compound, but it is normally difficult to distinguish enantiomers or to get the ACs for chiral compounds containing only light atoms. Because of a phase delay or abnormal dispersion, the interference pattern depends not only on interatomic distances but also on their relative positions in space, thus allowing to define the AC of moleculas containing heavy atoms. The AC of a compound can be obtained by determining the relative configuration at the position of interest against a reference compound or a substituent with known AC. A typical example is the X-ray crystallography taken after the introduction of a chiral auxiliary with known absolute configuration. In this case the absolute configuration can be determined using the chirality of the auxiliary introduced as an internal reference. For a molecule without a heavy atom, the AC can be determined by attaching another chiral moiety of known configuration to the sample.

Thus, the AC of the phosphor atoms in the quinoline salt of (+)-(*R*)-*O*-ethyl-*O*-phenylphosphorothioic acid and the brucine salt of (−)-(*S*)-ethyl phenylthiophosphonous acid were determined by X-ray single-crystal diffraction analysis (Scheme 1.9) [92, 96].

In X-ray crystallography, the Flack parameter is used to estimate the AC which is determined by single-crystal X-ray structural analysis. This parameter, introduced by H. D. Flack, became one of a standard set of values being checked for structures with noncentrosymmetric space groups. This approach determines the absolute structure of a noncentrosymmetric crystal. The Flack parameter can be calculated during structural refinement using the following equation: $I(hkl) = (1-x)[F(hkl)]^2 + x[F(-h-k-l)]^2$, where x is the Flack parameter, I is the square of the scaled observed structure factor, and F is the calculated structure factor [97]. The value x determined for all data usually should lie between 0 and 1. In the case when the value of the Flack parameter is near 0, the absolute structure determined by structural analysis is evidently correct; if the value is near 1, then the inverted structure is correct. However, if the value is near 0.5, the crystal should be racemic or twinned. The technique is most effective when the crystal contains both lighter and heavier atoms. Light atoms usually show only a small anomalous dispersion effect. In this case, the Flack parameter can refine to a physically unrealistic value (<0 or >1) and has no meaning.

For example, Scheme 1.9 shows an example of X-ray structural analysis of the molecule for which the Flack parameter was refined to −0.05(8), which is close to 0, that is, the determined AC (R_P, *S*) is correct. Crystallographic data should to be registered in the Crystallographic center, and the registration number should be obtained without which it is not possible to publish the article. The crystallographic data (excluding the structure factors) for the structure were deposited at the Cambridge Crystallographic Data Centre, as supplementary publication no. CCDC 195666 [95].

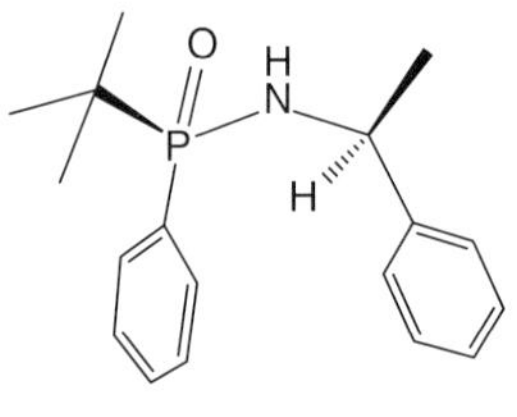

Scheme 1.9 Example of single-crystal X-ray structural analysis of *t*-Bu(Ph)P(O)NHCH(Me)Ph [95].

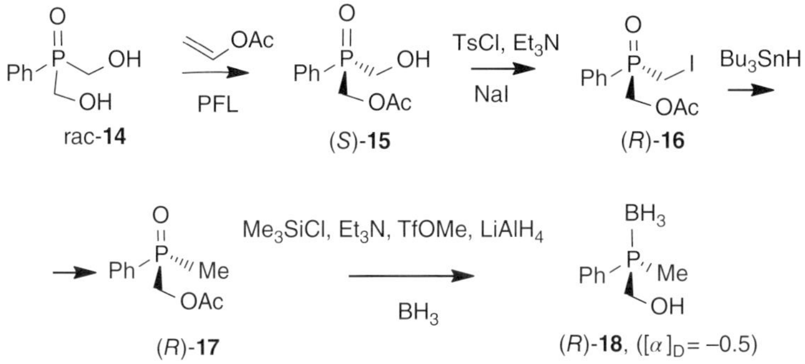

Scheme 1.10

1.4.2 Method of Chemical Correlation

AC can be determined relatively by chemical correlation, or by comparison of the optical rotation or CD spectrum of the compound in question with that of reference compounds with known AC. Although this method is frequently used, a careful selection of reference compounds is necessary for reliable analysis [98, 99]. For example, the biocatalytic acetylation of prochiral bis(hydroxymethyl) phenylphosphine oxide **14** in the presence of lipase PFL (*Pseudomonas fluorescences* lipase) led to the formation of chiral compound (*S*)-**15** in 50% yield and with 79% ee. The AC of compound (*S*)-**15** was determined by chemical correlation (Scheme 1.10). To this end, alcohol (*S*)-**15** was converted to iodide (*R*)-**16** which was then reduced to phosphinate (*R*)-**17** whose transformation resulted in a borane complex of phosphinic acid (*R*)-**18** with known AC [100].

Mastalerz *et al.* [99] have established the ACs of a number of optically active phosphonic analogs of serine, α-chloroalanine, phenylalanine, tyrosine, and 2-aziridinephosphonic acid via chemical correlations with phosphonic analogs of alanine or aspartic acid of known configuration. For example, by conversion of (−)-PheP **19** to TyrP **21**, the configuration of TyrP as (*R*)-(−) was established. The conversion was accomplished by nitration, reduction, and diazotization. Specific rotations indicate that the *p*-nitro **20**, and *p*-amino congeners of TyrP **21** also have the (*R*)-(−) configuration (Scheme 1.11).

1.4.3 The Assignment of Absolute Configuration by NMR

The ACs of secondary alcohols are frequently determined by an advanced Mosher method developed by Kusumi *et al.* [101, 102]. In this case, the ACs of chiral

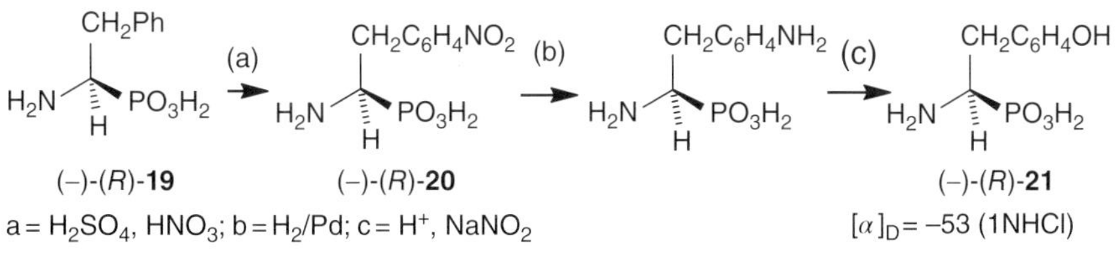

Scheme 1.11 Determination of absolute configurations by chemical correlation.

auxiliaries, such as α-methoxy-α-trifluoromethylphenylacetic acid (MTPA) or α-methoxyphenylacetic acid (MPA)], are known, and the preferred conformation of the esters formed with chiral secondary alcohols and MTPA or MPA acid is rationalized. Moreover, the phenyl group generates the effect of magnetic anisotropy because of the ring current that is induced by an external magnetic field, and therefore the NMR signals of alcohol shielded by the phenyl group are displaced to a higher magnetic field. By observing the ^{1}H NMR or ^{31}P anisotropy effect, we can determine the AC of the chiral compound. This method is particularly convenient because it does not require special purification of compounds [101, 102]. This method involves converting of chiral hydroxyphosphonates to their corresponding MTPA esters, followed by NMR analysis of the resulting derivatives. Mosher proposed that the carbonyl proton and ester carbonyl, as well as the trifluoromethyl group of an MTPA moiety, lie in the same plane. Calculations on this MTPA ester demonstrate that the proposed conformation is just one of two stable conformations. As a result of the diamagnetic effect of the benzene ring, the NMR signals of protons of the (*R*)-MTPA ester should appear upfield relative to those of the (*S*)-MTPA ester. Therefore, for $\Delta\delta = \delta S - \delta R$, protons on the right side of the MTPA have positive values ($\Delta\delta > 0$) and protons on the left side of the plane have negative values ($\Delta\delta < 0$).

> *Determination of the absolute configuration of hydroxyphosphonates (or secondary alcohols) by the Mosher-Kusumi method (Scheme* 1.13*).* To a solution of 0.1 mmol of any hydroxyphosphonate in CH_2Cl_2, 0.2 mmol of anhydrous pyridine is added. Then 0.1 mmol of (*R*)-(−)-MTPA-Cl is added to the first sample (marked A), and 0.1 mmol of (*S*)-(+)-MTPA-Cl to the second sample (marked B). The progress of the reaction is monitored by thin-layer chromatography, by eluting with a solvent (usually, hexane-ethyl acetate in a 4 : 1 ratio) and visualizing the plate using ultraviolet light or using any other appropriate method. The MTPA esters will have a higher R_f value compared to the starting alcohol. The product can be isolated by partitioning the reaction mixtures between diethyl ether and water, followed by extraction of the aqueous phase with ether. The ether phases are dried and the residues are dissolved in $CDCl_3$ for NMR analysis. Then ^{1}H NMR spectra of both esters are recorded and the chemical shift differences ($\Delta\delta = \delta S - \delta R$) between the diastereomers **22** and **23** are calculated by gathering the positive and negative $\Delta\delta$ values together. The AC of the hydroxyphosphonate can then be determined by using the formula shown below with $\Delta\delta < 0$ and $\Delta\delta > 0$ replaced by the actual structure fragments that exhibit positive and negative differences, respectively.

Kakisawa *et al.* [102] described the application of a modified Mosher's method to the N-MTPA derivatives of amino acid esters and acyclic amines, indicating that this method may be generally used to determine the ACs of the α-carbons of amino compounds. For the assignment of AC of hydroxy-, aminophosphonates, as well as of secondary alcohols or amines, the model of the target molecule should be constructed and it should be confirmed that all assigned nuclei with positive and negative $\Delta\delta$ values are actually found on the right and left sides of the MTPA plane, respectively (Scheme 1.12).

The absolute values of $\Delta\delta$ must be proportional to the distance from the MTPA moiety. Evidently, to correlate the information obtained from the NMR spectra with the AC, a detailed understanding of the structure as well as the strength and direction of

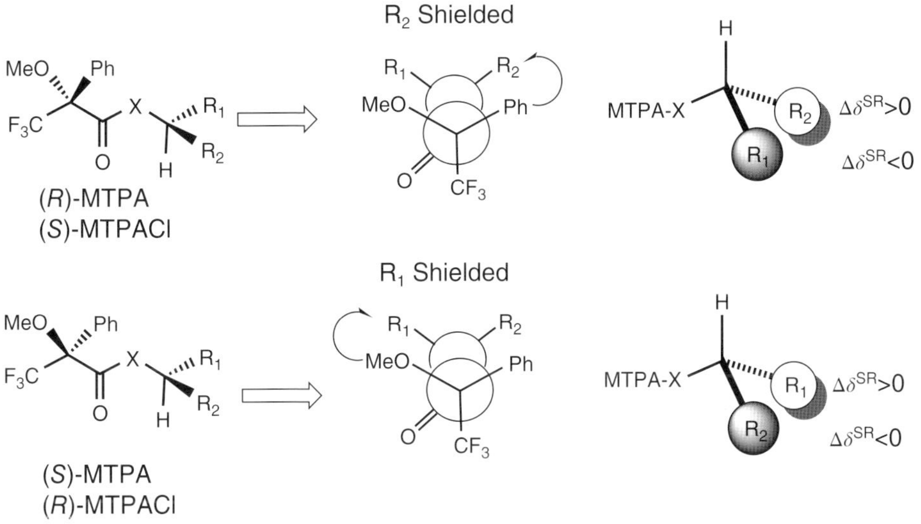

Scheme 1.12 Model to predict the absolute configurations of secondary alcohols, hydroxyphosphonates (X = O), primary amines or aminophosphonates (X = NH).

the anisotropic effect of Ph on R_1 and R_2 in each conformation is necessary. When these conditions are fulfilled, the model should indicate the correct AC of the target compound.

The AC of many α-hydroxyphosphonates was determined by derivatization with MTPA and application of ^{1}H, ^{19}F, and ^{31}P NMR spectroscopy [32, 33, 39]. Chemical shifts δ_P of derivatized (*S*)-hydroxyphosphonates **22** are usually in downfield relationship to signals of the corresponding (*R*)-hydroxyphosphonates **23**. The difference in chemical shifts δ_P (0.40 – 1.09 ppm) allows us to determine ACs of hydroxyphosphonates (Scheme 1.13).

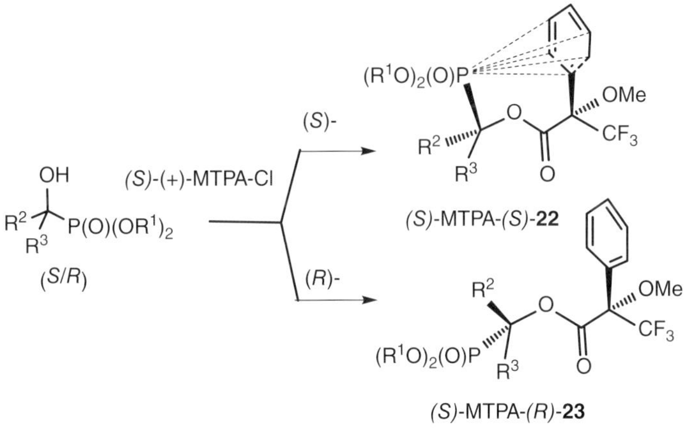

Scheme 1.13 Determination of configuration of hydroxyphosphonates by means of MTPA.

The conformation model of Mosher's esters derived from α-hydroxyphosphonates shows that the trifluoromethyl group and the carbonyl hydrogen at the C=O group (R_2 = H, D) are eclipsed by the carbonyl oxygen. The phosphorus atom in the (*R*)-MTPA ester is shielded by the phenyl group if the chiral alcohol has an (*R*)-configuration at C-1 relative to alcohol having an (*S*)-configuration. Therefore, the chemical shift of the phosphorus atom in the ^{31}P NMR spectra of the (*R*)-MTPA derivatives of (*R*)-hydroxyphosphonates will be upfield when compared with those of the (*S*)-alcohol.

The ^{31}P NMR spectra of (*R*)-MTPA esters with (*S*)-hydroxyphosphonates **24** confirmed that their signals δ_P are indeed downfield and that signals δ_P of (*R*)-hydroxyphosphonates **25** derivatized with (*R*)-MTPA esters are upfield. The shift differences were within 0.28–0.50 ppm (Table 1.2) [61]. In another work, Hammerschmidt and Wuggenig [103] have analyzed the resolution of phosphonates catalyzed by enzymes and confirmed Mosher's assumption that signals δ_P of (1*R*,2*R*)-esters and (*R*)-MTPA should be upfield in relation to (*S*)-MTPA esters (Scheme 1.14). The different orientation of the aromatic shielding cone effect in the diastereomeric derivatives leads to a selective shielding or deshielding of the R^1 or R^2 substituents at the asymmetric center. The AC of α-hydroxyphosphonates can be determined by ^{1}H and ^{31}P NMR spectroscopy of the mandelate ester derivatives. The observed chemical shifts allow assignment of the AC of the hydroxyphosphonates depending on the position of the phenyl group in the compounds and its shielding effect. Thus, the ^{1}H NMR spectra of (1*S*,2*R*)-diastereomers showed a downfield shift of signals for the *O*-methyl protons relative to the parent alcohol. The AC of compounds was additionally determined by

Table 1.2 Chemical shifts (δ_P) in the ^{31}P NMR spectra of Mosher's esters (Scheme 1.13).

Mosher ester			δ_P (ppm)		Δδ	References
R^1	R^2	R^3	(*S*)	(*R*)	[δ(*S*)−δ(*R*)]	
Et	H	Me	22.21	21.77	0.44	[61]
Pr	H	Me	21.9	21.5	0.40	[61]
i-Pr	H	Et	19.33	18.93	0.40	[61]
Bu	H	Me	19.80	19.39	0.41	[61]
C_5H_{11}	H	Et	19.90	19.43	0.47	[61]
Et	H	Et	19.85	19.37	0.48	—
$PhCH_2CH_2$	H	*i*-Pr	17.42	17.01	0.41	[61]

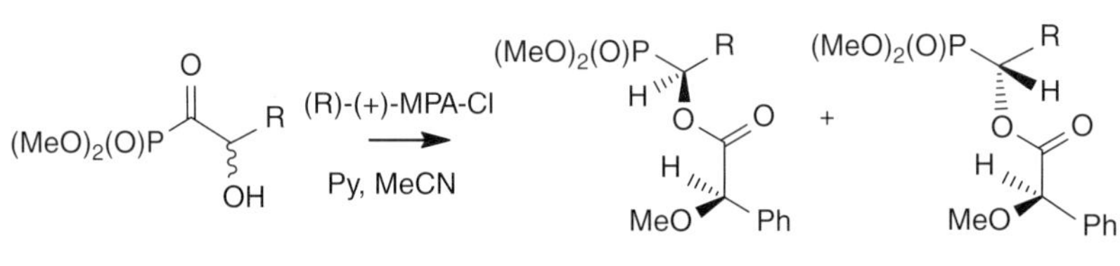

(1*S*,2*R*)-**24** (1*R*,2*R*)-**25**

Scheme 1.14 Determination of absolute configuration by derivatization of hydroxyphosphonates with MPA.

X-ray analysis. Kozlowski and Ordóñez [63, 64] determined the enantiomeric purity and the AC of α-hydroxy phosphonates **24** using esterification with mandelic acid. The spatial relationship between R^1/R^2 and the aryl ring is correlated to the observed chemical shift change. The R^1 substituent of the (*R*,*R*)-diastereomer is at higher field than the R^2 substituent. Inversely, R^2 in the (*S*,*R*)-derivative shifts more upfield relative to R^1. The upfield or downfield shifts of the signals for the methyl-group protons in the ^{1}H NMR spectra of compounds depend on the arrangement of the phenyl group in the molecule and its shielding effect (Scheme 1.14).

Blazewska *et al.* [104, 105] have described the assignment of the AC of hydroxy- and aminophosphonates by their double derivatization with commercially available naproxen as CDA. They have shown that the diethoxyphosphoryl group is dishielded in esters of (*R*)-α-aminophosphonates with (*S*)-naproxen and in esters of (*S*)-1-aminophosphonates with (*R*)-naproxen. The correlation between the spatial arrangement around the stereogenic carbon center and the signs of the $\Delta\delta_{RS}$ allows determination of the AC of hydroxy- and aminophosphonates by comparison of the ^{1}H and ^{31}P NMR spectra of the (*R*)- and (*S*)-naproxen ester or amide derivatives. The proposed model is also valid for 2-aminophosphonates on consideration of the change in the order of the initial substituent. The analysis can be carried out by ^{1}H or ^{31}P NMR; however, the determination of AC is essentially simplified in case of ^{31}P NMR. This method was applied for the determination of the AC for a series of hydroxyphosphonates (Scheme 1.15).

Yokomatsu *et al.* [106] determined the AC of dihydroxyphosphonates **26** by converting them into the cyclic acetonides [107, 108]. The dihedral angles between HCCP were calculated by the MOPAC semiempirical program. On the basis of these calculations and the phosphorus version of the Karplus equations, a large vicinal proton–phosphorus coupling constant ($^3J_{PH} = 17.2$ Hz) was expected for *trans*-**27**, while a small coupling constant ($^3J_{PH} = 1.7$ Hz) was assumed for the *cis*-isomer. Careful analysis of the ^{1}H NMR spectrum of the *cis*- and *trans*-isomers established the vicinal coupling constant to be 10.1 and 9.8 Hz, respectively, suggesting their *trans*-stereochemistry. The (*S*)-AC of compounds was confirmed also by means of ^{31}P NMR analysis of their (*R*)-MTPA esters. The δ_P chemical shifts of (*S*)-diastereomers in the low field of the ^{31}P NMR spectra were assigned to the (*S*)-configuration. Hence the results obtained by the two alternative methods coincided and consequently the AC of initial compounds **26** was unambiguously established as (1*S*,2*S*) (Scheme 1.16).

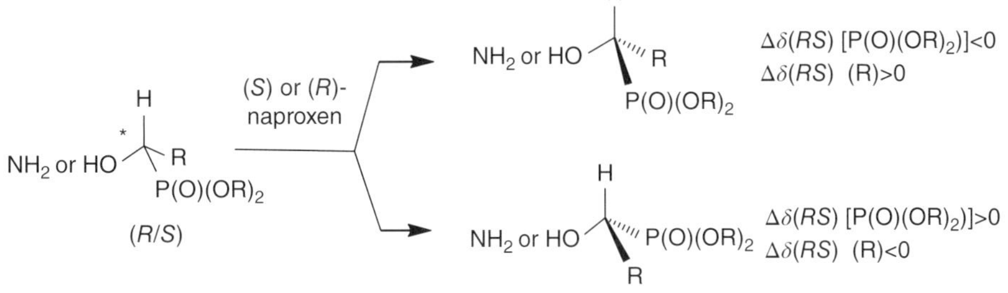

Scheme 1.15 Determination of absolute configuration of hydroxy- or aminophosphonates, using naproxen as CDA.

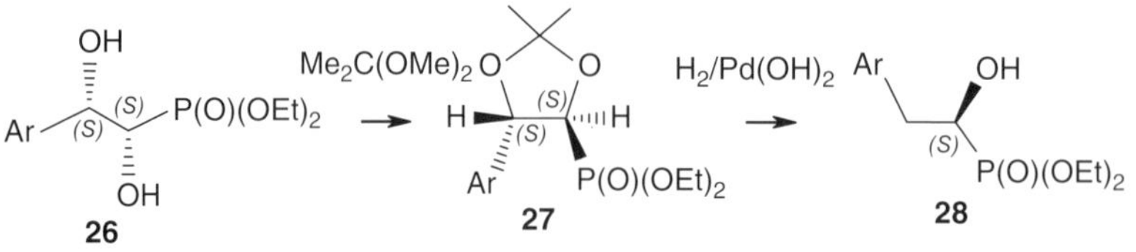

Scheme 1.16 Modification of dihydroxyphosphonate for determination of absolute configuration Ar = 3-MeOC_6H_4, 4-ClC_6H_4, furyl, 1-naphthyl.

The AC of a series of 3,3′-disubstituted-MeO-BIPHEP derivatives was determined by the ^{1}H NMR spectra of the methoxyl group when the 3,3′-disubstituted-MeO-BIPHEP derivative was mixed with (−)-(2*R*,3*R*)-dibenzoyltartaric acid ((−)-DBTA) (1 : 2) and its NMR spectrum was run in $CDCl_3$. The chemical shift of the methoxyl group in the S_{ax} enantiomer occurred at higher field than that in the corresponding R_{ax} enantiomer [109].

The CD method is successfully used for determination of the AC of hydroxyphosphonates [106, 110, 111]. CD involves circularly polarized light. Left-hand circular (LHC) and right-hand circular (RHC) polarized light represent two possible spin angular momentum states for a photon, and so CD is also referred to as dichroism for spin angular momentum. It is exhibited in the absorption bands of optically active chiral molecules. For example, Wynberg and co-workers [110] used the CD spectra to determine the AC of a number of chiral α-hydroxy phosphonates. The enantiomers having (*S*)-configuration show a negative Cotton effect at 225 nm. Yokomatsu *et al.* [106] used the CD spectra to determine the AC of 1,2-dihydroxyphosphonates. It was found that α,β-dihydroxy phosphonates having the (1*S*,2*S*)-configuration show a positive Cotton effect at 210 – 230 nm, whereas (1*R*,2*R*)-isomers show a negative Cotton effect at these wavelengths (Scheme 1.17). Keglevich determined successfully the AC of 3-phospholene oxides by CD spectroscopy using quantum chemical calculations for the analysis of the spectra [81].

1.5 Asymmetric Induction and Stereochemistry

1.5.1 Asymmetric Induction

Asymmetric induction in stereochemistry describes the preferential formation in a chemical reaction of one enantiomer or diastereoisomer over the other as a result of the influence of a chiral inductor present in the substrate, reagent, catalyst, or reaction medium [1, 6]. There are three types of asymmetric induction: (i) internal asymmetric

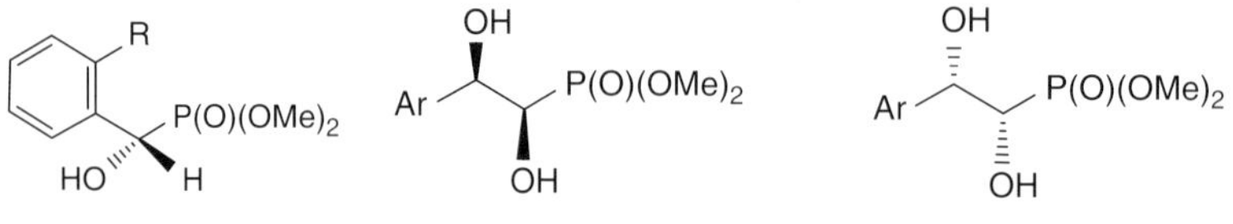

R = H, F, Cl, Br, NO_2, NH_2 (1*R*,2*R*)-**29** (1*S*,2*S*)-**30**

Ar = 3-MeOC_6H_4, 4-ClC_6H_4, Furyl, 1-Naphthyl

Scheme 1.17 Examples of 1,2-dihydroxyphosphonates the absolute configuration of which was defined by means of CDs-spectra.

induction, which uses a chiral center bound to the reactive center through a covalent bond and remains so during the reaction, (ii) relayed asymmetric induction, which uses a chiral information that is introduced in a separate step and then removed in a separate chemical reaction, and (iii) external asymmetric induction, in which chiral information is introduced in the transition state through a chiral catalyst or chiral auxiliary. A chiral auxiliary is a chemical compound or unit that is temporarily incorporated into an organic synthesis in order to control the stereochemical outcome of the synthesis. Asymmetric induction can involve chiral features in the substrate, reagent, catalyst, or environment and it works by making the activation energy required to form one enantiomer lower than that of the opposing enantiomer [112]. Asymmetric induction can occur intramolecularly when given a chiral starting material. This behavior can be exploited, especially when the goal is to make several consecutive chiral centers to give a specific enantiomer of a specific diastereomer. When two reactants of a reaction are stereogenic, the stereogenic elements of each reactant may operate either in concert (matched pair) or in opposition (mismatched pair). This phenomenon is known [113–115] as double asymmetric induction or double diastereoselection. See Section 3.5.

1.5.2 Asymmetric Synthesis

This involves a reaction that selectively creates one configuration of one or more new stereogenic elements by the action of a chiral reagent or auxiliary, acting on heterotopic faces, atoms, or groups of a substrate. The stereoselectivity is primarily influenced by the chiral catalyst, reagent, or auxiliary, despite any stereogenic elements that may be present in the substrate.

1.5.3 Asymmetric Transformation

This involves the conversion of a racemic mixture of stereoisomers into a single stereoisomer or a mixture in which one isomer predominates. An "asymmetric transformation of the first kind" involves such a conversion without separation of the stereoisomers, whereas in an "asymmetric transformation of the second kind," separation, such as an equilibration, is accompanied by selective crystallization of one stereoisomer.

1.5.4 An Enantioselective Reaction

An enantioselective reaction is one in which one enantiomer is formed in preference to the other, in a reaction that creates an optically active product from an achiral starting material, using either a chiral catalyst, an enzyme, or a chiral reagent. An important variant is the *kinetic resolution*, in which a pre-existing chiral center undergoes reaction with a chiral catalyst, an enzyme, or a chiral reagent such that one enantiomer reacts faster than the other and leaves behind the less reactive enantiomer, or in which a pre-existing chiral center influences the reactivity of a reaction center elsewhere in the same molecule.

1.5.5 Enantioselective Synthesis

Enantioselective synthesis (or asymmetric synthesis), is defined by IUPAC as "a chemical reaction in which one or more new elements of chirality are formed in a substrate molecule and which produces the stereoisomeric (enantiomeric or diastereoisomeric) products in unequal amounts." Enantioselective synthesis is a key process in modern chemistry and is particularly important in the field of pharmaceuticals, as the different enantiomers or diastereomers of a molecule often have different biological activity.

Enantiomers possess identical enthalpies and entropies, and hence should be produced in equal amounts by an undirected process – leading to a racemic mixture. The solution is to introduce a chiral feature which will promote the formation of one enantiomer over another via interactions at the transition state. Reactions giving unequal amounts of stereoisomers are called stereo differentiating and prefixed according to the nature of the substrate as enantiomer- and diastereomer-, enantiotopos- and diastereotopos-, and further enantioface- and diastereoface-differentiating reactions, according to whether stereoisomers, groups, or faces are differentiated. Note that the first two types cover substrate-selective transformations, while the last four product-selective ones. Izumi's classification is rather appealing because the conditions of selectivity can be defined very simply: enantio-differentiation requires chiral means, whereas diastereo-differentiation does not. The development of chiral catalysts is the most significant success in asymmetric synthesis in the last decades, which is capable to invoke transformation of achiral substrates in chiral products. Asymmetric catalysis, in which one molecule of chiral catalyst can yield many molecules of chiral product, has significant potential advantages over these older procedures. The catalysts are typically rendered chiral by using chiral ligands; however, it is also possible to generate chiral-at-metal complexes using simpler achiral ligands. Most enantioselective catalysts are effective at low concentrations making them well suited to industrial-scale synthesis, as even exotic and expensive catalysts can be used affordably. Compounds with a center of prochirality as well as a center of chirality can be transformed to a mixture of diastereomers without the interference of an additional source of chirality. In an enantioselective synthesis, the differentiation between the two enantiomers is possible by the same principle of diastereomeric transition states created under influence of additional sources of chirality (e.g., chiral catalyst). These methods are especially valuable for reactions in which two new stereogene centers are formed stereoselectively in one step as, for example, in the case of multistereoselectivity [114, 115].

1.6 Summary

This chapter provides general information concerning stereochemistry, nomenclature of chiral systems, determination of enantiomer composition, determination of AC as well as some other stereochemical terms used in organophosphorus chemistry. As the purpose of this book is asymmetric synthesis, the remaining chapters provide details of asymmetric syntheses of chiral organophosphorus compounds.

References

1 Rosen, J. (2008) Symmetry Rules: How Science and Nature Are Founded on Symmetry, The Frontiers Collection, Springer-Verlag, Berlin, Heidelberg, 312 pp.

2 Crosland, M.P. (1970–1980) Biot, Jean-Baptiste, in Dictionary of Scientific Biography, vol. **2**, Charles Scribner's Sons, New York, pp. 133–140.

3 Flack, H.D. (2009) Louis Pasteur's 1848 discovery of molecular chirality and spontaneous resolution, together with a complete review of his chemical and crystallographic work. *Acta Crystallogr., Sect. A: Found. Crystallogr.*, **A65** (5), 371–389.

4 Beckmann, E. (1905+) *Johannes Wislicenus*, vol. **37**, Berichte der deutschen chemischen Gesellschaft, Verlag Chemie, pp. 4861–4946.

5 Nobelprize.org (2015) Jacobus H. van't Hoff Biographical. Nobel Media AB 2014, http://www.nobelprize.org/nobel_prizes/chemistry/laureates/1901/hoff-bio.html (accessed 30 October 2015).

6 Fischer, E. (1890) Ueber die optischen Isomeren des Traubezuckers, der Gluconsäure und der Zuckersäure. *Ber. Dtsch. Chem. Ges.*, **23**, 2611–2624.

7 Bredig, G. and Fiske, P.S. (1912) Durch Katalysatoren bewirkte asymmetrische Synthese. *Biochem. Z.*, **46** (1), 7.

8 Marckwald, W. (1904) Ueber asymmetrische Synthes. *Ber. Dtsch. Chem. Ges.*, **37** (3), 349–354.

9 Pracejus, H. (1960) Organische Katalysatoren, LXI. Asymmetrische synthesen mit ketenen, 1 Alkaloid-katalysierte asymmetrische Synthesen von α–Phenyl-propionsaureestern. *Justus Liebigs Ann. Chem.*, **634** (1), 9–22.

10 Bijvoet, J.M., Peerdeman, A.F., and Van Bommel, A.J. (1951) Determination of the absolute configuration of optically active compounds by means of X-rays. *Nature*, **168** (4268), 271–272.

11 Dale, J.A. and Mosher, H.S. (1973) Nuclear magnetic resonance enantiomer reagents. Configurational correlations via nuclear magnetic resonance chemical shifts of diastereomeric mandelate, O-methylmandelate, and alpha, methoxy-.alpha, trifluoromethylphenylacetate (MTPA) esters. *J. Am. Chem. Soc.*, **95** (2), 512–519.

12 Sanders, J.K.M. and Williams, D.H. (1972) Shift reagents in NMR spectroscopy. *Nature*, **240**, 385–390.

13 Corey, E.J. and Ensley, H.E. (1975) Preparation of an optically active prostaglandin intermediate via asymmetric induction. *J. Am. Chem. Soc.*, **97** (23), 6908–6909.

14 Kunz, H. (2005) Leopold Horner (1911–2005): Nestor of preparative organic chemistry. *Angew. Chem. Int. Ed.*, **44** (47), 7664–7665.

15 (a) Horner, L. (1980) Dreibindige, optisch aktive Phosphorverbindungen, ihre Synthese, chemische Eigenschaften und Bedeutung fur die asymmetrische Homogenhydrierung. *Pure Appl. Chem.*, **52** (4), 843–858; (b) Horner, L., Siegel, H., and Buthe, H. (1968) Asymmetric catalytic hydrogenation with an optically active phosphinerhodium complex in homogeneous solution. *Angew. Chem. int. Edit.*, **7** (12), 942–942.

16 (a) Knowles, W.S. (2002) Asymmetric hydrogenations (Nobel Lecture). *Angew. Chem. Int. Ed.*, **41** (12), 1998; (b) Knowles, W.S. and Sabacky, M.J. (1968) Catalytic asymmetric hydrogenation employing a soluble, optically active, rhodium complex. *J. Chem. Soc., Chem. Commun.*, (22), 1445–1446

17 Noyori, R. (2003) Asymmetric catalysis: science and opportunities (Nobel Lecture 2001). *Adv. Synth. Catal.*, **345** (1-2), 15–32.

18 Schöllkopf, U., Hoppe, I., and Thiele, A. (1985) Asymmetric synthesis of α-aminophosphonic acids, I enantioselective synthesis of L-(1-aminoethyl)phosphonic acid by asymmetric catalytic hydrogenation of N-[1-(dimethoxyphosphoryl)ethenyl]formamide. *Liebigs Ann. Chem.*, **3**, 555–559.

19 Kagan, H.B. and Phat, D.T. (1972) Asymmetric catalytic reduction with transition metal complexes. I. Catalytic system of rhodium (I) with (−)-2,3-0-isopropylidene-2,3-dihydroxy-1,4-bis (diphenylphosphino)butane, a new chiral diphosphine. *J. Am. Chem. Soc.*, **94** (18), 6429–6433.
20 Crépy, K.V.L. and Imamoto, T.P. (2003) Chirogenic phosphine ligands, in *Topics in Current Chemistry*, vol. **229** (ed. J.-P. Majoral), Springer-Verlag, Berlin, Heidelberg, pp. 1–41.
21 Jugé, S. (2008) Enantioselective synthesis of *P*-chirogenic phosphorus compounds via the ephedrine-borane complex methodology. *Phosphorus, Sulfur Silicon Relat. Elem.*, **183** (2-3), 233–248.
22 McEwen, W.E. and Berlin, K.D. (1975) *Organophosphorus Stereochemistry*, Parts 2, John Wiley & Sons, Inc., New York, 320 pp.
23 Żymańczyk-Duda, E., Lejczak, B., Kafarski, P., Grimaud, J., and Fischer, P. (1995) Enantioselective reduction of diethyl 2-oxoalkylphosphonates by baker's yeast. *Tetrahedron*, **51** (43), 11809–11814.
24 Kiełbasinski, P. and Mikołajczyk, M. (2007) in *Future Directions in Biocatalysis* (ed. T. Matsuda), Elsevier, p. 159.
25 Pietrusiewicz, K.M., Holody, W., Koprowski, M., Cicchi, S., Goti, A., and Brandi, A. (1999) Asymmetric and doubly asymmetric 1,3-dipolar cycloadditions in the synthesis of enantiopure organophosphorus compounds. *Phosphorus, Sulfur Silicon Relat. Elem.*, **146** (1), 389–392.
26 Drabowicz, J., Kudelska, W., Lopusinski, A., and Zajac, A. (2007) The chemistry of phosphinic and phosphinous acid derivatives containing t-butyl group as a single bulky substituent: synthetic, mechanistic and stereochemical aspects. *Curr. Org. Chem.*, **11** (1), 3–15.
27 Börner, A. (ed.) (2008) Phosphorus Ligands in Asymmetric Catalysis: Synthesis and Applications, vol. **1**, John Wiley & Sons, Inc., 1546 pp.
28 Moss, G.P. (1996) Basic terminology of stereochemistry (IUAC Recommendations 1996). *Pure Appl. Chem.*, **68** (12), 2193–2222.
29 Mata, P., Lobo, A.M., Marshall, C., and Johnson, A.P. (1993) The CIP sequence rules: analysis and proposal for a revision. *Tetrahedron: Asymmetry*, **4** (4), 657–668.
30 IUPAC (2004) Provisional Recommendation, Chemical Nomenclature and Structure Representation Division. Nomenclature of Organic Chemistry. Preferred IUPAC Names, September, Chapter 9, p. 59.
31 Pirkle, W.H. and Finn, J. (1983) in *Asymmetric Synthesis*, vol. **1** (ed. J.D. Morrison), Academic Press, New York, p. 201.
32 Pirkle, W.H. and Hoover, D.J. (1982) NMR chiral solvating agents. *Top. Stereochem.*, **13**, 263.
33 Weisman, G.R. (1983) in *Asymmetric Synthesis*, vol. **1** (ed. J.D. Morrison), Academic Press, New York, pp. 153–172.
34 Parker, D. (1991) NMR determination of enantiomeric purity. *Chem. Rev.*, **91** (7), 1441–1451.
35 Li, Y. and Raushe, F.M. (2007) Differentiation of chiral phosphorus enantiomers by ^{31}P and ^{1}H NMR spectroscopy using amino acid derivatives as chemical solvating agents. *Tetrahedron: Asymmetry*, **18** (12), 1391–1397.
36 Uccello-Barretta, G., Pini, D., Mastantuono, A., and Salvadori, P. (1995) Direct NMR assay of the enantiomeric purity of chiral β-hydroxy esters by using quinine as chiral solvating agent. *Tetrahedron: Asymmetry*, **6** (8), 1965–1972.

37 Zymariczyk-Duda, E., Skwarczynski, M., Lejczak, B., and Kafarski, P. (1996) Accurate assay of enantiopurity of 1-hydroxy- and 2-hydroxyaikylphosphonate esters. *Tetrahedron: Asymmetry*, **7** (5), 1277–1280.

38 Maly, A., Lejczak, B., and Kafarski, P. (2003) Quinine as chiral discriminator for determination of enantiomeric excess of diethyl 1,2-dihydroxyalkanephosphonates. *Tetrahedron: Asymmetry*, **14** (8), 1019–1024.

39 Majewska, P., Doskocz, M., Lejczak, B., and Kafarski, P. (2009) Enzymatic resolution of α-hydroxyphosphinates with two stereogenic centres and determination of absolute configuration of stereoisomers obtained. *Tetrahedron: Asymmetry*, **20** (13), 1568–1574.

40 Kolodyazhnyi, O.I., Kolodyazhna, A.O., and Kukhar, V.P. (2006) A method for determining the optical purity of hydroxy phosphonates using cinchonidine as chiral solvating reagent. *Russ. J. Gen. Chem.*, **74** (8), 1342–1343.

41 Dunach, E. and Kagan, H.B. (1985) A simple chiral shift reagent for measurement of enantiomeric excesses of phosphine oxides. *Tetrahedron Lett.*, **26** (22), 2649–2652.

42 Demchuk, O.M., Swierczynska, W., Pietrusiewicz, M.K., Woznica, M., Wójcik, D., and Frelek, J.A. (2008) Convenient application of the NMR and CD methodologies for the determination of enantiomeric ratio and absolute configuration of chiral atropoisomeric phosphine oxides. *Tetrahedron: Asymmetry*, **19** (20), 2339–2345.

43 Pakulski, Z., Demchuk, O.M., Kwiatosz, R., Osinski, P.W., Swierczynska, W., and Pietrusiewicza, K.M. (2003) The classical Kagan's amides are still practical NMR chiral shift reagents: determination of enantiomeric purity of *P*-chirogenic phospholene oxides. *Tetrahedron: Asymmetry*, **14** (11), 1459–1462.

44 Ma, F., Shen, X., Ou-Yang, J., Deng, Z., and Zhang, C. (2008) Macrocyclic compounds as chiral solvating agents for phosphinic, phosphonic, and phosphoric acids. *Tetrahedron: Asymmetry*, **19** (1), 31–37.

45 Drabowicz, J., Pokora-Sobczak, P., Krasowska, D., and Czarnocki, Z. (2014) Optically active t-butylphenylphosphinothioic acid: synthesis, selected structural studies and applications as a chiral solvating agent. *Phosphorus, Sulfur Silicon Relat. Elem.*, **189** (7-8), 977–991.

46 Wenzel, T.J. (2007) Discrimination of Chiral Compounds Using NMR Spectroscopy, John Wiley & Sons, Inc., Hoboken, NJ.

47 Wenzel, T.J. and Chisholm, C.D. (2011) Assignment of absolute configuration using chiral reagents and NMR spectroscopy. *Chirality*, **23** (3), 190–214.

48 Rudzinska, E., Dziedzioła, G., Berlicki, Ł., and Kafarski, P. (2010) Enantiodifferentiation of α-hydroxyalkanephosphonic acids in 31P NMR with application of α-cyclodextrin as chiral discriminating agent. *Chirality*, **22** (1), 63–68.

49 Berlicki, Ł., Rudzinska, E., and Kafarski, P. (2003) Enantiodifferentiation aminophosphonic и aminophosphinic кислот acids with α– and β-cyclodextrins. *Tetrahedron: Asymmetry*, **14** (11), 1535–1539.

50 Berlicki, L., Rudzinska, E., Mucha, A., and Kafarski, P. (2004) Cyclodextrins as NMR probes in the study of the enantiomeric compositions of N-benzyloxycarbonylamino-phosphonic and phosphinic acids. *Tetrahedron: Asymmetry*, **15** (10), 1597–1602.

51 Malinowska, B., Młynarz, P., and Lejczak, B. (2012) Chemical and biocatalytical methods of determination of stereomeric composition of 1,4-di[(diethoxylphosphoryl) hydroxymethyl]benzene. *ARKIVOC*, IV, 299–313.

52 Rudzinska, E., Berlicki, Ł., Mucha, A., and Kafarski, P. (2007) Chiral discrimination of ethyl and phenyl N-benzyloxycarbonylaminophosphonates by cyclodextrins. *Tetrahedron: Asymmetry*, **18** (13), 1579–1584.

53 Rudzinska, E., Poliwoda, A., Berlicki, Ł., Mucha, A., Dzygiel, P., Wieczorek, P.P., and Kafarski, P. (2007) Enantiodifferentiation of N-benzyloxycarbonylaminophosphonic and phosphinic acids and their esters using cyclodextrins by means of capillary electrophoresis. *J. Chromatogr. A*, **1138** (1-2), 284–290.

54 Rudzinska, E., Berlicki, Ł., Mucha, A., and Kafarski, P. (2007) Analysis of pD-dependent complexation of N-benzyloxycarbonylaminophosphonic acids by α–cyclodextrin. Enantiodifferentiation of phosphonic acid pKa values. *Chirality*, **19** (10), 764–768.

55 Uzarewicz-Baig, M. and Wilhelm, R. (2016) Straightforward diastereoselective synthesis of *P*-chirogenic (1R)-1,8,8-trimethyl-2,4-diaza-3-phosphabicyclo [3.2.1] octane 3-oxides: application as chiral NMR solvating agents. *Heteroat. Chem*, **27** (2), 121–134.

56 Glowacki, Z., Topolski, M., Matczak-Jon, E., and Hoffmann, M. (1989) ^{31}P NMR enantiomeric purity determination of free 1-aminoalkylphosphonic acids via their diastereoisomeric Pd(I1) complexes. *Magn. Reson. Chem.*, **27** (10), 922–924.

57 Rockitt, S., Magiera, D., Omelanczuk, J., and Duddeck, H. (2002) Chiral discrimination of organophosphorus compounds by multinuclear magnetic resonance in the presence of a chiral dirhodium complex. *Phosphorus, Sulfur Silicon Relat. Elem.*, **177** (8-9), 2079–2080.

58 Andersen, K.V., Bildsee, H., and Jakobsen, H.J. (1990) Determination of enantiomeric purity from solid-state P MAS NMR of organophosphorus compounds. *Magn. Reson. Chem.*, **28** (1), 47–51.

59 Viswanathan, T. and Toland, A. (1995) NMR spectroscopy using a chiral lanthanide shift reagent to access the optical purity of l-phenylethylamine. *J. Chem. Educ.*, **72** (10), 945–946.

60 Dunina, V.V., Kuz'mina, L.G., Rubina, M.Y., Grishin, Y.K., Veits, Y.A., and Kazakova, E.I. (1999) A resolution of the monodentate P*-chiral phosphine $PBu^tC_6H_4Br$-4 and its NMR-deduced absolute configuration. *Tetrahedron: Asymmetry*, **10** (8), 1483–1497.

61 Hammerschmidt, F. and Li, Y.-F. (1994) Determination of absolute configuration of α-hydroxyphosphonates by ^{31}P NMR spectroscopy of corresponding Mosher ester. *Tetrahedron*, **50** (34), 10253.

62 Hammerschmidt, F. and Vollenkle, H. (1989) Absolute Konfiguration der (2-Amino-1-hydroxyethy1)phosphonsaure aus Acanthamoeba castellanii (Neff) - Darstellung der Phosphonsaure-Analoga von (+)- und (−)-Serin. *Liebigs Ann. Chem.*, **6**, 577–583.

63 González-Morales, A., Fernández-Zertuche, M., and Ordóñez, M. (2004) Simultaneous separation and assignment of absolute configuration of γ-amino-β-hydroxyphosphonates by NMR using (S)-methoxyphenylacetic acid (MPA). *Rev. Soc. Quim. Mex.*, **48**, 239–245.

64 Kozlowski, J.K., Rath, N.P., and Spilling, C.D. (1995) Determination of the enantiomeric purity and absolute configuration of α-hydroxy phosphonates. *Tetrahedron*, **51** (23), 6385–6396.

65 Glowacki, Z. and Hoffmann, M. (1991) ^{31}P NMR enantiomeric excess determination of 1-hydroxyalkylphosphonic acids via their diastereoisomeric phosphonodidepsipeptides. *Phosphorus, Sulfur Silicon Relat. Elem.*, **55** (1-4), 169–173.

66 Glowacki, Z. and Hoffmann, M. (1991) ^{31}P NMR non-equivalence of the diastereoisomeric phosphonodidepsipeptides. Part III. *Phosphorus, Sulfur Silicon Relat. Elem.*, **63** (1-2), 171–175.

67 Devitt, P.G., Mitchell, M.C., Weetmann, J.M., Taylor, R.J., and Kee, T.P. (1995) Accurate in-situ P-31(H-1) assay of enantiopurity in alpha-hydroxyphosphonate esters using a diazaphospholidine derivatizing agent-zeta. *Tetrahedron: Asymmetry*, **6** (8), 2039.

68 Blazewska, K. and Gajda, T. (2002) (S)-Naproxen® and (S)-Ibuprofen® chlorides—convenient chemical derivatizing agents for the determination of the enantiomeric excess of hydroxy and aminophosphonates by ^{31}P NMR. *Tetrahedron: Asymmetry*, **13** (7), 671–674.

69 Zymanczyk-Duda, E., Lejczak, B., Kafarski, P., Grimaud, J., and Fischer, P. (1995) Enantioselective reduction of diethyl 2-oxoalkylphosphonates by baker's yeast. *Tetrahedron*, **51** (43), 11809–11814.

70 Brunel, J.M., Pardigon, O., Maffei, M., and Buono, G. (1992) The use of (4R,5R)-dicarboalkoxy 2-chloro 1,3,2-dioxaphospholanes as new chiral derivatizing agents for the determination of enantiomeric purity of alcohols by ^{31}P NMR. *Tetrahedron: Asymmetry*, **3** (10), 1243–1246.

71 Alexakis, A., Mutti, S., Normant, J.F., and Mangeney, P. (1990) A new reagent for a very simple and efficient determination of enantiomeric purity of alcohols by ^{31}P NMR. *Tetrahedron: Asymmetry*, **1** (7), 437–440.

72 Anderson, R.C. and Shapiro, M.J. (1984) 2-Chloro-4(R),5(R)-dimethyl-2-oxo-1,3,2-dioxaphospholane, a new chiral derivatizing agent. *J. Org. Chem.*, **49** (7), 1304–1305.

73 Johnson, C.R., Elliott, R.C., and Penning, T.D. (1984) Determination of enantiomeric purities of alcohols and amines by a phosphorus-31 NMR technique. *J. Am. Chem. Soc.*, **106** (17), 5019–5020.

74 Hulst, R., Zijlstra, R.W.J., Feringa, B.L., de Vries, K., ten Hoeve, W., and Wynberg, H.A. (1993) New ^{31}P NMR method for the enantiomeric excess determination of alcohols, amines and amino acid esters. *Tetrahedron Lett.*, **34** (8), 1339–1342.

75 Gorecki, Ł., Berlicki, Ł., Mucha, A., Kafarski, P., Lepokura, K.S., and Rudzinska-Szostak, E. (2012) Phosphorylation as a method of tuning the enantiodiscrimination potency of quinine—an NMR study. *Chirality*, **24** (4), 318–328.

76 Hulst, R., Kellogg, R.M., and Feringa, B.L. (1995) New methodologies for enantiomeric excess (ee) determination based on phosphorus NMR. *Reel. Trav. Chim. Pays-Bas.*, **114** (4–5), 115–138.

77 Kolodiazhnyi, O.I., Demchuk, O.M., and Gerschkovich, A.A. (1999) Application of the dimenthyl chlorophosphite for the chiral analysis of amines, amino acids and peptides. *Tetrahedron: Asymmetry*, **10** (9), 1729–1732.

78 Schurig, V. and Nowotny, H.P. (1990) Gas chromatographic separation of enantiomers on cyclodextrin derivatives. *Angew. Chem. Int. Ed. Engl.*, **29** (9), 939–957.

79 Dougherty, W., Liotta, F., Mondimore, D., and Shum, W. (1990) A convenient gas chromatographic method for the optical purity determination of chiral epoxy alcohols. *Tetrahedron Lett.*, **31**, 4389–4390.

80 Degenhardt, C.E.A.M., Verweij, A., and Benschop, H.P. (1987) Gas chromatography of organophosphorus compounds on chiral stationary phases. *Int. J. Environ. Anal. Chem.*, **30** (1-2), 15–28.

81 Bagi, P., Fekete, A., Kállay, M., Hessz, D., Kubinyi, M., Holczbauer, T., Czugler, M., Fogassy, E., and Keglevich, G. (2014) Resolution of 1-n-butyl-3-methyl-3-phospholene 1-oxide with TADDOL derivatives and calcium salts of O,O'-dibenzoyl-(2R,3R)- or O,O'-di-*p*-toluoyl-(2R,3R)-tartaric acid. *Chirality*, **26** (3), 174–182.

82 Spruit, H.E.T., Trap, H.C., Langenberg, J.P., and Benschop, H.P. (2001) Bioanalysis of the enantiomers of (±)-sarin using automated thermal cold-trap injection combined with two-dimensional gas chromatograph. *J. Anal. Toxicol.*, **25** (1), 57–61.

83 Ahuja, S. (ed.) (1997)Chiral Separations: Application and Technology, American Chemical Society, Washington, DC.

84 Caccamese, S., Failla, S., Finocchiaro, P., and Principato, G. (1998) Separation of the enantiomers of a-hydroxybenzylphosphonate esters by enantioselective HPLC and determination of their absolute configuration by circular dichroism. *Chirality*, **10** (1-2), 100–105.

85 Pirkle, W.H. and Pochapsky, T.C. (1989) Considerations of chiral recognition relevant to the liquid chromatography separation of enantiomers. *Chem. Rev.*, **89** (2), 347–362.

86 Pirkle, W.H., House, D.W., and Finn, J.M. (1980) Broad spectrum resolution of optical isomers using chiral high-performance liquid chromatographic bonded phases. *J. Chromatogr. A*, **192** (1), 143–158.

87 Fischer, C., Schmidt, U., Dwars, T., and Oehme, G. (1999) Enantiomeric resolution of derivatives of a-aminophosphonic and α-aminophosphinic acids by high-performance liquid chromatography and capillary electrophoresis. *J. Chromatogr. A*, **845** (1-2), 273–283.

88 Zarbl, E., Lämmerhofer, M., Hammerschmidt, F., Wuggenig, F., Hanbauer, M., Maier, N.M., Sajovic, L., and Lindner, W. (2000) Direct liquid chromatographic enantioseparation of chiral α– and β–aminophosphonic acids employing quinine-derived chiral anion exchangers: determination of enantiomeric excess and verification of absolute configuration. *Anal. Chim. Acta*, **404** (2), 169–177.

89 Chen, H., Lu, X.-Y., Gao, R.-Y., Zhou, J., and Wang, Q.-S. (2000) Separation of chiral phosphorus compounds on the substituted β-cyclodextrin stationary phase in normal-phase liquid chromatography. *Chin. J. Chem.*, **18** (4), 533–536.

90 Pirkle, W.H., Brice, L.J., Widlanski, T.S., and Roestamadji, J. (1996) Resolution and determination of the enantiomeric purity and absolute configurations of α-aryl-α-hydroxymethanephosphonates. *Tetrahedron: Asymmetry*, **7** (8), 2173–2176.

91 Okamoto, Y., Honda, S., Hatada, K., Okamoto, I., Toga, Y., and Kobayashi, S. (1984) Chromatographic resolution of racemic compounds containing phosphorus or sulfur atom as chiral center. *Bull. Chem. Soc. Jpn.*, **57** (6), 1681–1682.

92 West, C., Cieslikiewicz-Bouet, M., Lewinski, K., and Gillaizeau, I. (2013) Enantiomeric separation of original heterocyclic organophosphorus compounds in supercritical fluid chromatography. *Chirality*, **25** (4), 230–237.

93 Szyszkowiak, J. and Majewska, P. (2014) Determination of absolute configuration by ^{31}P NMR. *Tetrahedron: Asymmetry*, **25** (2), 103–112.

94 Harada, N. and Nakanishi, K. (1983) Circular Dichroic Spectroscopy. Exciton Coupling in Organic Stereochemistry, University Science Books, Mill Valley, CA, and Oxford University Press, Oxford.
95 Kolodiazhnyi, O.I., Gryshkun, E.V., Andrushko, N.V., Freytag, M., Jones, P.G., and Schmutzler, R. (2003) Asymmetric synthesis of chiral N-(1-methylbenzyl)aminophosphines. *Tetrahedron: Asymmetry*, **14** (2), 181–183.
96 Dou, S.Q., Zheng, Q.T., Dai, J.B., Tang, C.C., and Wu, G.P. (1982) The crystal structure and absolute configuration of quinine salt of (+)-o-ethyl o-phenyl phosphorothioic acid. *Acta Phys. Sin.*, **31** (4), 554–560.
97 Flack, H.D. (1983) On enantiomorph-polarity estimation. *Acta Crystallogr., Sect. A: Found. Crystallogr.*, **A39** (6), 876–881.
98 Hooft, R.W.W., Straver, L.H., and Spek, A.L. (2010) Using the t-distribution to improve the absolute structure assignment with likelihood calculations. *J. Appl. Crystallogr.*, **43**, 665–668.
99 Kowalik, J., Zygmunt, J., and Mastalerz, P. (1983) Determination of absolute configuration of optically active. 1 -aminoalkanephosphonic acids by chemical correlations. *Phosphorus Sulfur Rel. Elem.*, **18** (1-3), 393–396.
100 Kielbasinski, P., Zurawinski, R., Albrycht, M., and Mikolajczyk, M. (2003) The first enzymatic desymmetrizations of prochiral phosphine oxides. *Tetrahedron: Asymmetry*, **14** (21), 3379–3384.
101 Seco, J.M., Quinoa, E., and Riguera, R. (2004) The assignment of absolute configuration by NMR. *Chem. Rev.*, **104** (1), 17–117.
102 Kusumi, T., Fukushima, T., Ohtani, I., and Kakisawa, H. (1991) Elucidation of the absolute configurations of amino acids and amines by the modified Mosher's method. *Tetrahedron Lett.*, **32** (25), 2939–2942.
103 Hammerschmidt, F. and Wuggenig, F. (1999) Enzymes in organic chemistry. Part 9:[1]Chemo-enzymatic synthesis of phosphonic acid analogues of l-valine, l-leucine, l-isoleucine, l-methionine and l-α-aminobutyric acid of high enantiomeric excess. *Tetrahedron: Asymmetry*, **10** (9), 1709–1745.
104 Blazewska, K., Paneth, P., and Gajda, T. (2007) The assignment of the absolute configuration of diethyl hydroxyand aminophosphonates by ^{1}H and ^{31}P NMR using naproxen as a reliable chiral derivatizing agent. *J. Org. Chem.*, **72** (3), 878–887.
105 Błazewska, K.M. and Gajda, T. (2009) Assignment of the absolute configuration of hydroxy- and aminophosphonates by NMR spectroscopy. *Tetrahedron: Asymmetry*, **20** (12), 1337–1361.
106 Yokomatsu, T., Yamagishi, T., Suemune, K., Yoshida, Y., and Shibuya, S. (1998) Enantioselective synthesis of *threo*-α,β-dihydroxyphosphonates by asymmetric dihydroxylation of 1(*E*)-alkenylphosphonates with AD-mix reagents. *Tetrahedron*, **54** (5-6), 767–780.
107 Yokomatsu, T., Suemune, K., Murano, T., and Shibuya, S. (1996) Synthesis of (α,α–difluoroallyl)phosphonates from alkenyl halides or acetylenes. *J. Org. Chem.*, **61** (20), 7207–7211.
108 Yokomatsu, T., Kato, J., Sakuma, C., and Shibuya, S. (2003) Stereoselective synthesis of highly-functionalized cyclohexene derivatives having a diethoxyphosphoryldifluoromethyl functionality from cyclohex-2-enyl-1-phosphates. *Synlett*, **10**, 1407–1410.
109 Gorobets, E., Parvez, M., Wheatley, B.M.M., and Keay, B.A. (2006) Use of ^{1}H NMR chemical shifts to determine the absolute configuration and enantiomeric purity for

enantiomers of 3,3'-disubstituted-MeO-BIPHEP derivatives. *Can. J. Chem.*, **84** (02), 93–98.
110 Smaardljk, A.A., Noorda, S., van Bolhuls, F., and Wynberg, H. (1985) The absolute configuration of α-hydroxyphosphonates. *Tetrahedron Lett.*, **26** (6), 493–496.
111 Kielbasinski, P., Omelanczuk, J., and Mikolajczyk, M. (1998) Lipase-promoted kinetic resolution of racemic, P-chirogenic hydroxymethylphosphonates and phosphinate. *Tetrahedron: Asymmetry*, **9** (18), 3283–3287.
112 Morrison, J.D. and Mosher, H.S. (1976) *Asymmetric Organic Reactions*, Prentice Hall, Englewood Clifis, NJ.
113 Izumi, A. and Tai, S. (1977) *Stereodifferentiating Reactions*, Kodansha Ltd., Academic Press, Tokyo, New York, 334 pp.
114 Kolodiazhnyi, O.I. (2003) Multiple stereoselectivity and its application in organic synthesis. *Tetrahedron*, **59** (32), 5953–6018.
115 Kolodiazhnyi, O.I. and Kolodiazhna, A.O. (2016) Multiple stereoselectivity in organophosphorus chemistry. *Phosphorus, Sulfur Silicon Relat. Elem.*, **191** (3), 444–458.

2

Asymmetric Synthesis of P-Chirogenic Phosphorus Compounds

2.1 Introduction

P-chiral phosphorus compounds are used in many areas of chemistry including biologically active pharmaceuticals, agrochemicals, and ligands for transition metal complexes. Many methods were developed to prepare enantiomerically pure P-chirogenic phosphorus compounds, including classical optical resolution via diastereoisomers, chemical kinetic resolution, enzymatic resolution, chromatographic resolution, and asymmetric catalysis. In the last few years, enormous success has been achieved in the asymmetric synthesis of organophosphorus compounds and many articles devoted to the synthesis of chiral organophosphorus compounds have been published. Complexes with transition metals containing PAMP, DIPAMP, DIOP, and CHIRAPHOS ligands are widely used for the asymmetric formation of C–H and C–C bonds [1].

Phosphorus can form bonds with many other elements. Phosphorus has a valency of 3 or 5 and has coordination numbers from 1 to 6. It has also empty d-orbitals that readily accept electrons from any good donors. These properties of phosphorus give an extra edge to organophosphorus chemistry. Trivalent tricoordinate and pentavalent tetracoordinate organophosphorus compounds can exist in optically active states and are configurationally stable. Phosphorus atoms in low-coordinate valence states (mono- and dicoordinate trivalent phosphorus, tricoordinate pentavalent phosphorus) possess axial or planar symmetry and cannot be optically active. Pentacoordinate and hexacoordinate phosphorus compounds are conformationally unstable, although some of them have been obtained in optically active form. Chiral pentacoordinate and hexacoordinate phosphorus compounds are important intermediates in asymmetric syntheses and therefore their stereochemistry has been studied in detail (Scheme 2.1).

This chapter points out significant advances in the asymmetric synthesis of P-chiral organophosphorus compounds with many applications in stereoselective synthesis and asymmetric catalysis. Asymmetric addition and cycloaddition reactions, reductions and oxidation, including metal-catalyzed and non-metal biocatalytic methods, are described, in addition to synthetic approaches via nucleophilic substitution of appropriately substituted precursors.

Asymmetric Synthesis in Organophosphorus Chemistry: Synthetic Methods, Catalysis and Applications, First Edition. Oleg I. Kolodiazhnyi.

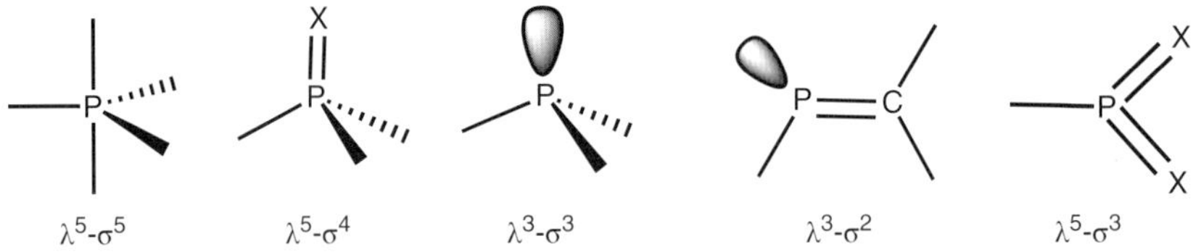

Scheme 2.1 Coordinated states of phosphorus.

2.2 Low-Coordinated Phosphorus Compounds

Low-coordinated phosphorus ligands can form various combinations of bonds owing to the presence of the phosphorus atom lone pair and a reactive π- and π*-system. Indeed, the ability of low-coordinated phosphorus compounds to form π-complexes leads to a number of important synthetic developments in coordination chemistry and applications in homogeneous catalysis [2–6]. Some of the most interesting combinations of bonds and complexes are presented in Scheme 2.2.

Prochiral structures of low-coordinated phosphorus are interesting as objects for asymmetric synthesis and as ligands. Stereoselective addition of optically active alcohols and amines ((−)-menthol, (−)-menthylamine, and (−)-α-phenylethylamine) to λ^3-iminophosphines possessing a trigonal configuration were described by Mikolajczyk, Markovski *et al.* [7, 8]. The highest stereoselectivity (34% de) was observed in the case of the reaction of λ^3-iminophosphines with (−)-menthol [8] (Scheme 2.3).

In the presence of chiral (−)-*N*-dimethylmenthylamine or (−)-*N*-dimethyl-1-methylbenzylamine stereoselectivity of the reaction of λ^3-iminophosphines with

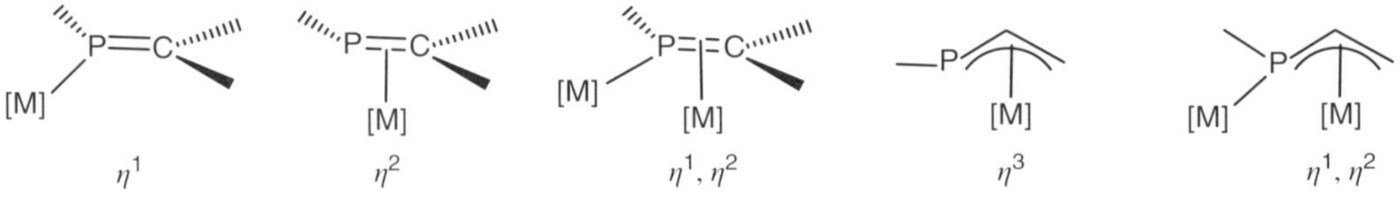

Scheme 2.2 Various types of low-coordinated phosphorus ligands.

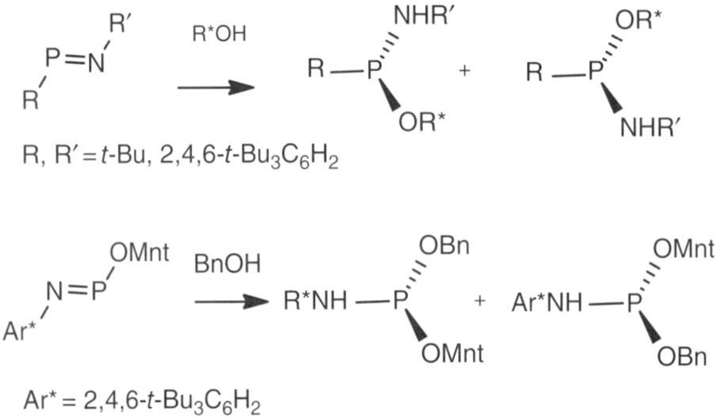

Scheme 2.3 Stereoselective addition of optically active alcohols to λ^3-aminophosphines.

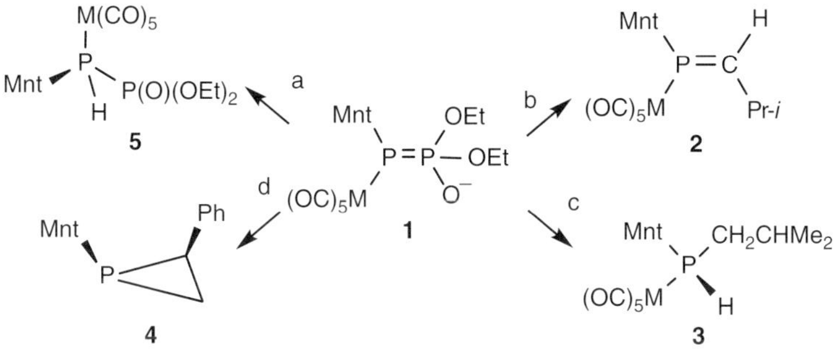

(a) BuLi; (b) -PrCHO; (c) H_2, CH_2Cl_2/L_2Ph^+; (d) (*S*)-phenyloxirane, diphos; M = Mo, W

Scheme 2.4 Chemical conversions of P-Wittig reagents.

methanol led to the formation of methoxyaminophosphine with 55% de. The stereoselectivity of these reactions was moderate: the highest level of stereoselectivity (34% de) was observed in the case of the reaction of 3-iminophosphines with (−)-menthol. However, in the presence of chiral tertiary amines [(−)-*N*-dimethylmenthylamine or (−)-*N*-dimethyl-1-methylbenzylamine] the reaction of 3-iminophosphines with methanol afforded methoxyaminophosphine with 55% ee. Good stereoselectivity (80% de) was obtained on the addition of benzyl alcoholate to P-menthoxy-3-iminophosphine [8]. The phosphaalkenes **1** (phospha-Wittig reagents) in the form of their tungsten and molybdenum complexes were used by Mathey for various syntheses (Scheme 2.4) [9, 10]. For example, L-menthyl phospha-alkene complexes **2** were prepared by the reaction of phospha-Wittig reagents **1** with aldehydes [9]. Catalytic hydrogenation of **1** using chiral rhodium complexes RhL_2^+ proceeded with high stereoselectivity; with the catalyst bearing L_2 = *diphos*, the diastereomeric yields of product **3** were above 90% de, and with L_2 = (−)-*chiraphos*, only one diastereomer of **3** with 100% de was obtained. The reaction of the anion of pentacarbonyl[(diethoxyphosphoryl) phosphine]tungsten complexes with optically pure (*S*)-styrene oxide led to the formation of corresponding optically active pentacarbonyl(phosphirane)tungsten complexes (24 : 76 dr) at 95 °C of complexes with *diphos* [bis(diphenylphosphino)ethane], to result in the optically active phosphirane **4** with inversion of the configuration at carbon (Scheme 2.5). The phospha-Wittig reaction was applied for the synthesis of phosphaalkenes **5**, which were transformed into complexes bearing a bulky group at the phosphorus atom. The reaction of **1** with (*R*)-(+)-phenyloxirane gives only two, (S_P,S_C)-**5** and (R_P,S_C)-**5**, of the four possible diastereomers. The other two isomers, (R_P, R_C)-**5** and (S_P, R_C)-**5**, were obtained from

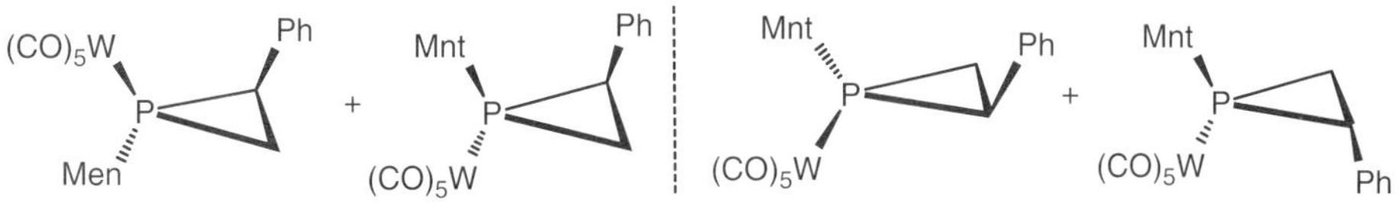

(S_P,S_C)-**5** (R_P,S_C)-**5** (R_P,R_C)-**5** (S_P,R_C)-**5**

Scheme 2.5 Optically active pentacarbonyl(phosphirane) tungsten complexes.

(*S*)-(−)-phenyloxirane, that is, the reaction is completely stereospecific with respect to the carbon configuration and total inversion at the oxirane carbon. The phosphirenes **5** were used as ligands in rhodium(I) complexes, which are potential hydrogenation catalysts [11, 12] (Scheme 2.5).

Interesting ligands with dicoordinated phosphorus were prepared on the basis of chiral oxazolines. The air-stable phosphaalkenes **8** is of considerable interest as a π-accepting ligands for asymmetric catalysis. The oxazoline intermediate **6** was prepared in two steps from L-valine. The treatment of **6** with *sec*-BuLi and TMEDA formed the desired carbanion. Claisen-type condensation of this anion with ethyl benzoate formed ketone **7** in 49% isolated yield. The ketone **7** via the phospha-Peterson reaction was then converted into phosphaalkene **8**, which was purified by crystallization in *n*-pentane and characterized crystallographically. The phosphaalkenes **8** were used as ligands in iridium complexes for asymmetric catalysis of hydrogenation, allylic alkylation, and hydroformylation [13, 14] (Scheme 2.6).

Certain axially dissymmetric phospha-allenes can be resolved on enantiomers [15–18]. Separation of enantiomers of axially dissymmetric 1,3-bis(2,4,6-tri-*t*-butylphenyl)-1,3-diphospha-allene **9** was attained by high performance liquid chromatography (HPLC) with a chiral (+)-poly(triphenylmethy1 methacrylate) column [16]. The absolute configuration of the optically pure (−)-phospha-allene **9** was determined by means of circular dichroism (CD) spectra. It was found that compound **9** racemizates under radiation with visible light; however, the rate of racemization decreased in the dark (Scheme 2.7). These phenomena are very similar to those observed for the systems of double-bonded phosphorus compounds in low coordination state such as diphosphaallene which resulted in racemization on photolysis or diphosphenes and phosphaethylenes of D_h symmetry which resulted in isomerization on photolysis. The racemization of the 1-phosphaallene **9** probably involves either rotation around the P=C (or C=C) bond or inversion at the phosphorus atom.

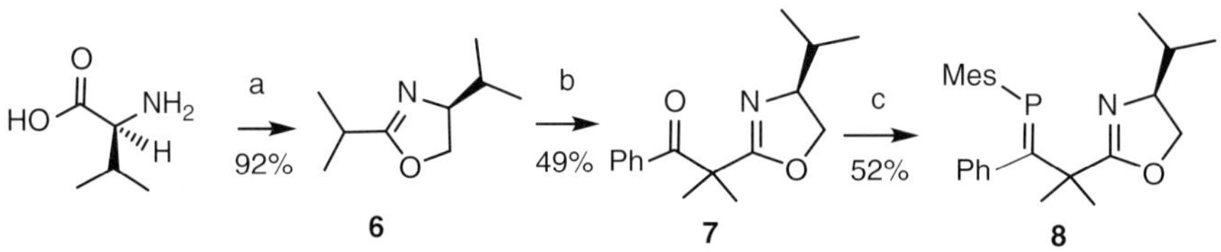

6 **7** **8**

(a) $NaBH_4$/ Me_2CCOOH; (b) *s*-BuLi/TMEDATHF/PhCOOEt; (c) $MesP(SiMe_3)Li$

Scheme 2.6 Synthesis of phosphaalkene ligands.

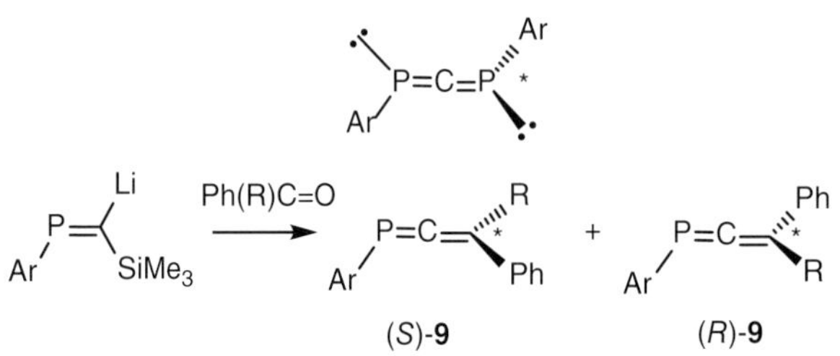

(*S*)-**9** (*R*)-**9**

R = H, Ph; Ar = 2,4,6-*t*-$Bu_3C_6H_2$

Scheme 2.7 Enantiomers of phospha-allenes **9**.

A number of chiral heterocycles containing dicoordinated phosphorus have been synthesized and studied as chiral ligands. The separation of enantiomers was attained by means of chromatography. Effective catalysts, and also phosphaferrocene–oxazoline ligands were developed on a basis of ferrocene [19]. Planar chiral bis-phosphine of the type (−)-**11** were successfully used in asymmetric hydrogenation catalyzed by rhodium dehydroaminoacids [19, 20] (Schemes 2.8 and 2.9). The group of Ganter *et al.* [10, 21] has reported that the formyl ferrocene could be employed as a convenient precursor for the synthesis of various enantiomerically pure ligands. Very efficient phosphaferrocene-based ligands such as **10**, **11**, and the mixed phosphaferrocene–oxazoline ligand **12** have been designed by the group of Fu in the $Fe(\eta^5C_5Me_5)$ series [20, 22, 23]. The resolution was carried out by chiral HPLC. Another approach for the synthesis of phosphanyl-substituted phosphaferrocenes through the use of an enantiomerically pure chiral phospholyl ligand has also been developed [20] (Scheme 2.8).

Interesting applications of low-coordinate phosphorus ligands in catalysis have been attained with phospha- and diphosphaferrocenes. These ligands, which combine a ferrocenyl backbone with unusual electronic properties of a dicoordinate P atom, can accommodate many transition metal centers and oxidation states. For example, Fu *et al.* reported that the planar chiral bisphosphine (−)-**11** was successfully employed in the rhodium-catalyzed asymmetric hydrogenation of dehydroamino acids (Scheme 2.9) [18, 22]. The [Rh(COD)(**13**)(+)][PF_6] complex catalyzed the enantioselective isomerization of allylic alcohols to the corresponding aldehydes in high yield and with good enantiomeric excess (ee) (Scheme 2.10) [22].

Enantiomeric antipods of atropisomeric phosphinine (*R*)-**14**/(*S*)-**14** were separated by using HPLC on a chiral stationary phase with *n*-hexane as eluent (Scheme 2.11) [20, 23]. The enrichment of one enantiomer and a subsequent investigation into its racemization kinetics revealed a barrier for internal rotation of $\Delta G_{298} = 109.5 \pm 0.5\,\text{kJ mol}^{-1}$, which is in excellent agreement with the theoretically predicted value of $\Delta G_{298} = 116\,\text{kJ mol}^{-1}$.

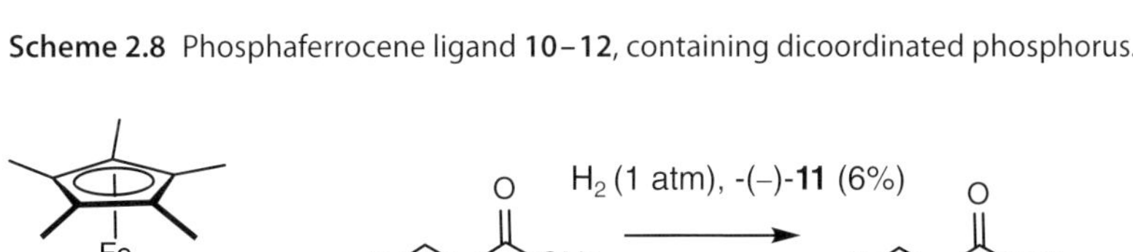

Scheme 2.8 Phosphaferrocene ligand **10–12**, containing dicoordinated phosphorus.

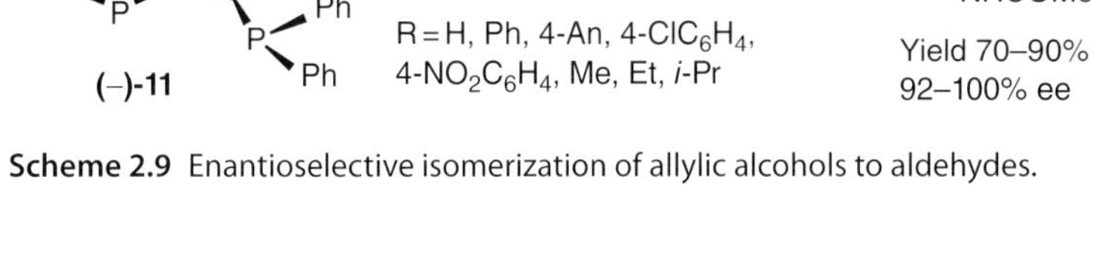

Scheme 2.9 Enantioselective isomerization of allylic alcohols to aldehydes.

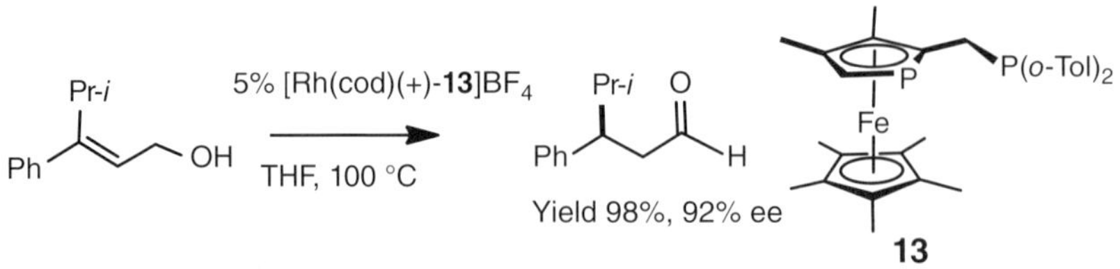

Scheme 2.10 Catalytic enantioselective isomerization of *E*-allylic alcohols to aldehydes.

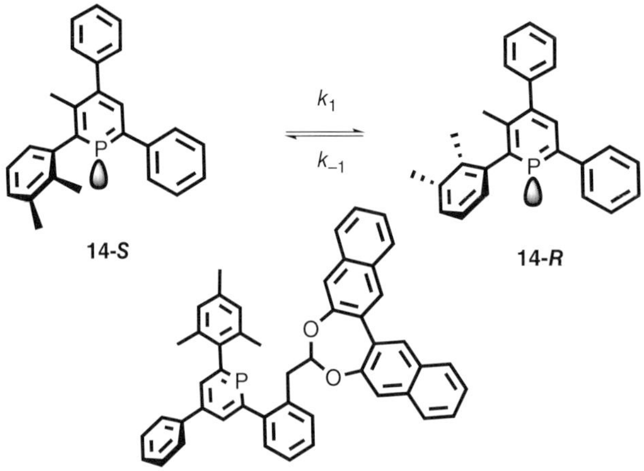

Scheme 2.11 Atropisomeric phosphinines (*S*)-**14** and (*R*)-**14**.

Further analysis with UV and CD spectroscopies and density functional theory (DFT) calculations led to the determination and assignment of the absolute configurations of both enantiomers. The successful separation of the enantiomers confirmed their reasonably high energy barrier for internal rotations. The assignment of the absolute configurations of enantiomers was achieved by comparing the experimental CD spectra of both enantiomers with theoretical spectra from optimized structures [24–26].

The stereochemistry of the reaction at the prochiral pentavalent tricoordinate phosphorus atom, possessing planar structure, was also studied [27, 28] (Scheme 2.12). Thus the addition of *tert*-butylamine to the metathioiminophosphate **15**, generated by

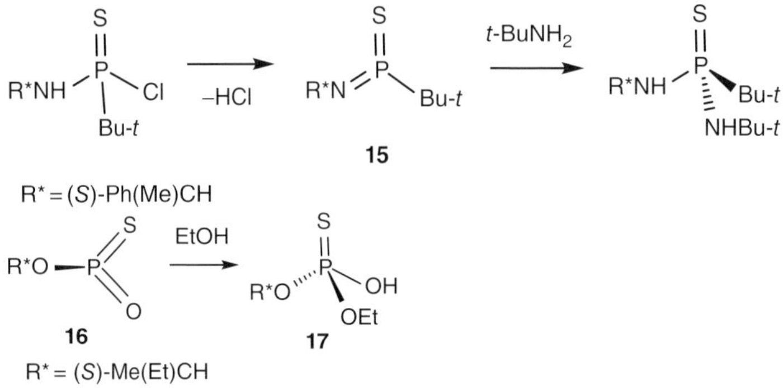

Scheme 2.12 Stereochemistry of reactions of pentavalent tricoordinated phosphorus compounds.

dehydrochlorination of N-substituted amide chlorophosphate, proceeded stereoselectively with the formation of a diastereomer mixture, the ratio of which increased with the decrease of the solvent polarity from 57 : 43 (MeCN) up to 80 : 20 (cyclohexane). In other cases, addition of an alcohol to prochiral (*S*)-*sec*-butoxy metathiophosphonate **16** possessing a planar–trigonal configuration led to the formation of an equimolar mixture of diastereomers **17** (Scheme 2.12). L. Queen applied low-coordinated phosphorus compounds for modification of silica gels, adamant, zeolites, and titanium dioxides, which showed high activity as solid phases in columns for HPLC [29, 30].

Hence, low-coordinate organophosphorus compounds are interesting starting compounds for asymmetric synthesis. However, their stereochemistry has not been sufficiently studied. Looking into the future, one can expect a wide application of low-coordinate organophosphorus compounds as starting reagents in asymmetric synthesis.

2.3 Trivalent Tricoordinated Phosphorus Compounds

Chiral trivalent organophosphorus compounds play a key role in phosphorus stereochemistry. The basic application of chiral tertiary phosphines consists in their use as ligands in catalysts for asymmetric synthesis. Significant efforts have therefore been devoted to working out convenient methods for the synthesis of chiral phosphines [31–33]. P-Chirogene trivalent organophosphorus compounds can be obtained by various methods: asymmetric synthesis, kinetic resolution of racemates, and chromatographic methods. Asymmetric synthesis is the most effective method for the preparation of chiral trivalent phosphorus compounds. Successful development of the asymmetric version of catalytic reactions with application of chiral complexes of transition metals depends on design and synthesis of new chiral ligands and also of new chiral phosphines. One of the effective roots to chiral tertiary phosphines is reactions of nucleophilic substitution at the trivalent phosphorus atom.

2.3.1 Configuration Stability of P(III)-Compounds

Configurational stability of chiral phosphorus compounds attracts considerable attention. A trivalent phosphorus atom bonded to three substituents in a pyramidal geometry and possessing one nonbonding electron pair may spontaneously undergo inversion of configuration [34]. Such a process of pyramidal atom inversion must involve passage through a transition state (TS) **A** in which the nonbonding pair possesses pure *p* character and the bonds from the central atom to the substituents are sp^2 [31] (Scheme 2.13).

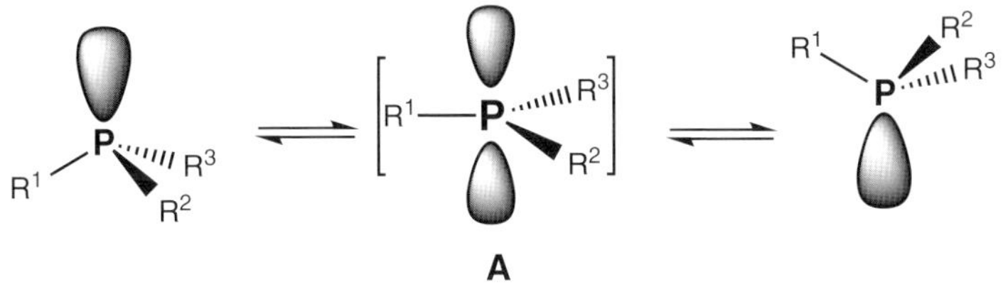

Scheme 2.13 Inversion of configuration in trivalent phosphorus compounds.

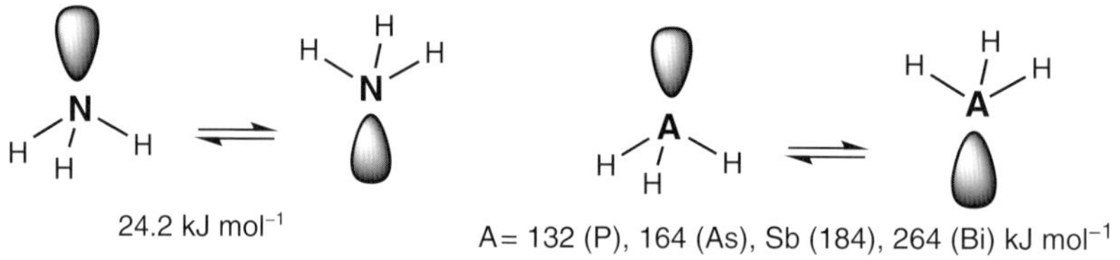

Scheme 2.14 Configuration stability of chiral compounds of five element groups.

Trivalent phosphorus compounds are more configurationally stable than nitrogen compounds. Racemization of trivalent phosphorus compounds depends strongly on their structure, and firstly on the electron-accepting substituents at the phosphorus atom, which decrease the configurational stability. The barrier of inversion in acyclic phosphines is about 150 kJ mol^{-1}, whereas the barrier of inversion in acyclic amines is about 30 kJ mol^{-1}. The barrier of inversion in phosphines depends on the electronegativities of the substituents bonded to the phosphorus atom. However, in some cases, compounds bearing electron-accepting groups at the phosphorus atom are racemized. For instance, electron-accepting groups in the para-position of phenyl rings in arylphosphines reduce the barrier of inversion. Moreover, chiral chlorophosphines are conformationally labile compounds existing as an equilibrium racemic mixture of (*R*)- and (*S*)-enantiomers. Although calculations indicate substantial pyramidal stability at phosphorus in halophosphines of the type R^1R^2PX, attempts to isolate enantiomerically pure chlorophosphines have been unsuccessful [31, 32, 35–42] (Scheme 2.14). Thus *tert*-butylphenylchlorophosphine of 49.4% ee lost its optical activity over 20 h in the polarimeter cell [39].

Jugé *et al.* [43] carried out experimental and computational studies on the configurational stability and racemization of chlorophosphines **18**. Pure chlorophosphines are configurationally stable; however, traces of acids, for example, HCl, which are almost unavoidable in experimental conditions, lead to easy racemization. It was found that HCl acts as a catalyst for inversion at the P-center. The mechanism of the racemization is explained by the phosphorus nucleophilic attack on H, with a concerted backside attack by the chlorine on the phosphorus center. The reaction intermediate, as indicated by the gas-phase computation, is an achiral pentacoordinated phosphorus with two chlorine atoms in the axial position (Scheme 2.15).

The esters of chiral trivalent phosphorus acid R_2POR are more stable than chlorophosphines and can be isolated as enantiomerically pure compounds. However, racemizations of P-chiral phosphorus esters R_2POR occur with measurable rates at room temperature. Acids catalyze racemization of chiral phosphonic esters and may involve not only the pyramidal inversion but also exchange of the ester group. Tertiary alkylaryl- and diarylphosphines are more or less configurationally stable. They have a barrier of pyramidal inversion of 30–35 kcal mol^{-1} and may be obtained as individual

18

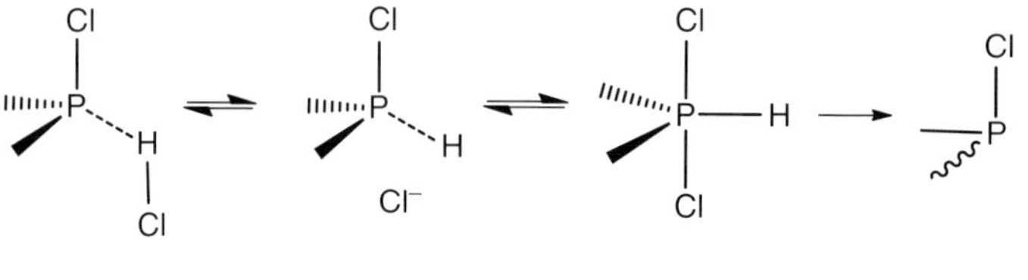

Scheme 2.15 Thermodynamically controlled racemization of tertiary phosphines.

enantiomers at ambient temperature. However, at high temperatures, they are easily racemized. Mechanisms other than pyramidal inversion, such as ligand exchange, have been also observed in the stereomutation of trivalent phosphorus, for example, a rapid phosphorus inversion, accelerated by $(p-p)_\pi$ and $(p-d)_\pi$ conjugation. The reduction in barrier height was explained by the presence of d orbitals on an adjacent substituent, as in the diphosphine and the silylphosphines [41]. The thermodynamically controlled pyramidal inversion of tertiary phosphines is useful synthetically. The facile acid-catalyzed racemization of secondary phosphines involves formation of an achiral phosphonium ion and often renders isolation of enantiomers difficult [44, 45]. Monochlorophosphoramidates and other configurationally unstable phosphorus compounds were employed in DYKAT (Dynamic Kinetic Asymmetric Transformation) and similar stereodynamic synthetic strategies [44].

Radosevich *et al.* [45] reported that the pyramidal inversion of trivalent phosphines is catalyzed by single-electron oxidation. Specifically, P-stereogenic (aryl)methylphenyl phosphines are shown to undergo rapid racemization at ambient temperature when exposed to catalytic quantities of a single-electron oxidant in solution [46]. For example, enantioenriched (R_P)-**19** (99% ee) is configurationally stable in solution at ambient temperature, with an experimentally determined barrier to thermal inversion of 31.4 kcal mol^{-1}. However, according to the proposed outer-sphere ET mechanism, the easy racemization of **19** can be attained by catalytic oxidation with organic aminium oxidant [P-An_3N][PF_6], (P-Tol_3N)PF_6, (Cp_2Fe)PF_6, Co$(OTf)_2$, or copper(II) triflate (Scheme 2.16).

2.3.2 Asymmetric Nucleophilic Substitution at P (III)

The most frequently encountered reactions in organic phosphorus chemistry are nucleophilic substitution reactions. The mechanism and steric course of S_NP reactions have been intensively studied. In the overwhelming majority of cases, S_N2 nucleophilic substitution at chiral tricoordinate trivalent phosphorus results in inversion of configuration that assumes the formation of pentacoordinate intermediate A, containing attacking and leaving groups in apical positions, in spite of the extension of the steric strain due to arrangement of the four-membered cycle in a diequatorial position [47–52]. Stereochemical consequences of nucleophilic substitution at chiral phosphorus may be different in cyclic systems. Ring size in particular can considerably affect stereochemical results (Scheme 2.17) [47, 48].

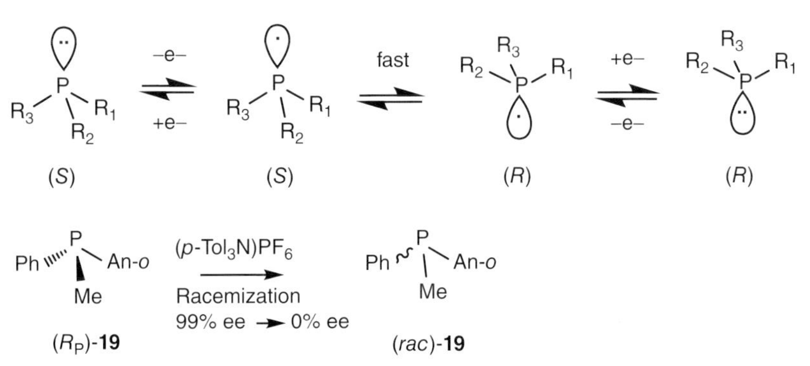

Scheme 2.16 Examples of catalytic racemization of P(III) compounds.

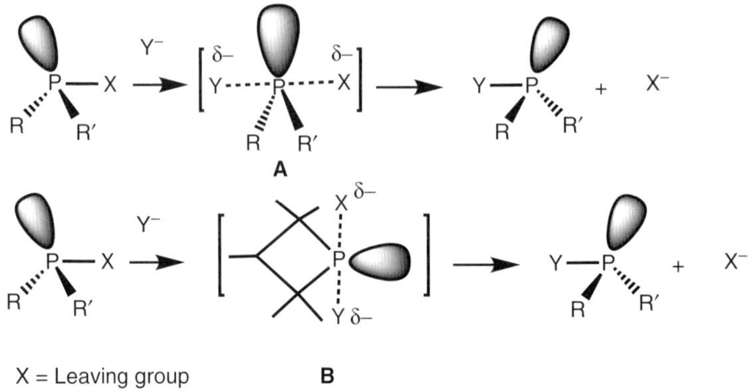

Scheme 2.17 Mechanism and stereochemistry of S_N2P reactions.

On the basis of theoretical calculations, Bickelhaupt found that the nucleophilic substitution at the tricoordinate phosphorus centers of the model reactions X + PH_2Y (S_N2@P3) is characterized by single-well reaction profiles with a stable, hypervalent transition complex (TC), similar to nucleophilic substitution at silicon (S_N2@Si). Differences between S_N2@P3 and S_N2@P4 are minor. However, the S_N2@P4 substitution (unlike S_N2@P3) exhibits a characteristic behavior, which is also observed in S_N2@Si: introducing sufficient steric bulk around the central atom causes the appearance of pre- and post-barriers that separate the TS from the reactant and product complexes [53, 54]. The reaction of prochiral dimenthyl phosphonite with alkyllithium at low temperature proceeded with formation of P-resolved menthyl alkylphenylphosphinites **20** of high diastereomeric purity (90–96% de). The second substitution gives enantiomerically enriched tertiary phosphines **20** (73–79% ee, Scheme 2.18) [55, 56].

The reaction of racemic isopropylphenylchlorophosphines with ortho-metallated (*R*)-[1-(dimethylamino) ethyl]naphthalene palladium(II) complex in dichloromethane produces the pair of (R,R_P)- and (R,S_P)-diastereomeric complexes **22** in a 78 : 22 ratio (Scheme 2.19) [35]. Free, configurationally stable (*R*)-**23** was isolated from this complex in 93% ee by treatment of the complex with bis(diphenylphosphino)ethane (dppe). Substitution of the P-chloride in the (R,R_P)-diastereomer **22** by methoxide proceeds with complete inversion at phosphorus [35] (Scheme 2.19).

Dahl [49] has reported that reactions of nucleophilic substitution at trivalent phosphorus proceed with inversion at the phosphorus atom. Very clean inversion was observed when the nucleophile is a strong base such as RLi or MeO- and the leaving group is OR, NR_2, or Ph. In other cases, the products initially formed with inversion, but they isomerized during the reactions.

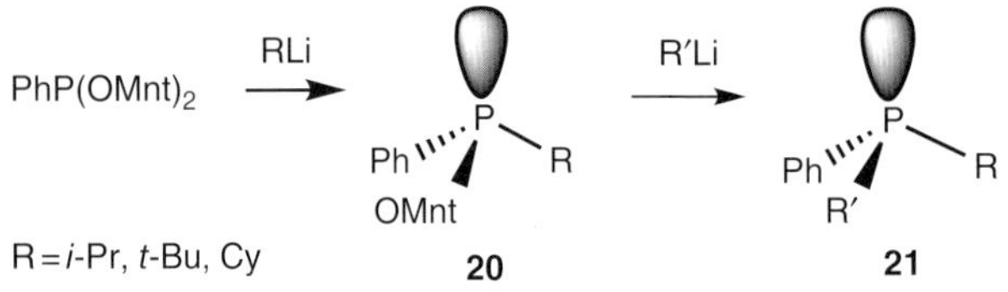

Scheme 2.18 Reaction of dimenthyl phosphonite with lithium alkyles.

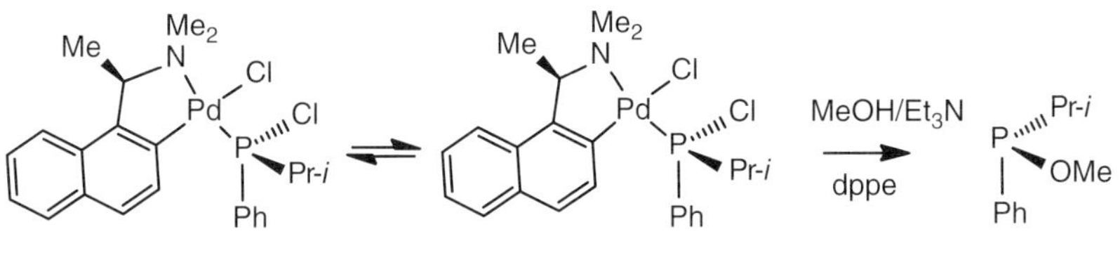

Scheme 2.19 Stereochemistry of reaction of the (R,S_P)-**22** complex with methanol.

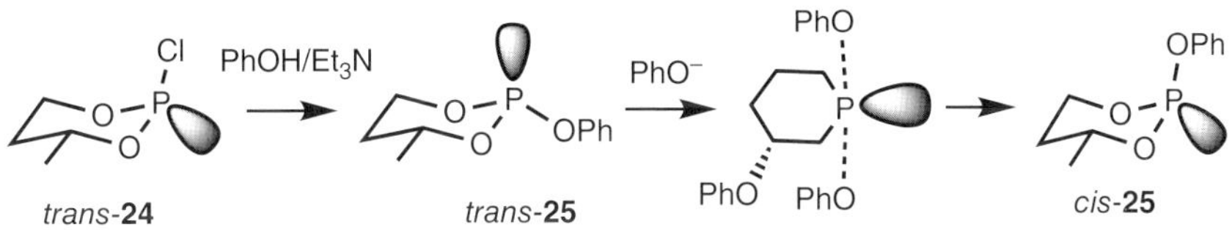

Scheme 2.20 Nucleophilic substitution at the trivalent phosphorus atom.

The stereochemical result of nucleophilic substitution at the trivalent phosphorus atom sometimes depends on reaction conditions. Thus the reaction of *trans*-**24** with an excess of phenol and triethylamine at room temperature led with retention of configuration to the *trans*-phenylphosphite 25 (84% de). However, addition of phenol to the chlorophosphine to avoid an excess of phenoxide ion during the reaction exclusively provides *cis*-phenylphosphite. This result is explained by an S_N2P mechanism and a second attack, following an antiperiplanar pathway, which produces the thermodynamically more stable *trans*-diastereomer **25** [57] (Scheme 2.20).

The effective methods of asymmetric synthesis, which use the reaction of nucleophilic substitution at P(III) atom allow to obtain various chiral phosphinic ligands. For example, such as presented below the diastereoselective reactions of secondary alcohols and amines with chlorides of trivalent phosphorus.

2.3.2.1 Secondary Alcohols as Chiral Auxiliaries

Optically active secondary alcohols (L-menthol, *endo*-borneol, glucofuranose derivatives, and others) serve as cheap and accessible chiral auxiliaries for the preparation of enantiopure organophosphorus compounds [58–64]. Diastereoisomers of menthyl phosphinite boranes [62, 65] are among the most popular chiral organophosphorus reactants in organic synthesis. In 1967, Mislow *et al.* [61] reported the synthesis of P-chiral ligands via resolution of diastereomeric menthylphosphinates and subsequent addition of Grignard reagents; the reduction of the phosphine oxide with trichlorosilane yielded the chiral phosphine (Scheme 2.21). Representative examples of chiral alkoxyphosphonates were prepared from easily available natural chiral precursors such as L-menthol or glucofuranose, obtained from readily available D-glucose. The unsymmetrically substituted menthyl phosphinates (R_P/S_P)-**26** could be separated into their diastereoisomers by recrystallization or column chromatography. Buono *et al.* [58, 59] reported that nucleophilic substitution of the alkoxy group of the H-phosphinates **27** with organolithiums reagents proceeds stereospecifically with inversion of configurations at phosphorus to give a wide range of P-stereogenic tertiary phosphine oxides **28** or secondary phosphine oxides **29** on quenching the reaction mixture with alkyl halides or water, respectively [59, 63]. The enantiopure (S_P)- and

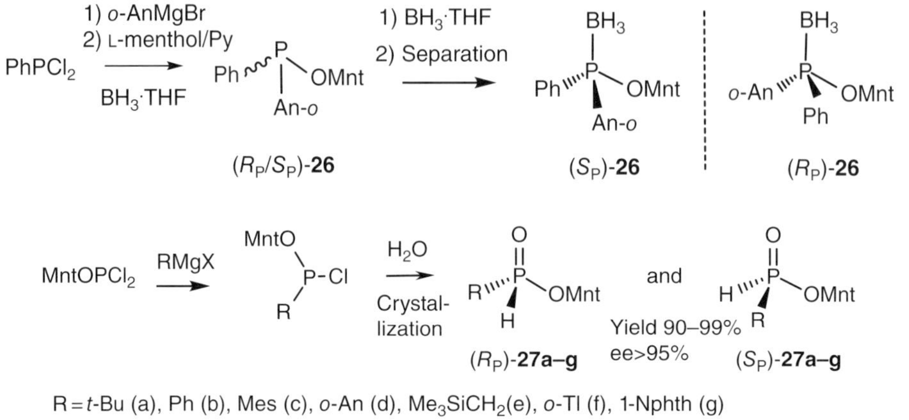

Scheme 2.21 Preparation of chiral menthyl phosphinate boranes.

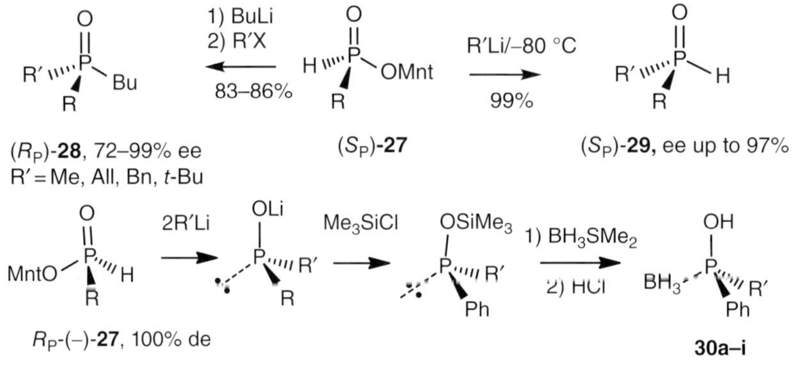

Scheme 2.22 Preparation of chiral *H*-phosphinates and *H*-phosphonates.

(R_P)-diastereomers of (−)-menthylmesitylphosphine **27c** were isolated by fractional crystallization of an (R_P)/(S_P)-mixture from acetonitrile containing a trace of sodium acetylacetonate as a proton scavenger (yield 66.2%, 97% de). The crystal and molecular structure of (S_P)-**27c** (R = Mes) has been defined (Schemes 2.21 and 2.22).

(S_P)-(−)-tert-Butylphenylphosphine oxide [59][1]

a) A solution of PhMgBr (200 mmol) in 100 ml of THF (tetrahydrofuran) was added dropwise at −70 °C to 40 g of dichloro-(−)-menthylphosphine in 300 ml of pentane and the reaction mixture was stirred at room temperature. Then an aqueous solution of ammonium chloride (100 ml) was added. The organic phase was washed with aqueous sodium carbonate solution, filtered, dried by Na_2SO_4, and concentrated under reduced pressure to give menthyl phenylphosphinate. The residue was crystallized in hexane in the refrigerator to give the diastereomerically pure product (yield 22%, ^{31}P NMR 24.0, d, $^{1}J_{PH}$ = 480 Hz).
b) A solution of menthyl phenylphosphinate (10 mmol) in 40 ml of THF at −85 °C was added slowly to a solution of *tert*-butyllithium (20 mmol) in 30 ml of pentane.

1 All preparations were developed or reproduced and modified in laboratory of author.

The reaction mixture was stirred at −80 °C for 3 h. Then the temperature was raised up to −20 to −30 °C, the mixture was diluted with 50 ml of ether, and a saturated aqueous solution of ammonium chloride was added (5 ml). The organic layer was separated off and dried with Na_2SO_4. The solution was filtered and evaporated under vacuum. The residue was chromatographed on a column with silica gel (Et_2O/pentane/EtOH as eluent) (yield 72%, $[\alpha]_D^{20}$ −35 (c = 1, $CHCl_3$); mp 77–78 °C; δ_P 49, d, $^1J_{PH}$ = 45.0 Hz).

Phosphinous acid was converted into enantiomers of secondary (S_P)- or (R_P)-phosphine boranes **30a–i** by reduction with sodium borhydride. The transformation of readily available enantiopure *H*-menthylphosphinates **27** into chiral phosphinous acid boranes **30** permits the elaboration of bulky P-stereogenic secondary phosphine boranes. Taking advantage of the synthetic potential of these compounds, a broad range of hindered P-chiral tertiary phosphine boranes **28** were prepared with excellent ee's [58, 59] (Scheme 2.23 and Table 2.1).

Buono described the preparation of menthyl phosphinites **27** from $PhPCl_2$ and their separation into diastereoisomers. The subsequent reaction of (S_P)-phosphinite borane **27** with sodium hydride and with methyl iodide afforded the menthyl methylphenylphosphinite borane (R_P)-**33** with retention of configuration at phosphorus [63] (Scheme 2.24). The palladium-catalyzed ($Pd[PPh_3]_4$) coupling reaction of the

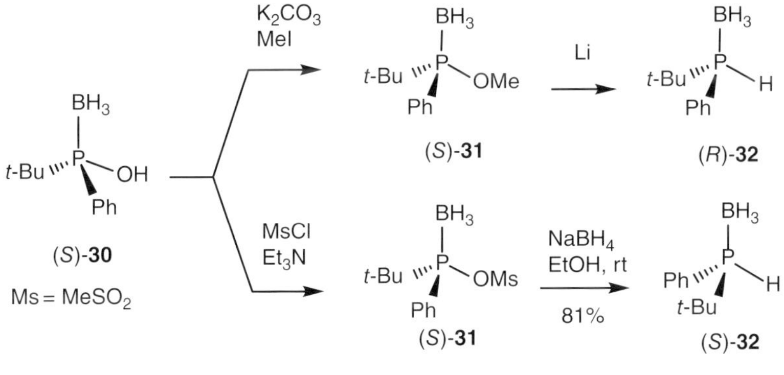

Scheme 2.23 Enantiopure secondary *H*-phosphines **31** (Table 2.1).

Table 2.1 Preparation of **30a–i** from menthyl *H*-phosphinates **27** (Scheme 2.23).

Compounds	R	Ar	Configuration	Yields (%)	ee (%)	References
30a	Me	Ph	(*S*)-(−)	78	95	[59]
30b	*n*-Bu	Ph	(−)	70	89	[59]
30c	*t*-Bu	Ph	(*S*)-(−)	70	84	[59]
30d	*t*-Bu	Tol	(*S*)-(−)	70	99	[58]
30e	*t*-Bu	1-Nphth	(*S*)-(−)	62	99	[58]
30f	2-Tol	Ph	(+)	75	97	[59]
30g	2-PhC_6H_4	Ph	(−)	85	95	[59]
30h	1-Nphth	Ph	(*R*)-(−)	75	99	[59]
30i	1-Furyl	Ph	(+)	72	80	[59]

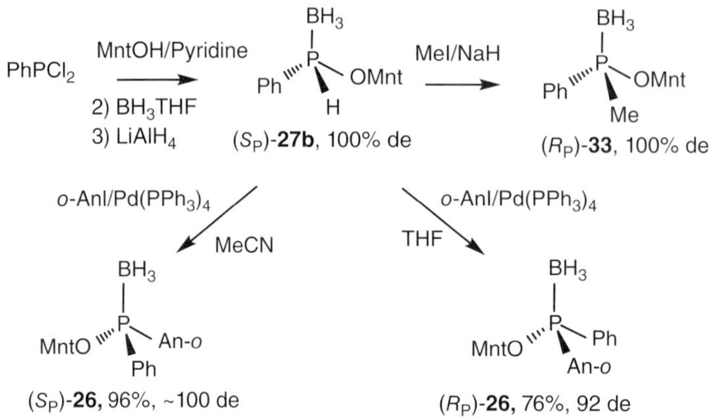

Scheme 2.24 Preparation of enantiopure (1*R*,2*S*,5*R*)-menthyl phosphinite boranes.

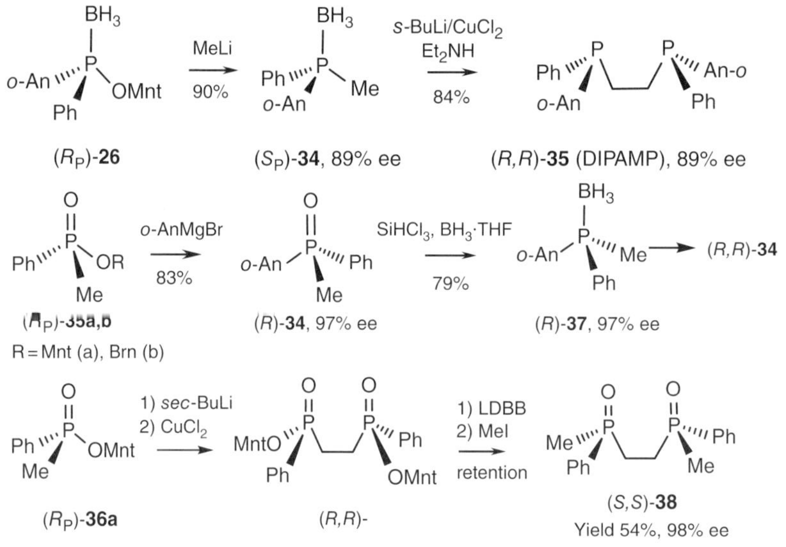

Scheme 2.25 Synthesis of DIPAMP analogs.

(S_P)-**27** with (S_P) and (R_P)-*o*-iodoanisole occurred either with complete retention of the configuration at the phosphorus atom or with almost complete inversion, depending on the solvent used (MeCN or THF). Both diastereoisomers (R_P)-**26** and (S_P)-**26** can thus be synthesized from a single starting diastereoisomer (S_P)-**27b** [58, 59].

Imamoto used menthyl phosphinite boranes for the synthesis of DIPAMP analogs, as shown in Scheme 2.25. The deboration of diborane by diethylamine afforded the (*R*,*R*)-DIPAMP **35** [61, 62]. Diastereoisomerically pure menthyl- or bornylphosphinates **36a,b** reacted with Grignard reagents to afford tertiary phosphine oxides **34** with inversion of absolute configuration at phosphorus [57–59]. Phosphine oxides **36** were reduced to tertiary phosphines or their boranes (*R*)-**37** by silanes. The subsequent treatment of phosphine oxides (*R*,*R*)-**36** with lithium 4,4′-di-*tert*-butyldiphenylide (LDBB) and methyl iodide furnished the diphosphine oxide (*S*,*S*)-**38** with retention of configuration at the phosphorus atom. This method is complementary to the nucleophilic substitution of the menthyl group, in which the substitution at the phosphorus atom occurs with inversion of configuration [57, 58].

The removal of the boranato group is easily achieved by treatment of phosphine borane with excess of diethylamine, DABCO, or by using certain acids. This step fully retains the configuration at the phosphorus atoms, in contrast to the stereochemical problems associated with reduction of oxides. For example, monodentate phosphines **40**, bearing 2-diphenyl group, were prepared as shown in Scheme 2.26 and used as effective ligands in complexes of palladium, catalyzing reactions of carbon–carbon bond formation.

(1*R*,2*S*,5*R*)-Menthyl-*H*-phosphinate **41** was prepared by reaction of anhydrous hypophosphorous acid with (1*R*,2*S*,5*R*)-menthol in the presence of trimethylorthoformate [66, 67]. The *H*-phosphinate **41** was stable and could be purified by distillation under vacuum. It entered into the Heck reaction with iodobenzene in the presence of a palladium complex to yield the diastereoisomerically enriched (S_P)-(−)-menthyl phenylphosphonite **27b**. The Todd–Atherton reaction of with CCl_4 and isopropylamine proceeded with the formation of amide **42**. Compound **41** also reacted with Schiff bases to give bis-amidophosphinates **43** (Scheme 2.27).

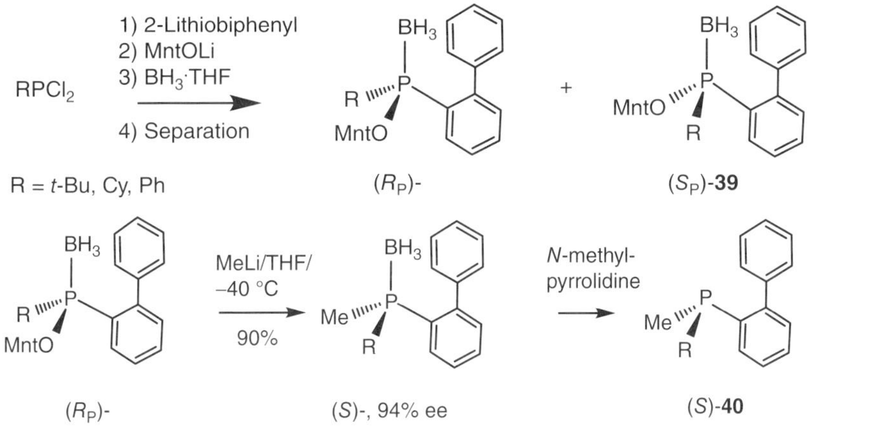

Scheme 2.26 Synthesis of monodentate phosphine ligands, bearing the 2-diphenyl group.

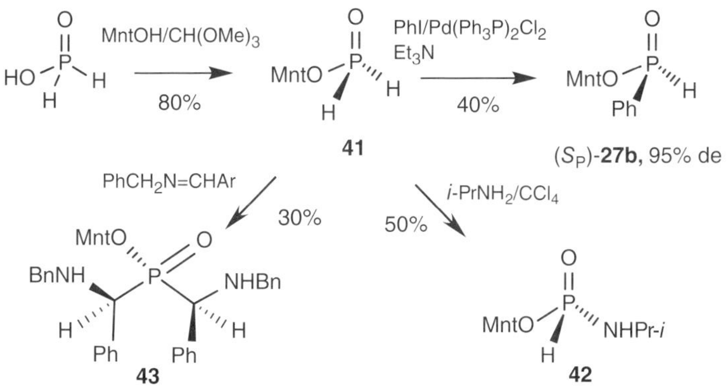

Scheme 2.27 (1*R*,2*S*,5*R*)-Menthyl-*H*-phosphinate **41** as starting chiral reactant.

(1R,2S,5R)-Menthyl-H-phosphinate **41**. Anhydrous hypophosphorus acid, 0.05 mol, in THF was treated consecutively with 0.06 mol of trimethyl orthoformate and 0.06 mol of (1*R*,2*S*,5*R*)-menthol. The mixture was kept for 3 h at 20 °C, and volatile compounds were removed *in vacuo*. The residue was a spectroscopically pure compound. It was additionally distilled under vacuum (yield 90% (before distillation), bp 60–70 °C (0.001 mm Hg), δ_P 10.5, dt, $^1J_{PH}$ = 557 Hz, $^3J_{HH}$ = 10 Hz).

(1R,2S,5R)-Menthyl (S)-phenylphosphonate. Bis(triphenylphosphine)palladium dichloride, 0.05 g, was added to a solution of 0.003 mol of menthyl phosphinate, 0.003 mol of iodobenzene, and 0.004 mol of triethylamine in 4 ml of acetonitrile. The mixture was heated for 2–3 h at 100 °C. The precipitate that formed was filtered off, the solvent was evaporated, and the residue was chromatographed on column with silica gel, eluent hexane–ethyl acetate (3 : 1). Compound **27b** was purified additionally by low-temperature crystallization from hexane. (R_f 3.2, yield 60%, $[\alpha]_D^{20}$ −21 (c = 4, benzene), δ_P ($CDCl_3$) 20.5, dd, $^1J_{HP}$ = 557 Hz, $^3J_{HH}$ = 13.0 Hz) [66, 67].

Montchamp has developed methods for the conversion of hypophosphorous acid and alcohols into various enantioenriched *H*-phosphonate diesters termed *organophosphorus chemistry without PCl$_3$*, which is interesting from the point of view of green chemistry (Scheme 2.28) [68–70]. Compound (R_P)-**44a** was prepared from hypophosphorous acid, paraformaldehyde, and L-menthol in 9% yield. Further cross-coupling of crystalline (R_P)-**44** with bromobenzene gave (R_P)-**45** in 68% yield. On the other hand, the cross-coupling of the mother liquor with bromobenzene and the crystallization of the resulting reaction mixture at room temperature led to (S_P)-**45** in 23% yield and with 97% de.

The compound (S_P)-**45** was also prepared from phenyl-*H*-phosphinic acid, L-menthol, and paraformaldehyde in 26% yield and with 95% de. These phosphorus synthons were functionalized into useful P-stereogenic compounds (Scheme 2.28) [71]. Diastereomer (R_P)-**44** is a versatile P-stereogenic building block for the preparation of chiral tertiary phosphines (Scheme 2.29) [71, 72].

Cross-coupling of (R_P)-**44** with arylhalogenides and Pd(OAc)$_2$ gives (S_P)-**46** in good yields, and subsequent oxidative cleavage delivers (R_P)-**47** in 81% yield. Compound (S_P)-**46** can be oxidized to form (R_P)-**47** stereospecifically in 81–95% de.

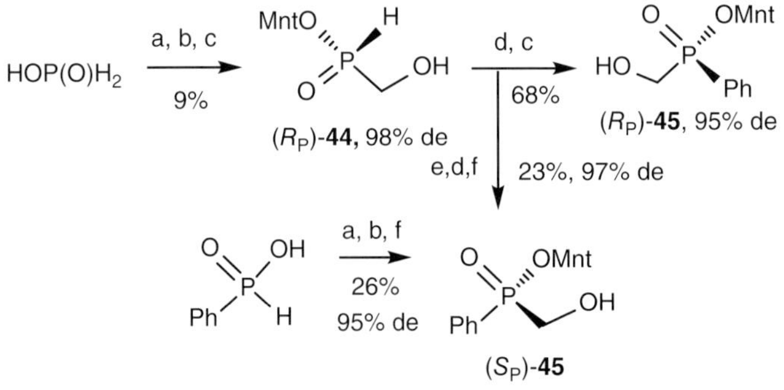

a = $(CH_2O)_n$,75 °C; b = L-Menthol, toluene, reflux; c = recrystallization at −18 °C
d = Pd(OAc)$_2$, Xantphos, DIPEA, PhBr, DMF/DME, 115 °C;
e = mother liquor, f = crystallization at room temperature

Scheme 2.28 Synthesis of P-stereogenic *H*-hydroxymenthylphosphonates.

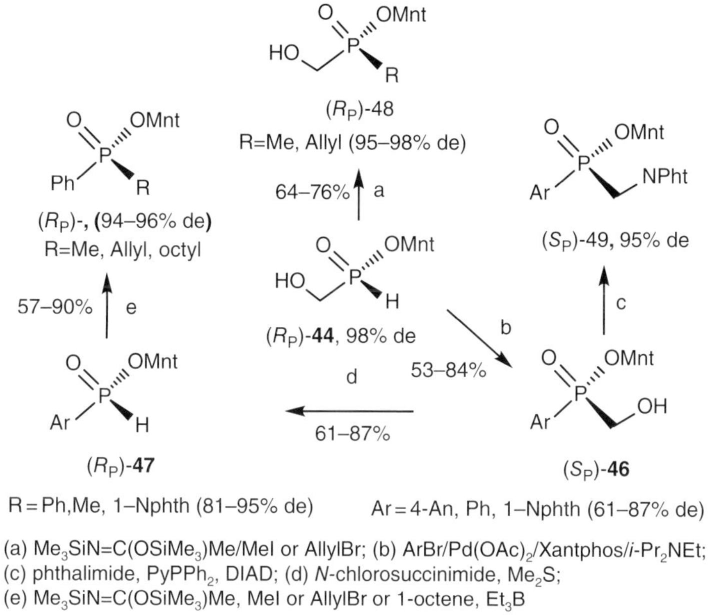

Scheme 2.29 P-Stereogenic *H*-phosphinate **44** as versatile P-stereogenic building blocks.

Therefore, cross-coupling of **44** followed by oxidation of **46** leads to P-configuration of phosphinates using L-menthol in all cases. The presence of the hydroxymethyl group in compounds **44** and **46** provides opportunities for functionalization as the carbon atom can be preserved if desired. Compound (R_P)-**44** can be viewed as a protected chiral equivalent of alkyl phosphinates $ROP(O)H_2$ as it can be stereospecifically alkylated to form **48**, or cross-coupled to form **46**, and the hydroxymethyl moiety can subsequently be cleaved to form *H*-phosphinates similar to **47**. For example, the Mitsunobu reaction of (S_P)-**46** with phthalimide gives (S_P)-**49** in 70% yield. In another interesting example, the stereoisomers of ethyl menthyl (methylthio)methylphosphonate **50** were oxidized with cumene hydroperoxide in the presence of a Sharplomplex to yield the corresponding (methylsulfinyl)methylphosphonates **51** (76–82% de). The sulfoxide **51** was obtained in a diastereomerically pure form (>98% de, upon recrystallization) and was shown to have an (R_P,S_S)-configuration (Scheme 2.30) [73]. Buono reported recently the synthesis of functionalized tertiary phosphine-boranes from readily accessible (hydroxymethyl) phosphine-boranes under mild conditions [74].

Glucofuranosyl phosphinites are an interesting alternative to menthyl phosphinites. The nucleophilic substitution at trivalent phosphorus of chlorophosphines with (−)-1,2:5,6-diisopropylidene- or (−)-1,2:5,6-dicyclohexylidene-D-glucofuranose

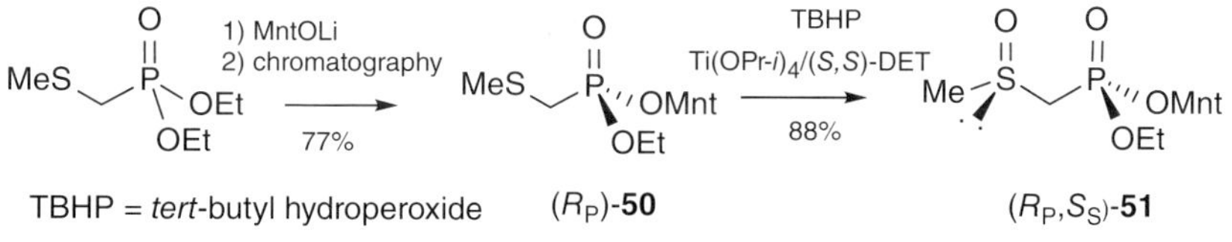

Scheme 2.30 Synthesis of ethyl menthyl (methylsulfinyl)methylphosphonates **51**.

proceeds with good stereoselectivity to afford enantiomerically pure phosphinites **52a** in good yields. Using different bases in the preparation of the phosphinites, it is possible to obtain either of the two diastereoisomers, (S_P)-**52** or (R_P)-**52**, with good diastereoselectivity [75–77]. The levorotatory (−)-(S_P)-phosphinites **52a** (or (S_P)-phosphinates) were obtained in the presence of triethylamine in toluene, and the dextrorotatory phosphinates (+)-(R_P)-**52a** (or (R_P)-phosphinates) were obtained in THF with pyridine as a base. The esters **52a,b** were converted into the corresponding tertiary phosphines (or phosphine oxides) (R_P)- or (S_P)-**53a,b** by reaction with organomagnesium (Scheme 2.31 and Table 2.2). Hii *et al.* reported that the diastereoisomeric

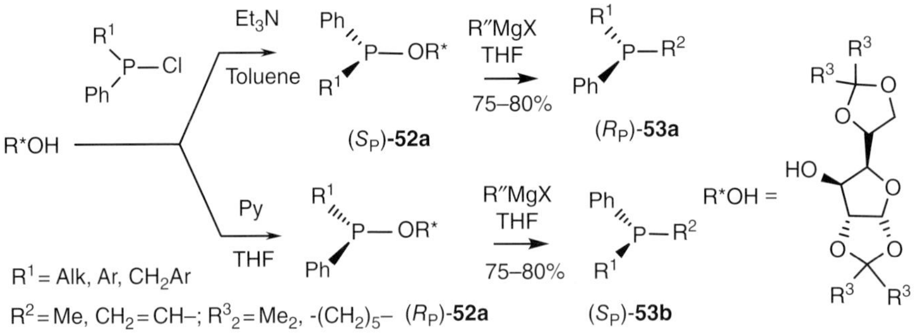

Scheme 2.31 Glucofuranosyl method for the synthesis of *P*-chiral tertiary phosphines.

Table 2.2 Reaction of racemic phenylphosphinic chlorides with glucofuranose (Scheme 2.31).

Compounds	R^1	CR^3_2	Base	Solvent	Yield (%)	(S_P):(R_P) 52
1	Me	CMe_2	Et_3N	Toluene	70	90 : 10
2	Et	CMe_2	Et_3N	Toluene	70	96 : 4
3	*i*-Bu	CMe_2	Et_3N	Toluene	70	95 : 5
4	Bn	CMe_2	Et_3N	Toluene	75	>99 : 1
5	Bn	CMe_2	Py	THF	70	25 : 75
5	Me	*c*-C_5H_{10}	Et_3N	Toluene	75	95 : 5
6	Me	*c*-C_5H_{10}	Et_3N	THF	70	95 : 5
7	Me	*c*-C_5H_{10}	Et_3N	CH_2Cl_2	70	87 : 13
8	Me	*c*-C_5H_{10}	Py	THF	70	30 : 70
10	Et	*c*-C_5H_{10}	Et_3N	Toluene	93	93 : 7
11	Et	*c*-C_5H_{10}	Py	THF	94	30 : 70
12	*i*-Pr	*c*-C_5H_{10}	Et_3N	Toluene	92	86 : 14
13	*i*-Pr	*c*-C_5H_{10}	Py	THF	90	40 : 60
14	Bn	*c*-C_5H_{10}	Et_3N	Toluene	95	90 : 10
15	Bn	*c*-C_5H_{10}	Py	THF	95	40 : 60
16	*o*-An	*c*-C_5H_{10}	Et_3N	Toluene	93	30 : 70
17	*o*-An	*c*-C_5H_{10}	Py	THF	94	55 : 45
18	1-Nphth	*c*-C_5H_{10}	Et_3N	Toluene	87	40 : 60
19	1-Nphth	*c*-C_5H_{10}	Py	THF	83	55 : 45

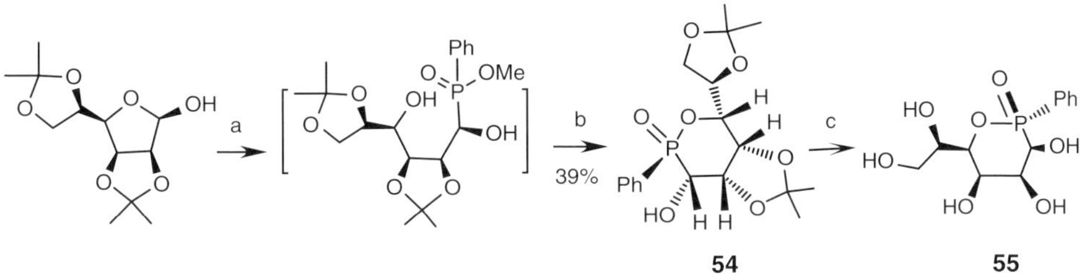

(a) = MeO(Ph)P(O)H, (b) = *t*-BuOK/THF; (c) = Amberlist 15, MeOH, r.t.

Scheme 2.32 Preparation of phosphinic analogous to C-arylglycosides.

ratio of **52** remained unchanged when the reactants were used in equimolar amounts, even at lower temperature [76].

The protected polyhydroxylated 1,2-oxaphosphinane **54** was prepared by a two-step sequence (phenyl-*H*-phosphinate addition on protected mannofuranose followed by intramolecular transesterification) on a gram scale. Deprotection of the di-isopropylidene derivative **54** using acidic cation-exchange resin afforded the free hydroxy organophosphorus heterocycle **55** analogous to C-arylglycosides (Scheme 2.32). The X-ray analysis allowed the absolute configuration of the newly created asymmetric centers of the diastereoisomer **54** to be assigned. Recrystallization in ethanol afforded the pure, fully deprotected arylphosphinosugar **55**. The phosphinosugar **55** analogous to C-arylated heptopyranose shows a boat structure **B** with P2(*S*), C3(*R*) absolute configuration (Scheme 2.32) [78].

(S_P)-1,2:5,6-Di-O-isopropylidene-α-D-glucofuranosyl benzylphenylphosphinite **52**. A solution of 0.02 mol of 1,2:3.5-diisopropylidene-D-glucofuranose in 5 ml of toluene was added dropwise to a solution of 0.02 mol of benzyl-phenylchlorophosphine and 3.5 ml of triethylamine in 10 ml of toluene cooled in an ice bath. The solution was stirred for 3 h at 0 °C and then allowed to stand at room temperature for 12 h under a nitrogen atmosphere. The precipitate of triethylamine hydrochloride was filtered off (yield 100%) and washed with 10 ml of ether. The filtrate was evaporated under reduced pressure. The residue was purified by flash chromatography (colorless oil, yield 85%, $[\alpha]_D$ −154 (c = 0.1, toluene), δ_P 124 ppm).

(R)-Methyl-benzyl-phenylphosphine oxide. To a solution of 0.01 mol of the phosphinite in 5 ml of diethyl ether cooled in an ice bath was added dropwise 0.011 mol of methyllithium in 5 ml of the same solvent. The solution was stirred for 0.5 h at 0 °C. Then the precipitate was filtered off, the solvent was removed under reduced pressure, and the residue was oxidized by *tert*-butylhypochlorite and then purified by crystallization in hexane (colorless prisms, yield 65%, mp 135 °C, $[\alpha]_D$ +50 (c = 0.1, EtOH), δ_P 41.22 ppm).

An advanced strategy for the synthesis of P-chiral gluco- and mannophosphonite boranes was developed on the basis of the addition of diethyl phosphonite borane to glucal-derived aldehyde, accompanied by cyclization and an ethyl/methyl exchange (Scheme 2.33). This direct P(III) strategy facilitates the preparation of various P-chiral phosphonite–boranes, further coupling reactions of which lead to the selective synthesis of two phostone dimers [79].

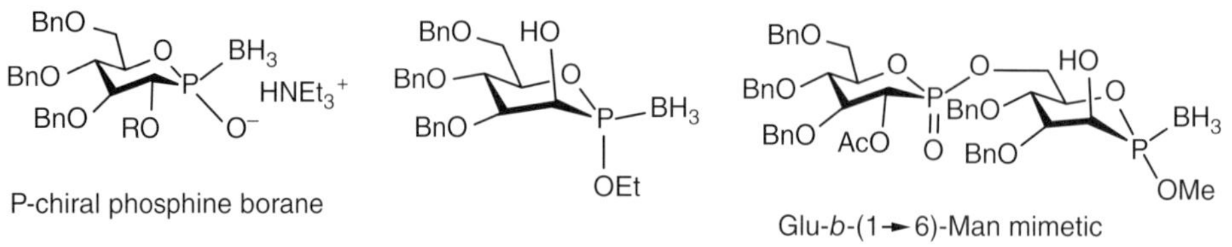

Scheme 2.33 P-Chiral glucophosphonite boranes and phostone–phostone dimer.

2.3.2.2 Optically Active Amines as Chiral Auxiliaries

The unsymmetrical chlorophosphines and chlorophosphine oxides react stereoselectively with chiral amines (amino acid esters or 1-methylbenzylamine) with the formation of enantioenriched aminophosphines **56a–d** and **57a–d** (R = *t*-Bu, R′ = Ph) (85–90% de), which after recrystallization were isolated as diastereoisomerically pure compounds. It was found that (*S*)-1-methylbenzylamine generates the (*R*)-configuration at the phosphorus atom, while, in contrast, (*R*) 1 methylbenzylamine gives rise to the (*S*)-configuration (Scheme 2.34) [80–93]. The stereoselectivity of the reaction depends on the reaction conditions: the reaction should be carried out at 20 °C, with a very slow addition of 1-methylbenzylamine and triethylamine to a benzene solution of chlorophosphine. The stereoselectivity of the reaction depends on the solvent (THF > hexane > toluene > benzene > diethyl ether), and on the organic bases (Et_3N > DABCO > DBU (1,8-diazobicyclo[5.4.0]undec-7-ene)). The mechanism of reaction includes the Berry pseudorotation of a pentacoordinate intermediate and an exchange of ligands at P(V) phosphorus as a result of which the thermodynamically most stable diastereomer is formed. The stability of the intermediate **b**, the apicophilicity of ligands as well as the asymmetric induction under the effect of the optically active 1-methylbenzylamine determine the stereochemical outcome of the reaction [81]. The effect of the reaction conditions on the diastereoisomeric ratio of products **c** shows thermodynamic control. For example, a decrease in the reaction temperature reduces the stereoselectivity, which is impossible with kinetic control, because temperature lowering leads to a deceleration in the establishment of equilibrium of pentacoordinated intermediate complex (Scheme 2.35).

Aminophosphines **56** are useful starting compounds for the preparation of enantiopure compounds. The treatment of aminophosphines **56** with borane in THF leads to the formation of the stable crystalline adduct **59** in quantitative yield. The BH_3 group of the phosphine boranes complex **59** can be easily removed by treatment with diethylamine to furnish the enantiopure aminophosphine (R_P)-**56** with almost quantitative yield. The

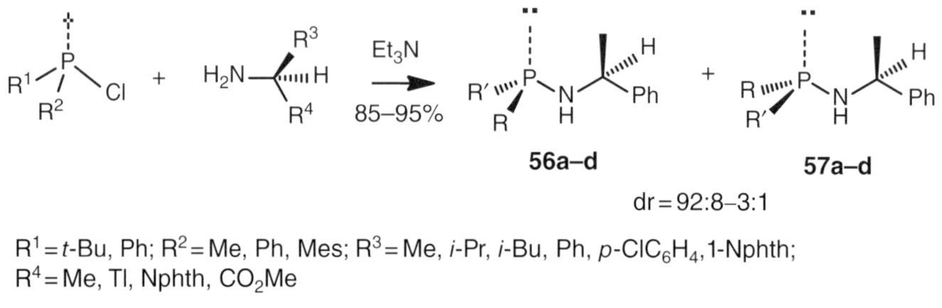

Scheme 2.34 The reactions of chlorophosphines with chiral primary amines.

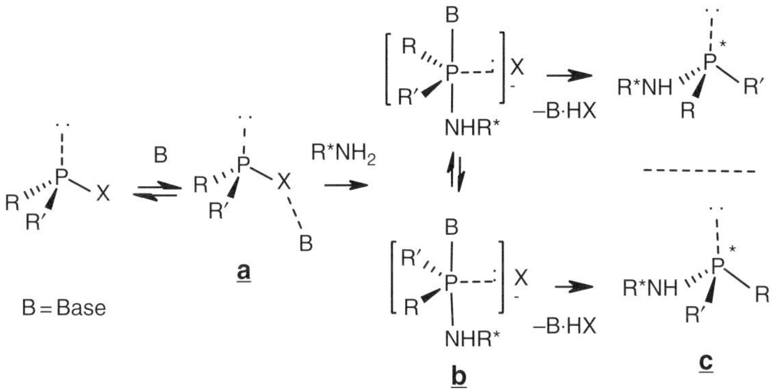

Scheme 2.35 The mechanism of asymmetric induction at the trivalent phosphorus atom.

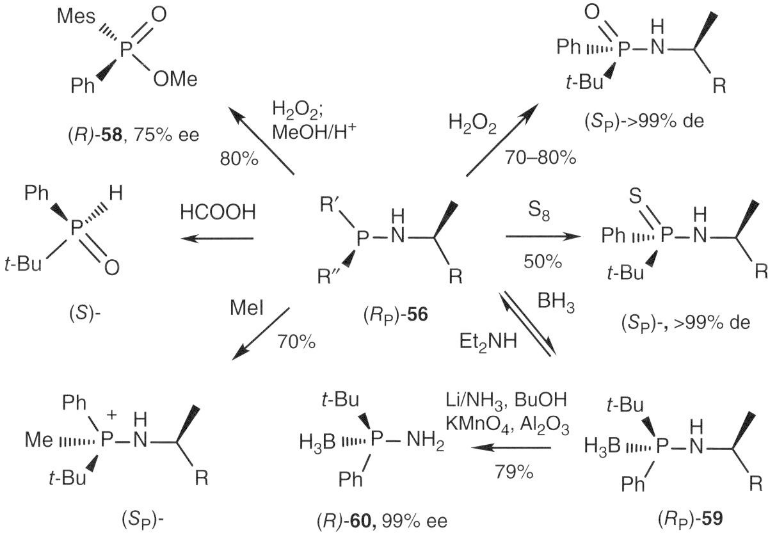

Scheme 2.36 Aminophosphines **56** as chiral starting reactants.

amino group of compounds was replaced by a methoxy group at the reflux in methanol containing sulfuric acid, with the formation of **58**. The acidolysis of **56** afforded enantiomers of *tert*-butylphenylphosphine oxide [83]. The deprotection of **59** was attained by treatment with lithium amide, leading to the formation of aminophosphine borane **60** (Scheme 2.36) [84–87]. The enantiopure aminophosphines **61** were applied as building blocks for the construction of chiral ligands. The reactivity of the amino group should permit further functionalization which can result in novel structures that preserve the original P-chirality. A straightforward application of these P-aminophosphines is the preparation of chiral aminodiphosphine (P–N–P) ligands. Thus, Riera and Verdaguer using the aminophosphines **60** have obtained the P–N–P- and P–N–S-ligands **62** and **63**, which are used in asymmetric catalytic hydrogenation [86, 87] (Scheme 2.37).

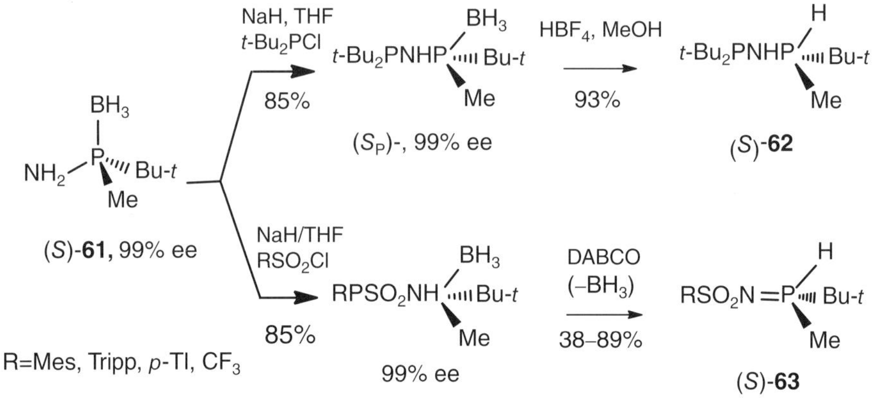

Scheme 2.37 Synthesis of PNP and PNS ligands **62** and **63.**

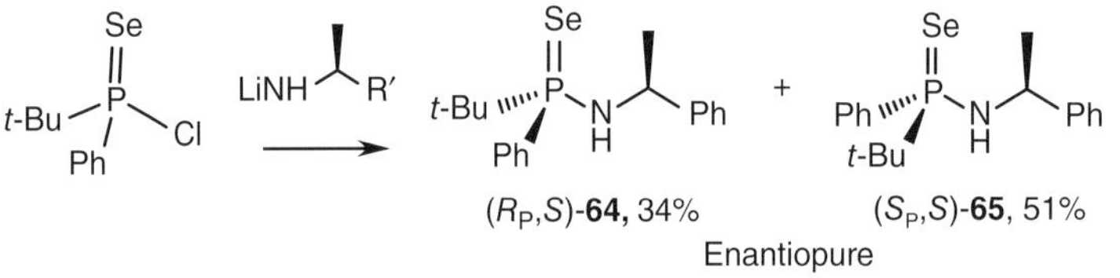

Scheme 2.38 Preparation of P-chiral phosphinoselenoic amides **64** and **65**.

N-(1-Phenylethyl)amino-tert-butylphenylphosphine. A solution of 0.025 mol of (*S*)-*N*-(1-methylbenzyl)amine and 0.026 mmol of triethylamine in 10 ml of benzene was added slowly with stirring over 3 h to a solution of 4.5 g (0.025 mmol) of *tert*-butylphenylchlorophosphine in 10 ml of benzene. The solution was stirred for 2–3 h at room temperature. Then, the reaction mixture was allowed to stand at room temperature overnight. The precipitate of triethylamine hydrochloride was filtered off and washed with 50 ml of diethyl ether. The filtrate was evaporated under reduced pressure and the residue was distilled *in vacuo*. The product was obtained as a colorless liquid (bp 140 °C (0.01 mmHg), yield 80%, borane complex, mp 140–141 °C (hexane), 95% ee, $[\alpha]_D^{20}$ +24.5 (c = 1, CH_2Cl_2) [81].

This method was also used for the preparation of P-chiral phosphinoselenoic amides (R_P,S)-**64** and (S_P,S)-**65** (Scheme 2.38). Enantiomerically pure amides **64, 65** were synthesized by the reaction of racemic phosphinoselenoic chlorides with opticallyactive lithium amides. Two diastereomers of (R_P,S)-**64** and (S_P,S)-**65** were separated by column chromatography on silica gel. The absolute configuration of phosphinoselenoic amide (R_P,S)-**64** was determined by X-ray analysis. Using this reaction, enantiomerically pure salts of phosphinoselenoic acid and P-chiral phosphinoselenoic chlorides were prepared [88, 89].

Ortiz *et al.* [90] described the ortho-directed lithiation of *P,P*-diphenylaminophosphazenes followed by electrophilic quenching as an efficient process for the preparation of P-chiral ortho-functionalized amidophosphinates **68** in good yields and diastereoselectivities. The usefulness of the method was shown with the preparation under mild reaction conditions of a variety of functionalized P-chiral compounds in high yield and excellent stereoselectivity, including phosphinic

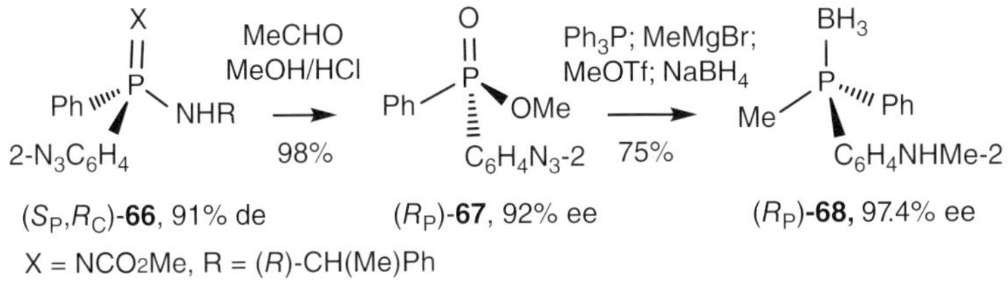

Scheme 2.39 Synthesis of *P*-chiral compounds **68** via *o*-desymmetrized aminophosphazenes.

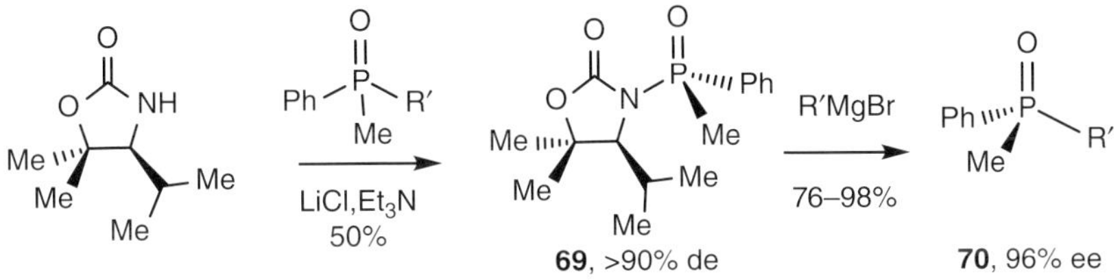

Scheme 2.40 Synthesis of chiral phosphine oxides **70**.

esters, amides, thioamides, phosphine oxides, and (2-aminophenyl)phosphine boranes (Scheme 2.39).

A range of chiral diaryl-methyl and alkyl-methylphenyl phosphine oxides **70** were synthesized under mild conditions with excellent enantioselectivity (>98 : 2 er) using the *N*-phosphinoyl oxazolidinone **69** derived from L-valine and methylphenyl phosphinic chloride. The use of lithium chloride and triethylamine allowed the phosphorylation of oxazolidinone, leading to the *N*-phosphinoyl oxazolidinone **69**. The methodology involves the highly stereoselective formation of P-chiral oxazolidinones **69** which were converted to the interesting phosphine oxides **70** by reaction with Grignard reagents (Scheme 2.40) [91].

Han *et al.* [92] reported a diastereoselective method for the synthesis of P-chiral phosphine oxides. The sequential nucleophilic substitution of 1,3,2-benzoxazaphosphinine-2-oxide **71**, bearing P–N and P–O bonds with metallorganics led to the formation of P-chiral phosphine oxides **72**. Cleavage of the P–O bond in **71** by treatment with MeMgCl led to the formation of (*S*)-**73** with inversion of configuration at P with 99 : 1 er. Diastereomerically pure (R_P)-**74** was obtained by crystallization, and its absolute configuration was confirmed by X-ray crystallographic analysis. The chiral versions of Buchwald-type ligands **74** were prepared effectively using this method. The method was also applied for the asymmetric synthesis of the bulky P-chiral phosphine oxides **75**, which were converted into many important P-chiral ligands (Scheme 2.41).

Nemoto and Hamada [93] have described the development of a new class of chiral phosphorus ligand – aspartic acid-derived P-chirogenic diaminophosphine oxides, DIAPHOXs – and their application to several Pd-catalyzed asymmetric allylic substitution reactions. Pd-catalyzed asymmetric allylic alkylation was initially examined in detail using diaminophosphine oxides **77**, resulting in the highly enantioselective construction of quaternary stereocenters. With the use of the Pd-DIAPHOX catalyst system, asymmetric allylic alkylation, asymmetric allylic amination, and enantioselective construction of quaternary carbons were achieved with high ee (up to 97–99% ee in many cases) (Scheme 2.42).

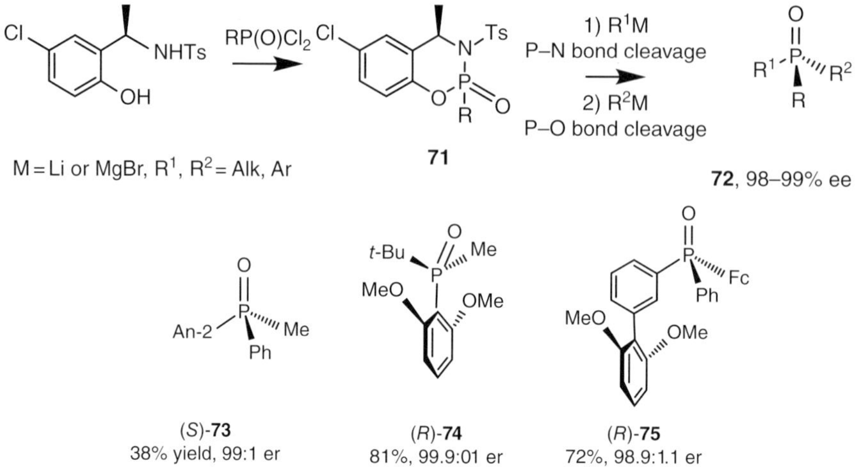

Scheme 2.41 Synthesis of P-chiral phosphine oxides.

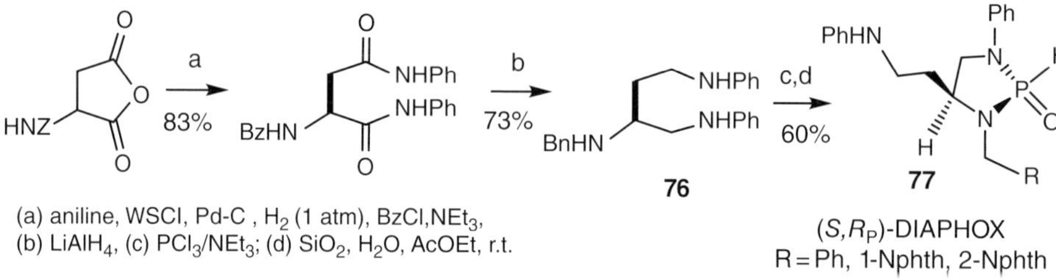

Scheme 2.42 Synthesis of (S,R_P)-Ph-DIAPHOX.

2.3.2.3 Ephedrine as Inductor of Chirality at P(III)

Jugé developed a powerful method (Jugé–Stephan method) [94–97] for the preparation of P-stereogenic phosphines based on the use of ephedrine as a chiral auxiliary. The key reactants in this methodology are 1,3,2-oxazaphospholidine boranes **78**, prepared by a one-pot reaction from bis(diethylamino)phenylphosphine and (−)-ephedrine, followed by protection with BH_3 [94–98]. Under these conditions, the reaction proceeds under thermodynamic control, to give products with diastereomeric ratio of approximately 95 : 5. The cyclization of the (−)-ephedrine leads stereoselectively to the preferential formation of the crystalline (R_P)-diastereoisomers **78** in approximately 90% de [93]. Enantiomeric antipodes of tertiary phosphines (S_P)-**79** and (R_P)-**79** were obtained from (+)- or (−)-ephedrine. The configuration at the P-atom is controlled by the configuration at the Ph-substituted C_1 of (+)-pseudoephedrine or (−)-ephedrine, respectively [98] (Scheme 2.43).

1,3,2-Oxazaphospholidines and 1,3,2-oxazaphospholidine boranes **78** react readily with electrophiles or nucleophiles to provide various chiral phosphorus compounds [97–106]. The acyclic phosphinite boranes **79** were obtained by reaction of oxazaphospholidine boranes **78** with alkyl lithiums or aryl lithiums in THF at −78 °C. Various substituents R^1 = *n*-alkyl, *c*-alkyl, aryl, or ferrocenyl, were introduced to aminophosphine boranes **79** in high yield (93–97%) and with high diastereoselectivity (dr > 98 : 2). The recrystallization of aminophosphine boranes **79** in propanol allowed to obtain diastereoisomerically pure compounds [97]. The reaction proceeded with

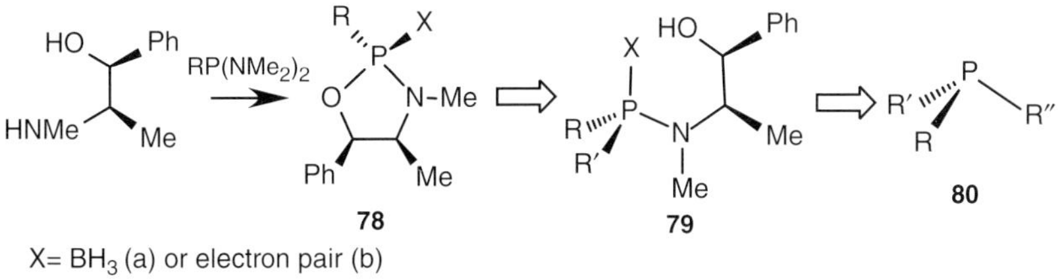

Scheme 2.43 Jugé–Stephan method.

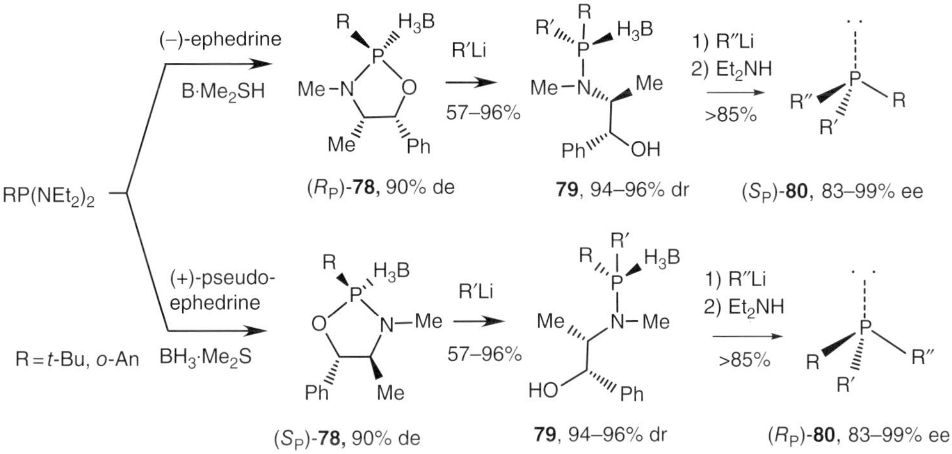

Scheme 2.44 Synthesis of (S_P)- and (R_P)-tertiary phosphines from (+)- or (−)-ephedrine.

retention of configuration at phosphorus, which was proved by the X-ray analysis of some phosphamide–boranes [102]. At the same time, the phosphinite borane **78** reacted with organo-lithium reagents at −70 °C with inversion of configuration on the P-center and with formation of tertiary phosphine borane **79** in good chemical yields and with very good ee's (85–100%). The decomplexation with DABCO or diethylamine yielded free tertiary phosphines with retention of absolute configuration [97] (Scheme 2.44). In addition to the "Jugé–Stephan method" [94], which is based on the nucleophilic ring opening of ephedrine-derived oxazaphospholidine boranes **78**, Bickelhaupt *et al.* [53] reported the stereodivergent ring opening of 2-phenyl oxazaphospholidines **81** (R = H or Me) with alkyl lithium reagents. N–H oxazaphospholidines **81** derived from both (+)-*cis*-1-amino-2-indanol and (−)-norephedrine provided inversion products with high stereoselectivity. In contrast, N–Me oxazaphospholidines **81** yielded ring-opening products with retention of configuration at the P-center. As a result, from a single amino alcohol auxiliary, both enantiomers of key P-stereogenic intermediates **82** were synthesized. The acid-catalyzed methanolysis of compounds **82** proceeded with inversion at the P-center to give the (*S*)- or (*R*)-methyl phosphinites **83** in very good yields and with high ee [99] (Scheme 2.45).

Reaction of oxazaphospholidine boranes prepared from ephedrine with organolithium compounds is a convenient method for the synthesis of chiral tertiary phosphines and ligands [94, 96, 98]. Thus, the phosphinite boranes **83** reacted with organo-lithium compounds at −70 °C with inversion of configuration at the P-center and with formation of tertiary phosphine borane **84** in good chemical yields and

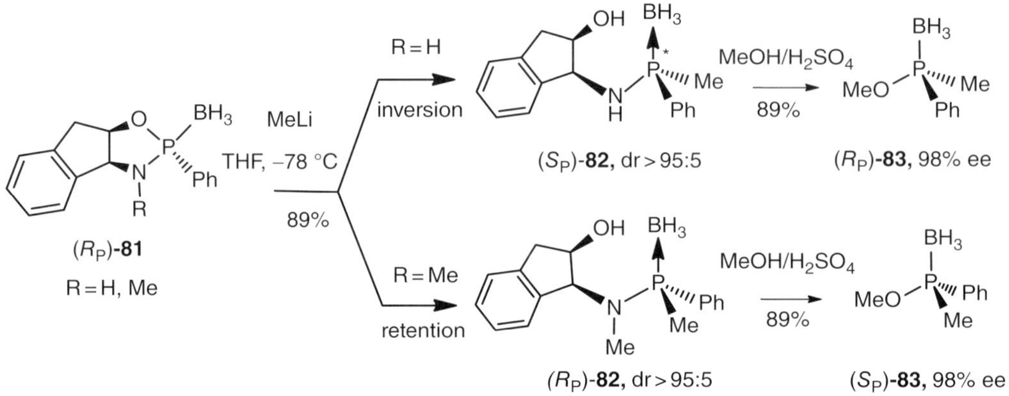

Scheme 2.45 Stereodivergent ring-opening of 2-phenyl oxazaphospholidines.

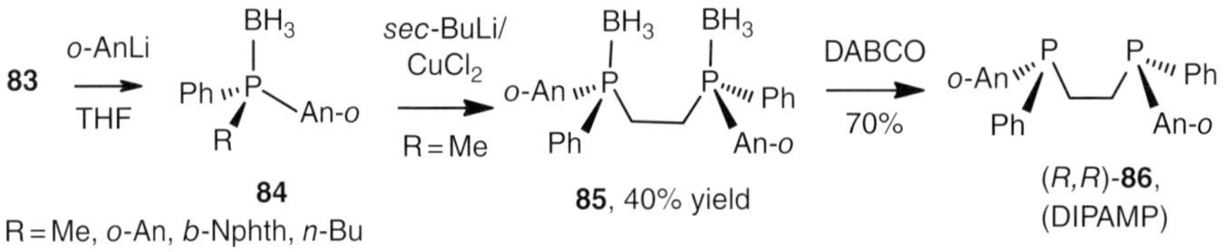

Scheme 2.46 The syntheses of PAMP·BH_3 **85** and DIPAMP from oxazaphospholidine borane **79**.

with very good ee's (85–100%). The phosphine borane **84** (R = Me) was coupled to the corresponding diphosphine diborane **85**, which was decomplexed by DABCO to afford in good yield the optically pure (*R*,*R*)-DIPAMP **86** with retention of absolute configuration at phosphorus (Scheme 2.46) [96, 97, 102].

The reaction of the aminophosphine boranes **79** with chlorophosphines afforded the corresponding aminophosphine-phosphinite boranes **87** in good yields (30–70%). Treatment of the borane complexes with DABCO or Et_2NH gave the enantiomerically pure aminophosphine-phosphinite ligands (AMPP) **88** in 70–90% yield [102–108], which provides an enormous potential for variation and ligand fine-tuning for a number of transition metal-catalyzed reactions [102, 106]. They were used in the Rh complex-catalyzed asymmetric hydrogenation of methyl α-acetamidocinnamate yielding the (*S*)-phenylalanine derivatives with 99% ee and in the rhodium-catalyzed asymmetric hydroformylation of vinyl arenes (Scheme 2.47) [105]. The stereoselective synthesis of P-chirogenic chlorophosphine boranes **89** was achieved by HCl acidolysis of the corresponding aminophosphine boranes **79**. The reaction resulted in P–N bond cleavage with inversion of the configuration at the phosphorus center, leading to the chlorophosphine boranes **89** with high to excellent enantiomeric purities (80–99% ee) (Scheme 2.48) [94, 109].

The stereoselective synthesis of the (*R*,*R*)- (or (*S*,*S*)-) ligands **92** was performed in several steps using the ephedrine methodology with (+)- or (−)-ephedrine, respectively. The key step of the synthesis is the methano bridge formation by reaction of the carbanion-derived methyl phosphine borane **90**, with the chlorophosphine borane **91**. The reaction of the (*S*)-chloro-phenyl-*m*-xylylphosphine borane **90** with MeLi afforded the corresponding (*R*)-methylphosphine borane **91** with inversion of configuration at the P-center. After deprotonation of the methylphosphine borane

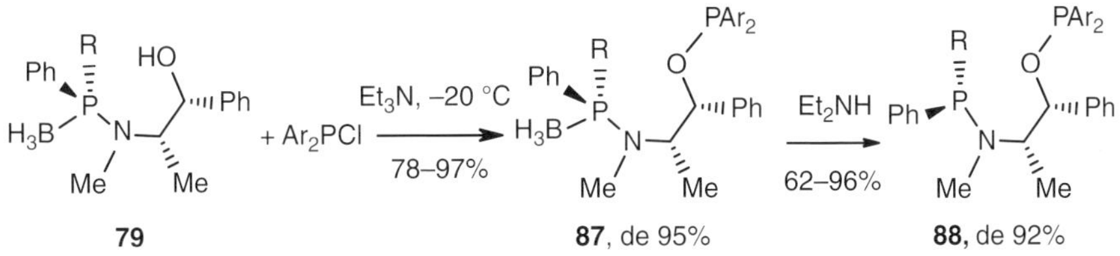

R	Ar	ee (%)
CH_3	Ph	75
CH_3	4-$CH_3C_6H_5$	71
CH_3	3,5-$(CF_3)_2C_6H_3$	32
n-Bu	Ph	75
n-Bu	4-$CH_3C_6H_5$	73
n-Bu	3,5-$(CF_3)_2C_6H_3$	46
1-Naphth	Ph	10

Scheme 2.47 Synthesis of aminophosphine phosphinite ligands.

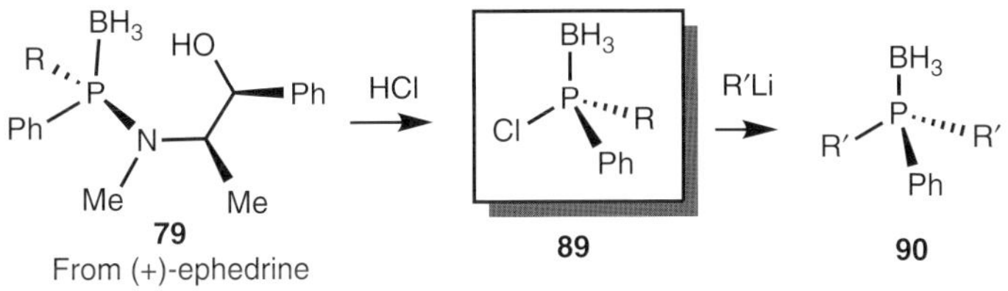

R	R	Yield (%)	ee (%)	Configuration
Me	*o*-An	85	90	(*R*)
o-An	Me	99	98	(*S*)
o-Tol	Me	87	98	(*S*)
2-Nphth	Me	61	85	(*S*)
o-PhPh	Me	45	99	(*S*)
c-C_6H_{11}	–	74	80	(*R*)

Scheme 2.48 The stereoselective synthesis of P-chirogenic chlorophosphine boranes **89**.

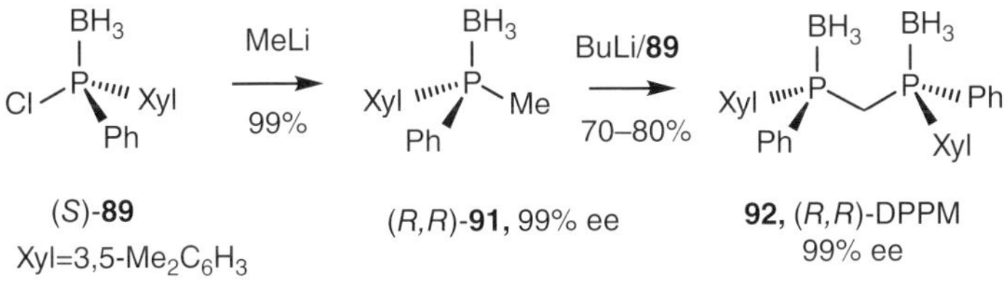

Scheme 2.49 Synthesis of the DPPM ligands **99**.

91 with *n*-BuLi, the reaction with the (*S*)-chlorophosphine borane **89** afforded the protected diphosphine diborane **92** in good yield (Scheme 2.49) [104].

The chlorophosphine boranes **89** were used as efficient starting reagents for the synthesis of various classes of P-chiral phosphorus compounds. Reactions of chlorophosphine boranes **89** with nucleophiles, such as carbanions, phenoxides, phenylthiolates, or amides, led to the formation of corresponding organophosphorus compounds **93–96** in yields of 53–99% and with up to 99% ee. This method was applied also for the preparation of various classes of symmetric and asymmetric P-chiral ligands useful for asymmetric reactions catalyzed by complexes of transition metals (Scheme 2.50) [99, 100, 108].

Both enantiomers of 2,3-bis(*tert*-butylmethylphosphino)-quinoxaline (QuinoxP*) **100**, 1,2-bis(*tert*-butylmethylphosphino)benzene (BenzP*) **101**, and 1,2-bis(*tert*-butylmethylphosphino)-4,5-(methylenedioxy)-benzene (DioxyBenzP*) **102** were prepared as stereochemically pure compounds in short steps from enantiopure (*S*)- and (*R*)-*tert*-butylmethylphosphine boranes **97** as shown in Schemes 2.51 and 2.52 [110]. The overall retention of configuration was rationalized by a reaction sequence involving attack of FcLi on the electrophilic phosphorus followed by pseudorotation and termination by chloride elimination [43, 110].

Ortho-lithiation of oxazaphospholidine borane **104** was carried out with diastereoselectivity of greater than 99%, affording a new and efficient way for

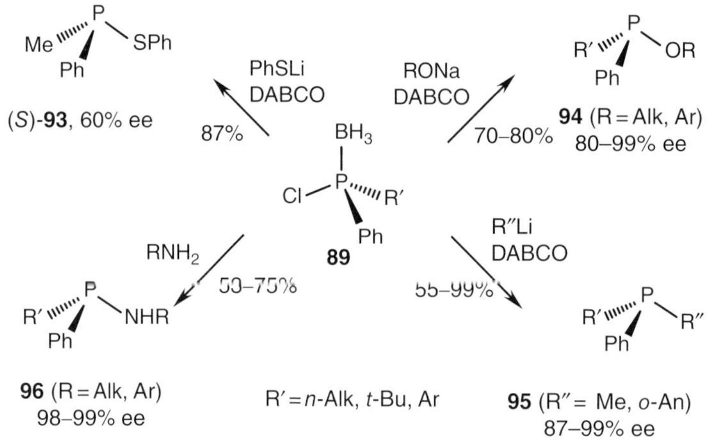

Scheme 2.50 The chlorophosphine boranes **97** as chiral starting reagents.

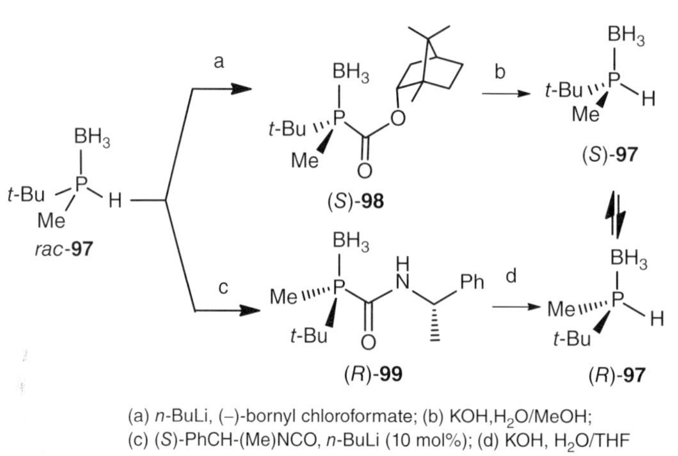

Scheme 2.51

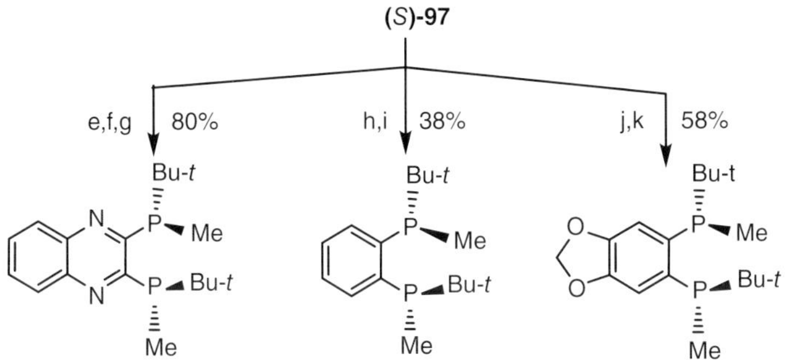

(*R*,*R*)-**100**, Quinox P **101**, Benz P (*R*,*R*)-**102**, Dioxy Benz P

(e) *n*-BuLi, THF, –80 °C, (f) 2,3-dichloroquinoxaline, –80 °C to r.t. (g) TMEDA, r.t.; (h) *n*-BuLi, THF, *o*-$C_6H_4Br_2$, (i) DABCO, THF, reflux; (j) *s*-BuLi , –80 °C, (k) 1,2- diiodo-4,5-(methylenedioxy)benzene

Scheme 2.52 The synthesis of P-chiral ligands **100–102**.

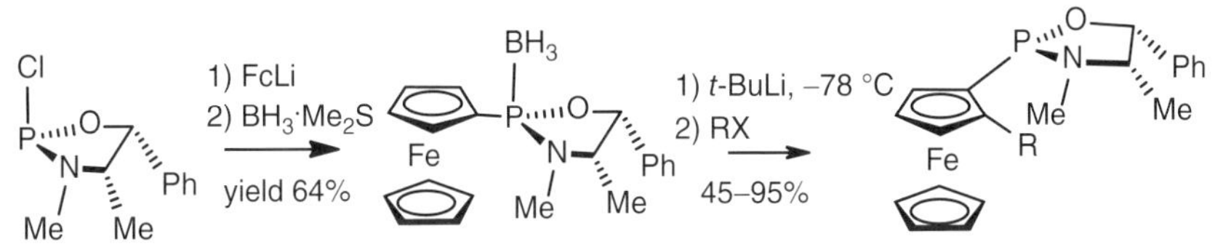

103 **104**, 99% de **105**, 99% de

RX = MeI, I_2, Me_3SiCl, Ph_2CO, $B(OMe)_3$, Ph_2PCl, Cy_2PCl, $(p\text{-}CF_3C_6H_4P)_2PCl$, $(p\text{-}An)_2$ PCl

Scheme 2.53 Ferrocene phosphine ligands **106**.

introducing planar chirality into the ferrocene backbone. Various electrophiles were used, showing the wide applicability of the new methodology and its potential to generate ligands **105** for use in asymmetric catalysis (Scheme 2.53) [107]. The P-stereogenic diphosphine ligands **108** were prepared with high diastereo- and enantiomeric purity by a multistep asymmetric synthesis starting from the oxazaphospholidine boranes **78**, via the phosphinite boranes **83**, and selective lithiation of the enantiomerically pure **107**. The ligands **107** were used as catalyst precursor for preparation of Pd-complexes **108**. The Pd complex **108** was tested as catalyst in allylic alkylation reactions as well as in hydrogenation reactions with enantioselectivity up to 97.7% ee (Scheme 2.54) [110–112].

The synthesis of P-chiral diarylphosphinocarboxylic acids **109** was achieved with excellent enantiopurity starting from the oxazaphospholidine boranes **78**. Amido- and amino-diphosphine ligands **110** containing an L-proline backbone were also derived from **78**. The catalytic activity of the ligands **110** was evaluated in the Pd-catalyzed allylic alkylation of 1,3-diphenylpropenyl acetate (Scheme 2.55) [113].

The P-chirogenic aminophosphane-phosphinite ligands (AMP*P) [100], supported on the upper rim of a calix[4]arene moiety **111**, were synthesized in two steps using the ephedrine methodology. Ligand **111** was used for the preparation of the corresponding rhodium complex [Rh(COD)-(AMP*P)]BF_4, which was tested for asymmetric catalyzed hydrogenation of various substrates with excellent enantioselectivities up to 98%. For example, the asymmetric hydrogenation of methyl α-acetamidocinnamates

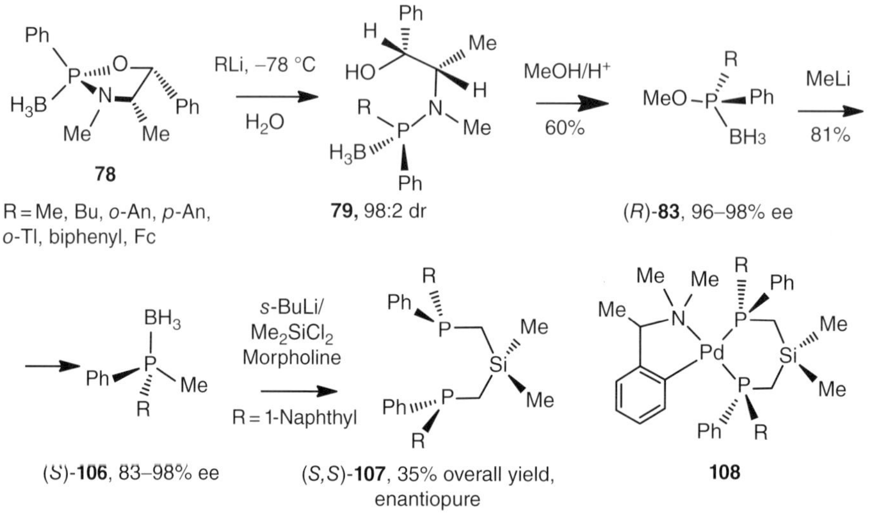

Scheme 2.54 Preparation of diphosphine ligands **107**.

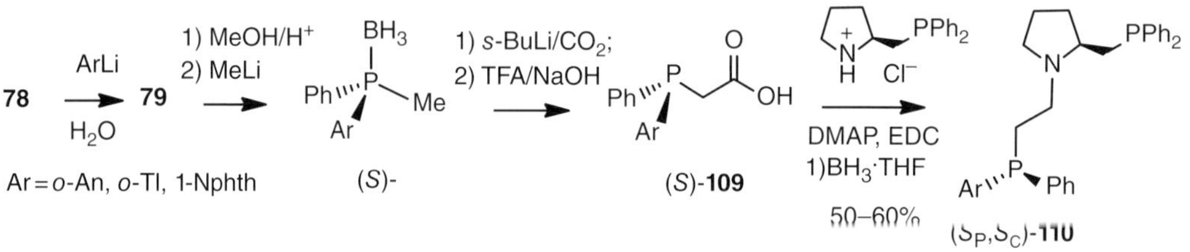

Scheme 2.55 Synthesis of P-chiral amido and amino diphosphines.

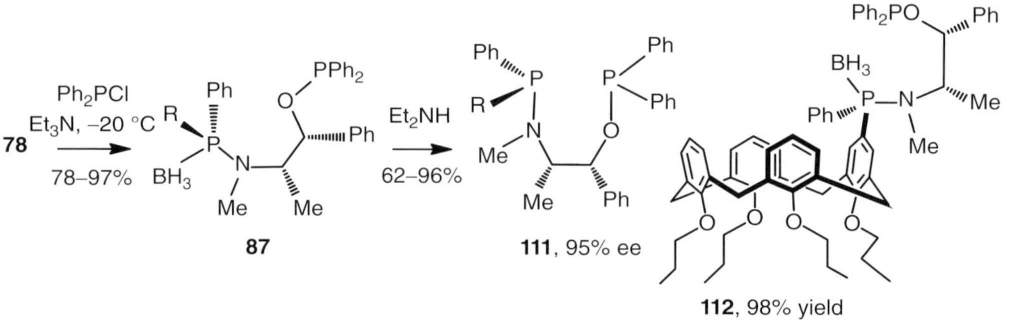

Scheme 2.56 Preparation of aminophosphin-phosphinite ligands.

catalyzed with these Rh complexes yielded (*S*)-phenylalanine derivatives with 99% ee. Investigation of modified P-chirogenic aminophosphane-phosphinite ligands **111**, bearing similar substituent on the P-chirogenic aminophosphane unit, demonstrates that the calix[4]arene substituent of the aminophosphane moiety plays a major role in better asymmetric induction (Scheme 2.56) [105].

By using this method, a number of tertiary phosphines **112**, including ferrocenyl and adamantyl phosphines, were synthesized in good yields (50–90%) and (80–98%) ee.

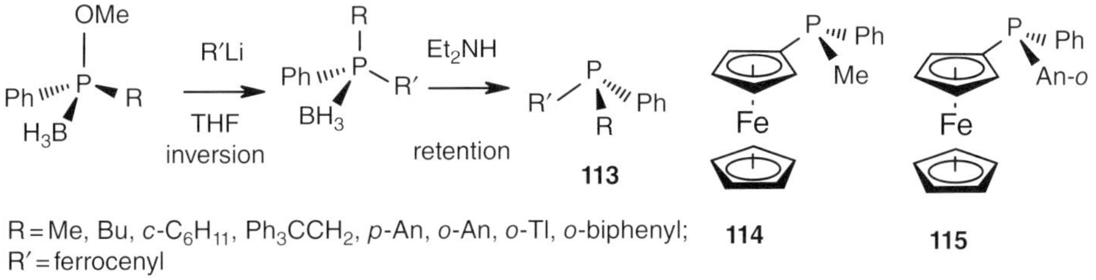

Scheme 2.57 Synthesis of ferrocene-substituted tertiary phosphines.

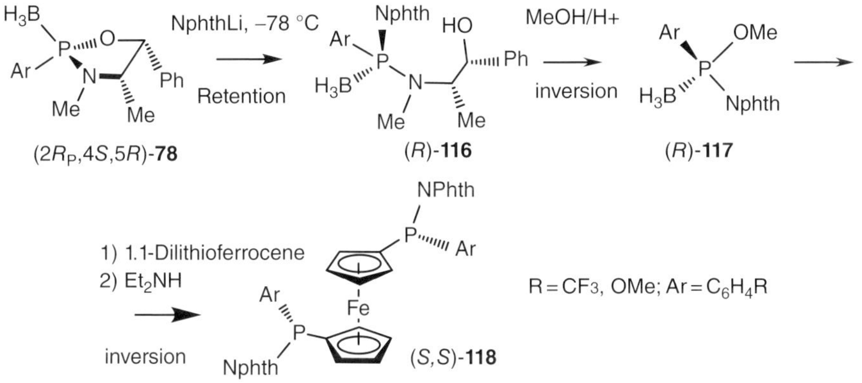

Scheme 2.58 P-Chiral 1,10-bis(1-naphthylphenylphosphino) ferrocenes.

The monodentate ferrocenyl phosphines **114** and **115** were evaluated as ligands in asymmetric catalytic reductive coupling of alkynes and aldehydes to give chiral allylic alcohols with good enantioselectivity in many cases and with complete (*E*)-selectivity in all cases (>98 : 2) (Scheme 2.57) [114–116]. The Jugé–Stephan route to P-chiral mono- and diphosphines gives easy access to a variety of structures [117]. The enantiopure P-chiral diphosphine 1,10-bis(1-naphthyl-phenylphosphino) ferrocene **118** and its electronically modified derivatives bearing methoxy or trifluoromethyl groups at the para-positions of the phenyl rings were investigated as ligands in rhodium-catalyzed asymmetric hydroformylation (Scheme 2.58) [118].

The synthesis of P-chiral diarylphosphinocarboxylic acids **119** was achieved with very good enantiopurity starting from the oxazaphospholidine boranes **78**. Amido- and amino-diphosphine ligands **120, 121** containing an L-proline backbone were also derived from **78**. The catalytic activities of the ligands **121** were evaluated in the Pd-catalyzed allylic alkylation reaction of 1,3-diphenylpropenyl acetate [113] (Scheme 2.59).

The diastereoselective synthesis of a P-chirogenic β-aminophosphine ligand **124** by carbon–carbon bond formation of the ethano bridge in a 3 : 1 ratio via reaction of an α-metallated P-chiral phosphine borane with a benzaldimine has been described. The major diastereoisomeric β-aminophosphine borane (S_P)-**123** is separated and decomplexed into the corresponding β-aminophosphine **124** under neutral conditions and without epimerization at reflux in EtOH [119] (Scheme 2.60).

Buono developed syntheses of chiral tertiary phosphine oxides starting from oxazaphospholidines **125**. The enantiomerically pure oxazaphospholidine (R_P)-**125**

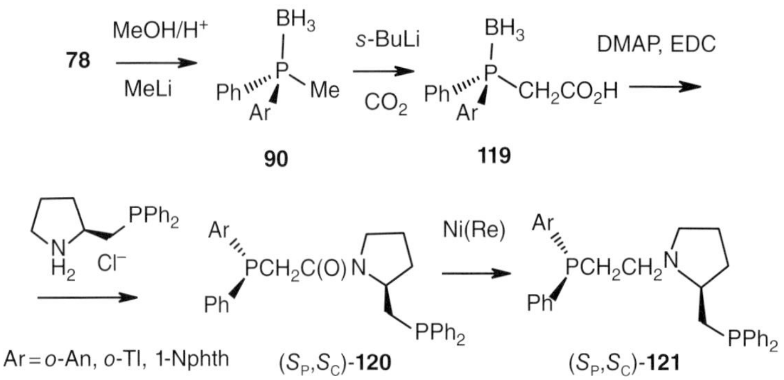

Scheme 2.59 Synthesis of P-chiral diarylphosphinocarbonic acids.

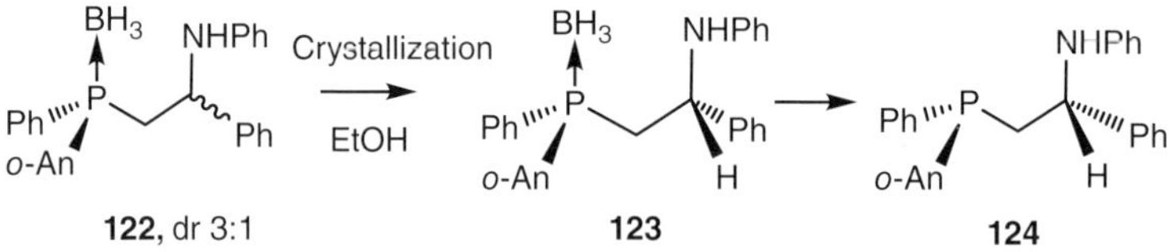

Scheme 2.60 Diastereoselective synthesis of P-chirogene β-aminophosphine ligands.

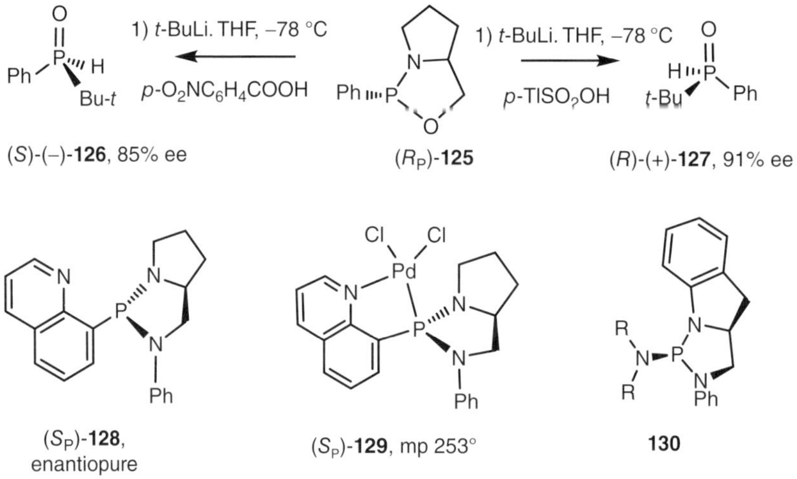

Scheme 2.61 Preparation of (*R*)- and (*S*)-*tert*-butylphenylphosphine oxides from (R_P)-oxazaphospholidine.

was prepared from $PhP(NMe_2)_2$ and (*S*)-(+)-prolinol. Subsequent treatment of **125** with a variety of acids followed by hydrolysis gave both enantiomers of *tert*-butylphenylphosphine oxide **127** with good yields and enantioselectivities up to 91% ee (Scheme 2.61) [120–125]. The opening of the oxazaphospholidine rings (R_P)-**125** with *tert*-butyllithium occurred with retention of absolute configuration on the phosphorus atom, affording the borane complex of aminophosphine (R_P)-**126** (Scheme 2.62). Buono *et al.* [121, 122] have also synthesized P-chiral diazaphospholidine and triaminophosphine ligands **128, 129**. The QUIPHOS-PN_5 **130** was prepared in two

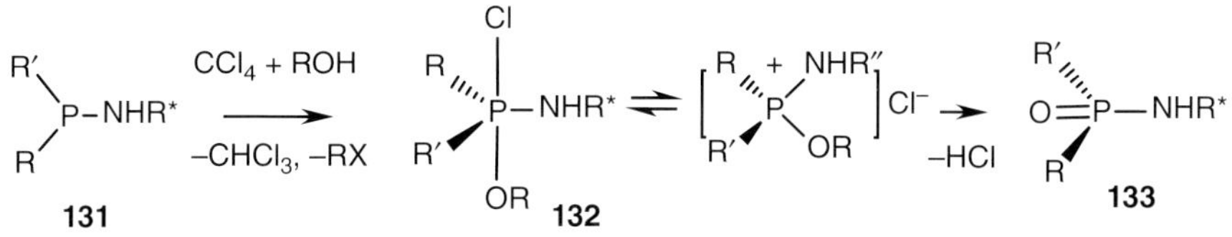

Scheme 2.62 Asymmetric oxidation of phosphines with methane tetrahalides.

steps from 8-bromoquinoline. The structure of compound **128** was confirmed by the X-ray analysis of a palladium(II) complex **129** bearing this ligand [122] (Scheme 2.61).

2.3.3 Asymmetric Oxidation of P(III) Compounds

The preparation of enantiopure phosphine oxides can be attained by asymmetric oxidation of phosphines. Thus the enantioselective oxidation of P-racemic tertiary phosphines and aminophosphines was realized by treatment with tetrahalomethanes and an alcohol. Oxidation of aminophosphines by these reactants resulted in the formation of aminophosphinates **133** in 80–85% yields and with 50–98% de (Scheme 2.62). The compounds **133** were purified by crystallization and obtained with greater than 99% de. It was shown that the reaction of **131** with CCl_4 and alcohol proceeded via the formation of the alkoxychlorophosphorane **132.** The pseudorotation of ligands at P(V) in **132** and Arbuzov rearrangement led to the formation of the most thermodynamically stable diastereomer (Scheme 2.62) [75, 82, 126].

Gilheany *et al.* [127, 128] reported the oxidation of tertiary phosphine **134** with polyhaloalkanes in the presence of L-menthol (the asymmetric Appel reaction). As a result, chiral phosphine oxide **135** was prepared with good ee. The chiral bis-phosphine oxide (*R,R*)-**136** was produced in 98% ee and the minor amount of meso-isomer was easily removed by recrystallization from benzene and enantiopure (>99.9% ee) bis-phosphine oxide **135** was isolated in yield of 73% (Schemes 2.63 and 2.64) [129].

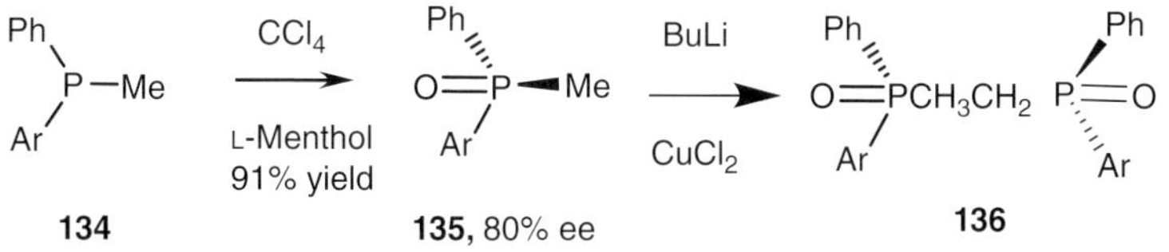

Scheme 2.63 Oxidation tertiary phosphines with polyhaloalkanes.

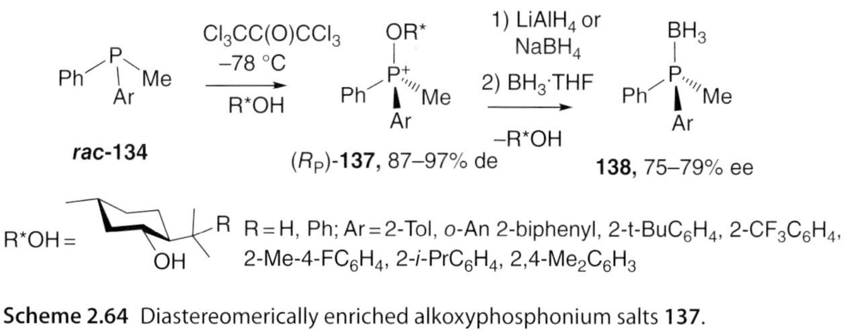

Scheme 2.64 Diastereomerically enriched alkoxyphosphonium salts **137**.

Gilheany *et al.* [130, 131] have also developed the preparation of diastereomeric alkoxyphosphonium salts **137** by reaction of tertiary phosphines with hexachloroacetone, and menthol. The reaction of **137** with $LiAlH_4$ or $NaBH_4$ gives the corresponding enantioenriched P-stereogenic phosphanes **138** in good yields and with moderate ee's. Both enantiomeric phosphine oxides were prepared from a single intermediate, (R_P)-alkoxyphosphonium chloride **139**, which was formed in the course of a selective dynamic kinetic resolution using a single enantiomer of menthol as the chiral auxiliary. Under conditions of Arbuzov-type reaction, this intermediate was converted with retention of the configuration at the phosphorus into tertiary phosphine (R_P)-**140**. Conversely, alkaline hydrolysis of the P–O bond led to the opposite enantiomer (S_P)-**140** (Scheme 2.65) [131–133]. The treatment of racemic P-stereogenic tertiary phosphines **134** with enantiomerically pure bis-phosphoryl or bis-thiophosphoryl disulfides under kinetic resolution conditions afforded enantiomerically enriched tertiary phosphine oxides **135** or phosphine sulfides **141** with 3.5–39% ee. However, the same reaction performed under dynamic kinetic resolution conditions (1 : 1 ratio of reagents) in the presence of chloride ions gave phosphinoyl chloride (R = *t*-Bu) with approximately 70% ee (Scheme 2.66) [134]. The oxidation of racemic phosphines with chiral dioxoruthenium(VI) porphyrins was described by Simonneaux *et al.* [135]. Dioxoruthenium(VI) complexes bearing optically active α-methoxy-α-(trifluoromethyl)phenylacetyl residues on both sides of the porphyrin plane were used for the oxygen atom transfer to phosphorus, with formation of optically active phosphine oxides. For example, the oxidation of racemic benzylmethylphenylphosphine resulted in the (*S*)-enantiomer **142** with 41% ee [134, 135] (Scheme 2.67).

2.3.4 Asymmetric Electrophilic Substitution at P(III)

The number of known stereoselective electrophilic reactions at phosphorus proceeding with high asymmetric induction is not very high and practically limited to chiral tricoordinate phosphorus compounds that on reaction with electrophilic

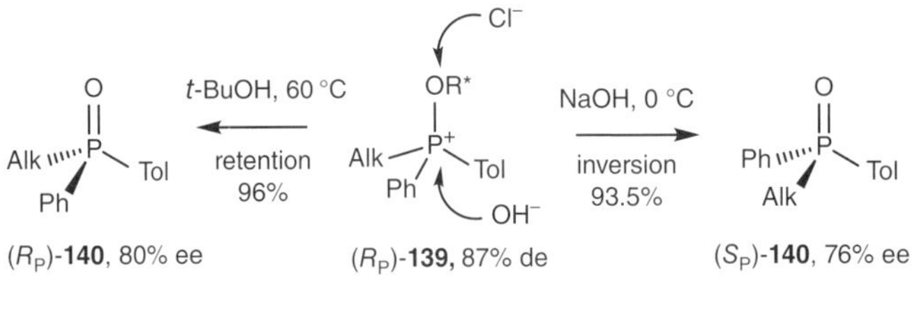

Scheme 2.65 DKR Oxidation of racemic tertiary phosphines.

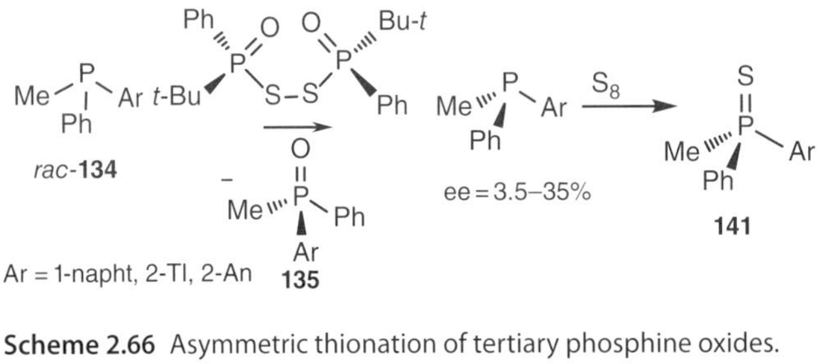

Scheme 2.66 Asymmetric thionation of tertiary phosphine oxides.

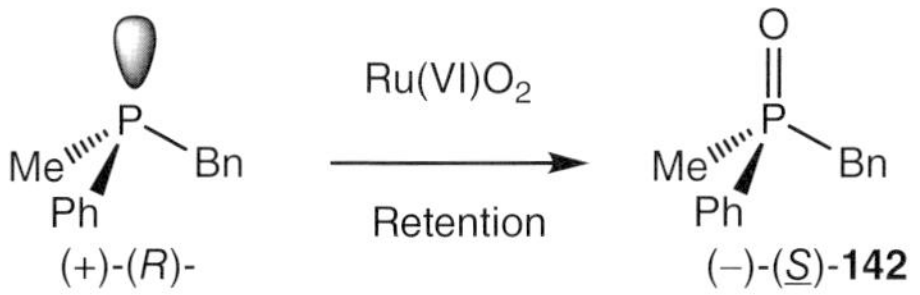

Scheme 2.67

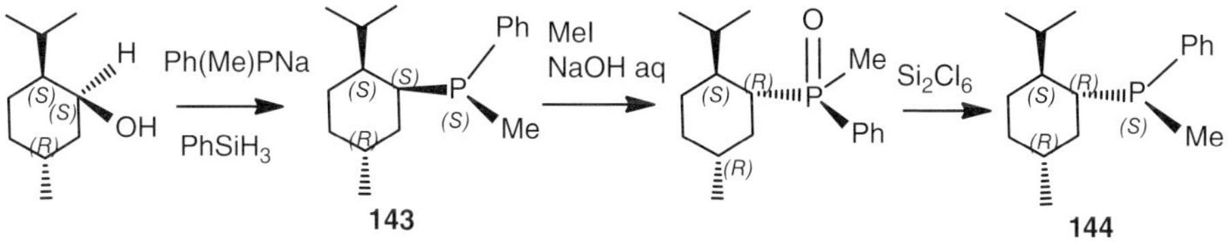

Scheme 2.68 Synthesis of chiral menthyl-methyl-phenyl phosphine.

reagents produce more stable tricoordinate derivatives. It is generally assumed that the electrophilic attack is directed on the lone-electron pair of phosphorus and results in retention of configuration. The alkylation of P-prochiral phosphides was utilized by several groups, but in almost all cases the asymmetric induction was low (Scheme 2.68) [136–142].

For example, the menthylmethylphenylphosphine **144** was prepared from neomenthylmethyl-phenylphosphine **143** by a method that is potentially general for the synthesis of phosphines and phosphine oxides having stereogenic groups at the chiral phosphorus [137]. A scalemic mixture of tertiary phosphines was oxidized to phosphine oxides, then separated by column chromatography. After reduction of phosphine oxides with phenylsilane the optically active tertiary phosphines **144** were obtained and used as ligands in complexes with $[Rh(COD)Cl]_2$. Mosher and Fisher noted that diastereomers of menthylmethylphenylphosphine **144** are stable at room temperature; however, they epimerizate at 120 °C to give a 70 : 30 equilibrium mixture of epimers (Scheme 2.68) [138, 139]. Burgess and co-workers performed the synthesis of stereochemically matched bis-phosphine ligands, representing DIOP–DIPAMP hybrids. Absolute configurations of chiral phosphine ligands were determined via single-crystal X-ray diffraction studies of molybdenum tetracarbonyl derivatives (Scheme 2.69) [140]. Reaction of lithium phosphides with the mesylate or the cyclic sulfate of (*R*,*R*)-2,4-pentanediol afforded, as general access to new chiral ligands, **145** based on the phospholane moiety. Mathey and co-workers reported that deprotonation and subsequent alkylation of (menthylphosphine)pentacarbonyltungsten with *i*-BuI gave two diastereomeric secondary phosphine complexes **146** in the ratio 7 : 3 (Scheme 2.70). Further alkylation of these complexes led to the preferential formation

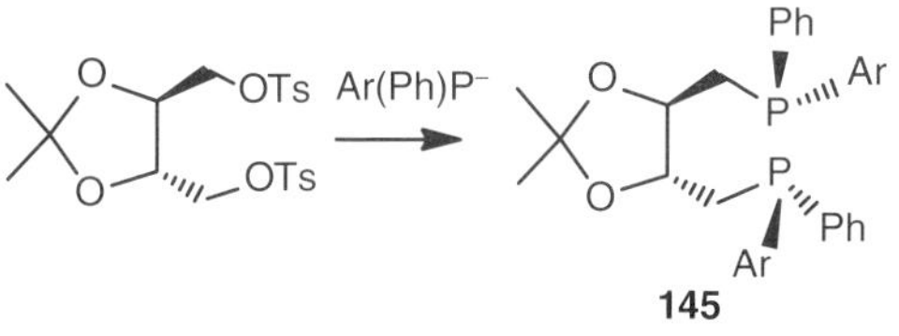

Scheme 2.69 Synthesis of enatiomerically pure bis-phosphine ligands.

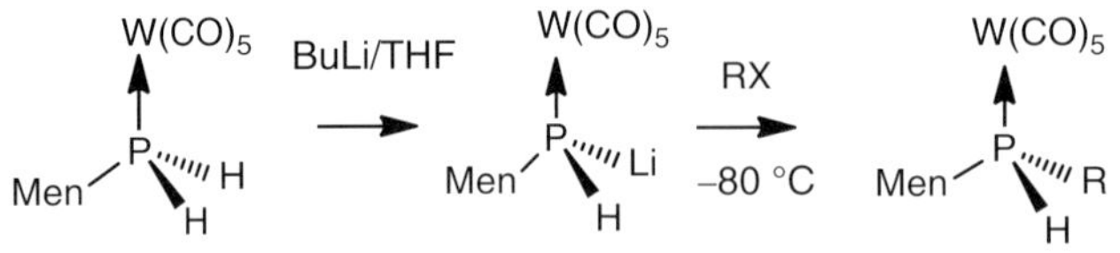

146

Scheme 2.70 The reaction of (menthylphosphine)pentacarbonyltungsten complex with alkyl halides.

of tertiary phosphines with a diastereomeric excess of 80%. Diastereomerically pure tertiary phosphines were isolated after crystallization and oxidative decomplexation. Different examples of alkylations and arylations of prochiral phosphides derived from tartaric acid have also been described by Nagel *et al.* [141, 142].

2.3.4.1 Asymmetric Michaelis–Arbuzov Reaction

The asymmetric version of the Michaelis–Arbuzov reaction has been studied by many authors. For example, Suga and co-workers reported that the reaction of cyclic phenylphosphonite **147**, prepared from (*S*)-1,3-butanediol, with alkylhalides resulted in the ring opening by cleavage of the primary P–O bond [143]. The same group utilized cyclic phosphonites **148** and **149** in a similar reaction, but here the enantiomeric purity of the resulting phosphine oxides was low (Scheme 2.71).

Mikolajczyk and Drabowicz have prepared the optically active (*R*)-methyl ethylphenylphosphinite **150** by reaction of racemic chlorophosphines with methanol in the presence of chiral (−)-*N*,*N*-dimethyl-(L-phenylethyl)amine at −70 °C in ether solution. The Michaelis–Arbuzov reaction of (*R*)-phosphinite **150** with methyl iodide proceeded with retention of absolute configuration at phosphorus affording the (+)-(*R*)-phosphine oxide **151** [144] (Scheme 2.72).

In many cases, the Arbuzov reaction proceeds with low stereospecificity. For example, Jugé found that the reaction of the diastereomerically pure (2*R*,4*S*,5*R*)-3,4-dimethyl-2,5-diphenyl-1,3,2-oxazaphospholidine **152** with alkyl halides proceeded with loss of chirality at phosphorus to afford the mixture of two diasteromers in 85 : 15 ratio. The reaction, monitored by NMR, showed the formation of phosphonium intermediates with the same diastereomeric ratio as that of the final products (Scheme 2.73) [145, 146].

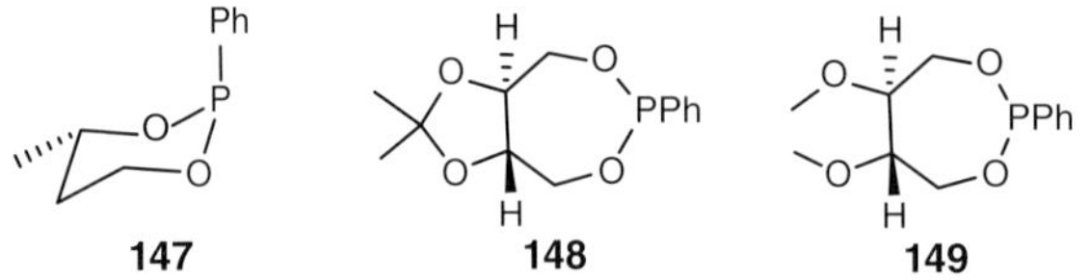

147 148 149

Scheme 2.71

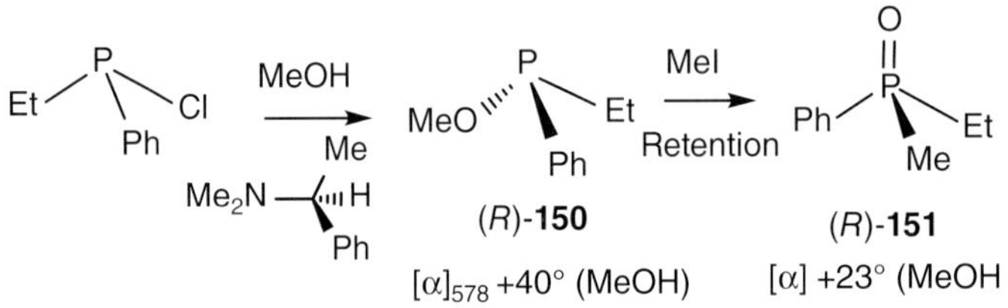

Scheme 2.72

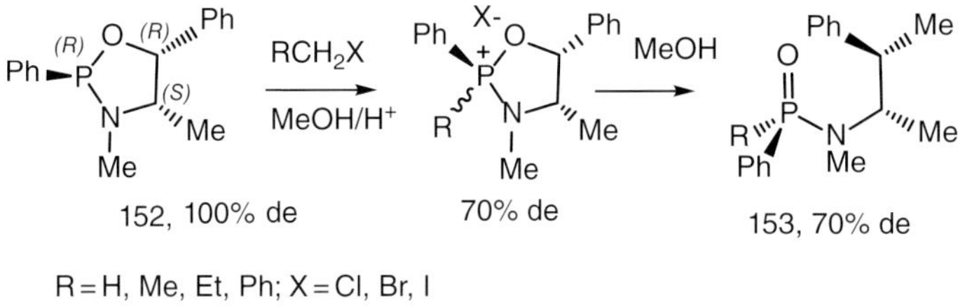

Scheme 2.73 The Arbuzov reaction of (R_P)-oxazaphospholidine **152** with alkyl halides.

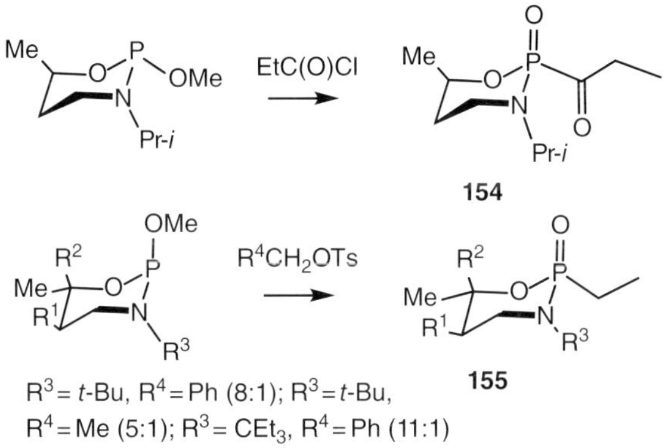

Scheme 2.74 Asymmetric Arbuzov reaction of 1,3,2-oxazaphosphorinanes with alkyl halides.

The Michaelis–Arbuzov reaction of propionyl chloride with 1,3,2-oxazaphosphorinanes led to the formation of a cyclic acylphosphonate **154** with 92 : 8 dr and in 81% yield. An alternative synthesis led predominantly to the *cis*-P-alkyloxazaphosphorinanes **155** (5 : 1–11 : 1 dr) (Scheme 2.74) [147–149].

Pietrusiewicz and co-workers found that the vinylphosphine reacted with (−)-menthyl homoacetate to afford a diastereomeric mixture of the Arbuzov products [150, 151]. Diastereomers were separated by crystallization from the crude reaction mixture. After recrystallization, (S_P)-stereoisomer was obtained with 100% de. Johnson and Imamoto, demonstrated that, prepared in the above manner, menthyl ester **156** could be readily separated into its diastereomers, which were separately subjected to hydrolysis and decarboxylation to give a tertiary phosphine oxide. This approach was successfully used to prepare a variety of phosphine oxides **157** (Scheme 2.75) [149, 152, 153].

The reaction of prochiral dimenthyl phenylphosphonite with alkyllithium at low temperature proceeds with formation of P-resolved menthyl alkylphenylphosphinites of high diastereomeric purity (90–96% de), The second substitution gives enantiomerically enriched tertiary phosphines (73–79% ee) [154].

2.4 Pentavalent P(IV)-Phosphorus Compounds

2.4.1 Introduction

Tetracoordinate organophosphorus compounds exhibit, in general, high configurational stability, although this depends on the structure of the compounds. Tertiary phosphine oxides are the most stable, esters of chiral phosphorus acids are also configurationally

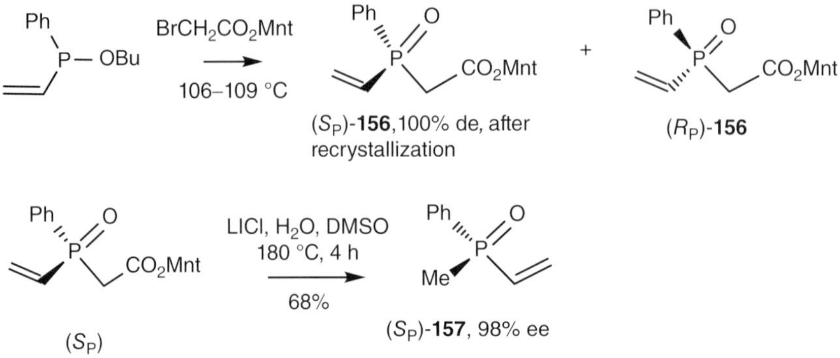

Scheme 2.75 Diastereoselective Arbuzov reaction.

stable, although they racemize slowly upon heating. Chlorides of phosphorus acids racemized at room temperature in the presence of nucleophilic agents. This is related to the formation of bipyramidal pentacoordinated phosphorus species that invert their configuration. A rapid phosphorus inversion accelerated by (p–d)π bonding has been implicated in the stereomutation of allylmethylphenylphosphine sulfide [146].

2.4.2 Nucleophilic Substitution Reactions

Nucleophilic substitution reactions at the tetracoordinate pentacovalent phosphorus atom are often highly stereoselective occurring with essentially complete inversion or retention of configuration. In other instances, the reactions show only marginal stereoselectivity, with the relative importance of the reaction pathways leading to products with inversion or retention of configuration being very much a function of the nucleophile, leaving group, and the reaction conditions [41]. In the overwhelming majority of cases, S_N2 nucleophilic substitution at the chiral tricoordinate trivalent phosphorus results in inversion of configuration that assumes the formation of a pentacoordinate intermediate, containing attacking and leaving groups in apical positions, in spite of the extension of the steric strain due to arrangement of the four-membered cycle in a diequatorial position. The compounds of pentacoordinate phosphorus have trigonal bipyramidal geometry, which can be formed via equatorial attack or via apical attack (Scheme 2.76). It is customary to consider that such reactions occur synchronously by an S_N2P mechanism involving a trigonal bipyramidal phosphorane intermediate that is formed by addition of the nucleophile (Nu) opposite the leaving group (L) occupying the apical position and which decomposes before any ligand pseudorotations have taken place. The result of group displacement (with retention or inversion) is dependent on the relative stability of the phosphorane intermediates and their rate of transformation [155]. As a rule, nucleophilic substitution at stereogenic tetracoordinate phosphorus results in inversion of configuration; however, retention takes place in some cases. For example, alkaline hydrolysis of acyclic *t*-butylphosphonium salts results in inversion of configuration [156]. In general, the stereochemical result (retention or inversion) of nucleophilic displacements at tetracoordinate phosphorus depends on the nature of the leaving group. The course of a displacement reaction at phosphorus, which involves the formation of a phosphorane intermediate, is determined by the energetics of the intermediates. Stereochemical and mechanistic aspects of nucleophilic substitution at

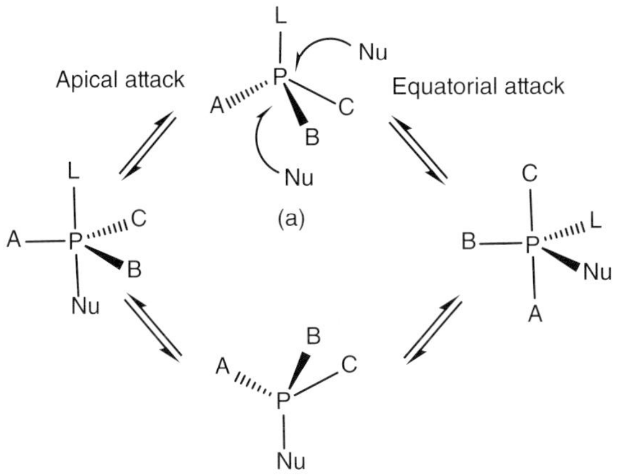

Scheme 2.76 S_N2P mechanism of nucleophilic substitution of P(IV) compounds.

the phosphorus in a six-membered ring, reported by Cremer [155], strongly suggest an S_N2 reaction mechanism.

Westheimer showed that all enzymatic reactions at phosphorus proceed with inversion and, therefore, occur without pseudorotation [157]. In fact, there is no unambiguous evidence that pseudorotation or adjacent attack at the P-atom is a process of significance in any biological system, and formal retention is rationalized by a multistereo process with an even number of inversions Ruedl and Bickelhaupt [54, 158, 159] have considered S_N2@P nucleophilic substitution at phosphorus by means of theoretical calculations and have concluded that increasing the coordination number of the central atom as well as the substituents' steric demand shifts the S_N2@P mechanism stepwise from a single-well potential (with a stable central TC) that is common for substitution at third-period atoms, via a triple-well potential (featuring a pre- and post-TS before and after the central TC), back to the double-well potential (in which pre- and postbarrier merge into one central TS) that is well-known for substitution reactions at carbon. The Walden inversion that accompanies nucleophilic substitution reactions proceeds as a concerted umbrella motion of the substituents at the central atom either via a labile TS, or a stable TC (Scheme 2.77).

Jennings *et al.* [51] have investigated degenerate nucleophilic substitution reactions of phosphonium salts (Scheme 2.78). They suggest a two-step mechanism of formation

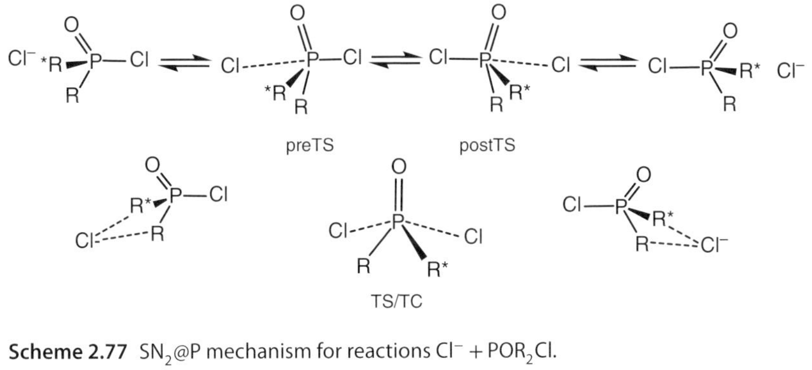

Scheme 2.77 SN_2@P mechanism for reactions $Cl^- + POR_2Cl$.

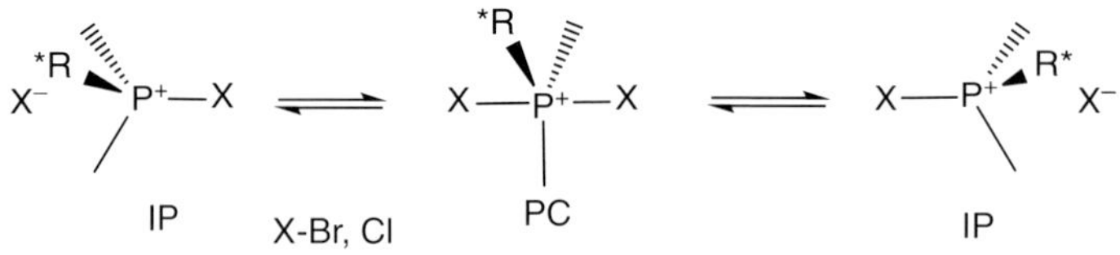

Scheme 2.78 Degenerate interconversion of ion pairs (IPs) via pentacoordinate intermediate.

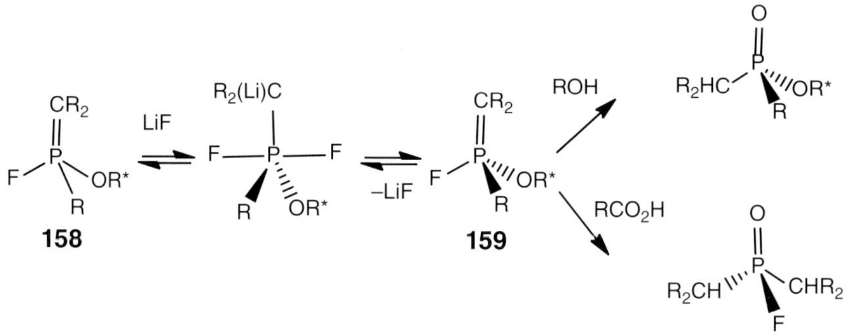

Scheme 2.79 Diastereoselective reaction of P-fluoroylides with chiral alcohols.

of a pentacoordinate dihalophosphorane via backside attack followed by dissociation, resulting in inversion of configuration at phosphorus. The experimentally determined barriers range from less than 9 kcal mol^{-1} to nearly 20 kcal mol^{-1}, ruling out a mechansm via Berry pseudorotation involving equatorial halides that diastereomeric halophosphonium salts exhibit two sets of signals in the NMR spectrum at low temperature and one set at room temperature. On the basis of this finding, the rates of epimerization at the phosphorus center and the respective energy barriers were quantified by using a combination of variable-temperature NMR and EXSY techniques (Scheme 2.78).

An interesting example of thermodynamically controlled asymmetric synthesis is the dehydrofluorination of alkoxyfluorophosphoranes bearing chiral ligands, resulting in a mixture of diastereomers of P-fluoroylides in a 1 : 1 ratio. The P-fluoroylides **158** reversibly add the lithium salt to form a fluorophosphorane intermediate that undergoes pseudorotation to provide the most thermodynamically stable diastereomer **159** [160]. Chiral phosphonium ylides and chiral P-stabilized carbanions are configurationally stable at phosphorus. This configurational stability persists through different electrophilic reactions [161, 162] (Scheme 2.79).

2.4.2.1 Nucleophilic Substitution at P(IV) with Chiral Alcohol

Nucleophilic substitution reactions at the tetracoordinate, pentacovalent phosphorus atom are often highly stereoselective occurring with essentially complete inversion or retention of configuration. In other instances, the reactions show only marginal stereoselectivity, with the relative importance of the reaction pathways leading to products with inversion or retention of configuration being very much a function of the nucleophile, the leaving group, and the reaction conditions [62, 163–173]. Further, in the case of P(III) compounds (Section 2.3.1), the synthesis of P-chirogenic phosphine oxides can be attained by nucleophilic substitution at P(IV) with chiral alcohol: L-menthol, *endo*-Borneol, Glucofuranose, and others (Scheme 2.80 and Table 2.3).

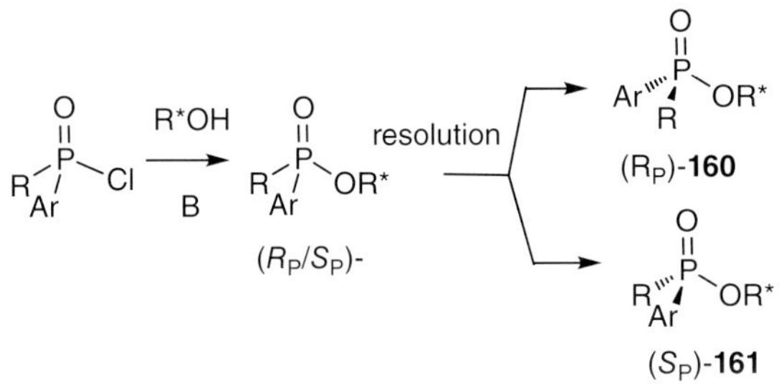

Scheme 2.80 Synthesis of enantiomeric (R_P)- and (S_P)-menthyl phosphinates.

Table 2.3 Reaction of racemic chlorophosphinates with chiral secondary alcohols (Scheme 2.80).

R	R*OH	B		Ratio	References
Me	GF1	Et_3N	Toluene	5	[174]
Me	GF2	Et_3N	Toluene	95 : 10	[175]
Me	GF1	Py	THF	97 : 3	[175]
Et	GF1	Et_3N	Toluene	96 : 4	[76]
Et	GF1	Py	THF	30 : 70	[76]
i-Pr	GF2	Et_3N	Toluene	86 : 14	[76]
i-Pr	GF2	Py	THF	40 : 60	[76]
i-Bu	GF1	Et_3N	Toluene	95 : 5	[174]
Bn	GF1	Et_3N	Toluene	~95::5	[176]
Bn	GF1	Py	THF	40::60	[176]
o-An	GF2	Et_3N	Toluene	30 : 70	[76]
o-An	GF2	Py	Et3N	55 : 45	[76]
Me	(1*S*)-Borneol	DMAP	Toluene	4 : 1	[41]
Me	(−)-Isopinocampheol	DMAP	Toluene	1 : 1	[41]
Me	(+)-Isopinocampheol	DMAP	Toluene	55 : 45	[175]
Me	(+)-Isoborneol	DMAP	Toluene	74 : 26	[175]
Me	(−)-Borneol	Et_3N	Toluene	75 : 25	[41]
Me	(+)-Isomentol	Et_3N	Toluene	13 : 87	[175]
Ph	(−)-Menthol	Et_3N	Toluene	4 : 1	[177]

GF1 = I.2:5.6-di-*O*-isopropylidene-α-D-glucofuranose and
GF2 = I.2:5.6-di-*O*-cyclohexy-lidene-α-D-glucofuranose

Unsymmetrically substituted menthyl phosphinates (R_PS_P)-**162** could be easily separated into their diastereoisomers (R_P)-**163** and (S_P)-**163** by recrystallization or column chromatography. Reaction of the racemic methyl(phenyl)phosphinic chloride with (1*S*)-borneol gave a 1 : 4 diastereomeric mixture of (1*S*)-bornyl (S_P)- and (R_P)-phosphinates, which were separated by column chromatography and reacted with (*o*-bromomethoxyphenyl)magnesium bromide to result in (*R*)-(*o*-methoxyphenyl)methyl-(phenyl)phosphine oxide. Diastereomeric phosphinate esters were also formed from racemic methyl(phenyl)phosphinic chloride with

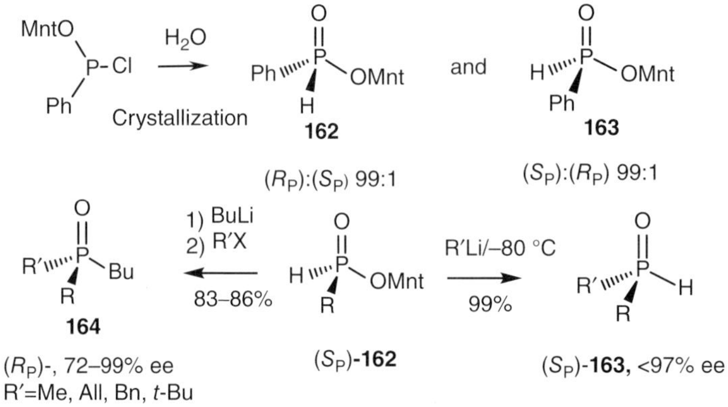

Scheme 2.81 Synthesis of chiral menthylphosphinates.

the terpene alcohols (−)-menthol, (−)-isopinocampheol, and (+)-isoborneol in the ratios 50 : 50, 50 : 50, and 74 : 26, respectively. Readily available enantiomerically pure *H*-menthyl phosphinates **162** permit the elaboration of bulky P-stereogenic secondary phosphine oxides (Scheme 2.81). Diastereoisomerically pure menthyl- or bornylphosphinates **165** reacted with Grignard reagents to afford tertiary phosphine oxides **166** with inversion of the absolute configuration at the phosphorus atom (Scheme 2.81) [97, 100].

These methodologies allow to obtain tertiary phosphines with retention of configuration at the phosphorus atom or with inversion of configuration. For example, Imamoto found that the subsequent treatment of phosphine oxides (*R*,*R*)-**167** with LDBB and methyl iodide furnished the diphosphine oxide (*R*,*R*)-**168** with retention of configuration [62].

This method is complementary to the nucleophilic substitution of the menthyl group, in which the substitution at the phosphorus atom occurs with inversion of configuration [178, 179]. Imamoto used menthyl phosphinite borane complexes for the synthesis of DIPAMP analogs, as shown in Scheme 2.82.

(S_P)-1.2:5.6-Di-O-isopropylidene-α-D-glucofuranosyl benzyl-phenylphosphinate. A solution of 0.02 mol of 1,2:3.5-diisopropylidene-D-glucofuranose in 5 ml of toluene was added dropwise to a solution of 0.02 mol of chloride of benzyl-phenylphosphinous

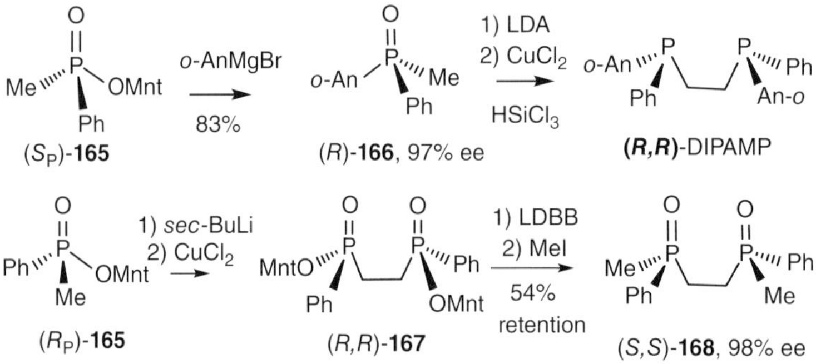

Scheme 2.82 Syntheses of DIPAMP, starting from (R_P)-phosphinates.

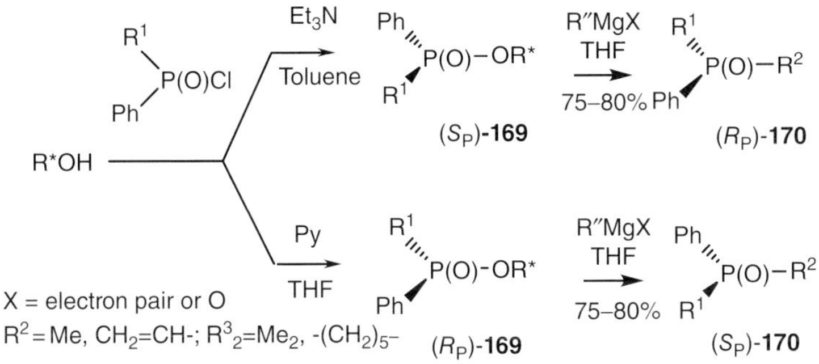

Scheme 2.83 Glucofuranosyle method for the synthesis of chiral phosphinates.

acid and 3.5 ml of triethylamine in 10 ml of toluene cooled in an ice-bath. The solution was stirred for 1 h at 0 °C, then allowed to stand at room temperature for a night under a argon atmosphere. The precipitate of trimethylamine hydrochloride was filtered off and washed with 5 ml of ether. After the solvent was evaporated under an reduced pressure, the residue was purified by crystallization in heptane (yield 75%, mp 175 °C, colorless prisms, $[\alpha]_D^{20}$ −47.5 ($c = 0.05$, acetone)).

Chlorides of dissymmetric phosphinous acids and chlorides of phosphonic acids, in particular the chloride of benzyl-phenylphosphinic acid, reacted stereoselectively with the derivatives of D-glucofuranose (R*OH) [174, 180]. The reaction proceeded in the presence of triethylamine in toluene solution to yield levorotatory (−)-(S_P)-phosphinates **169**, and dextrorotatory (+)-(R_P)-phosphinates **169** in the presence of pyridine in THF (Scheme 2.83) [176, 181, 182]. The phosphinate esters **169** were obtained with good to excellent yields after column chromatography, which also permitted the recovery of the excess chiral auxiliary (Table 2.3). The reaction of phosphinic acid chlorides with D-glucofuranose was performed using an equimolecular ratio of reagents with an excess of triethylamine in toluene at room temperature for 12–24 h. The yields of phosphinates were 70–75% and the diastereomeric excesses were 80–100%. The products were purified by crystallization from hexane. The ratio of diastereomers was not dependent on the excess of chlorophosphinate corresponding to thermodynamic control, unlike the reaction of trivalent phosphorus chlorides with glucofuranose proceeding under kinetic control. The stereoselectivity of the reaction was influenced by the nature of the base and the solvent. The highest stereoselectivity was achieved in toluene with triethylamine as base. The effect of the achiral base has been identified as the most important factor that determines the stereochemical outcome of the reaction [76, 175, 182]. Indeed, the selectivity was reversed in the presence of pyridine. Hii *et al.* [76] reported that, the diastereoisomeric ratio of phosphinites **169** remained unchanged when the reactants were used in equimolar amounts, even at lower temperatures.

The reaction of phosphinates (S_P)-**170** with vinyl magnesium bromide or alkyllithiums proceeded at −78 to −40 °C and led with inversion of configuration at the phosphorus atom to the formation of (R_P)-homochiral tertiary phosphine oxides **171**, which were used for the synthesis of phosphine ligands **172** [76] (Scheme 2.84).

(+)-(R_P)-Methylphenylvinylphosphine oxide. Vinylmagnesium bromide (1 M solution in THF, 2 equiv.) was added dropwise to a solution of the phosphinate ester (3 mmol)

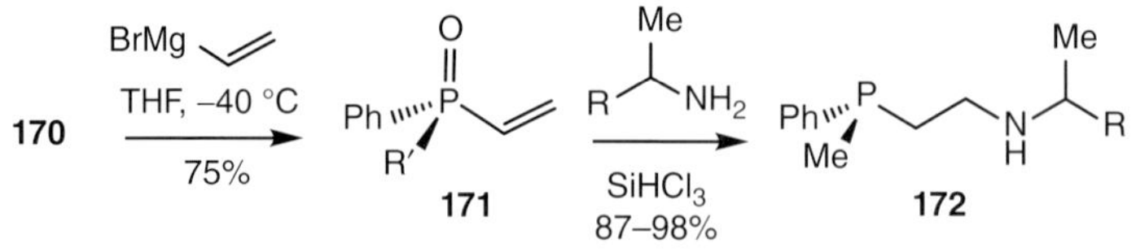

R = Ph, Pr, CH_2OH, CH_2OMe, CH(OH)Ph

Scheme 2.84

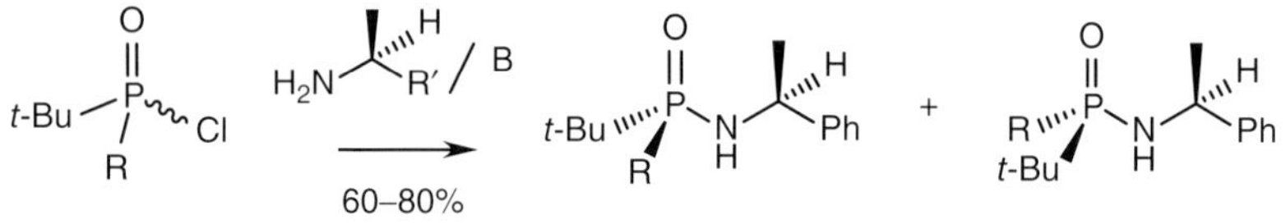

(R_P,S)-**173** (S_P,S)-**173**

B: = Et_3N, DABCO, or Py;
R = *t*-Bu; R′ = *i*-Pr, *i*-Bu, Ph; R = Ph, R′ = Mes

Scheme 2.85 Diastereoselective reaction of chiral amines with phosphinic chlorides.

in THF (20 ml) at −78 °C. The mixture was heated gradually to −40 °C and stirred until the reaction was complete (^{31}P NMR). The reaction mixture was quenched by a solution of 1 M aqueous NH_4Cl (100 ml) at 0 °C. Then the aqueous layer was extracted with CH_2Cl_2 (3 × 20 ml). The combined organic layers were dried (Na_2SO_4), filtered, and evaporated to furnish the crude product as an oil, which was purified by flash chromatography on silica using $CHCl_3$/acetone (8 : 2) as the eluting system. (80%, mp 79–81 °C, $[\alpha]_D^{20}$ +82° (*c* = 1.0, $CHCl_3$), δ_P ($CDCl_3$): +27.7) [76].

2.4.2.2 Nucleophilic Substitution at P(IV) with Chiral Amines

Optically active phosphorus(III) derivatives of *N*-(1-methylbenzyl)amine **174** are useful compounds for the synthesis of chiral organophosphorus compounds. It was found that the optically active 1-benzylethylamine reacts with the chlorides of phosphonic acids with the formation of diastereoisomerically enriched P-chiral aminophosphine oxides **174**. Enantiomerically pure amides **174** were synthesized by the reaction of racemic phosphinothio- and phosphinoselenoic chlorides with optically active amines (Scheme 2.85) [80, 86, 178]. The enantioselective synthesis of alkyl methylphosphonothioates **175** was achieved in high yields via BF_3 Et_2O-catalyzed alcoholysis of resolved phosphono-amidothioates (*R*)-**174** and (*S*)-**174** (Scheme 2.86). The diastereomers **175** were easily separated by fractional crystallization from a benzene–hexane mixture. The less soluble diastereomer was first recrystallized from a 1 : 5 mixture of benzene–hexane and then the more soluble diastereomer was recrystallized from a 1 : 7 mixture of benzene–hexane to give enantiopure amides (S_P,S) and (R_P,S)-**174** (100% de) [178].

A series of P-chirogenic phosphine oxides and phospine thioates, **176, 177** were prepared by nucleophinic reaction of L-prolinol derivatives with racemic chlorophosphine oxides. The reaction of phosphonic acid chloride with (*S*)-proline ethyl ester afforded a mixture of diastereomeric amides **176** in high stereoselectivity. The diastereomers were purified by column chromatography. Chiral organophosphorus was obtained by hydrolysis of **177** (Scheme 2.87).

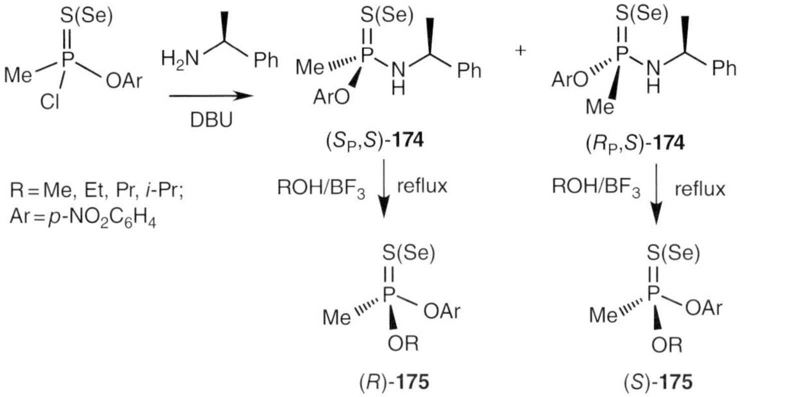

Scheme 2.86 Syntheses of derivatives of phosphonothionic and phosphonoselenic acids.

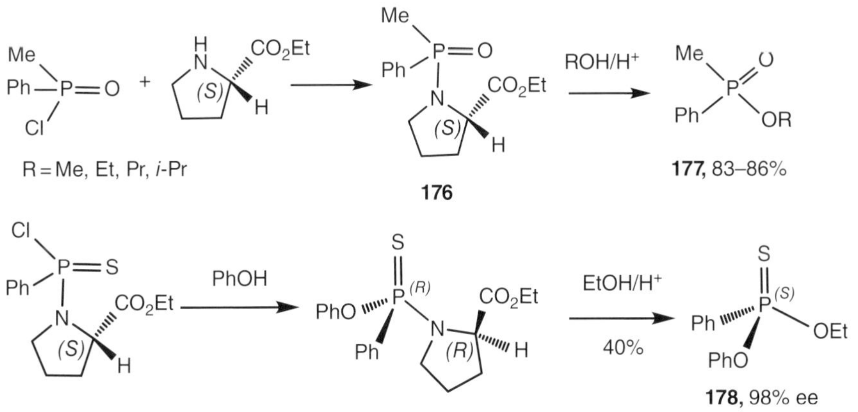

Scheme 2.87

Optically active alkyl methylphenylphosphinate **177** *(R = Et)*

a) A solution of L-proline ethyl ester (7 g, 0.05 mol) and trimethylamine (5 g, 0.05 mol) in THF (50 ml) was added dropwise to a solution of methylphenylphosphinyl chloride (8.9 g, 0.05 mol) in anhydrous tetrahydrofuran (50 ml) at 0 °C, with stirring. Then the reaction mixture was stirred overnight at room temperature. The mixture was then filtered and the solvent was evaporated under vacuum. The residue was dissolved in chloroform and the solution was washed with aqueous solution of $NaHCO_3$ and water. The organic phase was dried with Na_2SO_4 and evaporated to afford the diastereomeric mixture of (S,S_P) and (S,R_P)-**176** as a yellow oil (yield ~13.8 g (~98%)). The residue was chromatographed on column with SiO_2 (hexane–MeOH) to afford 150 mg of the isomers first and then 450 mg of diastereomers. Each isomer was additionally purified by microdistillation.
b) One of the diastereomers (0.01 mol) was allowed to stand overnight at room temperature in a 0.02 molar solution of sulfuric acid in ethanol (100 ml). The mixture was then neutralized with aqueous sodium hydrogen carbonate and evaporated in vacuum. The residue was extracted with ether (75 ml). The extract was dried with sodium sulfate and evaporated to give a yellow oil. The crude

product **177** was chromatographed on silica gel using benzene/ethanol mixture as eluent (yield 80%, bp 70 °C (0.01 mmHg), $[\alpha]_D$ +50 (MeOH)).

2.5 Chiral P(V) and P(VI) Phosphorus Compounds

Only a very limited number of X-ray single-crystal investigations of the absolute configurations of this type of phosphoranes has been reported. Vibrational circular dichroism (VCD) spectroscopy combined with the state-of-the-art DFT simulations has emerged in recent years as a powerful tool for assignment of absolute configurations and dominant conformations of chiral organic compounds and biomolecules.

Pentacoordinated phosphorus compounds have attracted attention as models of the TS in nonenzymatic and enzymatic phosphoryl transfer reactions [161, 177, 179, 183–192]. Very often, the compounds of tetracoordinate phosphorus are found in equilibrium with compounds of pentacoordinate phosphorus, for which pseudorotation with fluctuations of axial and apical bonds are characteristic (Scheme 2.88). For example, the enantioselective oxidation of racemic aminophosphines by reactant ROH/$CXCl_3$ proceeds via the formation of alcoxyhalophosphoranes which exist in equilibrium with alcoxyphosphonium salt and convert finally into amidophosphate. The equilibrium depends on substituents at the phosphorus atom. Alcoxyhalogenephosphoranes **178** containing a five-membered 1,3,2-oxaphospholane cycle effectively stabilizes the pentacoordinated intermediate that has been formed as a diastereomeric mixture in 94 : 6 ratio. The phosphoranes gradually converted at ambient temperature into amidophosphates (Scheme 2.88) [183].

Moriarty *et al.* [184] has isolated stable pseudorotamers of chiral monocyclic oxyphosphoranes containing five various substituents at the phosphorus atom.

Pentacoordinated diastereomers **179, 180** contain two sets of signals of equal intensity in ^{1}H, ^{13}C, and ^{31}P the NMR spectra belonging to two diastereomers of different absolute configuration of the phosphorus atom. The diastereomers **180** were separated by column chromatography and fractional crystallization (Scheme 2.89) [183, 184].

Various optically active compounds bearing chirality at the central pentacoordinate phosphorus atom have been prepared over the last few years [177, 183–189]. For example, Buono *et al.* [177, 185, 186] reported the chiral tricyclic pentacoordinated phosphorus compounds, "triquinphosphoranes." The triquinphosphoranes were

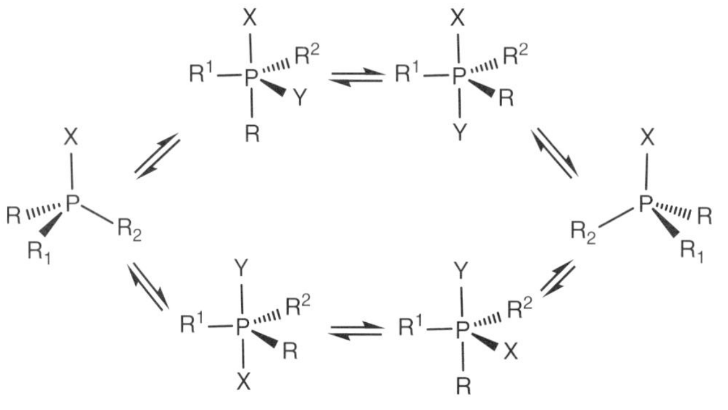

Scheme 2.88 P(IV) ⇌ P(V) equilibrium and pseudorotation of P(V) compounds.

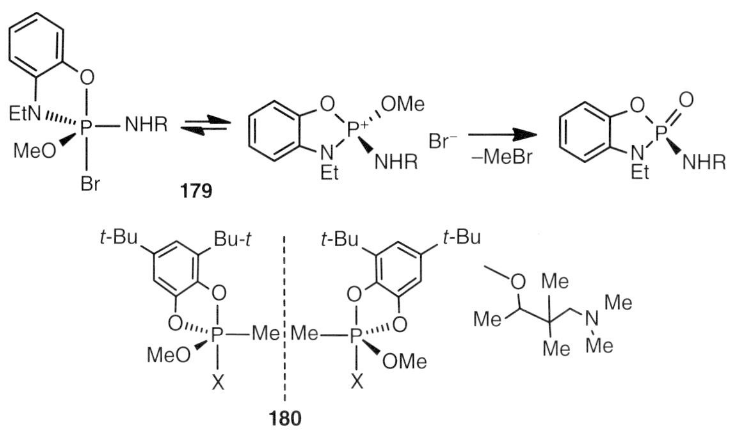

Scheme 2.89 Conversions of chiral five-membered cyclic phosphoranes.

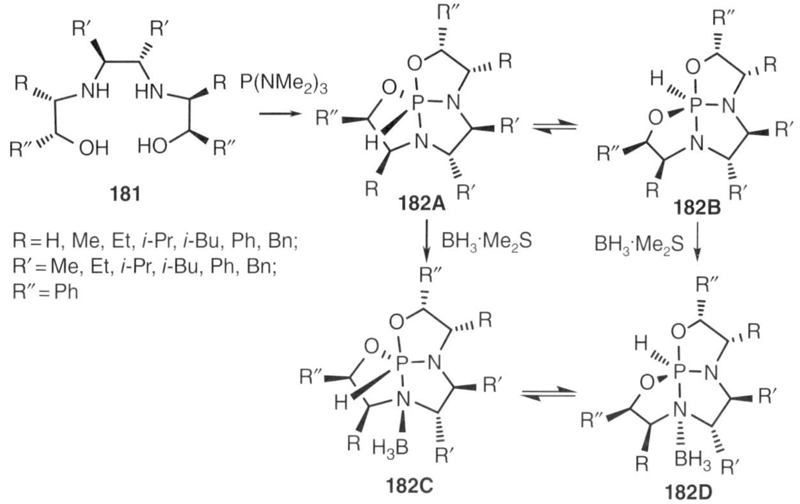

Scheme 2.90 Chiral spirooxyphosphoranes.

prepared from chiral enantiopure diamino diols that present a C_2-symmetry axis (Scheme 2.89).

Triquinphosphoranes reacted with borane to give two stable monoadducts **182A** and **182B** with opposite absolute configurations at the phosphorus center, which do not undergo epimerization. The ^{31}P and ^{13}C NMR data are consistent either with a low-energy single-step Berry pseudorotation process between the two possible diastereomeric trigonalbipyramidal structures TBP(R_P) and TBP(S_P) or with the chiral square-pyramidal structure (S_P) (Scheme 2.90).

The diastereoselectivity of this reaction depends strongly on the nature and the position of the substituents, the highest diastereomeric excesses being obtained with 4,9-diisopropyl (90% de) and 4,9-diisobutyl (86% de) compounds. The X-ray analysis of the major diastereomer revealed that it is close to an ideal TBP, exhibiting an (S_P) absolute configuration. Semiempirical AM1 MO calculations predict that TBP (R_P) and

TBP-(S_P) ground-state species are in rapid equilibrium through an S_P TS, the activation barrier being about 5 kcal mol^{-1}.

The highest diastereoisomeric excesses were obtained with 4,9-diisopropyl (90% de) and 4,9-diisobutyl (86% de) compounds. The X-ray structure of the major diastereoisomer revealed an (S_P) absolute configuration. Chiral triquinphosphoranes **183** reacted with trifluoroacetophenone, ketopantolactone, and aromatic aldehydes to afford pairs of diastereoisomeric hydroxyphosphoranes with diastereoselectivities up to 90% depending on the nature of the electrophile [177] (Scheme 2.91).

Chiral triquinphosphoranes undergo alkylthiolation reactions with methyl and *n*-butyl disulfides to give the corresponding chiral thiaphosphoranes in high yields (80–100%). The reaction with (t-BuS)$_2$ preferably gave the thiophosphoramide. An increased reactivity was observed when the experiments were conducted under UV irradiation (Scheme 2.92) [186].

The addition of chiral triquinphosphorane **193** to hexacarbonylmolybdenum led exclusively to complex **194**, in which the oxazaphospholidine ligand was coordinated to molybdenum and exhibited a single absolute configuration at the P atom. Compound **194** was isolated as colorless crystals (mp 140 °C) which were soluble in common

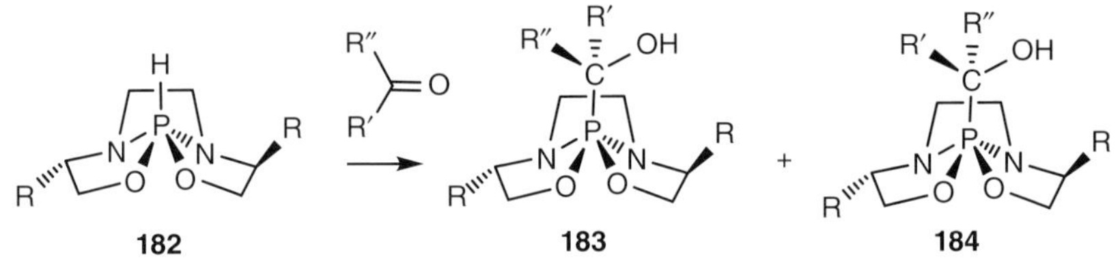

R	R′	R″	de (%)
Me	–C(CH$_3$)$_2$CH$_2$OC(O)–		86
i-Pr	–C(CH$_3$)$_2$CH$_2$OC(O)–		90
Bn	–C(CH$_3$)$_2$CH$_2$OC(O)–		84
i-Pr	H	Ph	32
i-Pr	H	Nphth	56
i-Pr	H	*p*-An	28

Scheme 2.91 Chiral triquinphosphoranes.

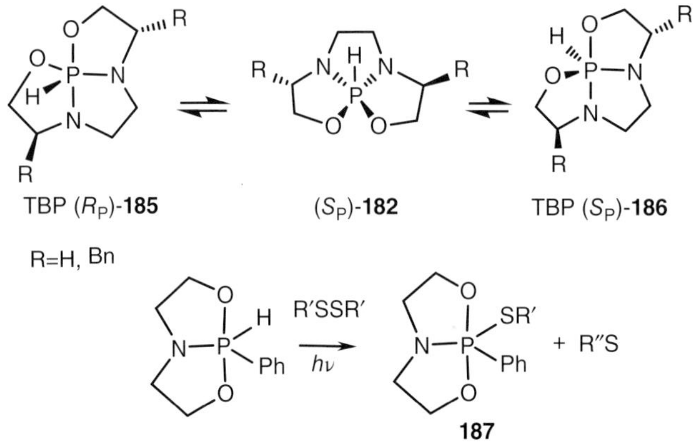

Scheme 2.92 Reactions of chiral triquinphosphoranes.

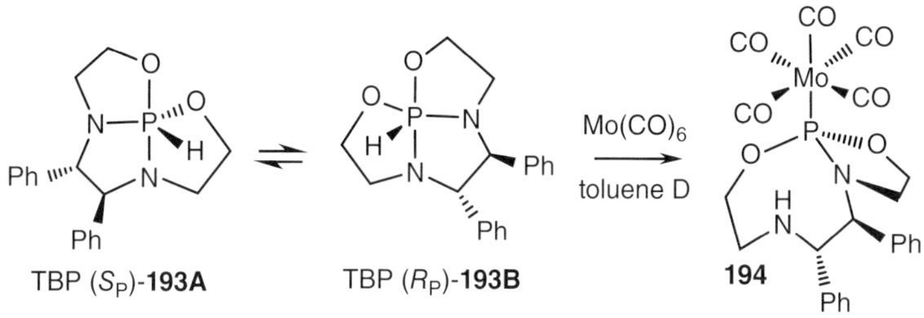

Scheme 2.93 Addition of chiral triquinphosphorane **193** to hexacarbonylmolybdenum.

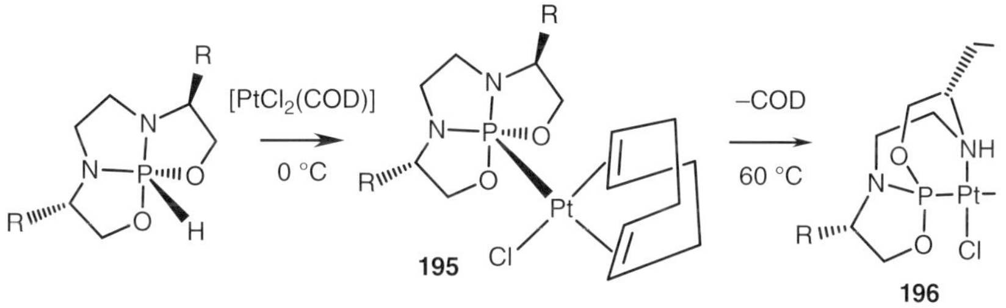

$[PtCl(COD)L]^{+}Cl^{-} + AgBF_4{}' \ [PtCl(COD)L]^{+}BF_4^{-} + AgCl$

Scheme 2.94 Coordination of a tricyclic hydrophosphorane to $[Pt(COD)Cl_2]$.

solvents. The structure of **194** was confirmed by X-ray analysis. The most prominent structural features are the nearly tetragonal structure adopted by the phosphorus atom with an (S_P) absolute configuration (Scheme 2.93) [186].

Complexation of the tricyclic hydrophosphorane with $[Pt(COD)Cl_2]$ at 0 °C led to the formation of the complex shown in Scheme 2.94. Gradual heating of the reaction solution to 20 °C led to opening of the phosphorane structure and to the formation of the complex. If the reaction was performed in the presence of $AgBF_4$, the corresponding BF_4 salt was isolated. The structure of the compound was determined by X-ray crystallography. The complex had a slightly distorted trigonal-bipyramidal geometry around the phosphorus, with the platinum fragment in the equatorial position. The Pt atom exhibits nearly square-planar coordination geometry (Scheme 2.94) [187].

Kojima reported optically active pentacoordinate phosphoranes with an asymmetry at the phosphorus atom. The treatment of diastereoisomeric (R_P)- and (S_P)-**188** with $LiAlH_4$ gave a pair of enantiomerically pure phosphoranes (R_P)-**189** and (S_P)-**189**, with asymmetry only at the phosphorus atom. The diastereoisomers were resolved as prisms and needles by recrystallization from MeOH-H_2O. The absolute stereochemistry of a chiral pentacoordinated phosphorus compound with the asymmetry on the phosphorus center was established by X-ray crystal-structure analysis. The phosphorane **188** was converted by reduction and esterification into optically active compounds without epimerization (Scheme 2.95) [188–190, 193].

The stereochemistry of the substitution reaction of chiral bicycle phosphorane **188** has depended upon the incoming nucleophile, thus implying the presence of hexacoordinate intermediates. The nucleophilic substitution reaction of SR compounds with alkyllithium reagents resulted in inversion of configuration, whereas that of OR compounds gave various ratios of inversion and retention products depending on

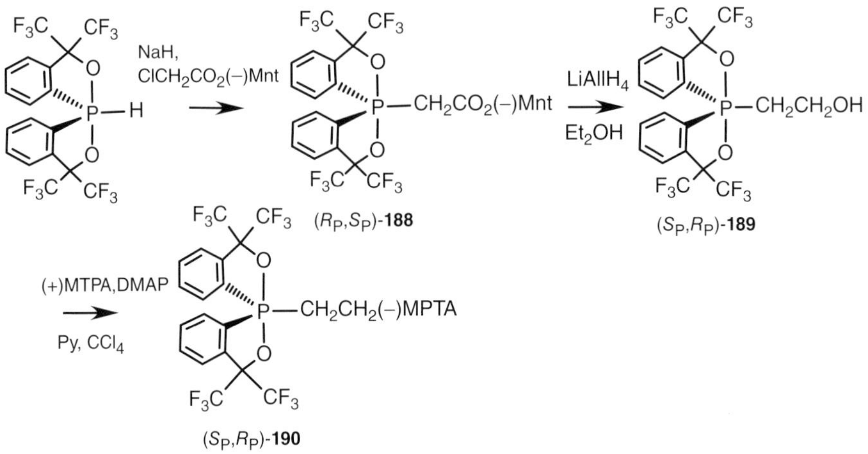

Scheme 2.95 Synthesis bicyclic optically active pentacoordinate phosphoranes.

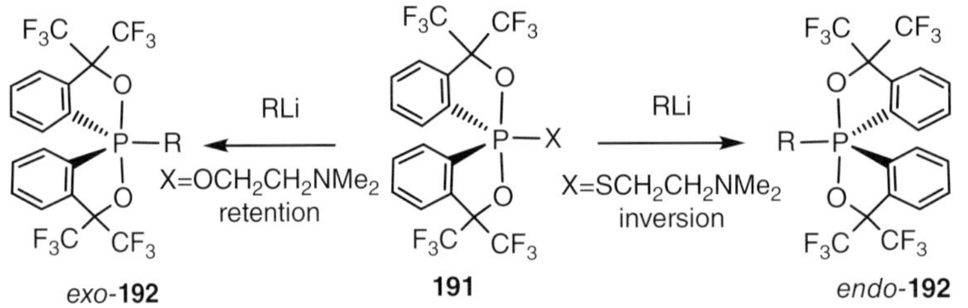

Scheme 2.96

the stereochemistry of the diastereomeric reactant phosphoranes and the solvent. However, the use of $OCH_2CH_2NMe_2$ as substituent lead to almost exclusive formation of the retention product. Retention of configuration indicates that the attack on pentacoordinate phosphorus had occurred from the rear side of a carbon atom to furnish hexacoordinate species from which extrusion of X had followed from the same face (Scheme 2.96).

The absolute configurations and stereochemistry of chiral pentacoordinated phosphorus compounds were studied by X-ray crystal analysis. The assignment of the absolute configurations and dominant conformations of diastereomers of chiral pentacoordinate phosphoranes were determined by VCD spectroscopy, in conjunction with DFT calculations [194, 195].

The octahedral geometry of pentavalent hexacoordinate phosphorus allows the formation of chiral phosphate anions by complexation with three identical, symmetrical, bidentate ligands [196]. Enantiopure anions of D_3-symmetry can be used in several fields of chemistry that involve chiral or prostereogenic cationic species. The resolution of enantiomeric cations, determination of their enantiomeric purity, and asymmetric synthesis of cationic species have been described [198]. Many organic compounds of hexacoordinate phosphorus are known and in some cases such compounds mimic intermediate products or TSs in nucleophilic displacements at the tetrahedral phosphorus. Kinetic studies of the decomposition of 3-hydroxypropyl triphenylphosphonium chloride catalyzed by ethoxide ions showed the presence of a

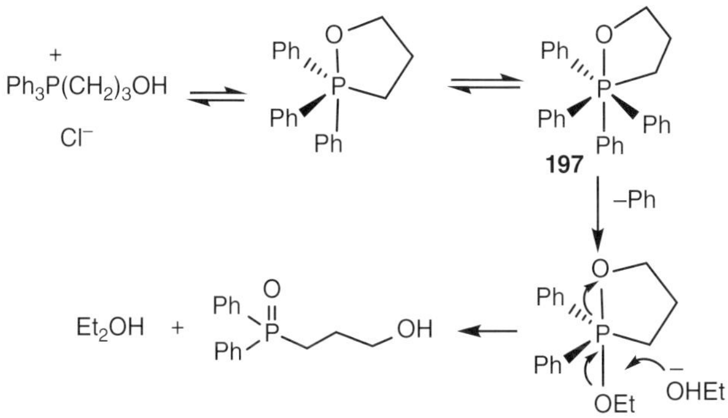

Scheme 2.97

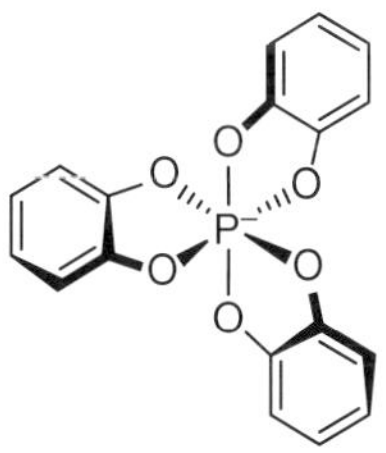

198, δ_P-83 ppm

Scheme 2.98 Chiral hexacoordinated phosphate anion.

hexacoordinate intermediate **197** or TS formed in the case of the attack of the ethoxide ion on a pentacoordinate intermediate (Scheme 2.97).

Hellwinkel and co-workers were the first to report the synthesis and resolution of a chiral hexacoordinated phosphate anion, the tris-biphenyl-2,2′-diyl phosphate(V) **198** [198, 199]. The optically active penta-arylphosphorane **198** has structure of the propeller-shaped anion spirophosphorane, constituting a mixture of two rapidly equilibrating pseudorotational isomers. Optically active phosphate(V) ions with a hexacoordinated phosphorus center were configurationally unstable. Koenig and Klaebe showed that on racemization of phosphate(V) ions, **198** is acid catalyzed. Determination of the kinetic parameters led to the suggestion of pseudorotation as the rate-determining step. Lacour and co-workers prepared the configurationally stable, enantiomerically pure tris (tetrachlorobenzenediolato) phosphate(V) anion from electron-poor tetrachlorocatechol (Scheme 2.98) [200,].

The nature of the solvent (MeOH, $CHCl_3$) plays a crucial role on the kinetics of epimerization and the position of the resulting equilibrium. For anions made with a (2*R*, 3*R*) tartaric backbone, a L configuration is always preferred in MeOH; the selectivity, obtained after a slow equilibration being independent of the nature of the ester alkyl chain (dr 3 : 1).However, in chloroform, the D diastereomer is rapidly obtained and the selectivity is best if the ester side chain is sterically demanding. The introduction of electron-withdrawing chlorine atoms on the aromatic nuclei of the catecholate ligands increases the configurational (and chemical) stability of the resulting tris(tetrachlorobenzenedio-lato) phosphate(V) derivative. This D_3-symmetric TRISPHAT anion was resolved by an association with a chiral ammonium cation [104]. A general one-pot process was developed for the preparation of C_2-symmetric anion

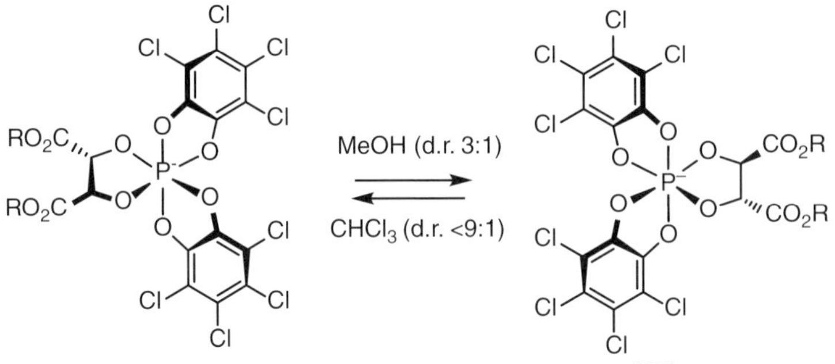

199D **199L**

Scheme 2.99 Epimerization of TARPHAT aniones initiated by the solvent.

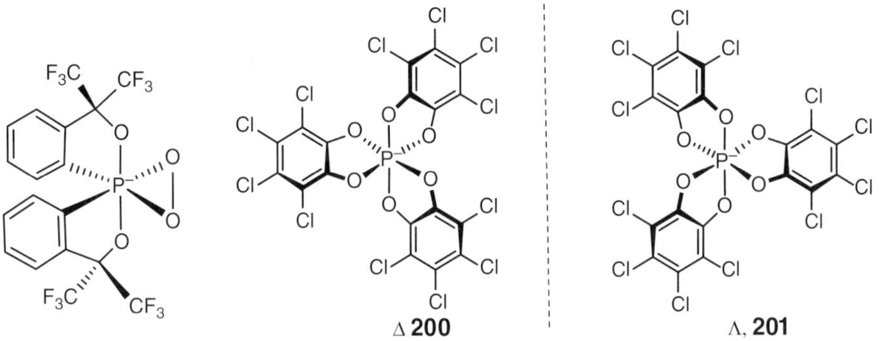

Δ **200** Λ, **201**

TRISPHAT

Scheme 2.100 Chiral P(VI) aniones.

containing enantiopure 2,20-dihydroxyl-1,10-binaphthyl (BINOL), hydrobenzoin, and tartrate-derived ligands **199** respectively. The C_1-symmetric anion was prepared similarly in two steps from methyl-α-D-mannopyranoside. All these anions were isolated as their dimethylammonium salts in good yields and chemical purity. The efficiency of TRISPHAT anions **200** as NMR chiral shift agents for chiral cations has been demonstrated over the last few years. Additions of ammonium salts of the D- or L-enantiomers of **200** and **201** to solutions of racemic or enantioenriched chiral cationic substrates have led to efficient NMR enantiodifferentiations [197]. Applications of anions such as **199** or **200** as NMR chiral shift reagents as resolving agents for organic and inorganic cations and as chiral auxiliaries in stereoselective processes have been reported (Schemes 2.99 and 2.100) [197,].

2.6 Summary

This chapter covers the synthesis and properties of P-chirogenic phosphorus compounds. It should be noted that despite the impressive progress achieved in the synthesis and studies of properties of P-chiral compounds, not all problems have been solved. The problem of the development of enantioselective methods giving easy access to both optical antipodes of chiral tertiary phosphines still remains. The creation of highly effective catalysts for the asymmetric synthesis of P-chirogenic compounds, or for the creation of chiral organophosphorus synthons, is an important problem, which is currently awaiting a solution. The actual problem is the resolution of enantiomers and purification of chiral tertiary phosphines and phosphine oxides. The exact structure and absolute configuration can only be successfully solved in a limited number of cases.

References

1 Morrison, J.D. and Mosher, H.S. (1976) *Asymmetric Organic Reactions*, American Chemical Society, Washington, DC, pp. 258–262.

2 Le Floch, P. (2006) Phosphaalkene, phospholyl and phosphinine ligands: new tools in coordination chemistry and catalysis. *Coord. Chem. Rev.*, **250**, 627–681.

3 Schoeller, W.W. and Lerch, C. (1986) Dicoordination and tricoordination at phosphorus. Ab initio study of bonding in bis(imino)phosphoranes and related compounds. *Inorg. Chem.*, **25** (4), 576–580.

4 Escudie, J., Ranaivonjatovo, H., and Rigon, L. (2000) Heavy allenes and cumulenes E=C=E′ and E=C=C=E′ (E = P, As, Si, Ge, Sn; E′ = C, N, P, As, O, S). *Chem. Rev.*, **100** (10), 3639–3696.

5 Escudie, J. and Ranaivonjatovo, H. (2007) Group 14 and 15 heteroallenes E=C=C and E=C=E′. *Organometallics*, **26** (7), 1542–1559.

6 Gates, D.P. (2005) Expanding the analogy between P=C and C=C bonds to polymer science. *Top. Curr. Chem.*, **250**, 107–126.

7 Markovski, L.N., Romanenko, V.D., and Ruban, A.V. (1987) Synthesis, structures and reactivities of dicoordinated phosphorus compounds. *Phosphorus, Sulfur Silicon Relat. Elem.*, **30** (1–2), 447–450.

8 Mikolajczyk, M., Omelanczuk, J., Perlikovska, W., Markovski, L.I., Romanenko, V.D., Ruban, A.V., and Drapailo, A.B. (1988) A new enantioselective asymmetric synthesis of tri-coordinate phosphorus compounds from di-coordinate λ^3-aryl(alkyl)iminophosphines. *Phosphorus, Sulfur Silicon Relat. Elem.*, **36** (1–4), 267.

9 de Vaumas, R., Marinetti, A., Ricard, L., and Mathey, F. (1992) Use of prochiral phosphaalkene complexes in the synthesis of optically active phosphines. *J. Am. Chem. Soc.*, **114** (1), 261–266.

10 Ganter, C., Brassat, L., and Ganter, B. (1997) Enantiomerically pure phosphaferrocenes with planar chirality. *Tetrahedron: Asymmetry*, **8** (15), 2607–2611.

11 Marinetti, A., Mathey, F., and Ricard, L. (1992) Phosphirane complexes from oxiranes via the "Phospha-Wittig" reaction. the case of optically active species. *Synthesis*, (1/2), 157–162.

12 Marinetti, A., Ricard, L., and Mathey, F. (1993) Synthesis of optically active phosphiranes and their use as ligands in rhodium(I) complexes. *Organometallics*, **12** (4), 1207–1212.

13 Dugal-Tessier, J., Dake, G.R., and Gates, D.P. (2008) Chiral ligand design: a bidentate ligand incorporating an acyclic phosphaalkene. *Angew. Chem. Int. Ed.*, **47** (42), 8064–8067.

14 Tang, W. and Zhang, X. (2003) New chiral phosphorus ligands for enantioselective. *Chem. Rev.*, **103** (8), 3029–3069.

15 Kolodiazhnyi, O.I. (1982) Premier heteroallene stable comportant un atome de phosphore. *Tetrahedron Lett.*, **23** (47), 4933–4936.

16 Yoshifuji, M., Toyota, K., Niitsu, T., Inamoto, N., and Okamoto, Y. (1986) The first separation of an optically active 1,3-diphospha-allene of axial dissymmetry. *J. Chem. Soc., Chem. Commun.*, (20), 1550–1551.

17 Yoshifuji, M., Toyota, K., Okamoto, Y., and Asakura, T. (1990) Separation of an optically active phosphaallene of pseudo axial dissymmetry. *Tetrahedron Lett.*, **31** (16), 2311–2314.

18 Caminade, A.M., Verrier, M., Ades, C., Paillous, N., and Koenig, M. (1984) Laser irradiation of a diphosphene: evidence for the first cis-trans isomerization. *J. Chem. Soc., Chem. Commun.*, (14), 875–876.

19 Qiao, S. and Fu, G.C. (1998) The first application of a planar-chiral phosphorus heterocycle in asymmetric catalysis: enantioselective hydrogenation of dehydroamino acids. *J. Org. Chem.*, **63** (13), 4168–4169.

20 Shintani, R., Lo, M.M. C., and Fu, G.C. (2000) Synthesis and application of planar-chiral phosphaferrocene-oxazolines, a new class of *P,N*-ligands. *Org. Lett.*, **2** (23), 3695–3697.

21 Ganter, C., Kaulen, C., and Englert, U. (1999) Cyclopentadienyl-substituted phosphaferrocenes: synthesis of a bis(phosphaferrocene) *P,P*-chelate ligand. *Organometallics*, **18** (26), 5444–5446.

22 Tanaka, K. and Fu, G. (2001) A versatile new catalyst for the enantioselective isomerization of allylic alcohols to aldehydes: scope and mechanistic studies. *J. Org. Chem.*, **66** (24), 8177–8186.

23 Ogasawara, M., Arae, S., Watanabe, S., Subbarayan, V., Sato, H., and Takahashi, T. (2013) Synthesis and characterization of benzo[*b*]phosphaferrocene derivatives. *Organometallics*, **32** (17), 4997–5000.

24 Muller, C., Pidko, E.A., Totev, D., Lutz, M., Spek, A.L., van Santena, R.A., and Vogta, D. (2007) Developing a new class of axial chiral phosphorus ligands: preparation and characterization of enantiopure atropisomeric phosphinines. *J. Chem. Soc., Dalton Trans.*, (46), 5372–5375.

25 Muller, C., Pidko, E.A., Staring, A.J.P.M., Lutz, M., Spek, A.L., van Santen, R.A., and Vogt, D. (2008) Developing a new class of axial chiral phosphorus ligands: preparation and characterization of enantiopure atropisomeric phosphinines. *Chem. Eur. J.*, **14** (16), 4899–4905.

26 Muller, C., Guarrotxena Lopez, L., Kooijman, H., Spek, A.L., and Vogt, D. (2006) Chiral bidentate phosphabenzene-based ligands: synthesis, coordination chemistry, and application in Rh-catalyzed asymmetric hydrogenations. *Tetrahedron Lett.*, **47** (12), 2017–2020.

27 Harger, M.J.P. (1988) Stereoselectivity, non-stereospecificity, and the lifetime of the thiometaphosphonimidate intermediate formed in the reactions of the diastereoisomers of a phosphonamidothioic chloride with t-butylamine. *J. Chem. Soc., Chem. Commun.*, (18), 1256–1257.

28 Freeman, S. and Harger, M.J.P. (1985) Stereohemistry of reaction of a phosphonamidic chloride with t-butylamine. *J. Chem. Soc., Chem. Commun.*, (20), 1394–1395.

29 Quin, L.D., Sadani, N.D., and Wu, X.P. (1989) Generation and trapping of O-alkyl metathiophosphates. *J. Am. Chem. Soc.*, **111** (17), 6852–6853.

30 Quin, L.D. (1992) Applications of Low-coordination Phosphorus Chemistry in the Chemical Modification of Surfaces, http:www.dtic.mil/dtic/tr/fulltext/u2/a261208.pdf (accessed 17 May 2016).
31 Valentine, D. Jr., (1984) in *Asymmetric Synthesis*, vol. **4**, Chapter 3 (eds J.D. Morrison and J.W. Scott), Academic Press, Orlando, FL, p. 263.
32 Kagan, H.B. (1985) in *Asymmetric Synthesis*, vol. **5** (ed. J.D. Morrison), Academic Press, Orlando, FL, pp. 1–39.
33 Horner, L. (1980) Dreibindige, optisch aktive Phosphorverbindungen, ihre Synthese, chemische Eigenschaften und Bedeutung fur die asymmetrische Homogenhydrierung. *Pure Appl. Chem.*, **52** (4), 843–858.
34 Lambert, J.B. (1971) in *Topics in Stereochemistry*, vol. **6** (eds N.L. Allinger and E.L. Eliel), John Wiley & Sons, Inc., New York, pp. 19–97.
35 Pabel, M., Willis, A.C., and Wild, S.B. (1995) Attempted resolution of free (±)-chlorophenylisopropylphosphine. *Tetrahedron: Asymmetry*, **6** (9), 2369–2379.
36 Schwerdtfeger, P., Laakkonen, L.J., and Pyykkö, P. (1992) Trends in inversion barriers. I. Group-15 hydrides. *J. Chem. Phys.*, **96**, 6807–6819.
37 Harvey, J.S. and Gouverneur, V. (2010) Catalytic enantioselective synthesis of P-stereogenic compounds. *J. Chem. Soc., Chem. Commun.*, **46** (40), 7477–7485.
38 Baechler, R.D. and Mislow, K. (1970) The effect of structure on the rate of pyramidal inversion of acyclic phosphines. *J. Am. Chem. Soc.*, **92** (10), 3090–3093.
39 Omelanchuk, J. (1992) The first stereoselective synthesis of chiral halogenophosphines: optically active *tert*-butyl(phenyl)chlorophosphine. *J. Chem. Soc., Chem. Commun.*, (23), 1718–1719.
40 Montgomery, C.D. (2013) Factors affecting energy barriers for pyramidal inversion in amines and phosphines: a computational chemistry lab exercise. *Chem. Educ.*, **90** (5), 661–664.
41 Kolodiazhnyi, O.I. (1998) Asymmetric synthesis of organophosphorus compounds. *Tetrahedron: Asymmetry*, **9** (11), 1279–1332.
42 Wolf, C. (2008) *Dynamic Stereochemistry of Chiral Compounds: Principles and Applications*, RSC Publishing, Cambridge, 512 pp.
43 Humbel, S., Bertrand, C., Darcel, C., Bauduin, C., and Jugé, S. (2003) Configurational stability of chlorophosphines. *Inorg. Chem.*, **42** (2), 420–427.
44 Hayakawa, Y., Hyodo, M., Kimura, K., and Kataoka, M. (2003) The first asymmetric synthesis of trialkyl phosphates on the basis of dynamic kinetic resolution in the phosphite method using a chiral source in a catalytic manner. *J. Chem. Soc., Chem. Commun.*, (14), 1704–1705.
45 Reichl, K.D., Ess, D.H., and Radosevich, A.T. (2013) Catalyzing pyramidal inversion: configurational lability of P-stereogenic phosphines via single electron oxidation. *J. Am. Chem. Soc.*, **135** (25), 9354–9357.
46 Rauk, A., Allen, L.C., and Mislow, K. (1970) Pyramidal inversion. *Angew. Chem. Int. Ed. Engl.*, **9** (6), 400–414.
47 De Bruin, K.E., Zon, G., Naumann, K., and Mislow, K. (1969) Stereochemistry of nucleophilic displacement at phosphorus in some phosphetanium salts. *J. Am. Chem. Soc.*, **91** (25), 7027–7030.
48 Kolodiazhnyi, O.I. (2015) in *Topics in Current Chemistry*, vol. **360** (ed. J.-L. Montchamp), Springer International Publishing, Switzerland, pp. 161–236.
49 Dahl, O. (1983) Mechanism of nucleophilic substitution at tricovalent phosphorus. *Phosphorus Sulfur Rel. Elem.*, **18** (1–3), 201–204.

50 Lewis, R.A. and Mislow, K. (1969) Direct configurational correlation of trialkyl- and triarylphosphine oxides. *J. Am. Chem. Soc.*, **91** (25), 7009–7012.
51 Jennings, E.V., Nikitin, K., Ortin, Y., and Gilheany, D.G. (2014) Degenerate nucleophilic substitution in phosphonium salts. *J. Am. Chem. Soc.*, **136** (46), 16217–16226.
52 van Bochove, M.A., Swart, M., and Bickelhaupt, F.M. (2006) Nucleophilic substitution at phosphorus (S_N2@P): disappearance and reappearance of reaction barriers. *J. Am. Chem. Soc.*, **128** (33), 10738–10744.
53 Zijlstra, H., León, T., de Cózar, A., Fonseca Guerra, C., Byrom, D., Riera, A., Verdaguer, X., and Bickelhaupt, F.M. (2013) Stereodivergent SN2@P reactions of borane oxazaphospholidines: experimental and theoretical studies. *J. Am. Chem. Soc.*, **135** (11), 4483–4491.
54 van Bochove, M.A., Swart, M., and Bickelhaupt, F.M. (2007) Nucleophilic substitution at phosphorus centers (S_N2@P). *ChemPhysChem*, **8** (17), 2452–2463.
55 Neuffer, J. and Richter, W.J. (1986) Optisch aktive phosphine durch asymmetrische substitution prochiraler, homochiral substituierter phosphonite. *J. Organomet. Chem.*, **301**, 289–297.
56 Darcel, C., Uziel, J., and Juge, S. (2008) in *Phosphorus Ligands in Asymmetric Catalysis*, vol. **3** (ed. A. Boerner), Wiley-VCH Verlag GmbH, Weinheim, pp. 1211–1233.
57 Cordillo, B., Garduno, C., Guadarrama, G., and Hernandez, J. (1995) Synthesis and conformational analysis of six-membered cyclic phenyl phosphites. *J. Org. Chem.*, **60** (16), 5180–5185.
58 Gatineau, D., Giordano, L., and Buono, G. (2011) Bulky, optically active P-stereogenic phosphineboranes from pure H-menthylphosphinates. *J. Am. Chem. Soc.*, **133** (28), 10728–10731.
59 Moraleda, D., Gatineau, D., Martin, D., Giordano, L., and Buono, G. (2008) A simple route to chiral phosphinous acid–boranes. *J. Chem. Soc., Chem. Commun.*, (26), 3031–3033.
60 Cain, M.F., Glueck, D.S., Golen, J.A., and Rheingold, A.L. (2012) Asymmetric synthesis and metal complexes of a C_3-symmetric *P*-stereogenic triphosphine, (*R*)-MeSi(CH_2PMe(t-Bu)$)_3$ (MT-Siliphos). *Organometallics*, **31** (3), 775–778.
61 Korpiun, O., Lewis, R.A., Chickos, J., and Mislow, K. (1968) Synthesis and absolute configuration of optically active phosphine oxides and phosphinates. *J. Am. Chem. Soc.*, **90** (18), 4842–4846.
62 Crépy, K.V.L. and Imamoto, T.P. (2003) in *Topics in Current Chemistry*, vol. **229** (ed. J.-P. Majoral), Springer-Verlag, Berlin, Heidelberg, pp. 1–41.
63 Xu, Q., Zhao, C.-Q., and Han, L.-B. (2008) Stereospecific nucleophilic substitution of optically pure H-Phosphinates: a general way for the preparation of chiral *P*-stereogenic phosphine oxides. *J. Am. Chem. Soc.*, **130** (38), 12648–12655.
64 Leyris, A., Bigeault, J., Nuel, D., Giordano, L., and Buono, G. (2007) Enantioselective synthesis of secondary phosphine oxides from (R_P)-(−)-menthyl hydrogenophenylphosphinate. *Tetrahedron Lett.*, **48** (30), 5247–5250.
65 Nemoto, T. (2008) Transition metal-catalyzed asymmetric reactions using *P*-chirogenic diaminophosphine oxides: DIAPHOXs. *Chem. Pharm. Bull.*, **56** (90), 1213–1228.
66 Kolodiazhnyi, O.I. (2005) (1R,2S,5R)-menthyl phosphinate and its properties. *Russ. J. Gen. Chem.*, **75** (4), 656–657.

67 Kolodiazhna, A.O. (2009) Asymmetric synthesis of hydroxyphosphonates and their derivatives with potential biological activity. PhD thesis. IBONCH, Kiev, 150 pp., http:disser.com.ua/content/351636.html (accessed 17 May 2016).
68 Montchamp, J.-L. (2013) Organophosphorus synthesis without phosphorus trichloride: the case for the hypophosphorous pathway. *Phosphorus, Sulfur Silicon Relat. Elem.*, **188** (1), 66–75.
69 Fisher, H.C., Prost, L., and Montchamp, J.-L. (2013) Organophosphorus chemistry without PCl_3: a bridge from hypophosphorous acid to H-phosphonate diesters. *Eur. J. Org. Chem.*, **35**, 7973–7978.
70 Bravo-Altamirano, K., Coudray, L., Deal, E.L., and Montchamp, J.-L. (2010) Strategies for the asymmetric synthesis of H-phosphinate esters. *Org. Biomol. Chem.*, **8** (24), 5541–5551.
71 Berger, O. and Montchamp, J.L. (2013) A general strategy for the synthesis of *P*-stereogenic compounds. *Angew. Chem. Int. Ed.*, **52** (43), 11377–11380.
72 Coudray, L. and Montchamp, J.-L. (2008) Green palladium-catalyzed synthesis of benzylic H-phosphinates from hypophosphorous acid and benzylic alcohols. *Eur. J. Org. Chem.*, **2008** (28), 4101–4103.
73 Cardellicchio, C., Naso, F., Annunziata, M., and Capozzi, M. (2004) A convenient route to the phosphorus and sulfur stereoisomers of ethyl menthyl (methylsulfinyl) methylphosphonate. *Tetrahedron: Asymmetry*, **15** (9), 1471–1476.
74 Lemouzy, S., Jean, M., Giordano, L., Hérault, D., and Buono, G. (2016) The hydroxyalkyl moiety as a protecting group for the stereospecific alkylation of masked secondary phosphine-boranes. *Org. Lett.*, **18** (1), 140–143.
75 Kolodiazhnyi, O.I., Sheiko, S., and Grishkun, E.V. (2000) C_3-symmetric trialkyl phosphites as starting compounds of asymmetric synthesis. *Heteroat. Chem.*, **11** (2), 138–143.
76 Oliana, M., King, F., Horton, P.N., Hursthouse, M.B., and Hii, K.K. (2006) Practical synthesis of chiral vinylphosphine oxides by direct nucleophilic substitution. Stereodivergent synthesis of aminophosphine ligands. *J. Org. Chem.*, **71** (6), 2472–2479.
77 Kolodiazhnyi, O.I. (2012) Recent developments in the asymmetric synthesis of *P*-chiral phosphorus compounds. *Tetrahedron: Asymmetry*, **23** (1), 1–46.
78 Cristau, H.-J., Monbrun, J., Schleiss, J., Virieux, D., and Pirat, J.-L. (2005) First synthesis of *P*-aryl-phosphinosugars, organophosphorus analogues of C-arylglycosides. *Tetrahedron Lett.*, **46** (21), 3741–3744.
79 Ferry, A., Malik, G., Retailleau, P., Guinchard, X., and Crich, D. (2013) Alternative synthesis of P-chiral phosphonite-borane complexes: application to the synthesis of phostone–phostone dimers. *J. Org. Chem.*, **78** (14), 6858–6867.
80 Kolodiazhnyi, O.I., Gryshkun, E.V., Andrushko, N.V., Freytag, M., Jones, P.G., and Schmutzler, R. (2003) Asymmetric synthesis of chiral *N*-(1-methylbenzyl)aminophosphines. *Tetrahedron: Asymmetry*, **14** (2), 181–183.
81 Gryshkun, E.V., Andrushko, N.V., and Kolodiazhnyi, O.I. (2004) Stereoselective reactions of chiral amines with racemic chlorophosphines. *Phosphorus, Sulfur Silicon Relat. Elem.*, **179** (6), 1027–1046.
82 Kolodiazhnyi, O.I., Andrushko, N.V., and Gryshkun, E.B. (2004) Stereoselective reactions of optically active derivatives of α-methylbenzylaminophosphine. *Russ. J. Gen. Chem.*, **74** (4), 515–522.

83 Gryshkun, E.V., Kolodiazhna, A.O., and Kolodiazhnyi, O.I. (2003) Synthesis of chiral tert-butylphenylphosphine oxide. *Russ. J. Gen. Chem.*, **73** (4), 1823–1824.

84 Reves, M., Ferrer, C., Leon, T., Doran, S., Etayo, P., Vidal-Ferran, A., Riera, A., and Verdaguer, X. (2010) Primary and secondary aminophosphines as novel *P*-stereogenic building blocks for ligand synthesis. *Angew. Chem. Int. Ed.*, **49** (49), 9452–9455.

85 Len, T., Parera, M., Roglans, A., Riera, A., and Verdaguer, X. (2012) Chiral *N*-phosphino sulfinamide ligands in rhodium(I)-catalyzed [2+2+2]-cycloaddition reactions. *Angew. Chem. Int. Ed.*, **51** (28), 6951–6955.

86 Grabulosa, A., Doran, S., Brandariz, G., Muller, G., Benet-Buchholz, J., Riera, A., and Verdaguer, X. (2014) Nickel(II) and Palladium(II) complexes of the small-bite-angle P-stereogenic diphosphine ligand MaxPHOS and its monosulfide. *Organometallics*, **33** (3), 692–701.

87 Brun, S., Parera, M., Pla-Quintana, A., Roglans, A., León, T., Achard, T., Solà, J., Verdaguer, X., and Riera, A. (2010) *P*-stereogenic secondary iminophosphorane ligands and their Rhodium(I) complexes: taking advantage of NH/PH tautomerism. *Tetrahedron*, **66** (46), 9032–9040.

88 Kimura, T. and Murai, T. (2005) Enantiomerically pure *P*-chiral phosphinoselenoic chlorides: inversion of configuration at the *P*-chirogenic center in the synthesis and reaction of these substances. *J. Chem. Soc., Chem. Commun.*, (32), 4077–4079.

89 Kimura, T. and Murai, T. (2004) *P*-Chiral Phosphinoselenoic chlorides and optically active *P*-Chiral phosphinoselenoic amides: synthesis and stereospecific interconversion with extrusion and addition reactions of the selenium atom. *Chem. Lett.*, **33** (7), 878–879.

90 Casimiro, M., Roces, L., Garcıa-Granda, S., Iglesias, M.J., and Ortiz, F.L. (2013) Directed ortho-lithiation of aminophosphazenes: an efficient route to the stereoselective synthesis of P-chiral compounds. *Org. Lett.*, **15** (10), 2378–2381.

91 Adams, H., Collins, R.C., Jones, S., and Warner, C.J.A. (2011) Enantioselective preparation of *P*-chiral phosphine oxides. *Org. Lett.*, **13** (24), 6576–6579.

92 Han, Z.S., Goyal, N., Herbage, M.A., Sieber, J.D., and Qu, B. (2013) Efficient asymmetric synthesis of P-chiral phosphine oxides via properly designed and activated benzoxazaphosphinine-2-oxide agents. *J. Am. Chem. Soc.*, **135** (7), 2474–2477.

93 Nemoto, T. and Hamada, Y. (2007) Pd-catalyzed asymmetric allylic substitution reactions using *P*-chirogenic diaminophosphine oxides: DIAPHOXs. *Chem. Rec.*, **7**, 150–158.

94 Jugé, S. (2008) Enantioselective synthesis of *P*-chirogenic phosphorus compounds via the ephedrine-borane complex methodology. *Phosphorus, Sulfur Silicon Relat. Elem.*, **183** (2–3), 233–248.

95 Jugé, S. and Genet, J.P. (1989) Asymmetric synthesis of phosphinates, phosphine oxides and phosphines by Michaelis Arbuzov rearrangement of chiral oxazaphospholidine. *Tetrahedron Lett.*, **30** (21), 2783–2786.

96 Jugé, S., Stephan, M., Laffitte, J.A., and Genet, J.P. (1990) Efficient asymmetric synthesis of optically pure tertiary mono and diphosphine ligands. *Tetrahedron Lett.*, **31** (44), 6357.

97 Rippert, J., Linden, A., and Hansen, H.J. (2000) Formation of diastereoisomerically pure oxazaphospholes and their reaction to chiral phosphane-borane adducts. *Helv. Chim. Acta*, **83** (2), 311–321.

98 Johansson, J.M., Kann, N., and Larsson, K. (2004) (2R,4S,5R)-3,4-dimethyl-5-phenyl-2-[4-(trifluoromethyl)phenyl]-1,3,2-oxazaphospholidine(P-B)borane. *Acta Crystallogr.*, **E60**, o287–o288.

99 (a) Kaloun, E.B., Merdes, R., Genet, J.P., Uziel, J., and Jugé, S. (1997) Asymmetric synthesis of (S,S)-(+)-1,1′-bis-(methyl-phenyl-phosphino) ferrocene. *J. Organomet. Chem.*, **529**, 455–463; (b) Bayardon, J., Maronnat, M., Langlois, A., Rousselin, Y., Harvey, P.D., and Jugé, S. (2015) Modular P-chirogenic phosphine-sulfide ligands: clear evidence for both electronic effect and P-chirality driving enantioselectivity in palladium-catalyzed allylations. *Organometallics*, **34** (17), 4340–4358.

100 Leon, T., Riera, A., and Verdaguer, X. (2011) Stereoselective synthesis of *P*-stereogenic aminophosphines: ring opening of bulky oxazaphospholidines. *J. Am. Chem. Soc.*, **133** (15), 5740–5743.

101 Uziel, J., Darcel, C., Moulin, D., Bauduin, C., and Jugé, S. (2001) Synthesis of 3,5-disubstituted-1,2,4-oxadiazoles using tetrabutylammonium fluoride as a mild and efficient catalyst. *Tetrahedron: Asymmetry*, **12** (9), 1441–1443.

102 Maienza, F., Spindler, F., Thommen, M., Pugin, B., Malan, C., and Mezzetti, A. (2002) Exploring stereogenic phosphorus: synthetic strategies for diphosphines containing bulky, highly symmetric substituents. *J. Org. Chem.*, **67** (15), 5239–5249.

103 Bauduin, C., Moulin, D., Kaloun, E.B., Darcel, C., and Jugé, S. (2003) Highly enantiomerically enriched chlorophosphine boranes: synthesis and applications as *P*-chirogenic electrophilic blocks. *J. Org. Chem.*, **68** (11), 4293–4301.

104 Salomon, C., Dal Molin, S., Fortin, D., Mugnier, Y., Boere, R.T., Jugé, S., and Harvey, P.D. (2010) The first unpaired electron placed inside a C3-symmetry *P*-chirogenic cluster. *J. Chem. Soc., Dalton Trans.*, **39** (42), 10068–10075.

105 Khiri, N., Bertrand, E., Ondel-Eymin, M.J., Rousselin, Y., Bayardon, J., Harvey, P.D., and Jugé, S. (2010) Enantioselective hydrogenation catalysis aided by a σ-bonded calix[4]arene to a *P*-chirogenic aminophosphane phosphinite rhodium complex. *Organometallics*, **29** (16), 3622–3631.

106 Moulin, D., Darcel, C., and Jugé, S. (1999) Versatile synthesis of *P*-chiral (ephedrine) AMPP ligands via their borane complexes. Structural consequences in Rh-catalyzed hydrogenation of methyl α-acetamidocinnamate. *Tetrahedron: Asymmetry*, **10** (24), 4729–4743.

107 Ewalds, R., Eggeling, E.B., Hewat, A.C., Kamer, P.C.J., van Leeuwen, P.W.N.M., and Vogt, D. (2000) Application of *P*-stereogenic aminophosphine phosphinite ligands in asymmetric hydroformylation. *Chem. Eur. J.*, **6** (8), 1496–1504.

108 Darcel, C., Moulin, D., Henry, J.C., Lagrelette, M., Richard, P., Harvey, P.D., and Jugé, S. (2007) Modular *P*-chirogenic aminophosphane-phosphinite ligands for Rh-catalyzed asymmetric hydrogenation: a new model for prediction of enantioselectivity. *Eur. J. Org. Chem.*, (13), 2078–2090.

109 Moulin, D., Bago, S., Bauduin, C., Darcel, C., and Jugé, S. (2000) Asymmetric synthesis of *P*-stereogenic o-hydroxyaryl-phosphine (borane) and phosphine-phosphinite ligands. *Tetrahedron: Asymmetry*, **11** (19), 3939–3956.

110 Imamoto, T., Tamura, K., Zhang, Z., Horiuchi, Y., Sugiya, M., Yoshida, K., Yanagisawa, A., and Gridnev, I.D. (2012) Rigid *P*-chiral phosphine ligands with tert-butylmethylphosphino groups for rhodium-catalyzed asymmetric hydrogenation of functionalized alkenes. *J. Am. Chem. Soc.*, **134** (3), 1754–1769.

111 Grabulos, A., Muller, G., Ordinas, J.I., Mezzetti, A., Maestro, M.A., Font-Bardia, M., and Solans, X. (2005) Allylpalladium complexes with *P*-stereogenic monodentate phosphines. Application in the asymmetric hydrovinylation of styrene. *Organometallics*, **24** (21), 4961–4963.

112 Stoop, M., Mezzetti, A., and Spindler, F. (1998) Diphosphines containing stereogenic P atoms: synthesis of (S,S)-C,C′-tetramethylsilane-bis-(1-naphthylphenylphosphine) and applications in enantioselective catalysis. *Organometallics*, **17** (4), 668–675.

113 Lam, H., Horton, P.N., Hursthouse, M.B., Aldous, D.J., and Hii, K.K. (2005) The synthesis of potential Neramexane metabolites: cis- and trans-3-amino-1,3,5,5-tetramethylcyclohexanecarboxylic acids. *Tetrahedron Lett.*, **46** (44), 8145–8147.

114 Colby, E.A. and Jamison, T.F. (2003) *P*-chiral, monodentate ferrocenyl phosphines, novel ligands for asymmetric catalysis. *J. Org. Chem.*, **68** (1), 156–166.

115 Nettekoven, U., Widhalm, M., Karner, P.C.J., and van Leeuwen, P.W.N.M. (1997) Novel *P*-chiral bidentate phosphine ligands: synthesis and use in asymmetric catalysis. *Tetrahedron: Asymmetry*, **8** (19), 3185–3188.

116 Nettekoven, U., Widhalm, M., Kalchhauser, H., Kamer, P.C.J., van Leeuwen, P.W.N.M., Lutz, M., and Spek, A.L. (2001) Steric and electronic ligand perturbations in catalysis: asymmetric allylic substitution reactions using C2-symmetrical phosphorus-chiral (Bi)ferrocenyl donors. *J. Org. Chem.*, **66** (3), 759–770.

117 Maienza, F., Worle, M., Steffanut, P., Mezzetti, A., and Spindler, F. (1999) Ferrocenyl diphosphines containing stereogenic phosphorus atoms. Synthesis and application in the rhodium-catalyzed asymmetric hydrogenation. *Organometallics*, **18** (6), 1041–1049.

118 Nettekoven, U., Kamer, P.C.J., Widhalm, M., and van Leeuven, P.W.N.M. (2000) Phosphorus-chiral diphosphines as ligands in hydroformylation. An investigation on the influence of electronic effects in catalysis. *Organometallics*, **19** (22), 4596–4607.

119 Camus, J.M., Andrieu, J., Richard, P., Poli, R., Darcel, C., and Jugé, S. (2004) A *P*-chirogenic β-aminophosphine synthesis by diastereoselective reaction of the α-metallated PAMP–borane complex with benzaldimin. *Tetrahedron: Asymmetry*, **15** (13), 2061–2065.

120 Delapierre, G., Brunel, J.M., Constantieux, T., and Buono, G. (2001) Design of a new class of chiral quinoline–phosphine ligands. Synthesis and application in asymmetric catalysis. *Tetrahedron: Asymmetry*, **12** (9), 1345–1352.

121 Toselli, N., Fortrie, R., Martin, D., and Buono, G. (2010) New *P*-stereogenic triaminophosphines and their derivatives: synthesis, structure, conformational study, and application as chiral ligands. *Tetrahedron: Asymmetry*, **21** (9), 1238–1245.

122 Delapierre, G., Achard, M., and Buono, G. (2002) New *P*-stereogenic triaminophosphines and their derivatives: synthesis, structure, conformational study, and application as chiral ligands. *Tetrahedron Lett.*, **43** (22), 4025–4028.

123 Brunel, J.M. and Buono, G. (2002) in *Topics in Current Chemistry*, (ed. J.-P. Majoral), Springer, Berlin, Heidelberg, vol. 220, pp. 79–106.

124 Leyris, A., Nuel, D., Giordano, L., Achard, M., and Buono, G. (2005) Enantioselective synthesis of both enantiomers of tert-butylphenylphosphine oxide from (S)-prolinol. *Tetrahedron Lett.*, **46** (50), 8677–8680.

125 Ngono, C.J., Constantieux, T., and Buono, G. (2006) Diastereoselective synthesis of new *P*-Stereogenic (ortho-Hydroxyaryl)-diazaphospholidine–Borane complexes by

a totally stereoselective P–O to P–C migration rearrangement. *Eur. J. Org. Chem.*, **2006** (6), 1499–1507.

126 Kolodiazhnyi, O.I., Ustenko, S.N., Grishkun, E.V., and Golovatyi, O.R. (1996) Diastereoselective Rearrangements and Epimerisation of Organophosphorus Compounds. *Phosphorus, Sulfur, and Silicon*, **109–110**, 485–488.

127 Rajendran, K.V. and Gilheany, D.G. (2012) Identification of a key intermediate in the asymmetric Appel process: one pot stereoselective synthesis of *P*-stereogenic phosphines and phosphine boranes from racemic phosphine oxides. *J. Chem. Soc., Chem. Commun.*, **48** (80), 10040–10042.

128 Rajendran, K.V., Kennedy, L., and Gilheany, D.G. (2010) *P*-Stereogenic phosphorus compounds: effect of aryl substituents on the oxidation of arylmethylphenylphosphanes under asymmetric Appel conditions. *Eur. J. Org. Chem.*, (29), 5642–5649.

129 Bergin, E., O'Connor, C.T., Robinson, S.B., McGarrigle, E.M., O'Mahony, C.P., and Gilheany, D.G. (2007) Synthesis of *P*-stereogenic phosphorus compounds. asymmetric oxidation of phosphines under Appel conditions. *J. Am. Chem. Soc.*, **129** (31), 9566–9567.

130 Rajendran, K.V., Kudavalli, J.S., Dunne, K.S., and Gilheany, D.G. (2012) A U-turn in the asymmetric Appel reaction: stereospecific reduction of diastereomerically enriched alkoxyphosphonium salts allows the asymmetric synthesis of *P*-stereogenic phosphanes and phosphane boranes. *Eur. J. Org. Chem.*, **2012** (14), 2720–2723.

131 Rajendran, K.V., Kennedy, L., O'Connor, C.T., Bergin, E., and Gilheany, D.G. (2013) Systematic survey of positive chlorine sources in the asymmetric Appel reaction: oxalyl chloride as a new phosphine activator. *Tetrahedron Lett.*, **54** (51), 7009–7012.

132 Rajendran, K.V., Nikitin, K.V., and Gilheany, D.G. (2015) Hammond postulate mirroring enables enantiomeric enrichment of phosphorus compounds via two thermodynamically interconnected sequential stereoselective processes. *J. Am. Chem. Soc.*, **137** (29), 9375–9381.

133 Nikitin, K., Rajendran, K.V., Müller-Bunz, H., and Gilheany, D.G. (2014) Turning regioselectivity into stereoselectivity: efficient dual resolution of *P*-stereogenic phosphine oxides through bifurcation of the reaction pathway of a common intermediate. *Angew. Chem. Int. Ed.*, **53** (7), 1906–1909.

134 Perlikowska, W., Gouygou, M., Daran, J.C., Balavoine, G., and Mikołajczyk, M. (2001) Kinetic resolution of *P*-chiral tertiary phosphines and chlorophosphines: a new approach to optically active phosphoryl and thiophosphoryl compounds. *Tetrahedron Lett.*, **42** (44), 7841–7845.

135 Le Maux, P., Bahri, H., Simonneaux, G., and Toupet, L. (1995) Enantioselective oxidation of racemic phosphines with chiral oxoruthenium porphyrins and crystal structure of [5,10,15,20-tetrakis[o-((2-methoxy-2-phenyl-3,3,3-trifluoropropanoyl) amino)phenyl] porphyrinato] (carbonyl)(tetrahydrofuran) ruthenium(II) (alpha,beta,alpha,beta Isomer). *Inorg. Chem.*, **34** (18), 4691–4697.

136 Naylor, R.A. and Walker, B. (1975) New routes to optically active phosphorus compounds. Asymmetric alkylation of phosphide anions. *J. Chem. Soc., Chem. Commun.*, (1), 45–50.

137 Valentine, D. Jr., Blount, J.F., and Toth, K. (1980) Synthesis of phosphines having chiral organic groups ligated to chiral phosphorus. *J. Org. Chem.*, **45** (18), 3691–3698.

138 Fisher, C. and Mosher, H. (1977) Asymmetric homogeneous hydrogenation with phosphine-rhodium complexes chiral both at phosphorus and carbon. *Tetrahedron Lett.*, **18** (29), 2487–2490.
139 King, R.B., Bakos, J., Hoff, C.D., and Marco, L. (1979) Poly(tertiary phosphines and arsines). 17. Poly(tertiary phosphines) containing terminal neomenthyl groups as ligands in asymmetric homogeneous hydrogenation catalysts's. *J. Org. Chem.*, **44** (18), 3095–3100.
140 Burgess, K., Ohlmeyer, M.J., and Whitmire, K.H. (1992) Stereochemically matched (and mismatched) bisphosphine ligands: DIOP-DIPAMP hybrids. *Organometallics*, **11** (11), 3588–3600.
141 Nagel, U. and Krink, T. (1993) Neue optisch reine 3,4-Bis(phosphanyl)pyrrolidine mit Phenyl- und Anisylgruppen sowie deren Palladium- und Rhodiumkomplexe. *Chem. Ber.*, **126** (5), 1091–1100.
142 Nagel, U. and Bublewitz, A. (1992) Neue 1,2-Bisphosphanliganden mit vier stereogenen Zentren und zusatzlichen Methoxygruppen fur die asymmetrische katalytische Hydrierung. *Chem. Ber.*, **125**, 1061–1072.
143 Kato, T., Kobayashi, K., Masuda, S., Segi, M., Nakajima, T., and Suga, S. (1987) Asymmetric synthesis of phosphine oxides with the arbuzov reaction. *Chem. Lett.*, (12), 1915–1918.
144 Mikolajczyk, M., Drabowicz, J., Omelanczuk, J., and Fluck, E. (1975) Optically active trivalent phosphorus acids esters: an approach by asymmetric synthesis. *J. Chem. Soc., Chem. Commun.*, (10), 382–383.
145 Jugé, S., Wakselman, M., Stephan, M., and Genet, J.P. (1990) Evidence of the loss of chiralitv at the phosphonium step in the Arbuzov rearrangement. *Tetrahedron Lett.*, **31** (31), 4443–4446.
146 Kolodiazhnyi, O.I. (1998) in *Advances of Asymmetric Synthesis*, vol. **3** (ed. A. Hassner), JAI Press Inc., Stamford, CT, London, pp. 273–357.
147 Gordon, N.J. and Evans, S.A. (1993) Synthesis and enantioselective aldol reaction of a chiral 2-oxo-2-propionyl-1,3,2-oxazaphosphorinane. *J. Org. Chem.*, **58** (20), 5295–5297.
148 Denmark, S.E., Chatani, N., and Pansare, S.V.T. (1992) Asymmetric electrophilic amination of chiral phosphorus-stabilized anions. *Tetrahedron*, **48** (11), 2191–2208.
149 Bodalski, R., Rutkowska-Olm, E., and Pietrusiewicz, K.M. (1980) Synthesis and absolute configuration of (menthoxycarbonylmethyl)phenylvinyl phosphine oxide. *Tetrahedron*, **36** (16), 2353–2355.
150 Pietrusiewicz, M. (1996) Stereoselective synthesis and resolution of P-chiral phosphine chalcogenides. *Phosphorus, Sulfur Silicon Relat. Elem.*, **109–110** (1–4), 573–576.
151 Pietrusiewicz, K.M., Zablocka, M., and Monkiewicz, J. (1984) Optically active phosphine oxides. 2. Novel approach to enantiomeric dialkylphenylphosphine oxides. *J. Org. Chem.*, **49** (9), 1522–1526.
152 Imamoto, T., Sato, K., and Johnson, C.R. (1985) A new method for the synthesis of optically pure phosphine oxides. *Tetrahedron Lett.*, **26** (6), 783–786.
153 Johnson, C.R. and Imamoto, T. (1987) Synthesis of polydentate ligands with homochiral phosphine centers. *J. Org. Chem.*, **52** (11), 2170–2174.
154 Kolodyazhnyi, O.I., Sheiko, S., Neda, I., and Schmutzler, R. (1998) Asymmetric Induction In Arbuzov Reaction. *Russ. J. Gen. Chem.*, **68**, 1157–1158.

155 Tasz, M.K., Gamliel, A., Rodriguez, O.P., Lane, T.M., Cremer, S.E., and Bennett, D.W. (1995) Preparation, Reactions, and Stereochemistry of 4-tert-Butyl-1-chlorophosphorinane-Oxide and Derivatives. *J. Org. Chem.*, **60** (20), 6281–6288.
156 Lewis, R.A., Naumann, K., De Bruin, K., and Mislow, K. (1969) t-Butylphosphonium Salts: Nucleophilic Displacement at Phosphorus with Inversion of Configuration. *J. Chem. Soc., Chem. Commun.*, (17), 1010–1011.
157 Westheimer, F.H. (1968) Pseudo-rotation in the hydrolysis of phosphate esters. *Acc. Chem. Res.*, **1** (3), 70–78.
158 Merckling, F.A. and Ruedl, P. (1996) Diastereoselectivity in nucleophilic displacement reactions at phosphorus; isolation and characterization of a pentacoordinated intermediate. *Tetrahedron Lett.*, **37** (13), 2217–2220.
159 van Bochove, M.A., Swart, M., and Bickelhaupt, F.M. (2009) Stepwise walden inversion in nucleophilic substitution at phosphorus. *Phys. Chem. Chem. Phys.*, **11** (2), 259–267.
160 Kolodiazhnyi, O.I., Ustenko, S., and Golovatyi, O.R. (1994) Fluoroalkoxyphosphonium ylids. Epimerization and transformations. *Tetrahedron Lett.*, **35** (11), 1755–1758.
161 Kolodiazhnyi, O.I. (1996) C-element-substituted phosphorus ylids. *Tetrahedron*, **52** (6), 1855–1925.
162 Kolodiazhnyi, O.I. (1997) Methods of Preparation and Application In Organic Synthesis of C-substituted Phosphorus Ylides. *Russ. Chem. Rev.*, **66** (3), 225–254.
163 Pedronia, J. and Cramer, N. (2015) TADDOL-based phosphorus(III)-ligands in enantioselective Pd(0)-catalysed C–H functionalisations. *Chem. Commun.*, **51** (100), 17647–17657.
164 Xu, G., Li, M., Wang, S., and Tang, W. (2015) Efficient synthesis of P-chiral biaryl phosphonates by tereoselective intramolecular cyclization. *Org. Chem. Front.*, **2**, 1342–1345.
165 Gatineau, D., Nguyen, D.H., Hérault, D., Vanthuyne, N., Leclaire, J., Giordano, L., and Buono, G. (2015) H-adamantylphosphinates as universal precursors of P-stereogenic compounds. *J. Org. Chem.*, **80** (8), 4132–4141.
166 Copey, L., Jean-Gérard, L., Andrioletti, B., and Framery, E. (2016) Synthesis of P-stereogenic secondary phosphine oxides using α-D-glucosamine as a chiral precursor. *Tetrahedron Lett.*, **57** (5), 543–545.
167 Lemouzy, S., Nguyen, D.H., Camy, V., Jean, M., and Gatineau, D. (2015) Stereospecific synthesis of α- and β-hydroxyalkyl P-stereogenic phosphine–boranes and functionalized derivatives: evidence of the P=O activation in the BH_3-mediated reduction. *Chem. Eur. J.*, **21**, 15607–15621.
168 Orgué, S., Flores-Gaspar, A., Biosca, M., Pàmies, O., Diéguez, M., Riera, A., and Verdaguer, X. (2015) Stereospecific SN2@P reactions: novel access to bulky P-stereogenic ligands. *Chem. Commun.*, **51** (99), 17548–17551.
169 Chelouan, A., Recio, R., Álvarez, E., Khiar, N., and Fernández, I. (2016) Stereoselective synthesis of P-stereogenic *N*-phosphinyl compounds. *Eur. J. Org. Chem.*, **2016** (2), 255–259.
170 Kolodiazhnyi, O.I. (2002) Double asymmetric induction as method for the synthesis of chiral organophosphorus compounds. *Phosphorus, Sulfur Silicon Relat. Elem.*, **177** (8–9), 2111–2114.

171 Kolodiazhna, O.O., Kolodiazhna, A.O., and Kolodiazhnyi, O.I. (2011) Highly effective catalyst for the reaction of trialkylphosphites with C = X. *Electrophiles,* **186** (4), 796–798.

172 Nesterov, V.V. and Kolodiazhnyi, O.I. (2008) New method for the asymmetric reduction of ketophosphonates. *Phosphorus, Sulfur Silicon Relat. Elem.,* **183** (2–3), 687–688.

173 Kolodiazhnyi, O.I., Grishkun, E.V., Sheiko, S., Demchuk, O., Thoennessen, H., Jones, P.G., and Schmutzler, R. (1998) Chiral symmetric phosphoric acid esters as sources of optically active organophosphorus compounds. *Tetrahedron: Asymmetry,* **9** (10), 1645–1649.

174 Kolodiazhnyi, O.I., Grishkun, E.V., and Otzaluk, V.M. (1997) Asymmetric induction in reaction of asymmetric substituted phosphinoyl chlorides with 1,2:5,6-diisopropylidenegluco-furanose. *Russ. J. Gen. Chem.,* **67** (7), 1140–1142.

175 (a) Fernandez, I., Khiar, N., Roca, A., Benabra, A., Alcudia, A., Espartero, J.L., and Alcudia, F. (1999) A generalization of the base effect on the diastereoselective synthesis of sulfinic and phosphinic esters. *Tetrahedron Lett.,* **40** (10), 2029–2032; (b) Kolodyazhnyi, O.I., Sheiko, S., Neda, I., and Schmutzler, R. (1998) Asymmetric induction in reaction of thionation of aminophosphine diastereomeres. *Russ. J. Gen. Chem.,* **68** (7), 1212–1213.

176 Kolodiazhnyi, O.I. and Grishkun, E.V. (1996) Asymmetric induction in the reaction of nonsymmetrical phosphinic and phosphinous acid clorides with derivatives of d-glucofuranose. *Phosphorus, Sulfur Silicon Relat. Elem.,* **115** (1–4), 115–124.

177 Marchi, C. and Buono, G. (1999) Asymmetric addition of chiral triquinphosphoranes on activated compounds. *Tetrahedron Lett.,* **40** (52), 9251–9254.

178 Kimura, T. and Murai, T. (2005) Optically active *P*-chiral phosphinoselenoic amides: stereochemical outcome at the *P*-stereogenic center in the synthesis of these substances and their characterization. *Tetrahedron: Asymmetry,* **16** (22), 3703–3710.

179 Reddy, P.M. and Kovach, I.M. (2002) Synthesis of chiral 4-nitrophenyl alkyl methylphosphonothioates: $BF_3 \cdot Et_2O$-catalyzed alcoholysis of phosphonamidothioates. *Tetrahedron Lett.,* **43** (22), 4063–4066.

180 Kolodiazhnyi, O.I., Grishkun, E.V., Golovatyi, O.R., and Ustenko, S.N. (1996) Diastereoselective rearrangements and epimerization of organophosphorus compounds. *Phosphorus, Sulfur Silicon Relat. Elem.,* **109/110** (1–4), 485–488.

181 Kolodiazhnyi, O.I. (1995) Asymmetric induction in reaction of chlorophosphines with 1,2:5,6-glucofuranose. *Russ. J. Gen. Chem.,* **65** (11), 1926–1927.

182 Kolodiazhnyi, O.I. and Grishkun, E.V. (1996) Simple route to chiral organophosphorus compounds. *Tetrahedron: Asymmetry,* **7** (4), 967–970.

183 Kolodiazhnyi, O.I. (1995) Stereoselective oxidation of *N*-phosphor (III) substituted amino acids. *Tetrahedron Lett.,* **36** (22), 3921–3924.

184 Moriarty, R.M., Hiratake, J., Liu, K., Wendler, A., Awasthi, A.K., and Gilardi, R. (1991) Isolation and characterization of stereoisomers of pentacoordinated phosphorus. *J. Am. Chem. Soc.,* **113** (24), 9374–9376.

185 Marchi, C., Fotiadu, F., and Buono, G. (1999) The first example of coordination of a tricyclic hydrophosphorane to platinum(II). X-ray crystal structure of an unusual platinated phosphorane. *Organometallics,* **18** (5), 915–927.

186 Marchi, C. and Buono, G. (2000) Alkylthiylation of triquinphosphoranes by disulfides: an entry to chiral thiatriquinphosphoranes. *Tetrahedron Lett.,* **41** (17), 3073–3076.

187 Marchi, C. and Buono, G. (2000) Total diastereoselective opening of chiral hydridophosphorane "Triquinphosphoranes" by hexacarbonylmolybdenum: an entry to Mo(CO)5: chiral oxazaphospholidine complex. *Inorg. Chem.*, **39** (13), 2951–2953.

188 Mikhel, I.S., Bondarev, O.G., Tsarev, V.N., Grintselev-Knyazev, G.V., Lyssenko, K.A., Davankov, V.A., and Gavrilov, K.N. (2003) The first example of coordination of a tricyclic hydrophosphorane to platinum(II). X-ray crystal structure of an unusual platinated phosphorane. *Organometallics*, **22** (5), 925–930.

189 Kojima, S., Kajiyama, K., and Akiba, K.Y. (1994) Characterization of an optically active pentacourdinate phosphorane with asymmetry only at phosphorus. *Tetrahedron Lett.*, **35** (38), 7037–7040.

190 Bojin, M.L., Barkallah, S., and Evans, S.A. Jr., (1996) Reactivity of carbon anions r to pentacoordinated phosphorus: spirooxyphosphoranyl C-anions as valuable intermediates in olefination chemistry. *J. Am. Chem. Soc.*, **118** (6), 1549–1550.

191 Kojima, S., Takagi, R., and Akiba, K.Y. (1997) Excellent Z-selective olefin formation using pentacoordinate spirophosphoranes and aldehydes. Wittig type reaction via hexacoordinate intermediates. *J. Am. Chem. Soc.*, **119** (25), 5970–5971.

192 Kojima, S., Nakamoto, M., Yamazaki, K., and Akiba, K.-Y. (1997) Reactions of sterically rigid phosphoranes. *Tetrahedron Lett.*, **38** (23), 4107–4110.

193 Kojima, S., Kawaguchi, K., Matsukawa, S., and Akiba, K.-Y. (2003) Stereospecific stilbene formation from β-hydroxy-α,β-diphenylethylphosphoranes. Mechanistic proposals based upon stereochemistry. *Tetrahedron*, **59** (1), 255–265.

194 Yang, G., Xu, Y., Hou, J., Zhang, H., and Zhao, Y. (2010) Diastereomers of the pentacoordinate chiral phosphorus compounds in solution: absolute configurations and predominant conformations. *Dalton Trans.*, **39** (30), 6953–6959.

195 Yang, G., Xu, Y., Hou, J., Zhang, H., and Zhao, Y. (2010) Determination of the absolute configuration of pentacoordinate chiral phosphorus compounds in solution by using vibrational circular dichroism spectroscopy and density functional theory. *Chem. Eur. J.*, **16** (8), 2518–2527.

196 Favarger, F., Goujon-Ginglinger, C., Monchaud, D., and Lacour, J. (2004) Large-scale synthesis and resolution of TRISPHAT [Tris(tetrachlorobenzenediolato) Phosphate(V)] anion. *J. Org. Chem.*, **69** (24), 8521–8524.

197 Constant, S. and Lacour, J. (2003) New trends in hexacoordinated phosphorus chemistry, in *Topics in Current Chemistry*, vol. **250** (ed. J.-P. Majoral), Springer-Verlag, Berlin, Heidelberg, pp. 1–42.

198 Hellwinkel, D. (1965) Optisch aktive Tris-(biphenylen)-phosphate. *Angew. Chem.*, **77** (8), 378–379.

199 Hellwinkel, D. (1967) Verfahren zur Herstellung von sechsbindigen Phosphorverbindungen. 256. Patent DE1235913, p. 2, https://www.google.com/patents/DE1235913B?cl=fr (accessed 17 May 2016).

200 Lacour, J., Londez, A., Tran, D.-H., Desvergnes-Breuil, V., Constant, S., and Bernardinelli, G. (2002) Asymmetric synthesis and configurational stability of C2-symmetric hexacoordinated phosphate anions (TARPHATs) with predetermined chirality from tartrate esters. *Helv. Chim. Acta*, **85** (8), 1364–1381.

201 Lacour, J., Ginglinger, C., Grivet, C., and Bernardinelli, G. (1997) Synthesis and resolution of the configurationally stable tris(tetrachlorobenzenediolato) phosphate(v) ion. *Angew. Chem. Int. Ed. Engl.*, **36** (6), 608–610.

3

Phosphorus Compounds with Chiral Side-Chain Centers

3.1 Introduction

Organophosphorus compounds with chiral centers in the side chain play an important role in science and technology as chiral ligands, biologically active compounds, drugs, and finally as numerous classes of chiral compounds [1–4]. The C_2-symmetric atropisomeric diphosphines (BINAP (2,20-bis(diphenylphosphino)-1,10-binaphthyl), BINOL (2,20-dihydroxyl-1,10-binaphthyl), BIPHEMP, and MeO-BIPHEP) prove to be highly effective in many asymmetric transformations. Enormous success has been achieved in the use of C_2-symmetric atropisomeric diphosphine ligands in Rh- or Ru-catalyzed asymmetric hydrogenation reactions, such as JOSIPHOS, CATPHOS, PROPRAPHOS, BoPHOS, FerroTANE, and others. Phosphonic acids represent an important class of biologically active compounds which occur in nature. Many types of phosphonic acids have been found in hundreds of aquatic and terrestrial animals and microorganisms [5]. Typical representatives of natural hydroxyphosphonic acids are phosphonothrixin (PTX), dihydroxyphosphonic acid (FR-33289), hydroxy-2-aminoethylphosphonic acid (HO-AEP), 1,5-dihydroxy-2-oxopyrrolidinphosphonic acid (SF-2312), and others (Scheme 3.1) [1, 2]. There has been significant interest in (1*R*,2*S*)-(−)-(l,2-epoxypropyl) phosphonic acid, also known as *fosfomycin*, which is a cell-wall-active antibiotic isolated from the fermentation broth of *Streptomyces fradiae* or *Pseudomonas syringae.* Many of these compounds have attracted attention because of their antibacterial, antiviral, antibiotic, pesticidal, anticancer, and enzyme-inhibitor properties. Functionalized phosphonic acids, hydroxyl-, and aminophosphonates display high antibacterial, antiviral, anticancer activity, for example, the compound **4** inhibits HIV protease. Hydroxyphosphonates are efficient medicines against smallpox. Some phosphonates possessing potent antitumor activity are used for the treatment of cancer and are prospective drugs for the treatment of AIDS, for example, hydroxyalkyl-bis-phosphonates demonstrate antiproliferative activity in several human cancer cell lines with IC_{50} values in the l M range [4].

Bis-phosphonates are also used for the treatment of bony rarefication, hypercalcemia, and malignant tumors. The bis-phosphonate derivative of 3-azido-3-deoxythymidine (AZT, Zidovudine) has been registered as an anti-HIV drug and used in the treatment of AIDS [5]. Over the last 10 years, a significant number of methods for the asymmetric synthesis and practical application of chiral hydroxyphosphonates has been reported, which clearly show the theoretical interest and practical importance of hydroxyphosphonates.

Asymmetric Synthesis in Organophosphorus Chemistry: Synthetic Methods, Catalysis and Applications, First Edition. Oleg I. Kolodiazhnyi.

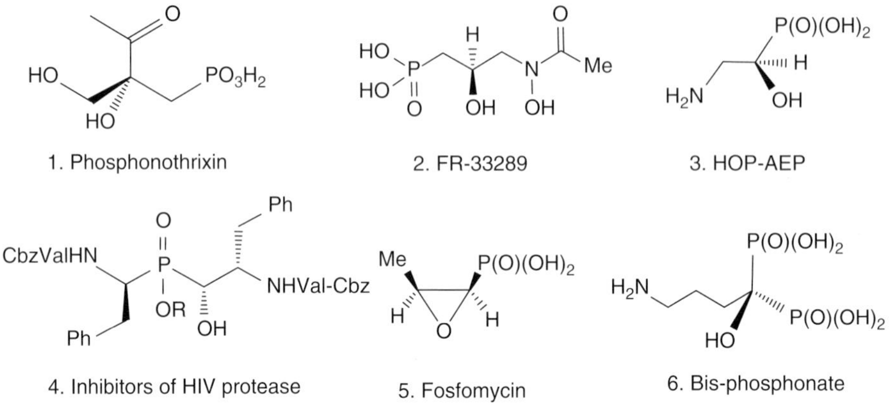

Scheme 3.1 Examples of biologically active phosphonates.

3.2 Asymmetric Induction in Side Chains

Chiral-functionalized phosphonic acids can be prepared by various methods such as kinetic resolution, chemoenzymatic synthesis, and asymmetric synthesis. The main method for the synthesis of functionalized phosphonates is the phosphonylation of carbonyl compounds: aldehydes, ketones, imines, and compounds with activated C=C bond. Several methods for asymmetric phosphonylations of carbonyl compounds are known, for example, the phospha-aldol addition (the Abramov reaction), the phospha-Mannich reaction, and reaction of phosphonate carbanions with aldehydes or ketones. The phospha-aldol reaction usually leads to the formation of hydroxyphosphonates, whereas the phospha-Mannich reaction is one of the most convenient methods for the synthesis of chiral α-aminoalkylphosphonates (Scheme 3.2) [6–10].

Metallocomplex, organocatalytic, and chemoenzymatic methods constitute a wide range of methodologies for the preparation of optically active functionalized side-chain phosphorus compounds [9]. These methods are discussed in the following chapters of

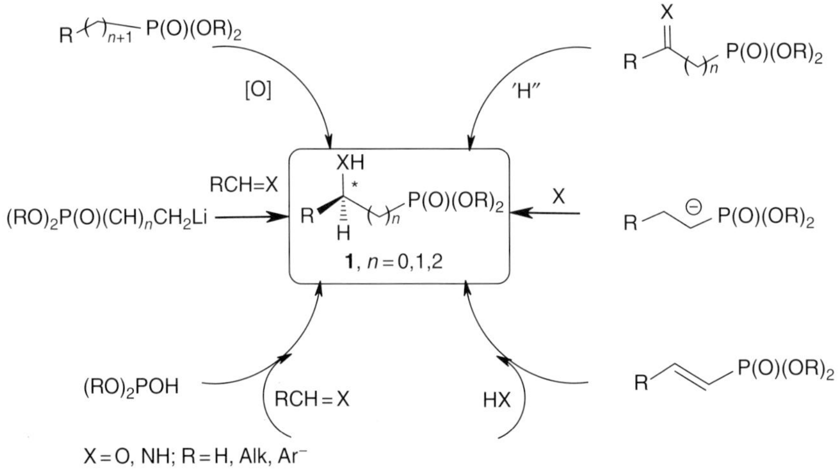

Scheme 3.2 Synthetic routes to chiral-functionalized phosphonates.

this monograph. Other useful methods for the synthesis of phosphonates with chiral centers in side-chain are the enantioselective reduction of ketophosphonates, enantioselective hydroxylation of phosphonate-stabilized carbanions, and [2,3]-sigmatropic Wittig rearrangements [5].

3.2.1 Transfer of Chirality from Phosphorus to Other Centers

The enantioselective reactions of phosphorus-stabilized carbanions, such as carbanions of α-lithiated phosphines, phosphonates, or phosphine oxides, have been studied extensively because of their broad synthetic utility. In fact, enantioselective reactions of α-lithiated phosphonates with carbonyl compounds give olefins having axial chirality. The chiral phosphorus moiety that induces optical activity can be easily removed from the molecule, thus presenting an additional advantage in the asymmetric synthesis of chiral organic compounds. On the basis of optically active organophosphorus compounds, different chiral organic compounds, in which asymmetric induction involved the transfer of chirality from a stereogenic phosphorus to a newly formed stereogenic center, were synthesized. All these reactions can be divided into two groups:

1) Reactions resulting in the formation of diastereomeric organophosphorus compounds containing the induced new stereogenic center as well as the optically active phosphorus atom. Here it is necessary to take into consideration the number of newly formed stereogenic centers, that is, to consider reactions proceeding with 1,2- or with 1,4-asymmetric induction.
2) Reactions in the course of which the formation of a new stereogenic center is accompanied by elimination of the phosphorus moiety.

3.2.1.1 *Chiral Phosphorus-Stabilized Anions*

The chemistry of phosphorus-stabilized anions in the areas of structure and reaction stereoselectivity has attracted the attention of many chemists [11–14]. Theoretical investigations of chiral phosphorus-stabilized carbanions have been performed by Cramer and co-workers [15]. The rotational coordinates about the P–C bond have been studied at the HF/3-21G$^{(*)}$ level. The locations of localized minima on the rotational coordinate were found to be dependent on opportunities for hyperconjugative stabilization.

Ab initio studies of the P–C bond rotation in phosphoryl- and thiophosphoryl-stabilized carbanions of five- and six-membered heterocycles showed a completely or nearly planar carbanion structure with its substituents parallel to the P=X axis (X=O, S). The transition-state (TS) structures have a strongly pyramidalized carbanion in which the lone pair (LP) is approximately perpendicular to the PdX bond. Isodesmic equations, bond length comparisons, and orbital interactions indicate a superior ground-state (GS) stabilization of the thioxo derivative and a favorable TS stabilization of the oxo species [16]. Leyssens and Peeters found the negative hyperconjugation effect in phosphorus-stabilized carbanions [17]. Spectroscopic analysis of lithiated cyclic phosphonates belonging to the 1,3,2-dioxaphosphorinane 2-oxide were examined by NMR spectroscopy and by single-crystal X-ray crystallography. They were characterized by free rotation and the lithium–carbon contact in the sp^2-hybridized anions was absent. The anions are most likely dimers linked through oxygen-lithium bridges. Cramer [15] and Denmark [18] studied NMR spectra and performed X-ray analyses of

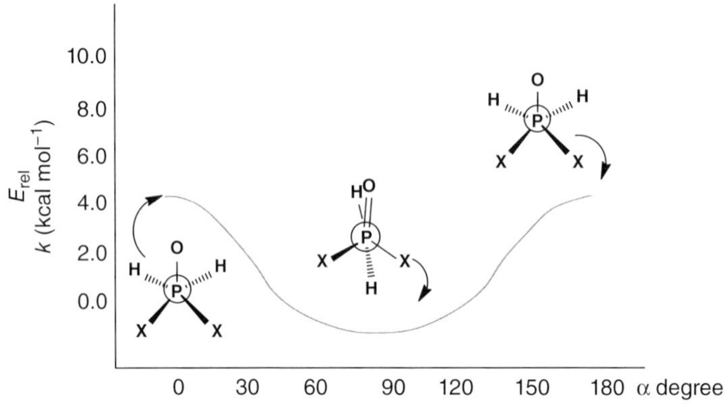

Scheme 3.3 Relative energy in kilocalorie per mole for $X_2P(O)CH_2^-$ (X = F) versus carbanion rotation angle α at the HF/3-21G$^{(*)}$ level [15].

phosphorus-stabilized carbanions **7** and **8**. For the phosphorus-stabilized carbanions, as shown in the Scheme 3.3, the local minimum occurs for the C $a = 90°$ structure and the barrier is described by the C $\alpha = 0°$ structure.

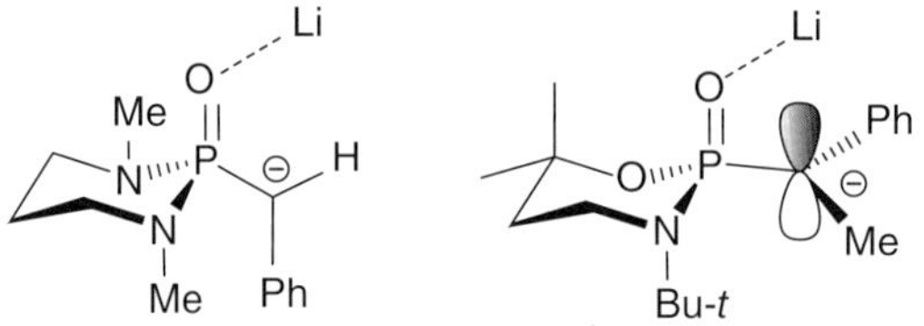

7

8

X-ray crystallography showed that the Li–C distance (3.88 Å) is greater than the sum of the van der Waals radii. The carbanion is planar and at nearly a 0° angle to the P=O to maximize P-type interaction and the preferred conformation of the anion is parallel [19]. The idealized anomeric stabilization of nitrogen- and carbon-bearing pairs of electrons leads to the familiar "pinwheel" conformation of methyl and benzyl groups. The chair conformation with the P=O group axially oriented is probable. The primary, hyperconjugative-type stabilization arises from $n \rightarrow \sigma^*_{P=O}$ donation which is best accommodated in a perfectly orthogonal conformation ($\theta = 90°$) [20]. The barrier to rotation about the P(1)–C(b) bond is very low (<8 kcal mol^{-1}). The lithium salts exist in general as dimers in tetrahydrofuran (THF) solution, and in the solid state there is no metal-carbon contact. The most remarkable feature of the anion is the pyramidality of the nitrogen, clearly disposing the methyl groups to axial and equatorial positions. The downfield shift of the ^{31}P NMR resonance is indicative of the polarization of the phosphoryl group to stabilize the anion [19]. The preferred conformation for P-carbanion is that which maximizes opportunities for hyperconjugative stabilization. The rate of racemization of phosphonate carbanions slows down in the presence of hexamethylphosphoramide (HMPA), because the HMPA increases the barrier of rotation around the P–C bond: $G^{\ddagger}_{205} = 9.8$ kcal mol^{-1} (THF), $G^{\ddagger}_{246} = 11.4$ kcal mol^{-1} (THF–HMPA solution) [21]. Aggarval analyzed the mechanism of racemization of phosphorus-substituted organolithium compounds and came to the conclusion that

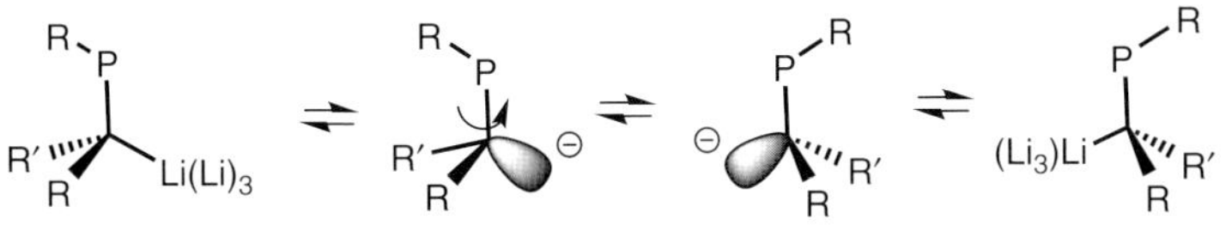

Scheme 3.4 Racemization of heteroatom-substituted organolithium compounds.

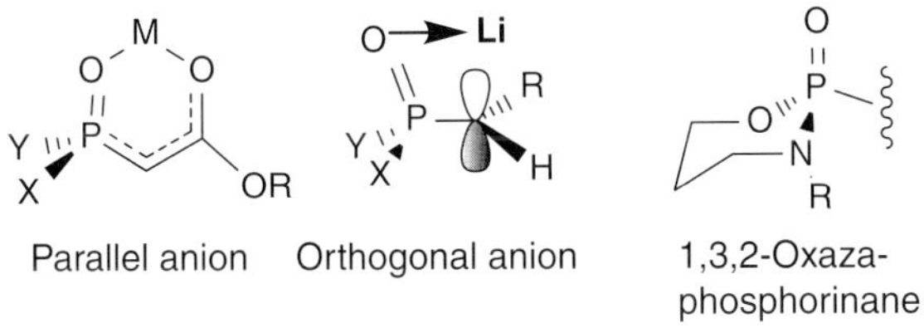

Scheme 3.5 Chiral P-stabilized carbanions.

rotation of the C–P bond plays a dominant role in racemization, which is reduced with increasing concentration of P-stabilized carbanions (Schemes 3.4 and 3.5) [22].

Reactions of chiral P-stabilized carbanions with carbonyl compounds possessing axial asymmetry (enantioselective Wittig or Horner–Wittig reactions) result in the formation of chiral alkenes with the elimination of the phosphorus moiety. Chiral P-stabilized carbanions are used for asymmetric olefination reactions [23], in asymmetric Michael addition reactions [24], in asymmetric reactions of alkylation and aminating [25–28], and others.

3.2.1.2 *1,2-Asymmetric Induction*

Chiral P-stabilized carbanions react with nucleophiles with the formation of organophosphorus compounds, bearing a newly formed chiral center formed as a consequence of 1,2- or 1,4-asymmetric induction. The discovery of new methods for 1,2- and 1.4-asymmetric induction in acyclic systems has been of keen interest in synthetic and theoretical organic chemistry. The 1,2-asymmetric induction includes alkylation, amination, carboxylation, and acylation reactions. A general method for the asymmetric synthesis of enantiomerically pure or enriched α-amino-α-alkylphosphonic acids was developed on the basis of amination or alkylation of chiral cyclic chloromethylphosphonamides, **9–11**, derived from (*R*,*R*)- or (*S*,*S*)-1,2-diaminocyclohexanes, and oxazophosphorinanes. Denmark and Amburgey studied enantioselective alkylations of P-carbanions (Schemes 3.6–3.8) [5, 29]. The quaternary stereocenter was created by alkylation of the various β-ketophosphonamidates as their potassium or sodium enolates. Stereoselective reduction resulted in diastereomers of β-hydroxyphosphonamidates **11** in the ratio 1 : 14–1 : 150 [11]. The high stereoselectivity of the reaction was explained by attack of the initially formed

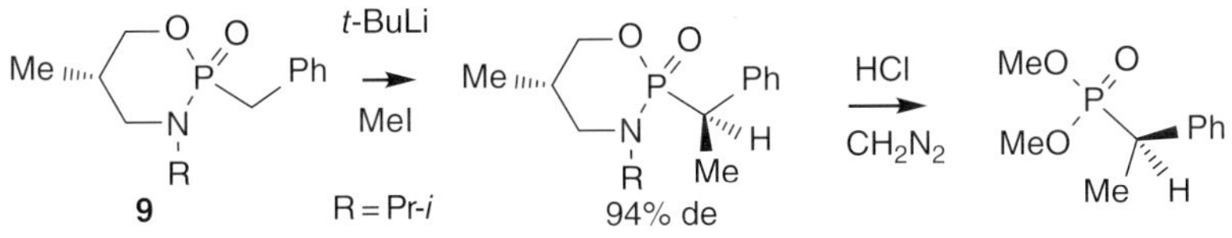

Scheme 3.6 Alkylation of 2-benzyl-6-methyl-1,3,2-oxazaphosphorinane 2-oxides **9**.

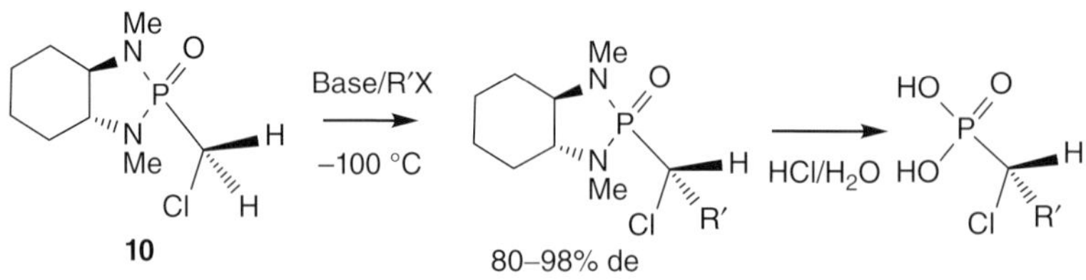

R = Alk, R′ = Alk; Base = BuLi or Et_2NLi

Scheme 3.7 Alkylation of bicyclic C_2-symmetric phosphondiamide **10**.

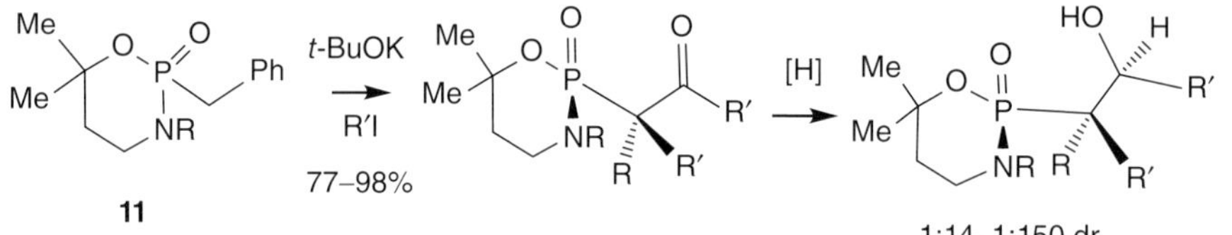

R = Bu-t; R = Me, R′ = Et; R = Et, R′ = Me

Scheme 3.8 Enantioselective alkylations of β-ketophosphonamidates **11**.

P-carbanion on the electrophile preferentially from the side facing the lone pair of one of the nitrogen atoms rather than the side facing the *N*-methyl group [30].

The aminations of carbanions **12a** derived from (*R*,*R*)-diaminocyclohexane by trisyl azide afforded the α-azidophosphonoamidates with good stereoselectivity (in average ~68–80% enantiomeric excess (ee)), and they were converted into aminophosphonic acid **13** as shown in Scheme 3.9 [13, 14]. In the best cases, (*S*)-aminophosphonic acids **13** were obtained with 92% ee. The deprotonation of oxazaphosphorinane carbanions **12b** with BuLi and treatment with NaN_3 and mild acid hydrolysis followed by hydrogenation of the azido group led to formation of the corresponding α-aminophosphonic acids **13** with 78–98% ee [13, 30]. The electrophilic amination of 1,3,2-oxazaphospholanes with azo-compounds proceeded with 52–83% diastereoselectivity in good agreement with theoretical calculations [29]. The synthesis of α- and β-aminophosphonic acids **13** based on the stereoselective addition of carbanions of chiral α-chloromethylphosphonamides **12a** to imines was also described [31]. The α-aminophosphonamides were obtained with greater than 95 : 5 diastereomeric ratio (dr) (Scheme 3.9).

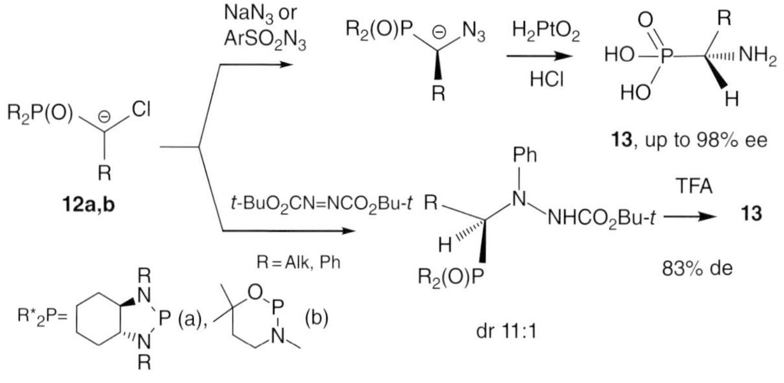

Scheme 3.9 Asymmetric synthesis of chiral aminophosphonic acids.

3.2.1.3 1,4-Asymmetric Induction

1,4-Addition of Phosphonate Carbanions Highly stereoselective asymmetric 1,4-addition reactions with cinnamate esters, which lead to the products with three and four contiguous stereogenic centers were described by Hanessian. An asymmetric addition of the P-carbanion to the *tert*-butyl cinnamate followed by addition of methyl iodide led to the formation of adducts **14** with 92 : 8 dr. Ozonolysis and sodium borohydride reduction afforded the hydroxy ester **15** as a single isomer. Treatment of **14** with trifluoroacetic acid (TFA) in dichloromethane led to the lactone containing three contiguous carbon substituents, the structure of which was unequivocally proved by single-crystal X-ray analysis [32]. Following the same protocol as above and quenching the enolate with methyl triflate, the same authors obtained the products **16** with 9 : 1 dr. Ozonolysis, reduction, and chromatographic separation gave the hydroxy ester as a single isomer **17** with four contiguous stereogenic centers [33] (Scheme 3.10). The creation of asymmetric carbon centers was attained in high stereoselectivity by conjugate additions of crotyl-, allyl-, and cinnamyl-derived anions to corresponding acceptors (enones, lactones, lactams, and α,β-unsaturated esters) followed by optional alkylation with formation of adducts. The resulting vinyl phosphonamide products bearing the chiral auxiliary were cleaved by ozonolysis to the corresponding aldehydes and the latter reduced to alcohols, respectively. Many highly functionalized, vicinally substituted compounds could be prepared by this method in good to excellent enantiopurity. Asymmetric conjugate additions using P-chiral phosphonamides with remarkable selectivity depending on the configuration of the P-stereogenic center were reported by Denmark [25] and Hanessian *et al.* [33, 34]. Thus, *cis*- or *trans*-orientation of the P-alkyl group relative to the *N*-alkyl group were obtained from (*R*)- or (*S*)-configurated phosphorus center, correspondingly. The addition of the Li-anion of *trans*-**18** to cyclic enones proceeded with a high level of stereocontrol, providing adducts **19** with up to 98% ee (Scheme 3.11).

Asymmetric 1,4-additions using phosphonamides is a convenient synthetic method, which was applied in a number of total syntheses of complex natural compounds (Scheme 3.12). For example, this method was used in the total synthesis of acetoxycrenulide [35, 36], berkelic acid [37], estrone [38], methyl jasmonate [39], and

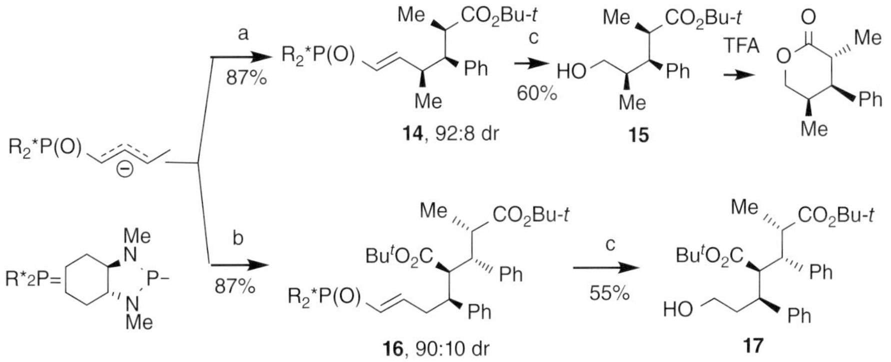

a = (1) PhCH=CHCO$_2$Bu-*t*, (2) MeI, −78 °C; **b** = PhCH=CHCO$_2$Bu-*t*, MeOSO$_2$CF$_3$, Py, −78 °C; **c** = (1) O$_3$; (2) NaBH$_4$

Scheme 3.10 Asymmetric 1,4-addition of the P-carbanion to the *tert*-butyl cinnamate.

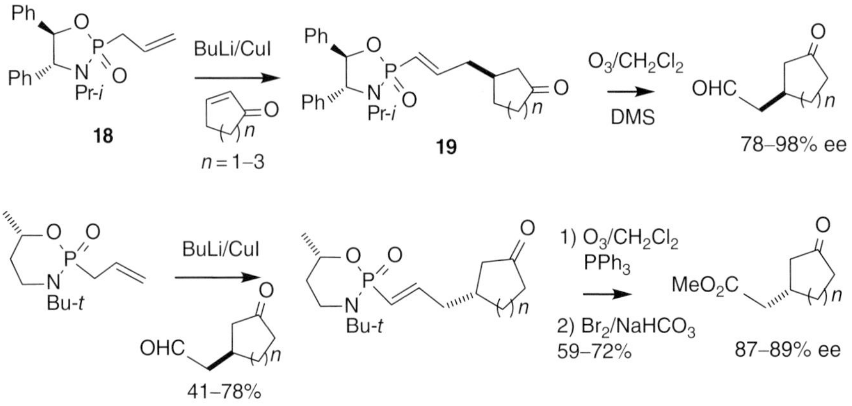

Scheme 3.11 Asymmetric conjugate additions of P-chiral phosphonamide to cyclic enones.

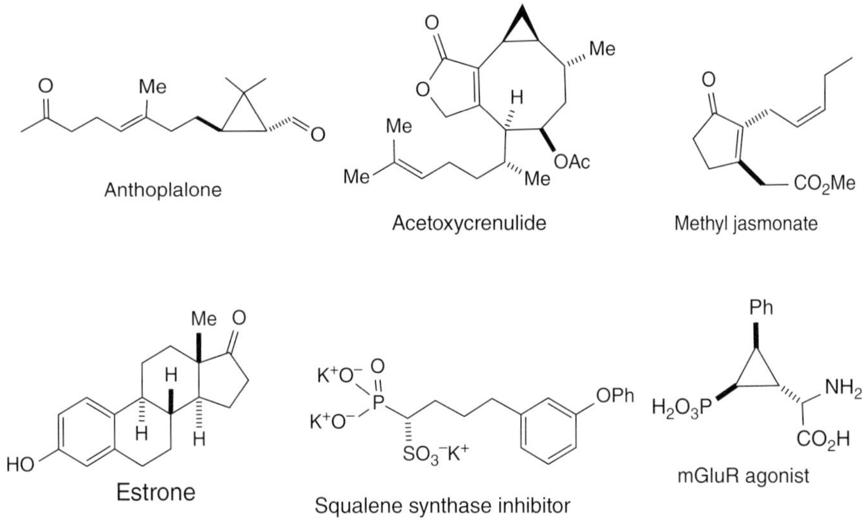

Scheme 3.12 Natural products and bioactive molecules synthesized using phosphonamide-based chemistry.

others [34]. Cyclopropanation with chloroallyl phosphonamide was used to construct cyclopropane fragments of anthoplalone [40]. Ottelione A and B [41] also employed this cyclopropanation methodology [42].

Reactions of anions derived from chiral nonracemic allyl and crotyl bicyclic phosphonoamides with unsaturated cyclic ketones, esters, lactones, and lactams take place at the position of the reagents and lead to diastereomerically pure or highly enriched products of conjugate addition. For example, chloromethyl phosphonamides as well as chloroallyl phosphonamides reacted with α,β-unsaturated esters to form cyclopropane products which can be converted to aminocyclopropylphosphonic acids (Scheme 3.13) [42]. The conjugate 1,4-addition of chloroallylphosphonamide *trans*-**20** to cyclic enones provided diastereomerically pure or highly enriched cyclopropane derivatives (88–90% de). The reaction gave the crystalline *endo,endo* isomer of the cyclopropane derivative **21** in 90% yield. Alternatively, utilization of the chloroallylphosphonamide *cis*-**20** with the same

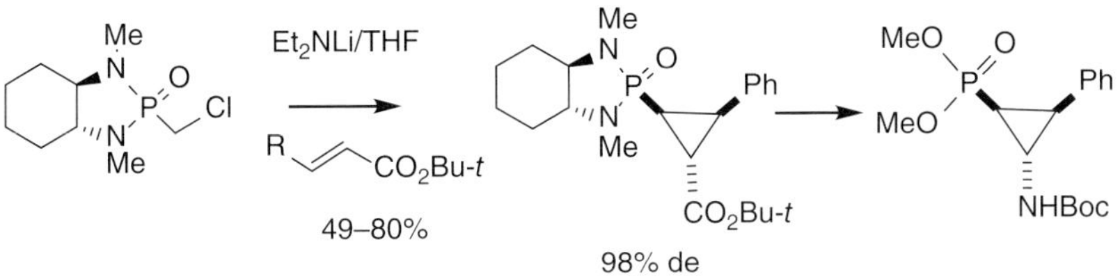

Scheme 3.13 Asymmetric cyclopropanation of chiral chloromethyl phosphonamide.

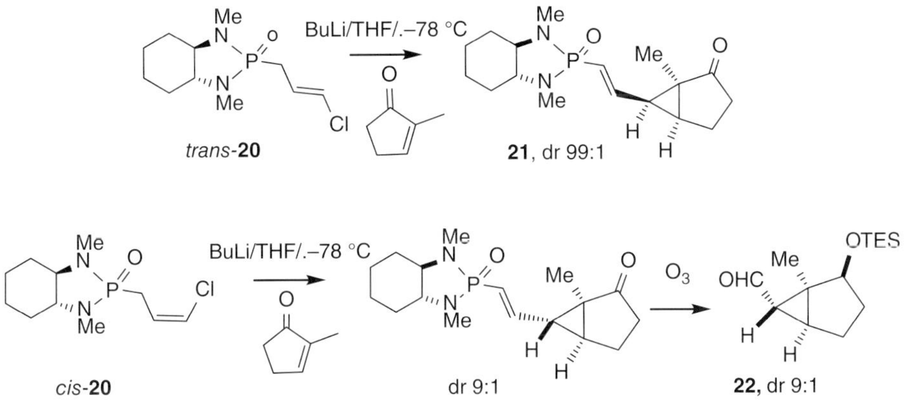

Scheme 3.14 1,4-Addition of 2-allyl-1,3,2-oxazaphosphorinane 2-oxide to cyclic enones.

enone led to the isomeric *exo,endo* product as the major isomer (>90 : 10). Stereoselective reduction of the carbonyl group ($NaBH_4$/MeOH), protection, and oxidative cleavage by ozonolysis afforded the aldehydes **22**, which constitute versatile cyclopropane chirons (Scheme 3.14) [32, 43, 44].

The cyclopropanation reaction was used in a wide range of substrates such as enones, lactones, lactams, and acyclic α,β-unsaturated esters. Oxidative cleavage led to products corresponding to the formal conjugative addition of an acetaldehyde or a propionaldehyde anion equivalent to unsaturated carbonyl compounds. The inclusion of HMPA was found to enhance the ratio of 1,4-addition and to improve the stereoselectivity in the case of 3-methylcyclopentanone. Mono-, di-, and trisubstituted cyclopentanones **23** were obtained as single diastereomers [42–44] (Scheme 3.15).

The stereocontrolled conjugate addition of anions derived from the chiral α-chlorophosphonoamides **12** to α-unsaturated esters led to the corresponding

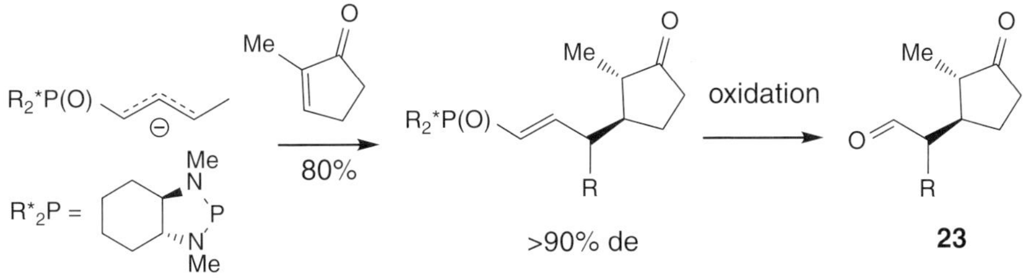

Scheme 3.15 Synthesis of substituted cyclopentanones **23**.

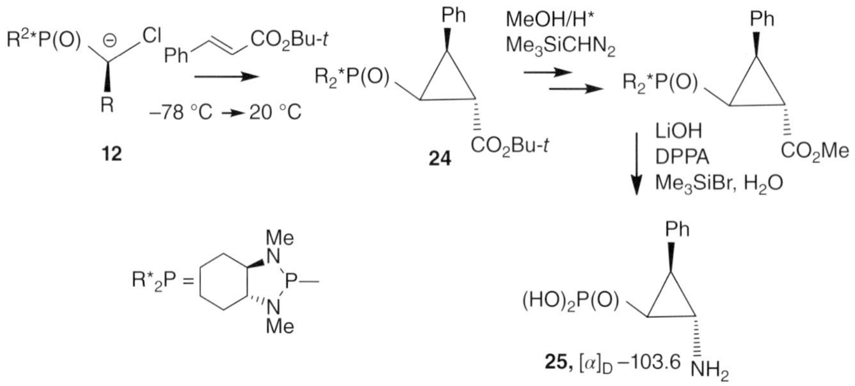

Scheme 3.16 3-Substituted cyclopropane 2-aminophosphonic acids **25**.

cyclopropane phosphonates **24** (dr 5 : 1 – 100 : 0). The obtained cyclopropanes were then converted into 3-substituted cyclopropane 2-aminophosphonic acids **25** (Scheme 3.16). Haynes has since described the enantioselective 1,4-addition of individual enantiomers of (*E*)-2-butenyl-*tert*-butylphenylphosphine oxide to 2-methyl-2-cyclopentenone and formulated a model for the asymmetric induction. *t*-Butyl(methyl)phenylphosphine oxide was lithiated with butyllithium and then treated with propylene oxide followed by BF_3-ether to provide a 1 : 2 mixture of diastereomers of the γ-hydroxyphosphine oxide **26**, which was converted into (*S*)- and (*R*)-allyl-*tert*-butylphenylphosphine oxides **27**. The lithiation of **27** and the treatment with 2-methylcyclopent-2-enone afforded the unsaturated diketone **28**, which was converted into the hydrindenone **29**, suitable for conversion into vitamin D analogs and their enantiomers (Scheme 3.17) [45, 46].

Claisen Rearrangement The carbanion-accelerated Claisen rearrangement (CACR) of allyl vinyl ethers has proved to be a reaction of interesting synthetic potential. The utility of various phosphonamide groups was examined in the context of the CACR [47, 48]. The *N*,*N*′-dibenzyl-1,3,2-diazaphospholidine group is the most optimal for the construction of the CACR precursor and the stereoselectivity of its rearrangement.

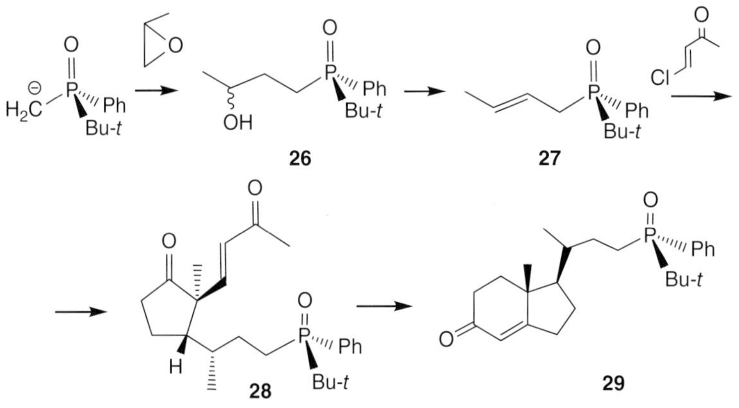

Scheme 3.17 Synthesis of hydrindenones **29**.

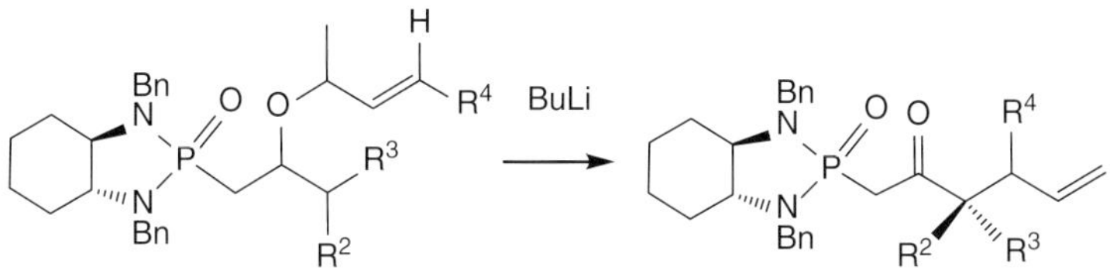

30 31

Scheme 3.18 Carbanion-accelerated Claisen rearrangement of allyl vinyl ethers **30**.

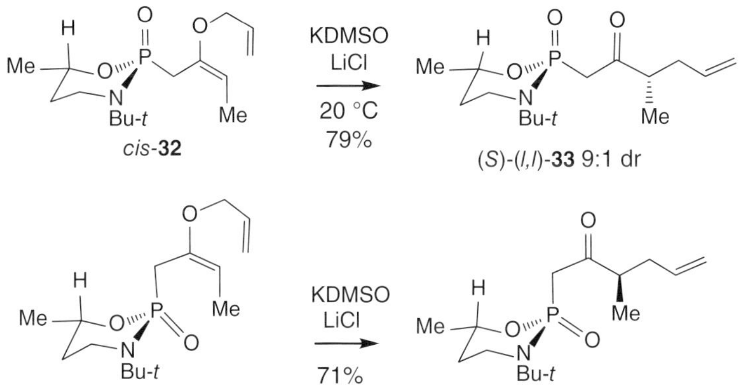

cis-**32** (*S*)-(*l*,*l*)-**33** 9:1 dr

trans-**32** (*S*)-(*u*,*u*)-**33** (*u*,*u*)/(*l*,*l*) = 92:8

Scheme 3.19 Claisen rearrangement of' cyclic *cis*- and *trans*-phosphonamides **32**.

For example, the treatment of the phosphonamides **30** with butyllithium leads to rearrangement into **31** with complete regioselectivity, in good yield, and with a high level of diastereoselectivity (>95% de) (Scheme 3.18). The CACR of allyl vinyl ethers bearing chiral 1,3,2-oxazaphosphorinanes takes place under extremely mild conditions (room temperature, 15 min) to afford γ,δ-unsaturated ketones with high levels of internal and relative diastereoselectivity. The relative diastereoselectivity with respect to the 1,3,2-oxazaphosphorinane ring was dependent upon the auxiliary structure and the reaction conditions (Scheme 3.19) [47]. All rearrangements of anions with lithium as the counterion gave excellent diastereoselectivities. In the absence of LiCl, there was no observed asymmetric induction in the anion-accelerated rearrangement. As the amount of LiCl was increased from 1 to 6 equiv., the dr's improved from approximately 2 : 1 to 9 : 1. The size of the nitrogen substituent was shown to be critical for high diastereoselectivity. Yamamoto described the catalytic enantioselective Claisen rearrangement of enolphosphonates for the synthesis of a wide range of α-ketophosphonate derivatives with contiguous tertiary and quaternary carbon centers in excellent yields and selectivities (70–90% yields and 90–95% ee). A plausible transition-state model **B** was proposed as the well-defined tetrahedral substrate/catalyst complex, in which bidentate chelation control results in excellent enantioface differentiation [48] (Scheme 3.20).

[2,3]-Wittig Rearrangement The [2,3]-Wittig rearrangement is a special class of [2,3]-sigmatropic rearrangements which involves oxycarbanions as the migrating

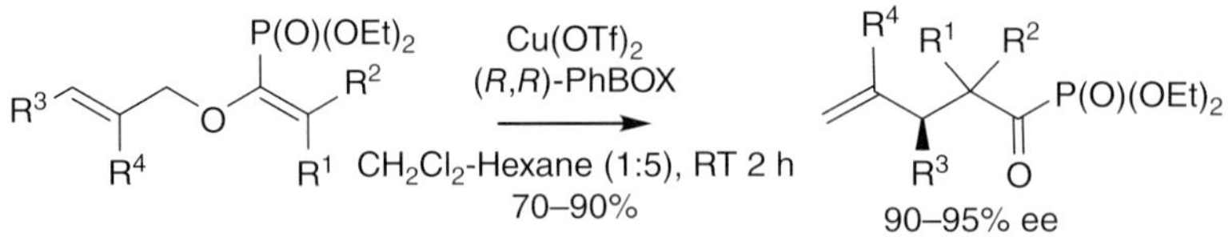

R^1	R^2	R^3	R^4	Yield (%)	ee (%)
Me	Me	Ph	$CH{=}CH_2$	90	97
$-(CH_2)_4-$		Ph	$CH{=}CH_2$	92	97
Me	Me	Me	$CH{=}CH_2$	90	90
Me	Me	1-Nphth	$CH{=}CH_2$	94	93
Me	Me	PhCH=CH	$CH{=}CH_2$	97	90
Me	Me	$4\text{-}C_6H_4Br$	$CH{=}CH_2$	94	96

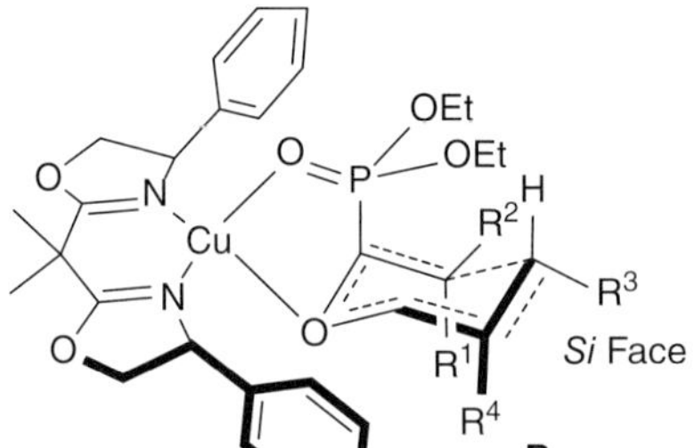

B

Scheme 3.20 Catalytic enantioselective Claisen rearrangement and transition state B.

terminus to afford various types of homoallylic alcohols. The most significant feature of the [2,3]-Wittig rearrangement is its ability for efficient diastereocontrol over the newly created stereogenic centers through the proper choice of the combination of substituents and substrate geometry. The [2,3]-Wittig rearrangement of chirally modified phosphonates proceeds with excellent diastereo- and enantioselectivity for allyloxymethyl and *Z*-2-butenyloxymethyl derivatives. Thus, deprotonation of 1,3,2-oxazaphosphorinate with butyllithium in THF at −70 °C generated the phosphorus-stabilized anion **34**, which underwent the [2,3]-Wittig rearrangement to afford a single diastereomer of hydroxy 3-butenyl-1,3,2-oxazaphosphorinanes **35** in good yield. The configuration of the hydroxyl-bearing stereocenters of compounds (*S*)-(+)- and (*R*)-(−)-**35** was determined by comparison of the sign of the specific rotation with compounds prepared by independent syntheses (Scheme 3.21) [49]. It was found that the configuration at the phosphorus controls the creation of the stereogenic centers in the side chain of newly formed rearranged product. Another interesting example of the [2,3]-Wittig rearrangement was described by Collignon for allyloxymethylphosphonate **34b** (Scheme 3.25). A diastereoselectivity of up to 90% was observed in the rearrangement of the lithiated derivative of the phosphonate **34b** (R* = MntO) on using the chiral dimenthylphosphinyl ester group as the stereodirecting auxiliary. After treatment with an excess of butyllithium in THF at −78 °C, dimenthyl allyloxymethylphosphonate **34b** underwent complete [2,3]-Wittig rearrangement giving, after low-temperature acid hydrolysis of the reaction mixture and subsequent work up, (1-hydroxy-3-buten)-1-yl phosphonate **35b**, isolated in 95% yield as a mixture of two diastereomers in a 96 : 4 ratio [50] (Scheme 3.21).

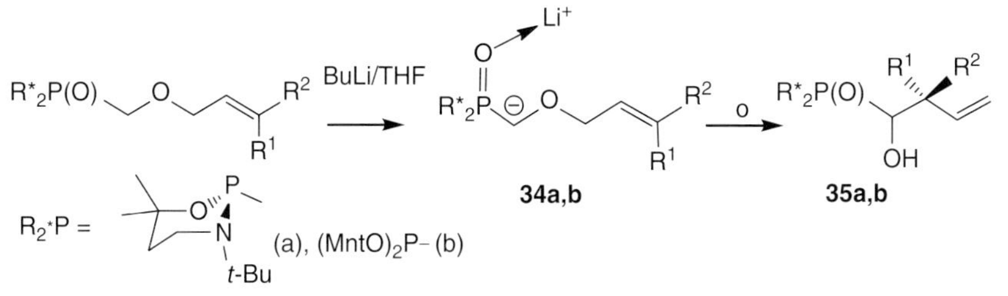

R^*_2	R^1	R^2	Yield (%)	dr	References
$-OC(Me)_2(CH_2)_2N(Bu\text{-}t)-$	H	H	87	>100:1	[49]
$-OC(Me)_2(CH_2)_2N(Bu\text{-}t)-$	Me	H	73	>100:1	[49]
$-OC(Me)_2(CH_2)_2N(Bu\text{-}t)-$	H	Me	81	1.6:1	[49]
$(MntO)_2$	Me	Me	95	96:4	[50]

Scheme 3.21 The [2,3]-Wittig rearrangement.

3.3 Enantioselective Olefination

Enantioselective olefination represents an interesting example of chemical properties of P-stabilized carbanions. The Horner–Wittig as well as the Wittig reactions between P-ylides or P-stabilized carbanions and aldehydes or ketones is an important and practical method for the construction of carbon–carbon double bonds [51, 52]. Different chirally modified phosphinates, phosphonates, phosphonamides, phosphinothionic amides, phosphine oxides, oxazaphosphorinanes, oxathiaphosphorinanes, and phosphoranes have been employed for asymmetric olefination with variable success. The first of these was performed by Bestmann, who employed chiral P-ylides for the synthesis of optically active allenes [51]. Since the Wittig-Horner olefination does not create a new sp^3 carbon center, efforts to develop an asymmetric version of the reaction were focused mainly on alkylidenecycloalkanes with axial chirality. Asymmetric carbonyl olefination may be accomplished by means of differentiation of enantiotopic carbonyls or desymmetrization of prochiral carbonyl compounds, which are highly dependent on the structure of the carbonyl compound [52]. The differentiation of enantiotopic carbonyls is based on differentiation of enantiotopic carbonyl groups in symmetrical molecules such as *meso*-compounds, and is therefore referred to as the desymmetrization of symmetric organic molecules (Scheme 3.22).

The desymmetrization concept was extended to intermolecular reactions by Fuji *et al.* [52]. For example, the chiral phosphonate **37** was used for desymmetrization of *meso*-diketone **36**. The selectivity of the reaction depended on the anion, temperature, solvent, and substituent (Scheme 3.23).

When symmetrically substituted monoketones are used as substrates, asymmetric carbonyl olefination causes desymmetrization to give olefinic products **40** with axial chirality. Thus, optically active (*ortho*-methoxyphenyl)phenylbutylphosphine oxide was used to create the new stereogenic carbon atom of the carbon framework, which was built up by stereoselective reaction leading to the formation of the chiral cyclic

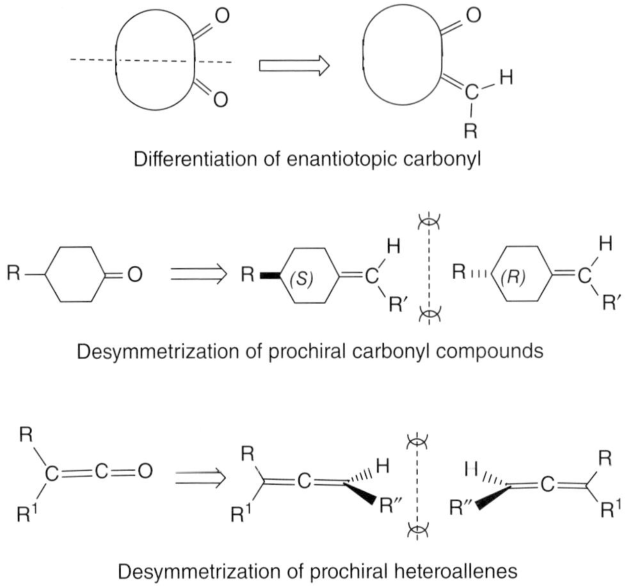

Scheme 3.22 General approaches for asymmetric carbonyl olefination.

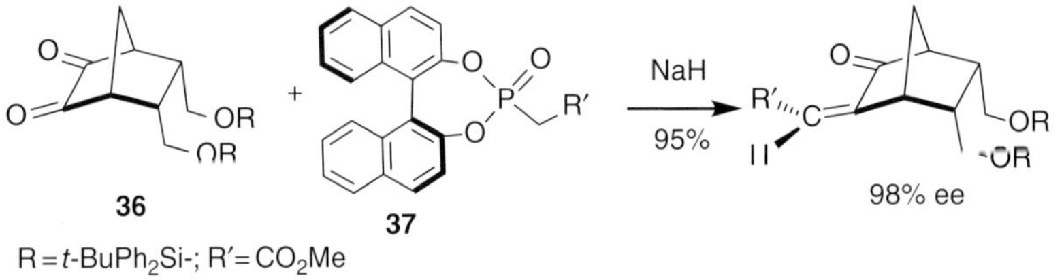

Scheme 3.23 Differentiation of enantiotopic carbonyls in the *meso*-diketone **36**.

phosphine oxide **38**. Elimination of the chiral auxiliary in a Horner–Wittig reaction gives compounds **39**. Optically active phosphine oxide **39** was converted into the optically active (*S*)-(+)-alcohol **40** [53] (Scheme 3.24).

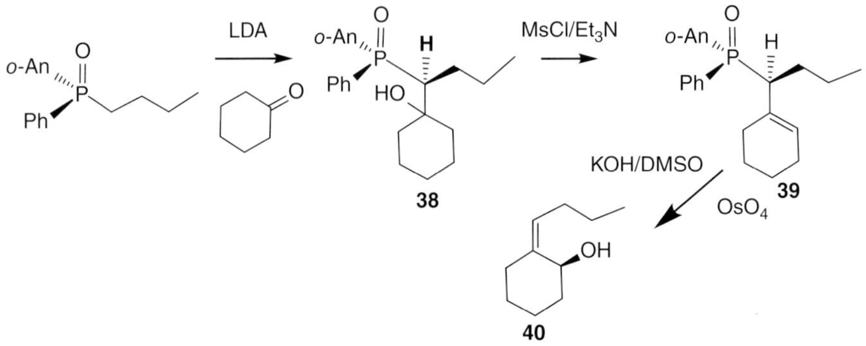

Scheme 3.24 Synthesis of optically active (*S*)-(+)- alcohol **40**.

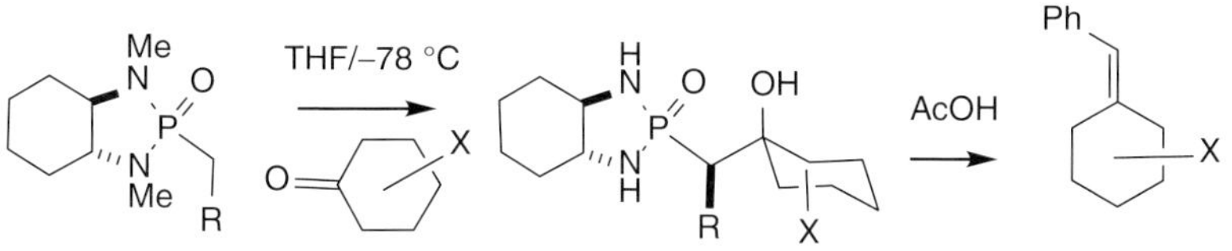

41

R = Me, Ph, vinyl; X = 4-*t*-Bu, 4-Me; 3-Me

42, 90–96% ee

Scheme 3.25 Preparation of axially chiral asymmetric alkenes **42**.

The *trans*-1,2-diaminocyclohexane phosphonates **41** were successfully applied for the preparation of axially chiral asymmetric alkenes (Scheme 3.25). Owing to the C_2 symmetry, the corresponding stabilized α-carbanions exhibited diastereofacial bias in the reaction with carbonyl electrophiles [54]. The treatment of alkylcyclohexanones with (*R,R*)- or (*S,S*)-chiral P-carbanions **41** led to enantiomerically pure allylidene, benzylidene, and propylidene alkylcyclohexanes **42**, which are suitable for incorporation in a liquid-crystal-based optical switch [55]. In the case of reagent **41**, the intermediate 2-hydroxyphosphonates were isolated and these were converted into the alkenes **42** [55, 56] (Scheme 3.25). Another chiral phosphoramidate **44**, which possesses stereogenic centers at both phosphorus and carbon atoms, was examined in asymmetric olefination reactions [57]. As with the phosphonic bis(amide) **41**, an additional elimination step to obtain the olefin **44** from the rather stable intermediate was required [11]. This conversion was best achieved by the action of trityl triflate. Overall, the process of asymmetric olefination using a combination of reagents of this type furnished asymmetric olefins with high enantioselectivity and in good chemical yield [58–63] (Scheme 3.26). The chiral P-stabilized phosphonate **37**, possessing optically active BINOL auxiliary, was effective as an asymmetric inducer in the desymmetrization of carbonyl compounds. It was found that the addition of zinc chloride increased the enantioselectivity and chemical yields [58] (Scheme 3.27). The chiral BINOL phosphonate **37** was also

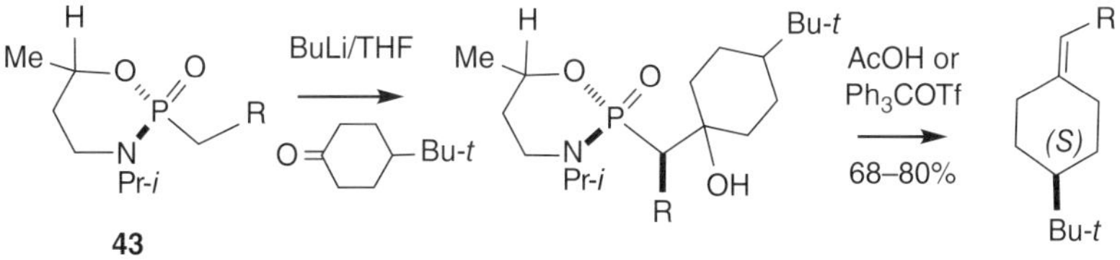

43

R = Me, *t*-Bu, Ph, CO_2Bu-*t*

(*S*)-**44**, 99% ee

Scheme 3.26 Asymmetric olefination of chiral phosphoramidates **44**.

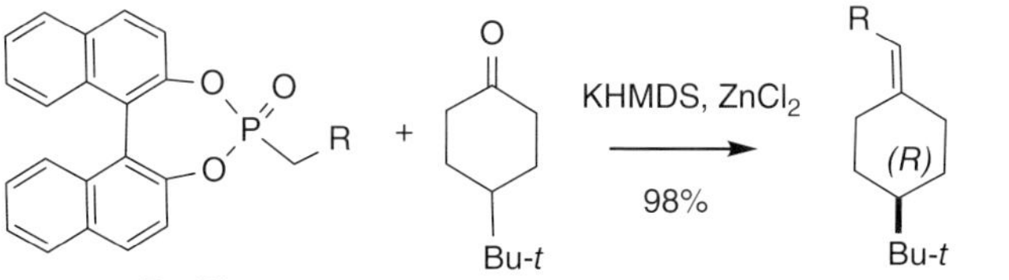

37, R = Ph

(*R*)-**44**, 98% ee

Scheme 3.27 BINOL phosphonate **37** as olefination reactant.

used for desymmetrization of bicyclo[3.3.0]octane derivatives **45a,b**. The observed stereochemistry of the alkene **44** was explained by considering an initial approach of the nucleophile to the W-shaped conformation of bicyclo[3.3.0]octanone **45b** in which steric interaction between the reagent and the substrate was minimized [55, 56] (Scheme 3.27). Olefination based on P-stabilized carbanions was employed in the construction of di- and trisubstituted double bonds in the total synthesis of polyoximic acid [58], jerangolid A [59], and ambruticin S [60]. Headley reported the application of P-stereogenic phosphines as chiral reagents in the asymmetric aza-Wittig reaction [61] (Schemes 3.28–3.30).

Hannessian used these reactants in total synthesis of the marine toxin (+)-acetoxycrenulide, which was isolated from the small brown seaweed of the family *Dictyotaceae* and from the sea hare (Scheme 3.30) [64, 65].

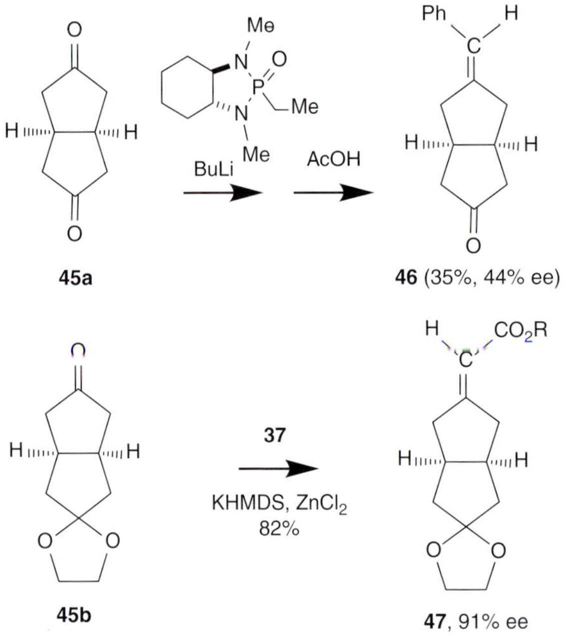

Scheme 3.28 The desymmetrization of bicyclo[3.3.0]octane derivatives **45a,b**.

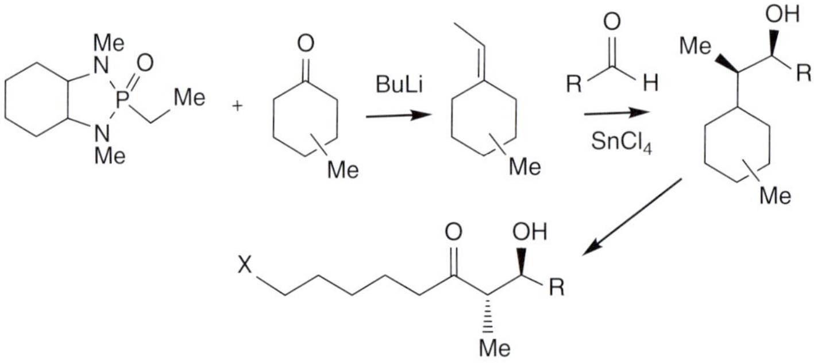

Scheme 3.29 Construction of chiral alicyclic cyclohexenes.

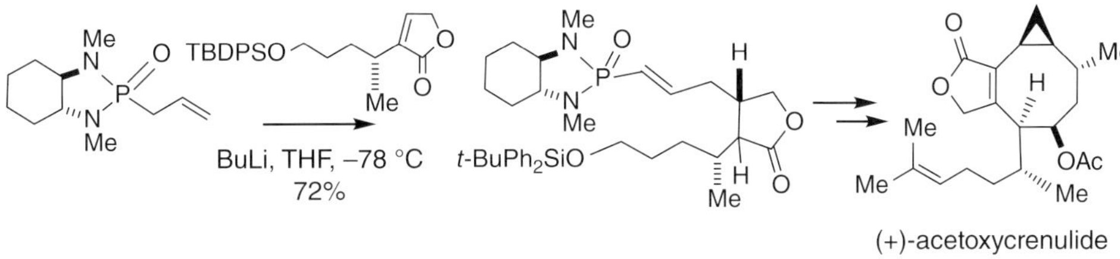

Scheme 3.30 Synthesis of the marine toxin (+)-acetoxycrenulide.

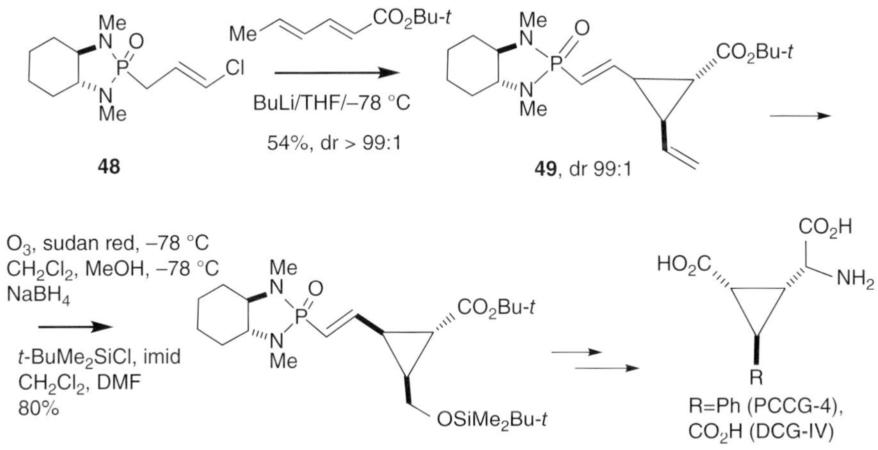

Scheme 3.31 Synthesis of constrained bioisosteres of L-glutamic acid **45**.

Pellicciari *et al.* reported the synthesis of constrained bioisosteres of L-glutamic acid DCG-IV. The phosphonocyclopropylamino acid DCG-IV was designed as an analog to PCCG-4 by replacing a carboxylic acid with a phosphonic acid moiety. The conjugate addition of the anion of **48** to (*E*)-*tert*-butyl cinnamate proceeded with excellent stereocontrol, and adduct **49** was isolated as a single diastereomer (Scheme 3.31) [42, 66, 67].

Many other examples of the total syntheses of natural products and bioactive compounds with application of P-carbanion technology have been described in the literature [35, 68–71]; also, for example, hexahydropyrrolo[2,3-*b*]indole alkaloids [63], various marine alkaloids [64], and prostacyclin analogs [65].

3.4 Stereoselective Addition of Phosphorous Nucleophiles to C=X Bonds

The rapid development of the chemistry and biology of phosphonic acid derivatives over the last decade has been determined by the development of highly effective methods for their preparation. Chiral phosphonic acids can be prepared by various routes. The main method for the synthesis of phosphonates is the phosphonylation of carbonyl compounds, mainly via the phospha-aldol reaction, phospha-Mannich reaction, or phospha-Michael reaction [72–79].

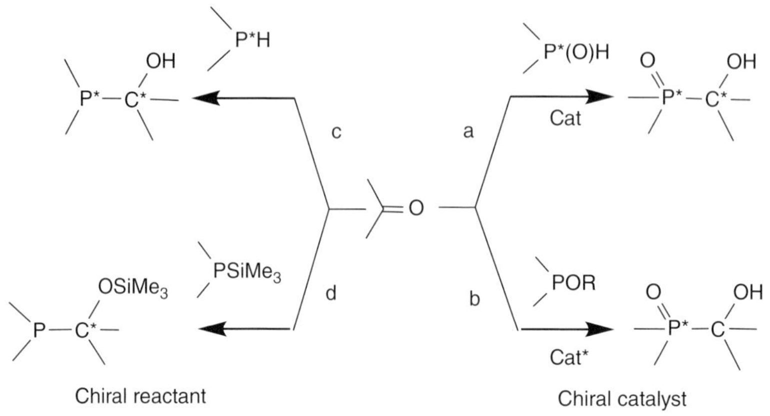

Scheme 3.32 Versions of phospha-aldol reaction.

Phospha-aldol reaction. Several types of phospha-aldol reactions are possible (Scheme 3.32): (a) the reaction of dialkylphosphites with carbonyl reagents proceeding in the presence of a base catalyst, which shifts the P(O)H⇌P–OH tautomeric equilibrium toward the H(O)-form; (b) the addition reaction of phosphoric acid esters to carbonyl compounds proceeding in the presence of proton-donating reagents or Lewis acids; (c) the addition of secondary phosphines, or (d) the addition of secondary silylphosphines to carbonyl compounds also represent versions of the phospha-aldol reaction.

The addition of diethylphosphite to carbonyl compounds under basic conditions was first described by Vasily Abramov in 1950 [73]. The asymmetric phospha-aldol reaction has been studied intensively, because α-hydroxyphospohonates are important components of enzyme inhibitors. The preparation of hydroxyphosphonates also used catalytic methods including metallocomplex catalysis (see Chapter 4), organo-catalysis (Chapter 5), and biocatalysis (Chapter 6), leading to the formation of functionalized molecules with high enantiomeric purity and, therefore having high potential in synthetic chemistry [72–75].

Phospha-Mannich-type reactions. The synthesis of enantioenriched α-aminophosphonates and aminophosphonic acids by means of asymmetric catalytic hydrophosphonylation of imines has attracted constant interest. A special case of the phospha-Mannich reaction is the Kabachnik–Fields reaction representing a one-pot, three-component procedure, which includes carbonyl compound, amine, and dialkyl phosphite. The three-component Kabachnik–Fields (phospha-Mannich) reaction involving the condensation of primary or secondary amines, oxo compounds (aldehydes or ketones), and >P(O)H species represents a good choice for the synthesis of α-aminophosphonates [76, 77]. Several types of phospha-Mannich reaction are possible (Scheme 3.33): (a) the reaction of dialkylphosphites with imino-compounds; (b) the addition reaction of phosphoric acid esters to imines proceeding in the presence of proton donating reagents or Lewis acids; (c) the addition of secondary phosphines; and (d) the addition of secondary silylphosphines to imines.

Phospha-Michael reaction. The phospha-Michael reaction is a nucleophilic addition of P(III)-phosphoric anions to an activated multiple bond and is one of the most useful methods for the formation of P–C bonds [78, 79], as P(III)-nucleophiles can be used as

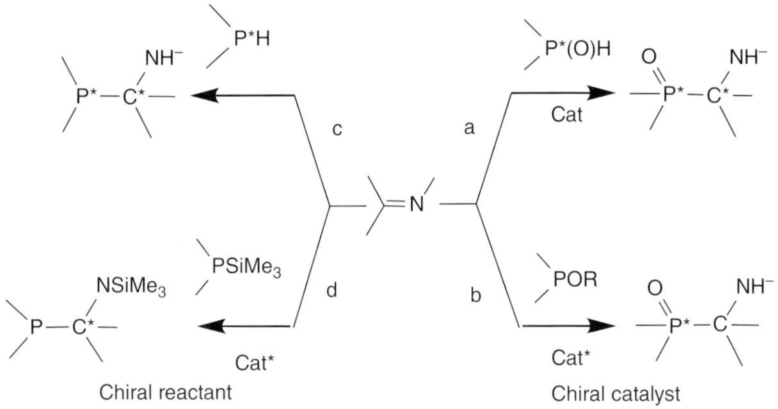

Scheme 3.33 Versions of phospha-Mannich reaction.

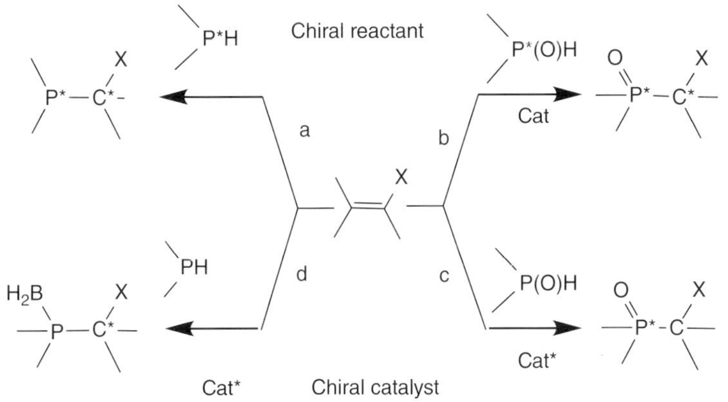

Scheme 3.34 Versions of phospha-Michael reaction.

anions of $R_2P(O)H$ or R_2PH compounds. A particular version of the phospha-Michael reaction is the Pudovik reaction that includes the addition of $R_2P(O)H$ to an activated C=C bond, catalyzed by Brønsted bases or Lewis acids. The reaction was described by Pudovik and Arbuzov [81, 82]. The asymmetric phospha-Michael reaction can be accomplished by two main routes: (a,b) substrate-controlled diastereoselective addition using chiral starting materials and chiral auxiliaries or (c,d) chiral catalytic enantioselective addition (Scheme 3.34).

The catalytic version of the phospha-Michael reaction can be organometallic and organocatalytic. Organometallic catalysis is the most convenient for primary and secondary phosphine additions to activated C=C bond with the formation of chiral tertiary phosphines [78, 79].

3.4.1 Phospha-Aldol Reaction

The main method for the synthesis of hydroxyphosphonates is the phosphonylation of carbonyl compounds, mainly via the phospha-aldol reaction (the Abramov reaction) [72–75]. The addition reaction of phosphorus acid esters with carbonyl compounds

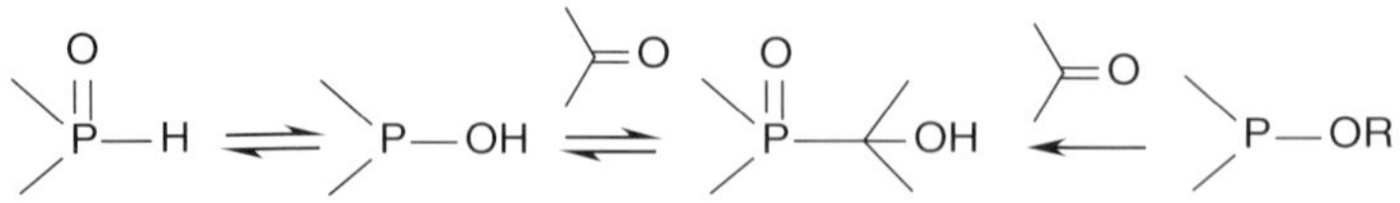

Scheme 3.35 Addition reaction of phosphonic acid esters to carbonyl compounds.

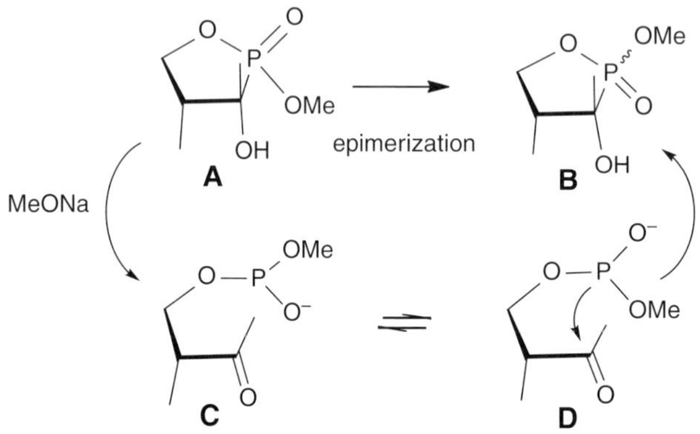

Scheme 3.36 Retro-Abramov reaction.

involves two steps: first, the formation of a P–C bond and, second, the cleavage of the ester function with the formation of the phosphonyl group. The first step of the addition reaction is reversible [83–85] (Scheme 3.35).

In the presence of strong bases, hydroxyalkylphosphonates dissociate with the formation of the dialkylphosphites and carbonyl compounds. This transformation is known as the retro-phospha-aldol reaction (or retro-Abramov reaction) [83]. Under the action of sodium methoxide in methanol, diastereomerically pure hydroxyphosphonates undergo racemization. Presumably, the reaction mechanism includes cleavage of the P–C bond and epimerization of the chiral P-anion, which brings about the formation of a mixture of diastereomers upon subsequent cyclization. For example, the diastereomerically pure 3,4-dimethyl-2-methoxy-2-oxo-1,2-oxaphospholan-3-ol in a solution of sodium methoxide in methanol afforded the diastereomeric mixtures **A** and **B** (Scheme 3.36) [86, 87, 88].

Chiral substrates in the asymmetric phospha-aldol reaction can be chiral phosphonites, having a stereogenic phosphorus atom, or phosphites, derived from chiral alcohols, amino alcohols, or amines. The addition of achiral aldehydes to chiral trialkylphosphites or chiral dialkylphosphites results, as a rule, in the formation of α-hydroxyphosphonates in good yields and moderate stereoselectivity [80, 89–98]. The addition of the chiral lithium derivatives of diamidophosphites **50** to aldehydes led to the formation of α-hydroxyalkylphosphonamides **51** in good yield and moderate stereoselectivity. However, the increase in the steric bulk of the alkyl groups in molecules of amidophosphite and aldehyde increased the stereoselectivity of the reaction (Scheme 3.37) [10, 87]. The nature of the chiral alkoxy groups at the phosphorus atom has a pronounced effect on the stereoselectivity of the phospha-aldol condensation. For example, high asymmetric inductions were obtained using a readily available TADDOL auxiliary. Subsequent racemization-free removal of the chiral auxiliary led to the desired α-hydroxyphosphonic acids in very good yields and high enantiomeric purity as the (*R*)-enantiomers [85]. The reaction of tris(glucofuranosyl)

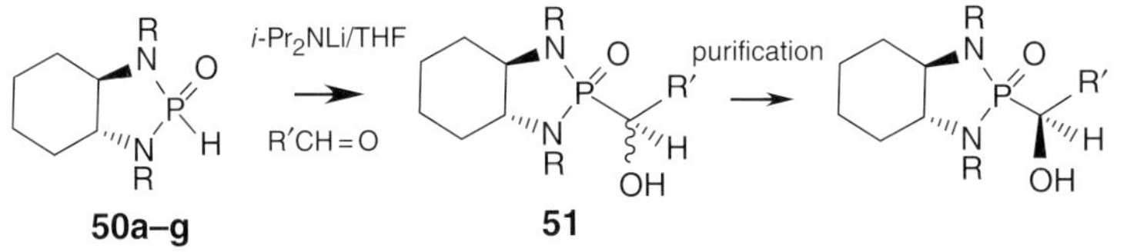

Compounds	R′	R	dr	References
51a	PhCH=CH	2-MeC_6H_4	1:1	[89]
51b	CH_2=CH	Bn	1.4:1	[10,89]
51c	PhCH=CH	l-Naphth-CH_2	1.5:1	[10,89]
51d	(*S*)-PhMeCH	Ph	1.8:1	[90]
51e	PhCH=CH	Bn	4.0:1	[10,89]
51f	PhCH=CH	*t*-BuCH_2	7.9:1	[80]
51g	Ph	*t*-BuCH_2	25:1	[80,89,90]

Scheme 3.37 Addition of chiral lithium diamidophosphite to aldehydes.

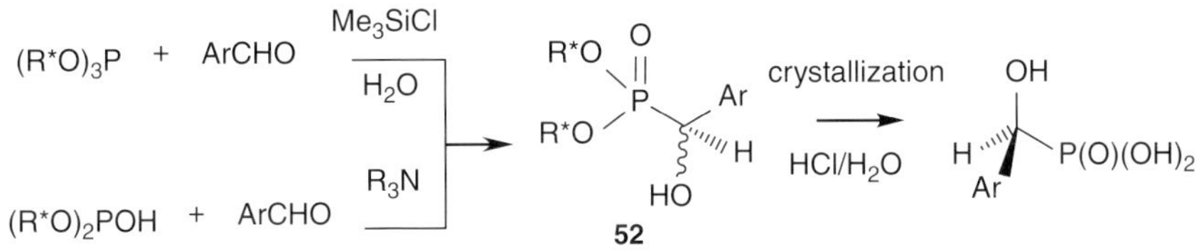

R*O = GF = (–)-1:2; 5:6-diisopropylidene-D-glucofuranosyl, Mnt = (1*S*,2*R*,5*S*)-Menthyl, Brn = *endo*-Bornyl

Scheme 3.38 Reaction of chiral secondary phosphites with aldehydes.

phosphite with benzaldehyde in the presence of chlorotrimethylsilane proceeded with good stereoselectivity to give hydroxy phosphonates with a diastereomeric excess (de) of 84%. Meanwhile, trimethyl and tribornyl phosphites react with aldehydes with low stereoselectivity, giving rise to mixtures of hydroxy phosphonate diastereomers **52**. The reaction of dimenthyl phosphite with aromatic and aliphatic aldehydes in the presence of 1,8-diazabicyclo[5.4.0]-undec-7-ene (DBU) as the catalyst has a moderate stereoselectivity. The diastereomers **52** were separated and, after hydrolysis, were converted into acids with enantiomeric purity 96–98% ee (Scheme 3.38).

Example of phospha-aldol reaction (Scheme 3.37) [80]. A solution of diisopropylamine (1.5 mmol) in THF (6 ml) was cooled to −60 °C, and *n*-butyllithium (1.2 mmol) was added. After 30 min, a solution of diamide **50** (R = Ph) (1.2 mmol) and then benzaldehyde (1.29 mmol) were added slowly to reaction mizture. After stirring for 4.5 h at −60 °C, the reaction mixture was quenched with aqueous ammonium chloride (1 ml) and diluted with $CHCl_3$ (50–60 ml). The solution was washed with H_2O (2 ml × 25 ml), dried (Na_2SO_4), and concentrated *in vacuo* to give the crude product (yield 90%, 25 : 1 dr). The product **51d**, recrystallized from ethyl acetate, was obtained in yield 49%, 94% de, mp 186–187 °C, δ_P 39.

(S)-(−)-1-Hydroxybenzylphosphonic acid. Aqueous 4 N HC1 (1 ml) was added to a solution of the phosphonamide **51d** (1 mmol) in dioxane (2 ml). The solution was stirred at room temperature and the reaction was monitored by ^{31}P NMR spectroscopy until complete. The solvent was evaporated *in vacuo* and the residue was dissolved

in water and passed through an ion exchange column (Amberlite IR-120(+)-I) eluting with water. The first 50 ml fraction was evaporated to give the phosphonic acid ^{31}P NMR (D_2O) δ_P 21,6. The phosphonic acid was dissolved in ethanol, and cyclohexylamine (0.1 ml) was added. The precipitated salt was collected by filtration (60%) (mp = >200 °C (MeOH, Et_2O); $[\alpha]_D$ ~13.8 (c = 0.1, aqueous MeOH); ^{31}P NMR (D_2O) δ_P 15.9 ppm).

Di[(1R,2S,5R)-menth-2-yl] 1-hydroxybenzylphosphonate **52**. Benzaldehyde (0.05 mol) and

a) 1.5 ml of trimethylchlorosilane were added to trimenthyl phosphite (0.05 mol) at 0 °C. The mixture was stirred for 1 h at 0 °C, then warmed to ambient temperature and left for 1–2 h [92]. The ^{31}P NMR spectrum showed the presence of only two signals at δ_P 20.36 and 20.08 ppm in the ratio of 3 : 1. The solvent was removed under reduced pressure. The residue was mixed with silica gel and was then washed with a 1 : 1 mixture of ethyl acetate-hexane. The solvent was removed under vacuum, the residue dissolved in hexane, and the solution placed in a refrigerator. After 2 days a crystalline product was obtained (yield: 60%, mp 139 °C, $[\alpha]_D{}^{20}$ 188.9 (c = 1, toluene), δ_P 23.71).
b) The solution of dimenthyl 1-phenylhydroxymethylphosphonate (1 g) in 50 ml of dioxane was placed into a flask and 25 ml of 6 N hydrochloric acid was added. Then the reaction mixture was left for 3–4 days at 80 °C and the course of hydrolysis was monitored by ^{31}P NMR spectroscopy. When the reaction was completed, the solvent was evaporated, and the residue was dissolved in alcohol and excess cyclohexylamine was added (about 1.5 ml). The precipitate of the dicyclohexylammonium salt of (*R*)-(−)-1-phenyl(hydroxymethyl)phosphonic acid was filtered off. The yield of dicyclohexylammonium salt of (1)-(*S*)-hydroxybenzylphosphonic acid was 70%, mp 226 °C, $[\alpha]_D{}^{20}$ −14.0 (aqueous MeOH).

The 2-triorganosiloxy-1,3,2-oxazaphospholidines undergo the Abramov reaction with benzaldehyde at room temperature to afford new esters in high yield and with good stereoselectivity. Recrystallization of diastereomeric mixtures from pentane afforded siloxyphosphonate esters **53–55** as white crystalline solids in up to 88% isolated yield and 95% isomeric purity [95, 96]. The reaction is kinetically controlled and the transfer of the silyl group to the oxygen is intramolecular, which results in retention of the relative configuration at the phosphorus atom (Scheme 3.39).

The reaction of achiral phosphites with chiral aldehydes proceeded in general also with moderate stereoselectivity [92–94]. For example, the reaction of dialkylphosphites with 2,3-*O*-substituted-D-glyceraldehyde, in the presence of such catalysts as triethylamine, lithium, or caesium fluorides, furnished diastereomeric mixtures of hydroxyphosphonates (*R*,*R*)- and (*S*,*R*)-**56** in the ratio of 45 : 55–35 : 55 (Scheme 3.40). Application of lithium diethylphosphonate allowed to improve the stereoselectivity of the reaction [98] and to obtain the phosphonates (1*R*,2*R*)- and (1*S*,2*R*)-**56** in a pure state after recrystallization. The reaction of dialkylphosphites with the Garner aldehyde in the presence of triethylamine resulted in the (1*R*,2*S*)-2-amino-1,3-dihydroxypropylphosphonate **57** in 80% de. At the same time, the reaction in the presence of titanium(IV) isopropoxyde as a catalyst yielded the diastereomeric mixture (1*S*,2*S*)-**57**/(1*R*,2*S*)-**57** in a ratio of 1 : 1, which was separated by column chromatography (Scheme 3.40) [99–101]. The addition of dialkyl phosphites to (*S*)-*N*,*N*-dibenzylphenylglycinal catalyzed by triethylamine resulted in (1*S*,2*S*)-1-hydroxy-2-aminophosphonate **59**, whereas the reaction of

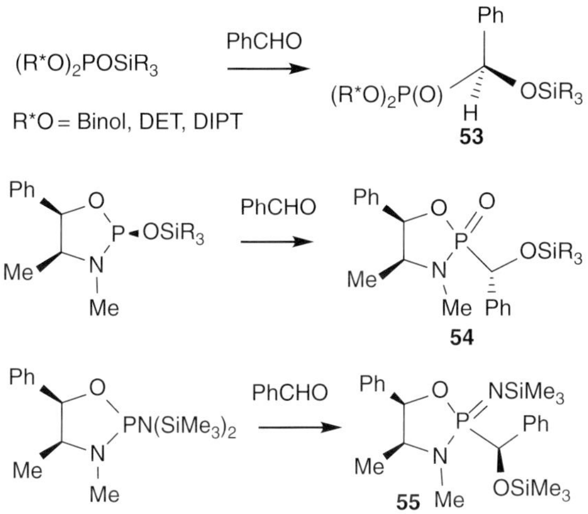

Scheme 3.39 Phospha-aldol reaction of chiral silylphosphites with aldehydes.

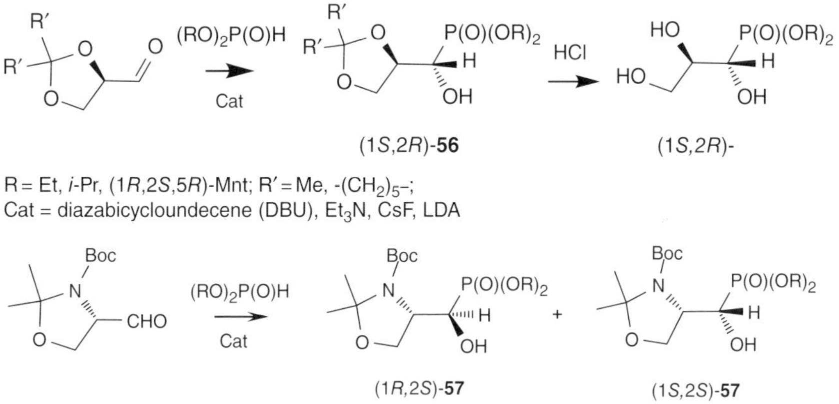

Scheme 3.40 Reaction of dialkylphosphites with chiral aldehydes.

dialkylphosphites with (*S*)-*N*-Boc-phenylglycinal under the same conditions, yielded the (1*S*,2*R*)-diastereomer **58** (Scheme 3.41)[102]. These results were rationalized by modeling of transition states in the addition of dialkyl phosphites to aldehydes. In the (*S*)-*N*-Boc-derivative, the intramolecular hydrogen bond stabilizes the conformation **A** and the dialkyl phosphite attacks the *Si* face of the carbonyl group, thus leading to the formation of the *syn*-adduct (model **A**). In the presence of triethylamine, the chelation is not possible, therefore the dialkyl phosphite preferentially attacks the *Re*-face of the carbonyl group in the (*S*)-aldehyde (model **B**). Addition of a metalated phosphite (Li, Mg, or Ti) to the (*S*)-aldehyde led to an increase in the *syn*-diastereomer due to the involvement of the chelated conformation **C** (Figure 3.1) . The reaction of (*S*)-*N*,*N*-dibenzylphenylglycinal with dimenthyl and dibornyl phosphite followed by hydrolysis afforded optically pure 2-amino-1-hydroxyalkylphosphonic acid (1*S*,2*S*)-**60**, whereas the addition of the same aldehyde to tris(trimethylsilyl) phosphite resulted in the (1*R*,2*S*)-**60**, which was isolated in the crystalline state (Scheme 3.42) [106]. Phosphonate analogs of natural anticancer substances, the taxoids, and fragments of taxoids

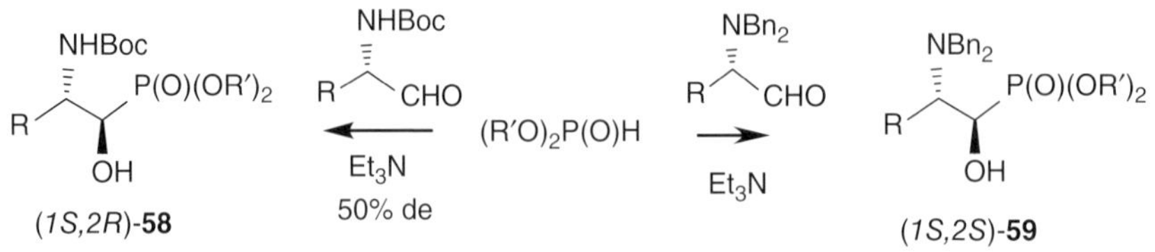

R′	R	X	Cat	Yield (%)	dr[a]	References
Me	Ph	H	Et_3N	62	56	[102]
Me	Ph	Li	—	100	14	[102]
Et	Bn	$SiMe_3$	$TiCl_4$	63	20	[103,104]
Et	Bn	$SiMe_2Bu$-*t*	$TiCl_4$	86	96	[103]
Et	Bn	$SiMe_3$	SnCl	51	48	[104]
Et	*t*-$BuOC_6H_4CH_2$	H	KF	82	60	[104]
Et	Bu-*i*	H	CsF	92	42	[104,107]

Scheme 3.41 Phosphonylation of chiral α-aminoaldehydes.

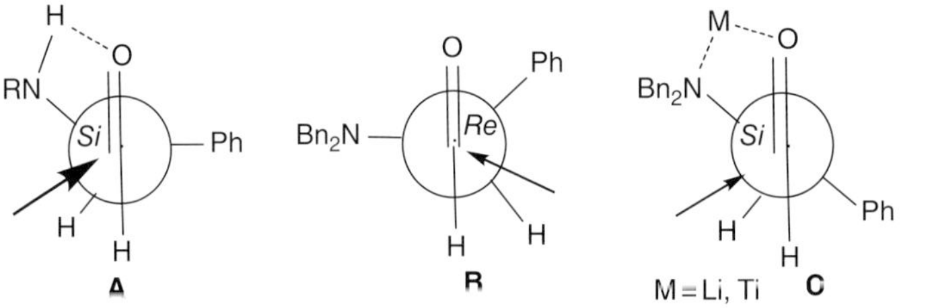

Figure 3.1 Modeling of transition states in the addition of dialkyl phosphites to aldehydes.

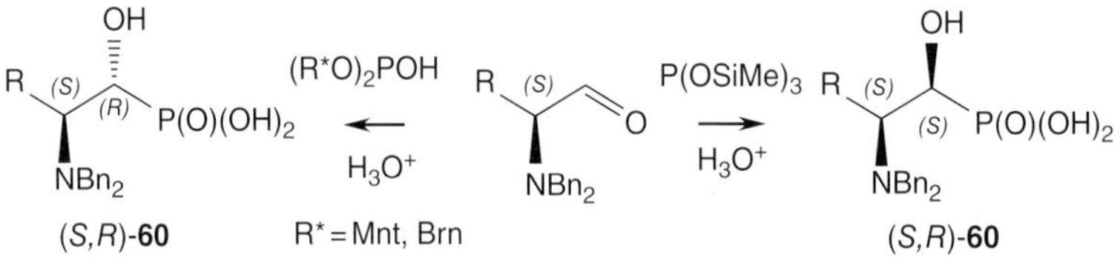

Scheme 3.42

have been synthesized with the use of this methodology [107]. The chiral phosphonate analogs of the paclitaxel side chain were obtained by reaction of diethylphosphite with chiral α-aminoaldehydes (Scheme 3.42) [102–106, 108, 109].

Di[(1R,2S,5R)-menthyl] (S)-[hydroxy[(4R)-2,2-dimethyl-1,3-dioxolan-4-yl]methyl] phosphonate (1S,2R)-**56**. Catalyst (1-2 drops of DBU or 25 mol% of aluminium-lithium bis(binaphthoxide) (ALB)) was added to a mixture of 0.01 mol of dimenthyl hydrogen phosphite and 0.01 mol of glyceraldehyde acetonide in 5 ml of THF at 0 °C. The resulting mixture was left to stand at this temperature for 12 h. The ^{31}P NMR spectrum of the reaction mixture showed the presence of two diastereomers (δ_P 19.98 and 20.9 ppm) in the above-mentioned ratio [*i.e.* in the ratio of 45 : 55–35 : 55 (Scheme 3.40)]. An optically pure (*S*,*R*) diastereomer was isolated by crystallization from acetonitrile (yield 50%, mp 98–100 °C, $[\alpha]_D^{20}$ −65 (c = 2, $CHCl_3$); ^{31}P NMR ($CDCl_3$): δ_P 20.9 ppm).

(1S,2R)-(1,2,3-Trihydroxypropyl)phosphonic acid. A solution of 0.005 mol of compound (1*S*,2*R*)-**56** in a 1 : 1 mixture of 40% hydrochloric acid and dioxane was left to stand at 80 °C for 48 h and then thoroughly evaporated in a vacuum. The residue was dissolved in 4 ml of ethanol, and 0.01 mol of cyclohexylamine was added. The cyclohexylammonium salt that precipitated was filtered off, yield 65%, mp > 200 °C (decomp.). ^{31}P NMR spectrum (CD_3OD), δ_P 18.1 ppm.

The phospha-aldol reaction of chiral aldehydes with dialkyl phosphites was used for the stereoselective syntheses of various biologically active compounds. For example, Patel *et al.* [110] prepared the tripeptidyl α-hydroxyphosphonates **61**, which are highly effective renin inhibitors (Scheme 3.43). Inhibitors of HIV-protease **4** [111, 112] were prepared as a mixture of three diastereomers in 3.4 : 1.7 : 1 ratio because of fast racemization of the chiral aldehyde under coupling conditions (Scheme 3.44). Optically active α-hydroxy and α-fluorophosphonates **63** and **64**, which represent interest as chiral synthetic blocks for preparing biologically active compounds and phosphorus

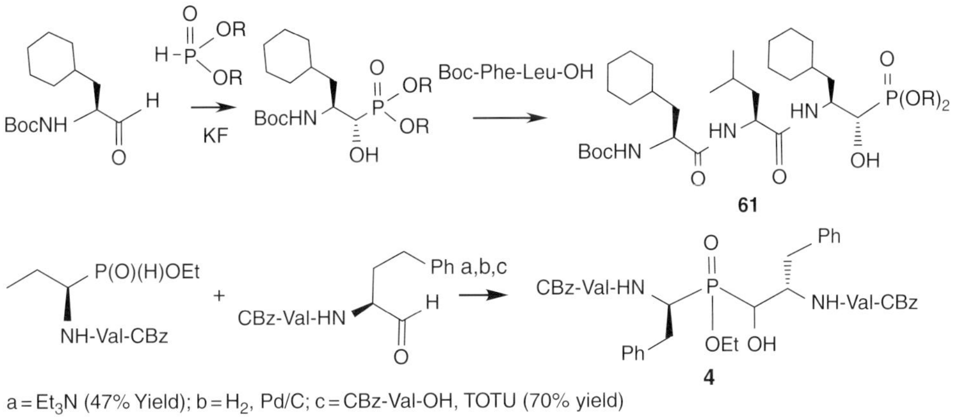

Scheme 3.43 Preparation of enantiomerically pure phosphonate analogs of tripeptides.

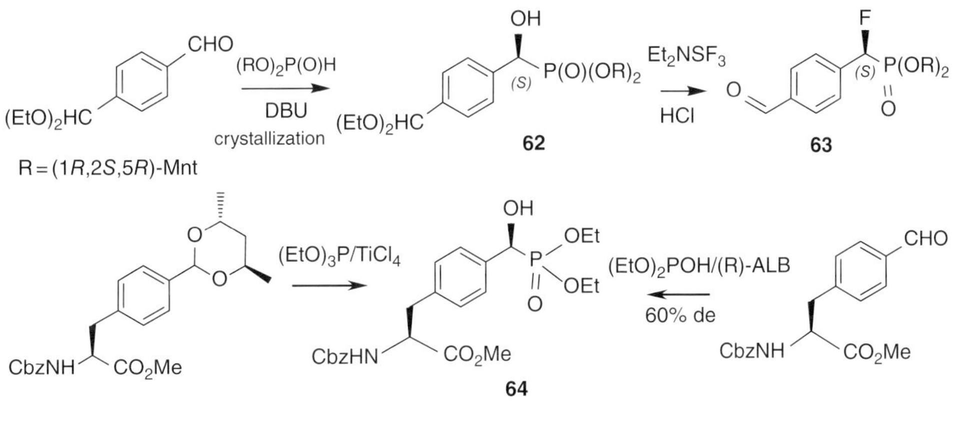

Scheme 3.44 Preparation of chiral synthetic blocks of bioactive compounds.

analogs of natural compounds were synthesized [113]. The stereoselective methods for the synthesis of the hydroxy phosphonate analog of tyrosine **64** based on the reaction of chiral pentane-2,4-diol acetal of protected 4-formyl-L-phenylalanine with triethyl phosphite in the presence of titanium tetrachloride and by the reaction of diethyl phosphite with methyl ester of *N*-benzyloxycarbonyl-4-formyl-L-phenylalanine in the presence of a chiral heterobimetallic catalyst ALB were developed (Scheme 3.43) [113].

Hydroxylmethylphosphonate **62** [110]. To a mixture of 3 mmol of (1*R*,2*S*,5*R*)-dimenthyl phosphite and 3 mmol of *p*-(diethoxymethyl)benzaldehyde was added two to three drops of DBU, and the mixture was left overnight at room temperature. The ^{31}P NMR spectrum of the reaction mixture showed two signals at δ_P 19.2 and 20.2 ppm belonging respectively to diastereoisomers (*S*)- and (*R*)- in a 35 : 65 ratio. The solvent was then removed, and the residue was recrystallized from acetonitrile to afford (*S*)-stereomer (yield 50%, mp 137–138 °C, $[\alpha]_D^{20}$ −68.7 ($c = 9$, $CHCl_3$)).

Fluoromethylphosphonate **63**. (Diethylamino) trifluorosulfurane, 4.5 mmol, was added dropwise to a solution of 3 mmol of hydroxymethylphosphonate (*S*)-**64** in methylene chloride at −80 to −85 °C under vigorous stirring. The reaction mixture was kept for 1 h at this temperature and then cooled to room temperature, washed with a sodium bicarbonate solution, and extracted with ethyl acetate. The extracts were evaporated, and the residue was purified by column chromatography (silica gel 60, Merck), R_f 0.33 (ethyl acetate–hexane, 1 : 7). ^{31}P NMR spectrum ($CDCl_3$), δ_P, ppm: δ 13.09 (J_{PF} 88.34 Hz). ^{19}F NMR spectrum ($CDCl_3$), δ, ppm: 198.45 d.d (J_{HF} 44.4 Hz, J_{PF} 88.4 Hz).

Chiral epoxy aldehydes **65** were obtained by asymmetric Sharpless epoxidation of 2-trimethylsilylalk-2-enols and by reaction with dialkyl phosphites yielded γ-hydroxy-β-oxophosphonates **66** with 96–97% ee [114] (Scheme 3.45). β-Substituted α-hydroxy phosphinates *syn*- and *anti*-**67** were prepared by lithium phenoxide-catalyzed reaction of protected α-hydroxy and α-aminoaldehydes with ethyl allylphosphinate [115] (Scheme 3.46). The phosphonylation of α-silyloxy aldehydes with silyl phosphites was accomplished by Bongini *et al.* [116, 117]. The stereoselectivity of

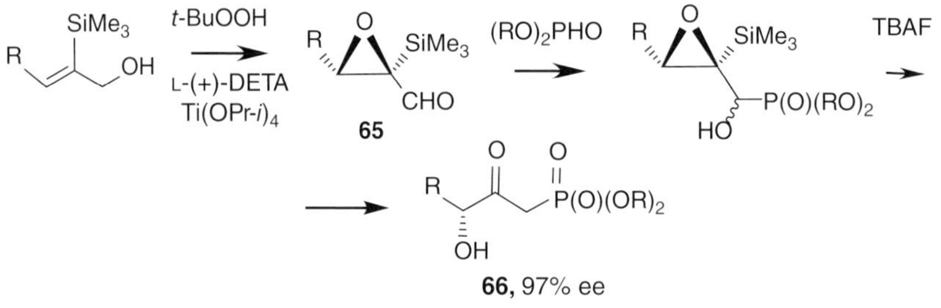

Scheme 3.45 Synthesis of chiral epoxyaldehydes by asymmetric Sharpless epoxidation.

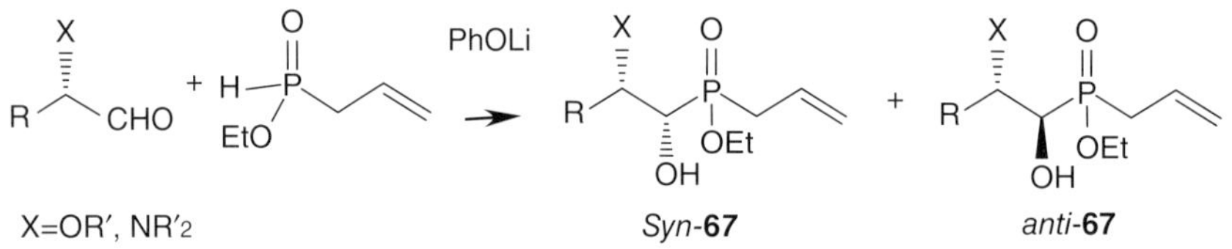

Scheme 3.46 Synthesis of β-substituted α-hydroxy phosphinates.

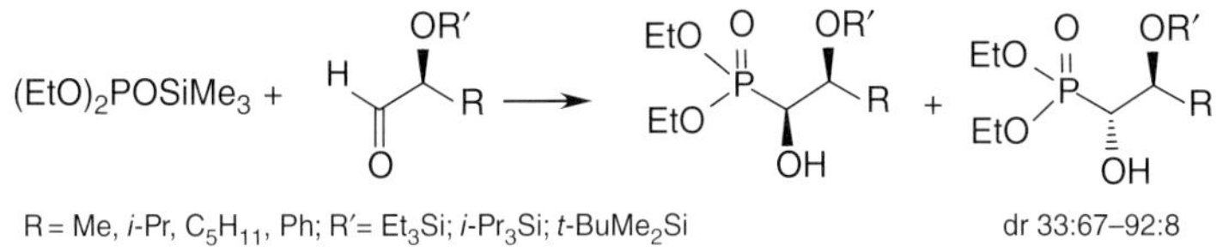

Scheme 3.47

the reaction increased with an increase in the bulk of the trialkylsilyl substituents in the phosphite and aldehyde molecules (dr 33 : 67 – 92 : 8) (Scheme 3.47).

The phospha-aldol reaction was used for the synthesis of a number of phosphorus analogs of natural compounds. For example, for the synthesis of fosfomycin **5**, which is an efficient antibiotic of natural origin was obtained by stereoselective addition of dibenzyl trimethylsilyl phosphite to *O*-triisopropylsilyl-(*S*)-lactaldehyde [118] (Schemes 3.48 and 3.49). The (1*S*,2*S*)-phosphothreonine **68** was prepared with good diastereomeric purity in a similar way by the reaction of diethyl trimethylsilylphosphite with silylated *N*-trimethylsilylimino-(*S*)-lactaldehyde [119]. The other phosphothreonine stereoisomers **68** were prepared by the addition of diethyl trimethylsilylphosphite to lactaldehyde followed by the Mitsunobu inversion of the corresponding α-hydroxy-β-silyloxyphosphonate. Diastereomerically pure (1*R*,2*S*)- and (1*S*,2*S*)-3-amino-1,2-dihydroxypropyl-phosphonates were prepared by a multistep synthesis, which includes the reaction of (*S*)-3-azido-2-benzyloxypropanal with dialkyl phosphites, chromatographic separation of the stereoisomers on a column with silica gel, and debenzylation with hydrogen in the presence of HPd/C [120] (Scheme 3.50). The stereoselective hydrophosphonylation of α-benzyloxyaldehydes catalyzed by titanium tetrachloride is used in the synthesis of the phosphonic analogs of amino hydroxy acids. Shibuya and co-workers [121, 122] used the phospha-aldol reaction to prepare β-benzyloxy-α-hydroxy phosphonates; their reactions with

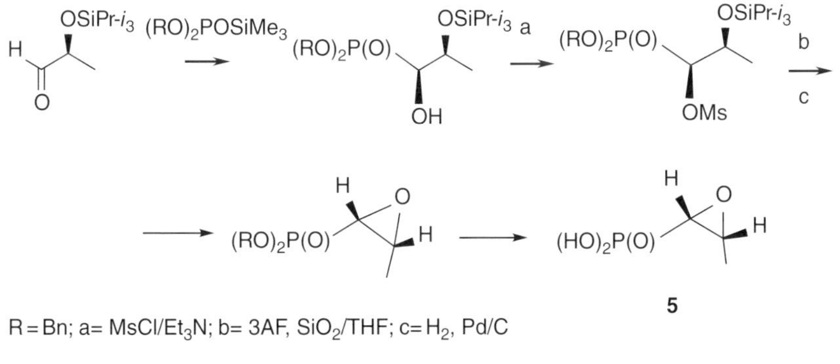

5

Scheme 3.48 Synthesis of fosfomycin **5**.

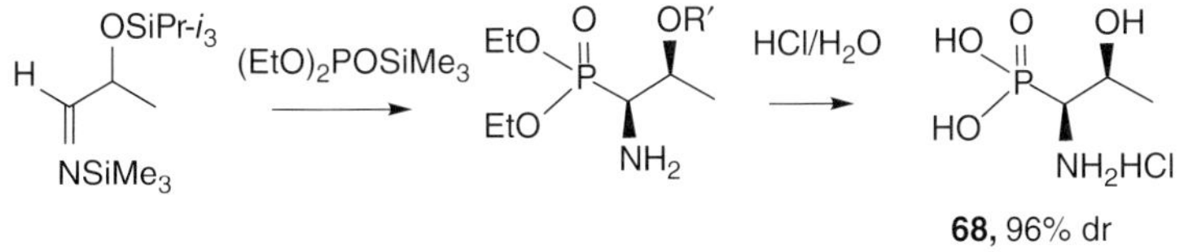

68, 96% dr

Scheme 3.49 Synthesis of (1*S*,2*S*)-phosphothreonine **68**.

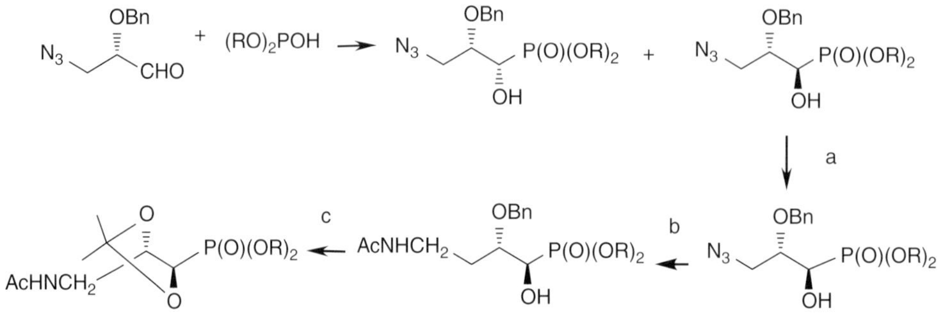

R = Me, Et; (a) Ac_2O, NEt_3, DMAP; (b) H_2, Pd-C; (c) $Me_2C(OMe)_2$, TosOH

Scheme 3.50 Diastereomerically pure 3-amino-1,2-dihydroxypropylphosphonates.

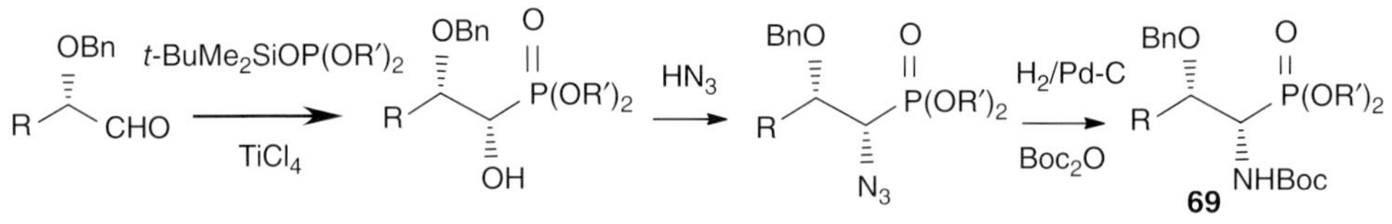

Scheme 3.51 Stereoselective hydrophosphonylation of α-benzyloxyaldehydes.

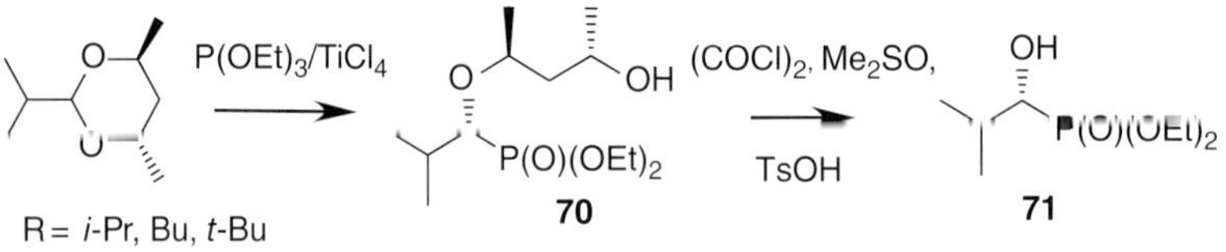

R = *i*-Pr, Bu, *t*-Bu

Scheme 3.52 Reaction of triethyl phosphite with homochiral (2*S*,4*S*)-pentanediol acetals.

hydrazoic acid, followed by reduction and protection, gave substituted α-amino β-hydroxyphosphonates **69** (Scheme 3.51). The reaction of triethyl phosphite with homochiral (2*S*,4*S*)-pentanediol acetals in the presence of titanium tetrachloride results in diastereoselective formation of alcohols **70** (dr = 93 : 7). The Swern oxidation followed by treatment with *p*-toluenesulfonic acid yields hydroxyphosphonates **71** (ee 95%) [123] (Scheme 3.52). Yamamoto and co-workers [122] used this strategy to prepare chiral (2*R*)-1- amino-1-deoxy-1-phosphinylglycerols.

The phospha-aldol reaction was applied to the preparation of various phosphonate carbohydrates: phosphorus-containing D-ribofuranose and D-glucopyranose, analogs of 5-deoxy-5-*C*-phosphinyl-D-xylose [124–134]. These compounds were converted into pseudo-carbohydrate nucleoside derivatives that possess biological activity (Scheme 3.53) [124–126]. Thiem and Guenter used the phospha-aldol reaction for the preparation of δ-phoston **72**, which was isolated as a stereochemically pure compound (Scheme 3.54) [130, 131]. Wroblewski [124–132] synthesized analogs of D-tetrulose, D-ribose, and some other carbohydrates containing phosphorus in the anomeric position (Scheme 3.55) [133]. The reaction of dimenthyl phosphite with galacto-hexodialdose proceeded with high stereoselectivity, owing to double asymmetric induction, resulting in the formation of C(6)-phosphorylated galactose derivative

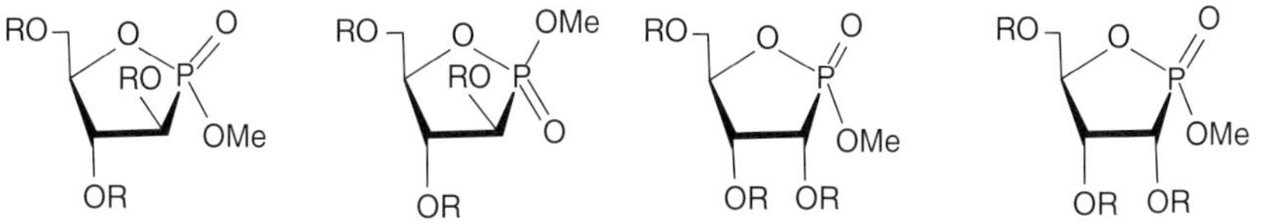

Scheme 3.53 Phosphonate analogs of carbohydrates.

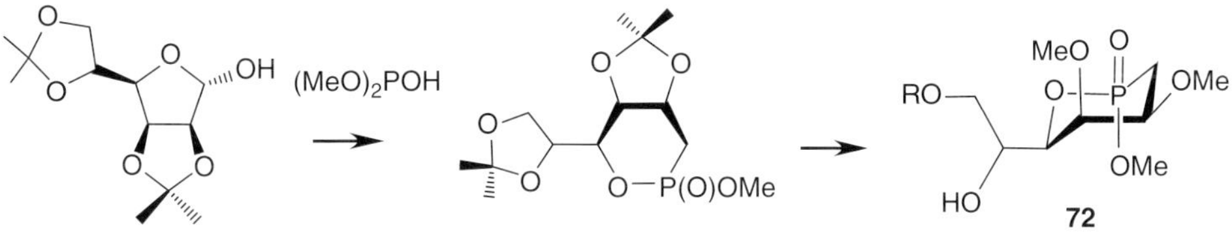

Scheme 3.54 Synthesis of the δ-phoston **72**.

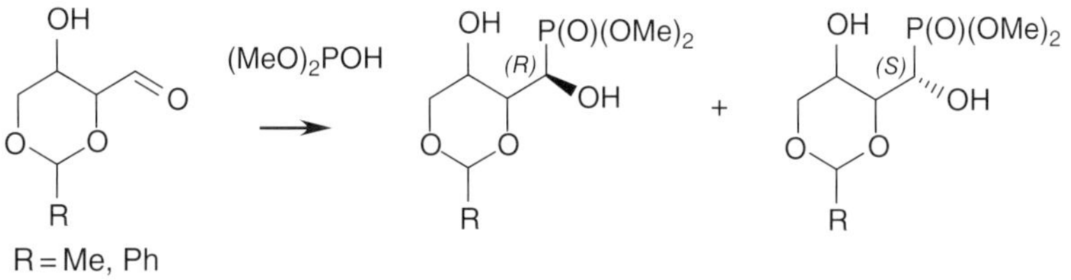

Scheme 3.55 The synthesis of D-erythritol and D-threitol derivatives.

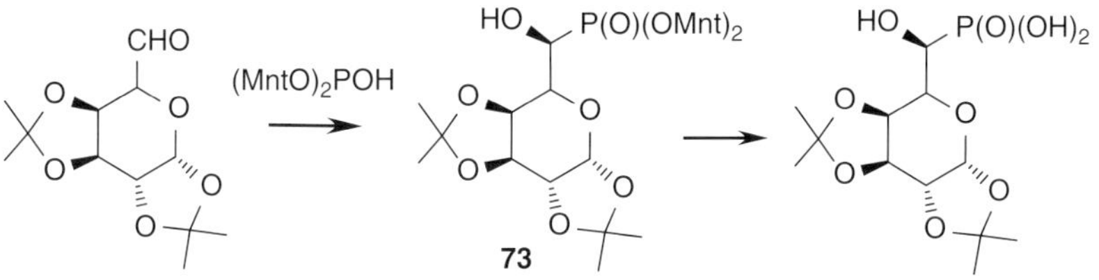

Scheme 3.56 Synthesis of C(6)-phosphorylated galactose derivative.

73 with 100% de. The adduct **73** was isolated as an optically pure crystalline compound (Scheme 3.56).

The addition of silylphosphines to chiral aldehydes proceeded with high diastereoselectivity to give optically pure tertiary α-trimethylsiloxyalkylphosphines. The diastereomeric purity of the addition products was 90–100% de (Scheme 3.57, Table 3.1) [134–137]. The reaction of bis(trimethylsilyl)phenylphosphine with aldehydes proceeded with very good stereoselectivity to give bis-hydroxyalkylphosphins. The subsequent treatment of this compound with borane resulted in an optically pure borane complex, which was stable to oxidation and hydrolysis by air and could be purified by column chromatography on silica gel.

Bis[2,2-dimethyl-1,3-dioxolan-4-yl(trimethylsiloxy)-methyl]phenylphosphine (Table 3.1) [136]. The solution of glyceraldehyde (2.2 mmol) in 5 ml of toluene was added slowly

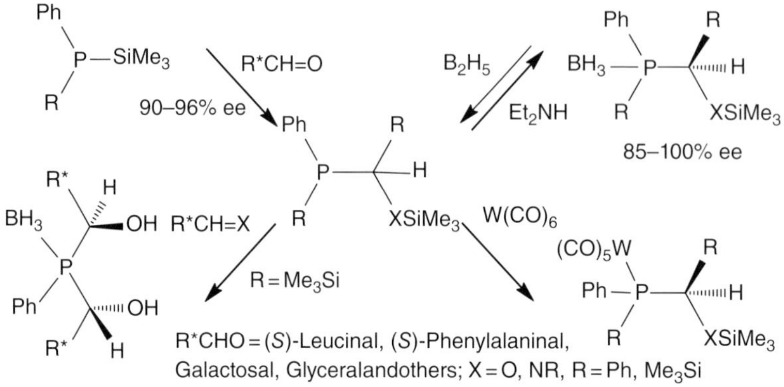

Scheme 3.57 Addition of silylphosphines to chiral aldehydes.

Table 3.1 Stereoselective reaction between Ph_2PSiMe_3 and R*CHO.

Compound	Solvent	*t* (°C)	Yield (%)	dr
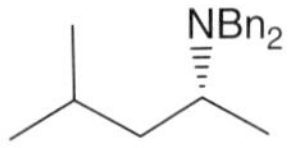	Toluene	−20 to +20	60	90 : 10
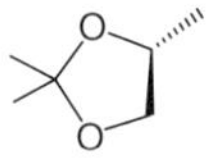	THF	−20	85	95 : 5
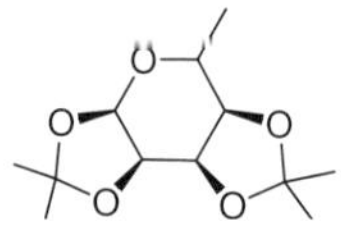	Toluene	20 to 0	90	~100 : 0

to $PhP(SiMe_3)_2$ (1.0 mmol) while cooling. The solution was left at this temperature overnight. The temperature was raised to +20 °C and in 1 h the solvent was evaporated to give tertiary phosphine, ^{31}P NMR ($CHCl_3$): δ −10.68 ppm. Then a solution of borane (1.1 mmol) in THF was added dropwise to a reaction solution (3 ml), with stirring at −20 °C. The reaction mixture was allowed to stand at room temperature. After 6 h the solvent was removed *in vacuo* to give tertiary phosphine and the residue was chromatographed on a column with silica gel, using a mixture of hexane-ethyl acetate (6 : 1) as eluent (yield 70%, R_f 0.37 (hexane-ethyl acetate 6 : 1), $[\alpha]^{20}_{D} = +23.6$ ($c = 3$, $CHCl_3$), NMR ^{31}P ($CHCl_3$): δ 27.4 ppm).

3.4.2 Phospha-Mannich Reaction

The asymmetric phospha-Mannich reaction is hydrophosphonylation of imines using chiral starting materials and chiral auxiliaries. This reaction afforded a convenient method for stereoselective preparation of aminophosphonic acids [138–142]. A special case of the phospha-Mannich reaction is the Kabachnik–Fields reaction, representing a one-pot, three component procedure, which includes carbonyl compound, amine, and

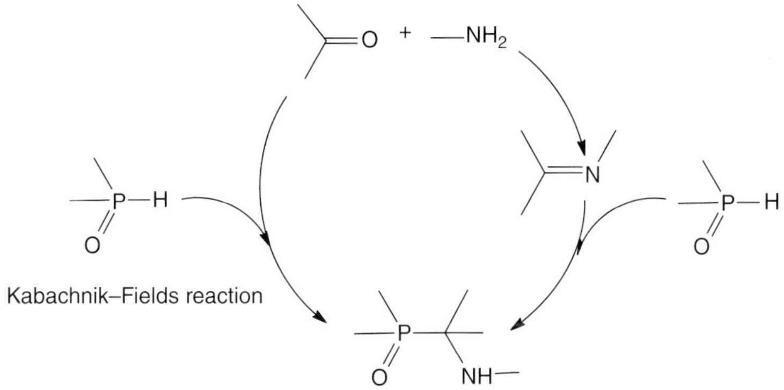

Scheme 3.58 Phospha-Mannich and Kabachnic–Fields reaction.

dialkyl phosphite. The first step in this reaction is the formation of an imine followed by an addition reaction of the phosphonate P–H bond into the C=N double bond with the formation of an α-amino phosphonate (Scheme 3.58).

The stereochemical behavior of the addition of dialkyl or diaryl phosphites to C=N bond of chiral Schiff bases such as (*R*) or (*S*)-*N*-α-methylbenzylimines was firstly described by Gilmore and McBride [143] and later was reported by many authors [144–146]. Diastereoselectivity of reported reactions varied from 2 : 1 to 9 : 1 of diastereoisomeric ratio. The diastereoisomeric aminophosphonates obtained were separated by chromatographic methods or crystallization and then were made to undergo hydrolysis to provide enantiopure, free aminophosphonic acids. Addition of dialkyl phosphite to Schiff bases can be performed at heating to 130–140 °C without catalyst. However, a high reaction temperature is not favorable for the preparation of chiral compounds.

Therefore, various types of catalysts and reaction conditions were studied. For example, Keglevich proposed the microwave-assisted solvent- and catalyst-free approach at moderate heating to 60–80 °C for the synthesis of α-aminophosphonates under conditions of the Kabachnik-Fields reaction [76].The presence of acidic catalysts including Lewis acids and protonic acids (*p*-toluene sulfonic acid) favors the reaction (Table 3.2). In this case, the addition takes place smoothly at room temperature. More nucleophilic sodium salts of dialkylphosphites also react with aldimines at moderate temperature. For example, the addition of diethylphosphites to (*S*)-*N*-aryl-α-methylbenzylamines at 130–140 °C, provided the diastereomeric mixture of (*S*,*S*)- and (*R*,*S*)-α-aminophosphonates in 70% yield and 9 : 1 dr [145]. However, the sodium salts of diethylphosphite added to (*R*)-*N*-aryl-α-methylbenzylamines in anhydrous ether at ambient temperature afforded the α-aminophosphonates in high yields and with very good diastereoselectivity [147]. In another example, the reaction of triethyl phosphite with chiral (*S*)- or (*R*)-aldimines catalyzed by TFA in toluene at room temperature afforded, after recrystallization, the diastereoisomerically enriched (*S*,*S*)-α-aminophosphonates in 44–57% yield and approximately 98% de [142] (Scheme 3.59).

[(1R,2S,5R)-Menth-2-yl]1-phenyl(benzylamine)methylphosphonate. Trimenthyl phosphite (3.5 g, 0.1 mol) was added to benzylbenzaldimine (1.8 g, 0.10 mol) at 0 °C and the reaction mixture was left for 12 h at 60–80 °C. ^{31}P NMR spectroscopy revealed

Table 3.2 Addition of H-phosphites to *N*-α-methylbenzylimines.

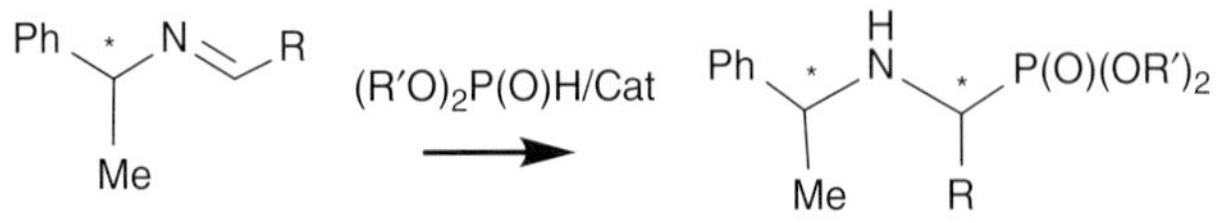

Starting reactants				Product			References
R	R′	Reaction conditions	Configuration	Yield (%)	*de* (%)	Configuration	
Ph	Et	140 °C, 1 h	*R*	64	50	*R,S*	[146]
Ph	Mnt	130–140 °C	*S*	60	92	*S,R*	[146]
Ph	Mnt	130–140 °C	*R*	64	50	*R,S*	[146]
Ph	Brn	130–140 °C	*S*	60	86	*R*	[146]
Ph	Brn	130–140 °C	*R*	60	60	*S*	[146]
Ph	Et	$[PyH]^+ClO_4^-$ 130 °C	*R*	65	50	S	[84]
Ph	Me	$BF_3{\cdot}OEt_2$, CH_2Cl_2	*S*	80	43	*R,S* + *S,S*	[144]
Ph	Et	$BF_3{\cdot}OEt_2$, CH_2Cl_2	*S*	80	61	*R,S* + *S,S*	[144]
Ph	Et	$AlCl_3$, CH_2Cl_2	*S*	77	70	*R,S* + *S,S*	[144]
2-C_6H_4OH	Et	NaH/Et_2O	*R*	63	25 : 2	*R,S*	[147]
2-C_6H_4OH	*i*-Pr	NaH/Et_2O	*R*	71	25 : 4	*R,S*	[148]
Ph	TMS	CH_2Cl_2, reflux	*R*	59	67	—	[149] [150]
Fer	TMS	CH_2Cl_2, reflux,	*R*	32	80	—	[149] [150]
Fer	TMS	CH_2Cl_2, reflux	*S*	32	100	—	[149] [150]
2-Thienyl	TMS	CH_2Cl_2, reflux	*R*	51	67	—	[149]
4-Pyridyl	TMS	CH_2Cl_2, reflux	*R*	58	5 : 6	—	[149]
c-Hexyl	TMS	CH_2Cl_2, reflux	*R*	68	60	—	[149]
2-Furyl	Bn	TFA-Me_3CN	*R*	70	2 : 1	*R,R*	[151]

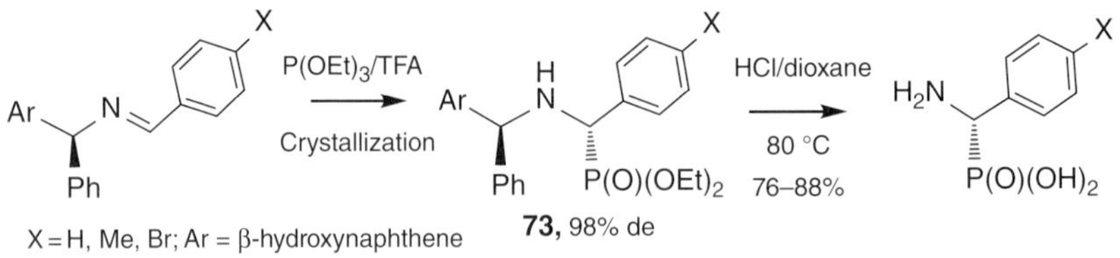

73, 98% de

Scheme 3.59 Phospha-Mannich reaction starting from triethylphosphite.

the presence of two signals at δ_P 21.91 and 21.67 in the 3 : 1 ratio. The reaction mixture was separated by flash chromatography (SiO_2, mixture of hexane-ethylacetate 2 : 1 (300 ml) and 1 : 1 (300 ml) as eluent). The solvent was removed under vacuum to afford the crystalline product in the residue, which was recrystallized from

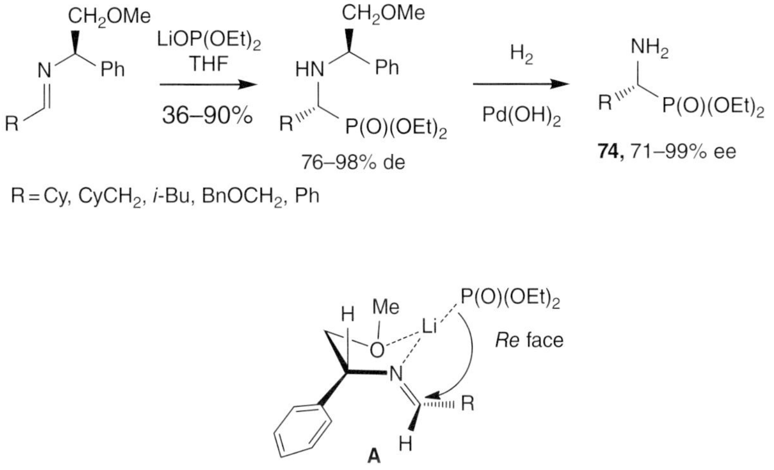

Scheme 3.60 Smith's synthesis of a series of α-amino phosphonates with high optical purities.

acetonitrile or from hexane to afford stereochemically pure adduct (yield 60%, mp 86–87 °C, $[\alpha]_D^{20}$ −57.9 (c = 1, toluene), NMR spectra ($CDCl_3$): δ_P 21.95).

(R)-1-Aminobenzylphosphonic acid. Hydrochloric acid (25 ml 6 N) was added to a solution of dimenthyl 1-phenyl-1-(*N*-benzylamino)-methylphosphonate (1 g) in 60 ml of dioxane and the reaction mixture was left for 2–3 days at 80 °C, the course of hydrolysis being monitored by ^{31}P NMR spectroscopy. When the reaction was completed, the solvent was removed under reduced pressure to result in 1-phenyl-1-(*N*-benzylamin) methylphosphonic acid, which was recrystallized from a mixture of water and alcohol (mp 219 °C, $[\alpha]_D$ 29.9 (c = 1, DMSO)). The hydrochloride of 1-phenyl-1-(*N*-benzylamino)-methylphosphonic acid (156 mg) was dissolved in 7 ml of water and Pd/C was added. Gaseous hydrogen was passed through the mixture at 20 °C till the end of reaction. The catalyst was filtered off and the solvent was evaporated. (*R*)-(−)1-phenyl-1-aminomethylphosphonic acid was obtained (yield, 70%; mp 226 °C; $[\alpha]_D^{20}$ 15.5 (c 0.32, in a 1 N NaOH solution)).

Smith and co-workers [152] reported the synthesis of a series of α-aminophosphonates with high optical purities. Lithium diethyl phosphonate ($LiPO_3Et_2$) was found to afford a fast reaction with chiral imines derived from corresponding enantiopure amine and aldehydes (Scheme 3.60). Yields of 36–81% and high diastereoselectivites (95–98% de) were observed when imines derived from aliphatic aldehydes were employed. However, the phenyl aldimine yielded adducts only with 76% de. The auxiliary groups were removed and α-aminophosphonates **74** were obtained without loss of enantiomeric purity. A transition state **A** was proposed to explain the stereo course of reaction. The phosphite anion was attacked from the *Re*-face to generate (*R*,*R*)-diastereomers.

Palacios *et al.* [153] reported the diastereoselective synthesis of a quaternary α-aminophosphonate using a chiral tartaric-acid-derived phosphonate. TADDOL-derived α-aminophosphonate was obtained by phospha-Mannich-type hydrophosphonylation of tosylimine using TADDOL phosphite. Selective *N*-chlorination of *N*-tosyl α-aminophosphonate with trichloroisocyanuric acid in CH_2Cl_2, followed by treatment with poly(4-vinylpyridine), gave α-ketiminophosphonate **75** in good yield (82%). α-Aminophosphonate **76** was obtained as a mixture of diastereoisomers in

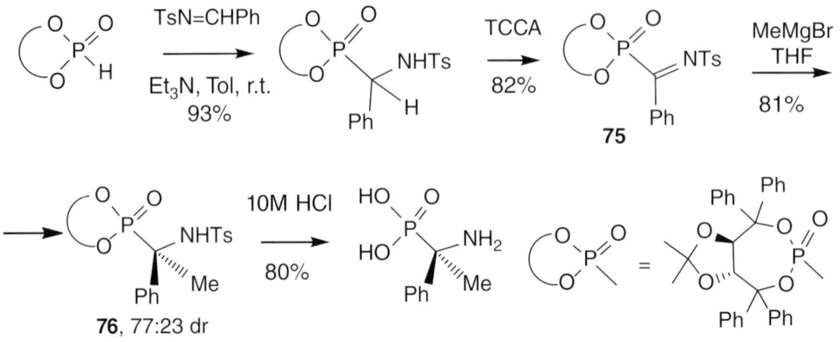

Scheme 3.61 Diastereoselective synthesis of a quaternary α-aminophosphonate.

yield 93% and 77 : 23 dr, which was separated by crystallization in diethyl ether. The hydrolysis of (*R*,*R*)-TADDOL phosphonate **76** with aqueous HCl at 100 °C afforded the optically enriched (*R*)-α-(*R*)-aminophosphonic acids (Scheme 3.61).

Nucleophilic addition of diethylphosphite to (*S*)-aldimines **77** obtained from (9*S*)-amino-deoxyquinine and 4-chlorbenzaldehyde, in toluene at 90 °C gave (*S*,*S*)-α-aminophosphonate **78** with moderate yield and excellent diastereoselectivity. However the imine of (9*R*)-configuration reacted with low yield and stereoselectivity. Stereochemical models explaining the observed high diastereoselectivity of hydrophosphonylation in one case and its lack in the other were discussed. The high diastereoselectivity of this transformation was attributed to the directing effect of the quinuclidine nitrogen. The separable products of the (9*R*)-configuration were obtained both in inferior yields and with no diastereoselectivity. The configuration of the products was established by means of NMR spectroscopy and density functional theory (DFT) calculations [154] (Scheme 3.62).

The addition of dialkyl phosphites to the *N*,*N*′-disalicylidene-1,2-diaminocyclohexane imines, in the presence of sodium hydride, led to the bis-aminophosphonates **79** in a high diastereoselectivity [147]. The diastereoselectivity of addition to *N*,*N*′-disalicylidene-(*R*,*R*)-1,2-diaminocyclohexane imines was explained by stabilization of their sodium salts by intramolecular coordination of a sodium cation by the phenolic oxygen and by the azomethine nitrogen. This stabilization enables the predominant formation of the (*RR*,*RR*) isomers of aminophosphonates **79** (Scheme 3.63).

The additions of diethyl and diisopropyl phosphites to salicylaldimine in anhydrous ether in the presence of sodium hydride led to the formation of diethyl and diisopropyl (*R*)-α-methylbenzylamino-(2-hydroxyphenyl)-methylphosphonates **81** [148] (Scheme 3.64). It is probable that the sodium dialkylphosphite is stabilized by intramolecular coordination of a sodium cation by the phenolic oxygen and by

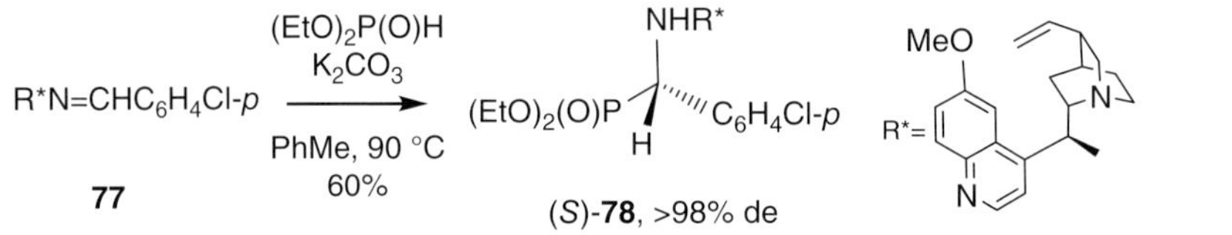

Scheme 3.62 Phospha-Mannich addition controlled by (9*S*)-amino-deoxyquinine auxiliary.

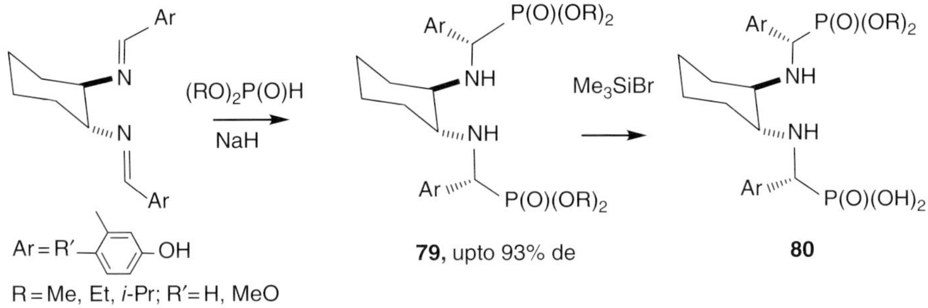

R	R[1]	Yield (%)	dr
H	Me	74	94:1:5
H	Et	95	92:1:7
H	*i*-Pr	72	95:4:1
OMe	Et	80	2:96:2
OMe	*i*-Pr	74	2:95:3
	Et	63	93:7

Scheme 3.63 Preparation of di(aminophosphonates) **79** and aminophosphonic acid **80**.

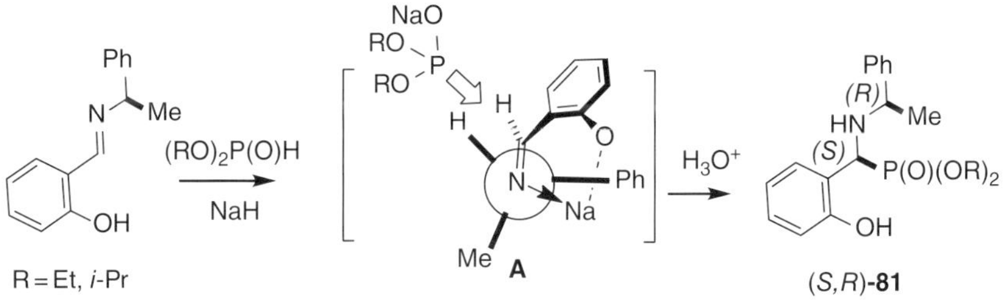

Scheme 3.64 Preparation of aminophosphonate **81**.

azomethine nitrogen. The sodium salt of *N*-salicylidene-(*R*)-α-methylbenzylamine would adopt the bis-conformation **A**. The attack of a phosphite molecule is then most easy from the side opposite to the phenyl ring and results in the major formation of the (R_S)-diastereoisomer of compound (*S*,*R*)-**81** [147].

Readily available chiral sulfinylimines represent convenient chiral reactants for the asymmetric phospha-Mannich reaction. Utilization of chiral sulfinimines allowed to obtain a number of α-aminophosphonic acids of high optical purity [155–162]. The chiral sulfinyl group not only activates the C=N bond of imines for the nucleophilic attack but also is an excellent inducer of chirality. After the nucleophilic addition, the sulfinyl group can be easily removed by treatment with trifluoroacetic acid. Within this context, Evans and co-workers [155] designed a synthetic approach to α-amino phosphonic acids involving the addition of metallo phosphites to enantiomerically enriched sulfinimines derived from sulfinate with good yields (93%) and enantioselectivity (up to 93–97% ee) (Scheme 3.66). Davis *et al.* [156] reported that the addition of the lithium salt of diethyl phosphite to enantiopure (*S*)-ketimines **82** gave the (S_S,R_C) and (S_S,S_C)-α-aminophosphonates in good yield and with approximately 98% de. Only in the case of the ketimine derived from 2-hexanone, the (S_S,R_C)

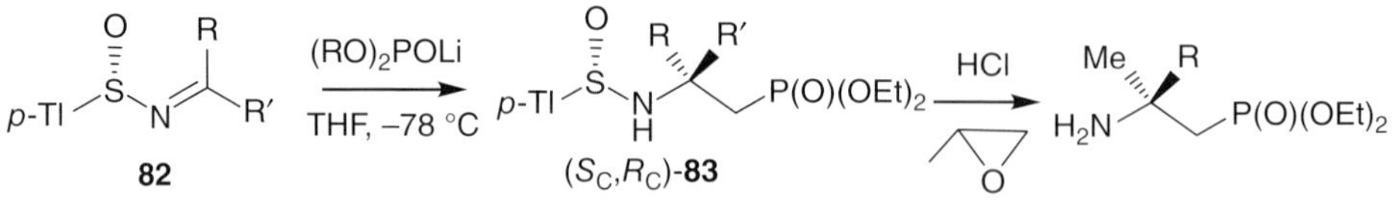

R	R′	Yield (%)	de (%)
Me	4-An	73	95
Me	4-Tl	91	>95
Me	Ph	92	>95
Et	Ph	93	>95
Me	4-$O_2NC_6H_4$	93	>95
Me	*t*-Bu	97	>95e
Me	*n*-Bu	71	64

Scheme 3.65 Addition of the lithium diethyl phosphite to enantiopure (*S*)-ketimines.

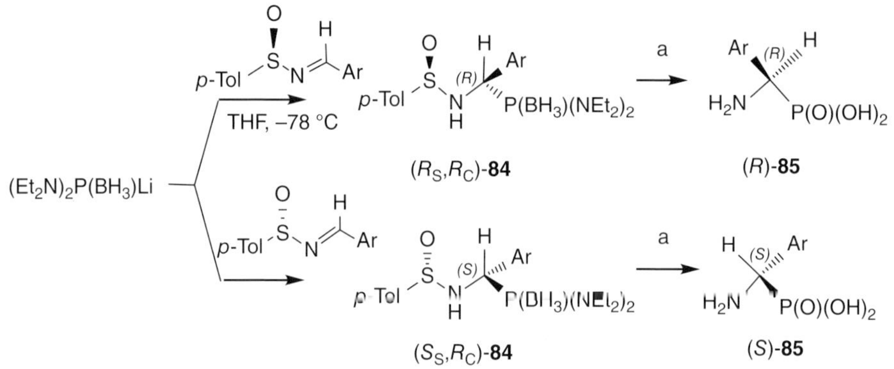

Ar = Ph, 2-Thienyl, 4-C_6H_4F, 4-C_6H_4Br, 4-$C_6H_4NMe_2$ a = HCl/AcOH/propylene oxide

Scheme 3.66 Addition of lithiated bis(diethylamino)phosphine borane complex to enantiopure sulfinimines.

and (S_S,S_C)-α-aminophosphonates **83** were obtained with 64% de. Cleavage of the *N*-sulfinyl auxiliary in the diastereoisomerically pure (S_S,R_C)-**83** with 10 N HCl at reflux followed by treatment with propylene oxide gave the enantiomerically pure (*R*)-α-aminophosphonic acids (Scheme 3.65). Mikołajczyk *et al.* [163] reported an efficient procedure for preparation of (*S*)-and (*R*)-α-aminophosphonic acids via addition of lithiated bis(diethylamino)phosphine borane complex to enantiopure sulfinimines (Scheme 3.66). The addition of the lithium salt of diethyl phosphite to (*S*)-sulfinylaldimine at 78 °C in dry THF gave the α-aminophosphonates (R_S,R_C)- and (S_S,R_C)-**84** with a 16 : 1 diastereoisomeric ratio. Hydrolysis of (R_C,S_S)- and (S_C,S_S)-**84** with 10 N HCl under heating afforded the optically pure α-aminophosphonic acid (*R*)-**85** in 71% yield and the optically pure α-aminophosphonic acid (*S*)-**85** in 75% yield.

Mikołajczyk and Łyżwa have also developed an approach to asymmetric synthesis of α-, β-, and γ-aminophosphonic acids using enantiopure *p*-toluenesulfinimines as key reagents [157, 164]. The addition of lithium diethylphosphite to enantiomeric pure sulfinimines proceeded with good enantioselectivity to afford the α-alkylaminophosphonates with high diastereoselectivity as shown in Scheme 3.67.

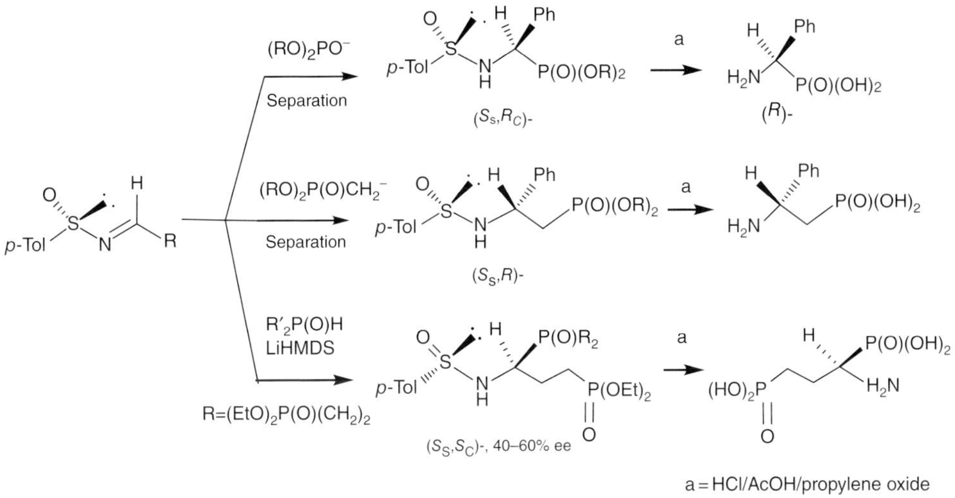

R$_2$POH	Adduct	dr
$(EtO)_2P(O)H$	R=$(EtO)_2P(O)$	9:1
$(Et2N)_2P(O)H$	R=$(Et_2N)_2P(O)$	6:1
$(Et2N)_2P(O)H$	R=$(EtO)_2P(O)CH_2$	3:1

Scheme 3.67 Asymmetric synthesis of α-, β-, and γ-aminophosphonic acids.

The methodology was applied for the synthesis of various structures of enantiomeric α- and β-aminophosphonic acids. The most interesting among them were the (*S*)-2-amino-4-phosphono-butanoic acid and phosphinothricin. The addition of the lithium salt of diethyl phosphite to (*S*)-sulfinylaldimine at low temperature in THF gave the α-aminophosphonates (R_C,S_S)- and (S_C,S_S)-**86** with a 16 : 1 diastereoisomeric ratio, which, after flash-chromatography and crystallization, afforded the diastereoisomerically pure (R_C,S_S)-**86**, which was obtained in 62% yield. Hydrolysis of (R_S,S_S)-**86** with 10 N HCl under heating afforded the optically pure (*R*)-α-aminophosphonic acid in 71% yield. On the other hand, the addition of lithium di-(diethylamino) phosphide borane complex to the (*S*)-sulfinylaldimine afforded the α-aminophosphaneborane (S_C,S_S)-**87** in 72% yield as a single diastereoisomer. Hydrolysis of (S_C,S_S)-**87** with HCl provided the optically pure (*S*)-α-aminophosphonic acid in 75% yield [157]. The nucleophilic phosphorus atom of the lithium dialkyl phosphites probably approaches the (*S*)-*trans*-conformation of C=N group from the diastereotopic π-face occupied by the sulfinyl oxygen atom to allow chelation of the lithium cation by the sulfinyl oxygen atom (Path A) and stabilization of the cyclic seven-membered transition state formed in this way. Among the four diastereotopic π-faces formed by (*S*)-*trans*-A and (*S*)-*cis*-B, that occupied by the sulfur lone electron pair in the more stable (*S*)-*cis*-B is the least hindered one. Therefore, approach of lithium aminophosphido–borane takes place from this π-face of (S)-*cis*-B (Scheme 3.68 and Figure 3.2) [159].

Analogously, enantiomerically pure (2-amino-4-phenylbuten-3-yl)phosphonic acids **88** were synthesized (Scheme 3.69). Reaction of (+)-(*S*)-sulfinylaldimine with the lithium salt of diethyl methanephosphonate afforded a mixture of diastereoisomeric adducts **87** in a 9 : 1 ratio from which pure major diastereoisomer (−)-(*R*)-**87** was isolated in 65% yield. Acidic hydrolysis of (−)-(S_S,R_C)-**87** gave corresponding

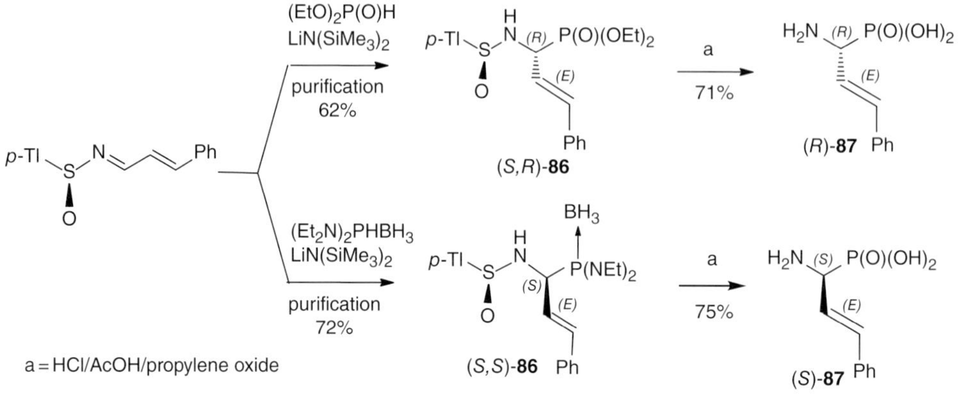

Scheme 3.68 Addition of lithium diethyl phosphite and lithium di-(diethylamino) phosphide borane complex to (*S*)-sulfinylaldimine.

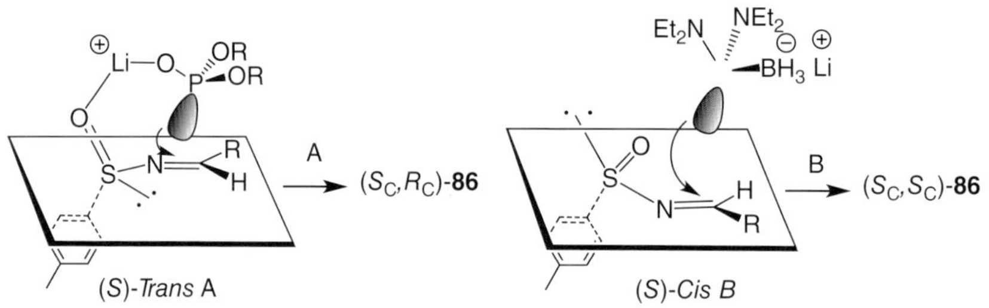

Figure 3.2 The transition-state models for the preferred addition of P- and C-nucleophiles to (+)-(*S*)-sulfinimine.

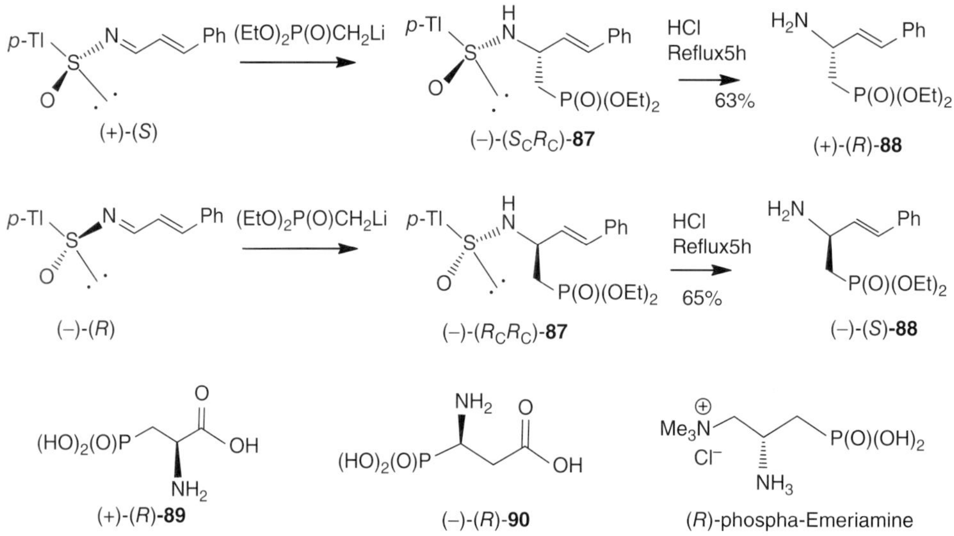

Scheme 3.69 Synthesis of enantiopure (2-amino-4-phenyl-buten-3-yl)phosphonic acid.

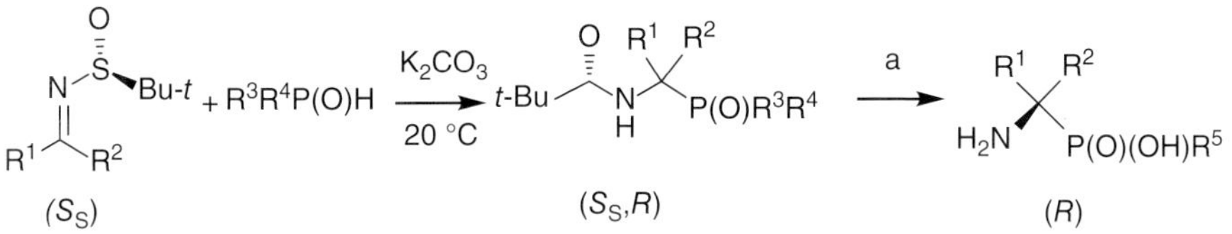

R^1 = Alk, Ar R^2 = Alk, H; R^3 = EtO R^4 = EtO, $CH(OEt)_2$, R^5 = H, OH a = HCl/AcOH/propylene oxide

Scheme 3.70

β-aminophosphonic acid (+)-(*R*)-**88** in 63% yield. Starting from (−)-(*R*)-sulfinimine the opposite enantiomer of acid **88** was similarly prepared [159]. The enantiopure (2-amino-4-phenyl-buten-3-yl)phosphonic acid was used for the synthesis of (+)-(*R*)-2-amino-3-phosphonopropanoic acid (phosphonoaspartic acid) **89** and (−)-(*R*)-3-amino-3-phosphonopropanoic acid **90**. The phosphonoaspartic acid **89** is a selective, potent modulator of the metabotropic excitatory amino acids receptor subtype. The phosphonic analog of emeriamine was also prepared. The P-Emeriamine, also called *aminocarnitine*, exhibits interesting pharmacological properties.

Chen and Yuan [161, 165] reported that the nucleophilic addition of dialkyl phosphites to *N-tert*-butylsulfinyl aldimines or ketimines occurring at room temperature with potassium carbonate as a base resulted in α-amino- and α-alkyl-α-amino-*N*-(*tert*-butylsulfinyl)phosphonates in good to excellent chemical yields (73–95%) and with good diastereoselectivity (72–95% de). The major diastereomers were separated and smoothly converted into enantiomers of α-amino- and α-alkyl-α-aminophosphonic acids [162] (Scheme 3.70). Competitive nucleophilic attack at the sulfur atom is minimized in the addition to *N-tert*-butylsulfinyl imines versus *N-p*-tolylsulfinyl imines owing to the greater steric hindrance and lower electronegativity of the *tert*-butyl group relative to the *p*-tolyl moiety. Evidently, substituents R and R′ have a crucial influence on the reaction rate and the diastereoselectivity.

Roschenthaler *et al.* developed asymmetric syntheses of fluorinated aminophosphonic acids using the sulfinimine methodology [166, 167]. Addition of diethyl lithiodifluoromethylphosphonate to enantiomerically pure aromatic, heteroaromatic, and aliphatic aldehyde-derived sulfinimines afforded *N*-sulfinyl α,α-difluoro-β-aminophosphonates **91** with generally good enantioselectivity and in high yields. The reaction with acetophenone-derived sulfinimine resulted in the formation of the addition product with high diastereoselectivity and in only moderate yield. A two-step deprotection involving treatment of diastereomerically pure *N*-sulfinyl α,α-difluoro-β-aminophosphonates with TFA in EtOH followed by refluxing with HCl, which was then followed by treatment with propylene oxide and ethanol, provided enantiopure α,α-difluoro-β-aminophosphonates and α,α-difluoro-β-aminophosphonic acids **92** (Scheme 3.71). Addition of dialkyl phosphites to (*S*)-*N-tert*-butanesulfinyl imine derived from fluoral afforded, under mild conditions, the corresponding *N-tert*-butanesulfinyl α-aminophosphonates in moderate to high yields and diastereoselectivity. The major diastereomers of *N-tert*-butanesulfinyl α-aminophosphonates **93** of (*S,S,S*) configuration were isolated and, after partial or complete deprotection, converted into enantiomerically pure phosphonotrifluoroalanine and its dialkyl esters [160] (Scheme 3.72).

N-Sulfinyl α,α-difluoro-β-aminophosphonates **91** [167]. LDA (1.30 mmol) was added to a solution of diethyl difluoromethylphosphonate (1.30 mmol) in THF (3 ml) at −78 °C.

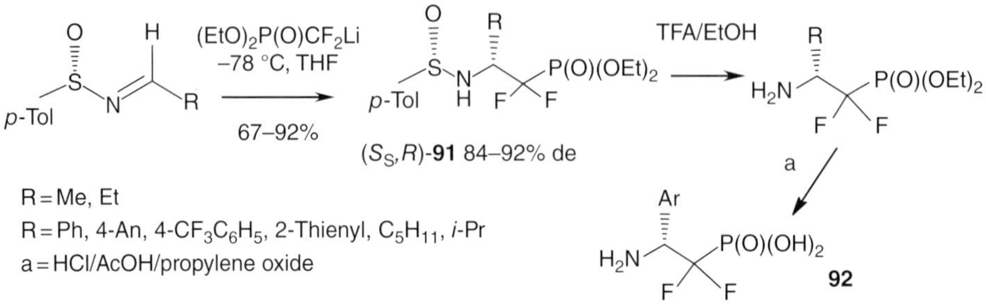

Scheme 3.71

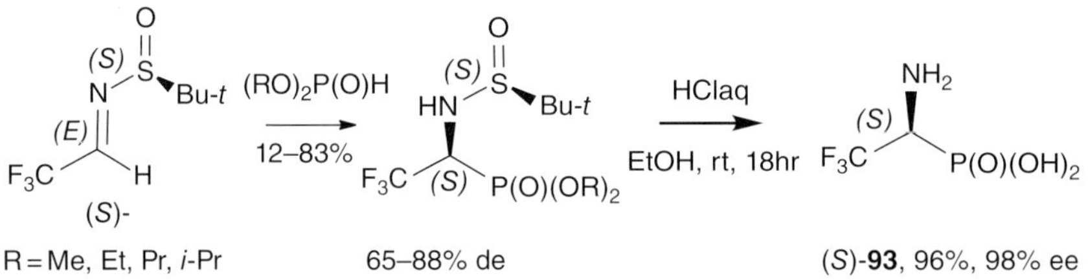

Scheme 3.72 Asymmetric syntheses of fluorinated aminophosphonic acids.

After 0.5 h, (*S*)-sulfinimine (1.00 mmol) in THF (1 ml) was added dropwise and the solution was stirred at 78 °C for 1 h. The reaction was then quenched by addition of saturated aqueous NH_4Cl (5 ml) and the solution was warmed to room temperature. After dilution with H_2O (2 ml), the solution was extracted with EtOAc (25 ml). The combined organic layers were washed with brine (5 ml) and dried ($MgSO_4$). Concentration under reduced pressure gave the crude phosphonate with 90% de. Crystallization from ether afforded (*Ss,R*)-**91** (yield 74% as a white solid; mp 95–97 °C; $[\alpha]_D^{20}$ +53.7 (c = 1.08, $CHCl_3$), ^{31}P NMR (121 MHz, $CDCl_3$): δ 5.9 (dd, J 103.6)).

The three-component Kabachnik–Fields reaction represents a good choice for the synthesis of α-aminophosphonates [168–170]. The Kabachnik-Fields reaction of benzaldehyde with (*S*)-α-methylbenzylamine and diethyl or dimethyl phosphite catalyzed by phenylphosphonic acid was proceeded at 80 °C under solvent-free conditions to afford the α-aminophosphonates (*R,S*)-**94** and (*S,S*)-**94** with 80 : 20 dr. The best yields and the highest diastereoselectivities of (*R,S*)-α-aminophosphonates were attained with (*S*)-3,3-dimethyl-2-butylamine as chiral catalyst (Scheme 3.73) [171]. The reaction of chiral dicarboxylate **95** with aromatic aldehydes and diethyl phosphoramidate in CCl_4 at −20 °C afforded the α-aminophosphonates **96** as a mixture of two diastereoisomers in good yield. With the acetophenone, the reaction furnished the quaternary α-aminophosphonate **96** in 62% yield and with 67 : 33 dr

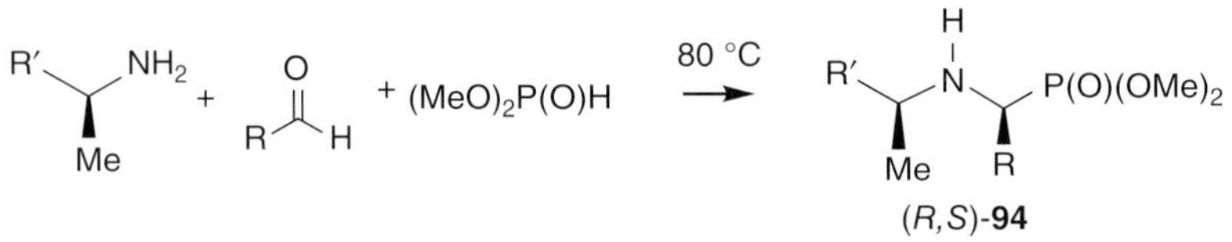

Scheme 3.73

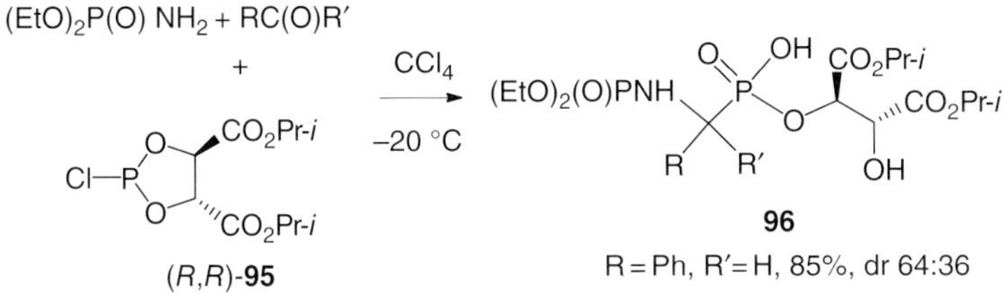

Scheme 3.74 Asymmetric syntheses of the α-aminophosphonates **96**.

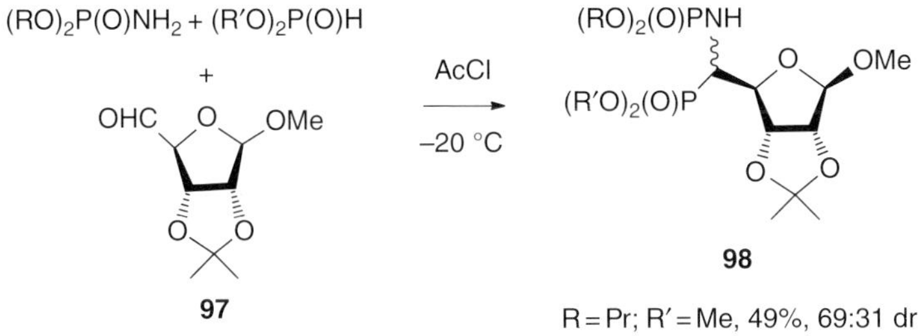

Scheme 3.75 Synthesis of dialkylphosphono-D-ribofuranosides **98**.

(Scheme 3.74). In a similar way, the three-component reaction of acetone with diethyl phosphoramidate and 2-chlorobenzo[*d*][1,3,2]dioxaphosphole under solvent- and catalyst-free conditions, followed by hydrolysis with hydrochloric acid, furnished the *N*-diethoxyphosphoryl-α-amino-α-methylethylphosphonate derivative in 80% yield [169]. Miao *et al.* [170] developed a convenient method for the diastereoselective synthesis of dialkylphosphono-D-ribofuranosides **98** under mild conditions. A one-pot reaction of a dialkyl phosphoramidate with a dialkyl phosphite and 1,4-furanoside **97** in acetyl chloride led to the formation of D-ribo-furanosides **98** in moderate yield and with 69 : 31 dr [170] (Scheme 3.75).

Palacios *et al.* [172] has described a simple asymmetric synthesis of 2*H*-azirin-2-phosphine oxides **99** from easily accessible oximes, using chiral amines immobilized on a polymer. These heterocycles are useful intermediates for the synthesis of α-ketamides and phosphorylated oxazoles. The key step is a solid-phase bound achiral or chiral amine-mediated Neber reaction of ketoxime tosylates derived from phosphine oxides. The reaction of 2*H*-azirines with carboxylic acids yielded phosphorylated ketamides. The ring closure of the ketamides with triphenylphosphine and hexachloroethane in the presence of triethylamine led to the formation of phosphorylated oxazoles (Scheme 3.76) [60, 61]. Alkaloids and solid-phase bound chiral amines were used as catalysts. The best results were obtained with quinidine (69–95% yields, up to 72% ee). In

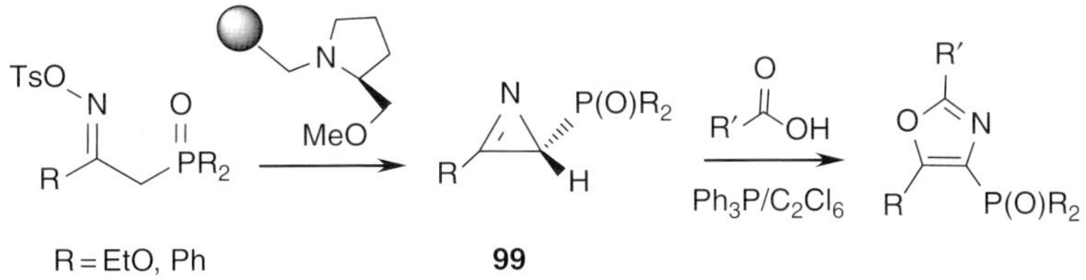

Scheme 3.76 Examples of diastereoselective phospha-Mannich additions.

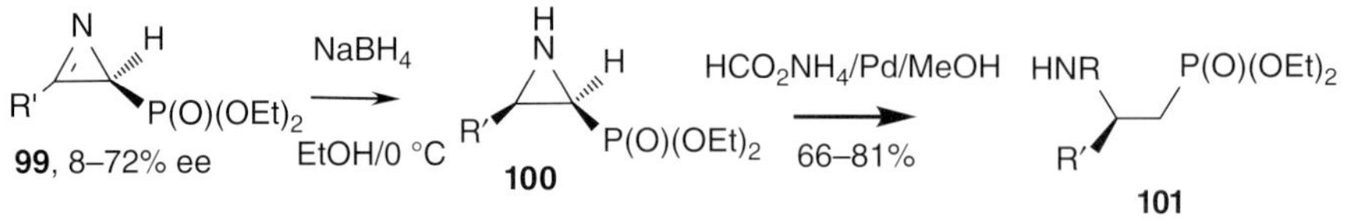

X = H, Ts; R = Me, Et, Ph; Base = , SP, QN, Q, HQN,

Scheme 3.77 Asymmetric synthesis of 2H-aziridine-phosphonates.

other cases, the stereoselectivity was either low or moderate. The subsequent reduction of 2*H*-azirines **99** with sodium borohydride in ethanol gave *cis*-aziridine-phosphonates **100** with moderate ee (up to 65% ee) (Scheme 3.77). The ring opening of enantiomerically enriched *N*-unsubstituted aziridine **99** by catalytic transfer hydrogenation with ammonium formate and palladium on carbon led to the formation the enantiomerically enriched β-aminophosphonates **101**. The absolute configuration of the β-aminophosphonates **116** was established by chemical correlation.

3.4.3 Phospha-Michael Reaction

The phospha-Michael reaction is one of the best methods for the formation of P–C compounds [173]. The asymmetric phospha-Michael reaction can be accomplished by several methods under control of internal or external asymmetric inductors. Internal asymmetric inductors represent chiral centers bound to the reactive center through a covalent bond. The external asymmetric inductors represent chiral organometallic catalysts or organocatalysts and the reaction proceeds under control of these catalysts.

Haynes and Yeung [174] used lithiated *tert*-butyl(phenyl)phosphane oxide in the additions to saturated and unsaturated carbonyl compounds. Nucleophilic lithiated P-chiral *tert*-butylphenylphosphine oxide reacted with aldehydes and α,β-unsaturated carbonyl compounds with the formation of chiral tertiary phosphines in good yields and with diastereoselectivities from low to high (33–98% de). In most cases, the reaction proceeded with retention of configuration at the phosphorus center. The diastereoselectivity of reaction was rationalized through consideration of the relative energies of diastereomeric five-membered transition states TS1 and TS2 involving OLi contacts, and which provide *ul* and *lk* products respectively, as depicted for the lithiated (*S*)-*tert*-butylphenylphosphine oxide (Scheme 3.78).

Asymmetric versions of phospha-Michael reactions mainly deal with substrate-controlled diastereoselective additions. For example, the TADDOL-derived chiral phosphite was added to the aromatic alkylidene malonates with good stereoselectivity [175]. The reaction was carried out with KOH as base supported on solid Fe_2O_3. The phosphonates **104** were obtained in good yields and with very good diastereoselectivities. The auxiliaries were then removed using Me_3SiCl/NaI to afford acids **105** without any epimerization or racemization (Scheme 3.79).

Stankevic reported the stereoselective Michael addition of phosphorus nucleophiles to the unsymmetrically substituted *tert*-butyl (1,4-cyclohexadienyl)phosphine oxide and its derivatives [176]. Yamamoto and co-workers [177] carried out a systematic study on the stereoselective addition of dimethyl phosphonate to the *E*-configured nitroalkenes. Two conditions of reaction were tested and it was found that the stereochemical outcomes were opposite. The first reaction condition (with Et_3N) predominantly gave the

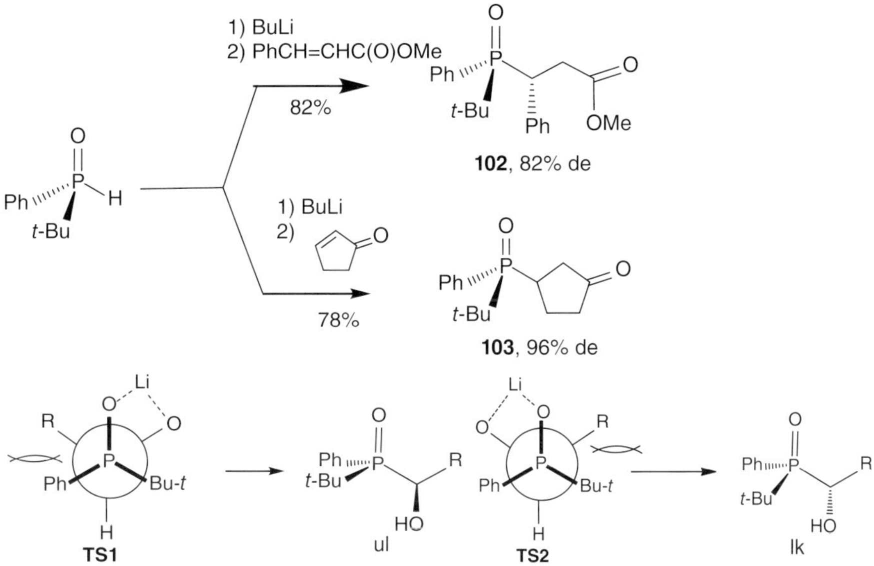

Scheme 3.78 Haynes's conjugate addition of P-chiral *tert*-butyl(phenyl)phosphine oxides.

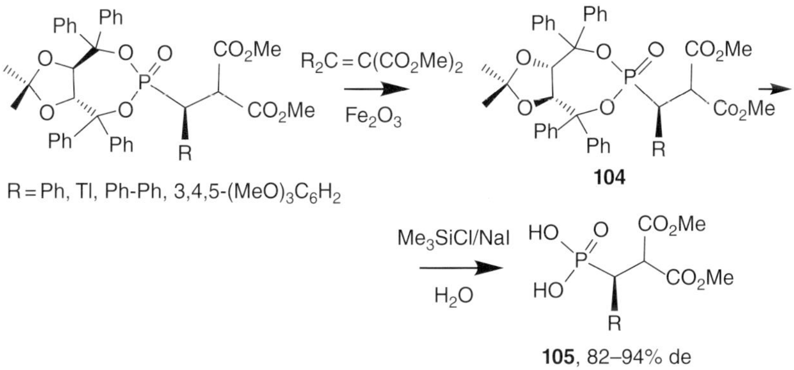

Scheme 3.79

stereoisomer (R)-**106**, whereas the stereoisomer (S)-**106** was obtained when the reaction mixture was heated to 100 °C in the absence of the base. However, the yields of products were moderate to good (55–94%) (Scheme 3.80).

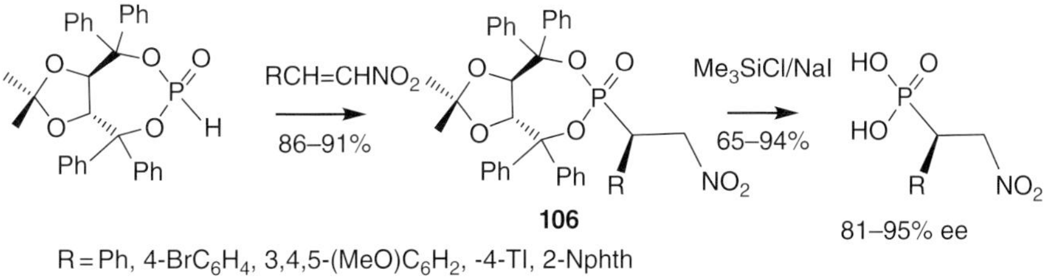

Scheme 3.80 Enders's asymmetric phospha-Michael reaction to nitroalkenes.

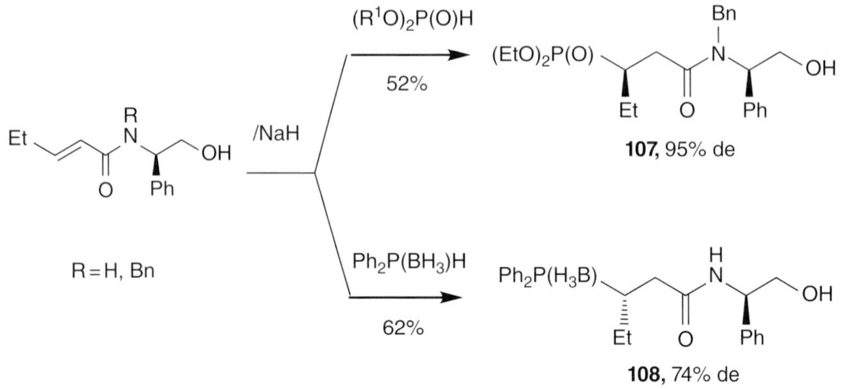

Scheme 3.81 The nucleophilic addition of dialkyl phosphites to unsaturated amides.

β-Amino alcohols are effective chiral auxiliaries in nucleophilic addition of dialkyl phosphites to unsaturated amides, imides, and oxazolines. The addition of unsaturated amides to sodium diethyl phosphite resulted in the formation of the respective 1,4-addition products **107** in moderate yields. The diastereoselectivity was excellent in cases of aliphatic acrylic amide moiety (90% de), but dropped significantly in the case of aromatic substituents (45% de). The final removal of the chiral auxiliary was accomplished using 8 N HCl [178] (Scheme 3.81). Similar to their work on dialkyl phosphites, Quirion and co-workers [179] also studied the 1,4-addition to unsaturated amides using $Ph_2P(BH_3)Li$. In this case, an optimization of the reaction revealed the necessity to use a primary amine as the chiral auxiliary (as opposed to phosphites, where a secondary amine is used). Interestingly, the nucleophilic attack of lithiated phosphite to the double bond now occurs from the opposite face to give the tertiary phosphane–boranes **108** in moderate yields and de's (Scheme 3.81).

An asymmetric Michael addition of phosphinic and aminophosphinic acid has been developed by Ebetino and co-workers [180]. The bis(trimethylsilyl) phosphinites reacted with the enantiopure acrylimides **109** yielding the addition products (*R*)-**110**. The treatment of silyl esters employing EtOH resulted in phosphinic acids in very good yields. The diphenylmethyl-substituted oxazolidinone gave much better diastereoselectivities than their benzyl analogs. The auxiliaries were cleaved successfully using LiOH (Scheme 3.82).

The Michael additions was used in the synthesis of phosphane ligands for late-transition metals. Among various nucleophiles, the phenylphosphanes Ph_2PH and $PhPH_2$ are most often used – either in form of their lithium salts (*vide infra*) or

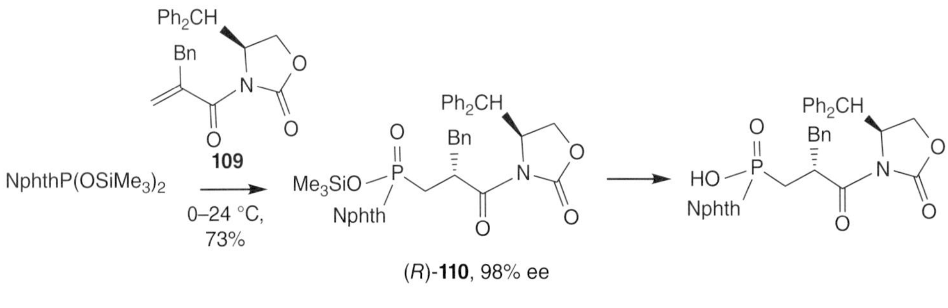

Scheme 3.82 Ebetino's asymmetric Michael addition of phosphinic and aminophosphinic acid.

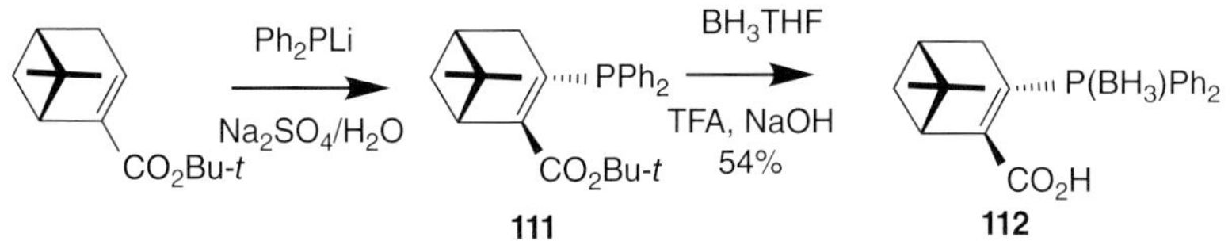

Scheme 3.83

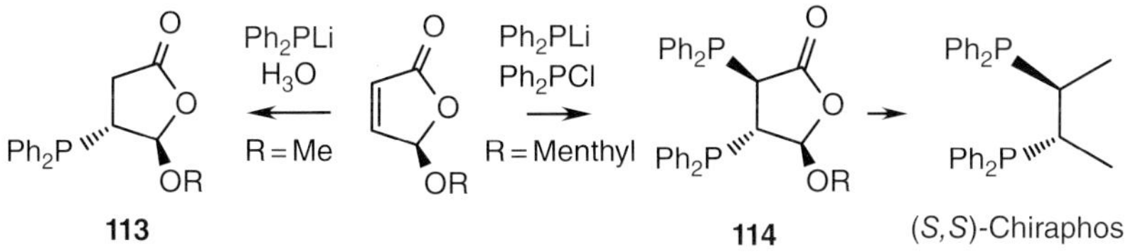

Scheme 3.84 Michael addition of lithio-diphenylphosphine to γ-butenolides.

in the presence of KOtBu as catalyst [181]. Helmchen and co-workers [182] utilized Ph_2PLi for the addition to (−)-(1*R*)-*tert*-butyl myrtenate (**102**) (Scheme 3.83). The reaction proceeded smoothly and diastereoselectively to give **111**, which was further transformed to the actual phosphane ligand **112** in four steps. This ligand was then tested in allylic substitution reactions with the cyclic alkenes. Good to very good yields of the substitution products along with good to excellent enantioselectivities were easily achieved in the case of six- and seven-membered rings.

Feringa and Jansen [183] reported the Michael addition of lithio diphenylphosphine to γ-butenolides (Scheme 3.84). The reaction with methoxy-2(5*H*)-furanone furnished the lactone **113** in high yield and with high diastereoselectivity in favor of the *trans*-isomer. Moreover, using the enantiomerically pure butenolide synthon (5*R*)-menthyloxy-2(5*H*)-furanone, the asymmetric Michael addition of lithio-diphenylphosphide was followed by trapping the intermediate with chlorodiphenylphosphine to afford lactone **114** as a single diastereoisomer. The enantiomerically pure (*S*,*S*)-Chiraphos was obtained from lactone **114** in an overall yield of 35%.

Phosphine-boranes can react as nucleophiles similarly to corresponding secondary phosphines. LeCorre and co-workers [184] have prepared bis-tertiary phosphine boranes **116** by addition of diphenylphosphine boranes to dienes **115**. Although **115** was reacted as a mixture of *E*/*Z* isomers, a single diastereomer **116** was obtained from the reaction of $Ph_2P(BH_3)Na$ and **115** in THF. However, the utilization of a stoichiometric amount of KOH yielded both diastereoisomers of **116**, which were separated by crystallization (Scheme 3.85).

Asymmetric versions of phospha-Michael reactions in most cases use substrate-controlled diastereoselective additions. For example, the substrate-controlled

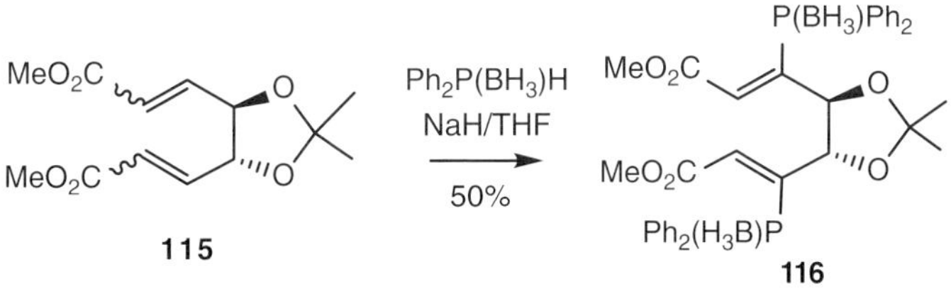

Scheme 3.85 Corre's stereoselective addition of phosphine-borane complex to bis-electrophile.

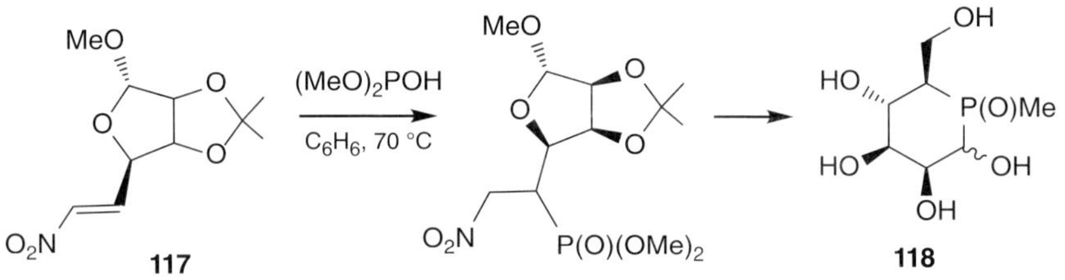

Scheme 3.86 Yamamoto's substrate-controlled diastereoselective addition of P-nucleophiles to unsaturated nitrosaccharides.

diastereoselective addition of phosphorus nucleophiles to unsaturated *Z*-nitroalkenes derived from sugar led to the formation of **118** with moderate stereoselectivities on heating to 70 °C [185]. This reaction demonstrated an interesting and accessible route to the preparation of sugar analogs **117** with a phosphorus atom in place of oxygen in the hemiacetal ring (Scheme 3.86). Yamashita and co-workers [186] described the diastereoselective addition of phosphorus nucleophiles to sugar derivatives of nitroalkenes catalyzed by Et_3N at heating to 90 °C (Scheme 3.87). The major product was L-idose derivatives **118** due to the steric hindrance caused by the 3-*O*-alkyl group of the sugar, as well as the R_2 and R_3 groups of the phosphorous compounds. The ratio of L-idose and D-glucose derivatives increased from 2 : 1 to 11 : 1 with different steric size of R_1-R_3. The highest dr = 11 : 1 was obtained when R^1 = Me, $R^2 = R^3$ = Ph, and X = O (Scheme 3.88). A one-step reaction of (R_P)-menthylphenylphosphine oxide with α,β-unsaturated aldehydes catalyzed by KOH at room temperature led to the formation of P,C-stereogenic 1,3-bis-phosphinylpropanes **122** as a single stereomer containing five stereogenic centers [186].

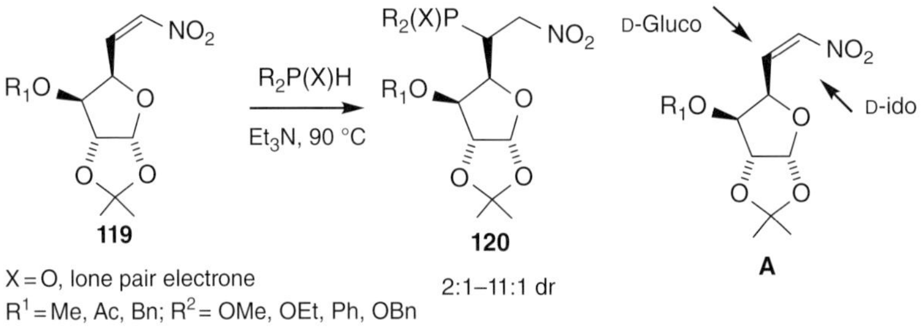

Scheme 3.87 Diastereoselective addition of phosphorus nucleophiles to nitroalkenes.

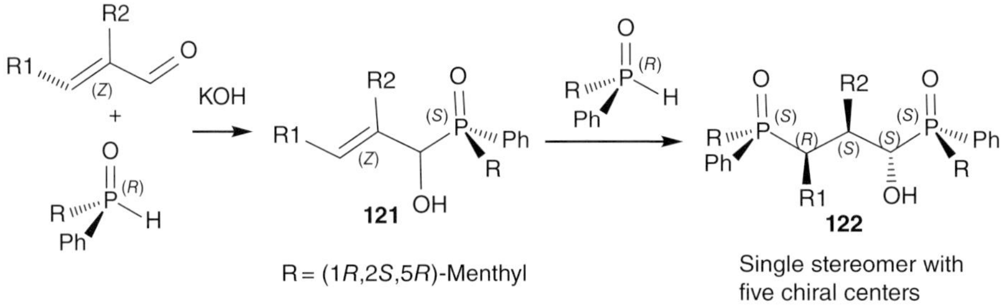

Scheme 3.88 Preparation of P,C-stereogenic 1,3-bis-phosphinylpropanes.

3.5 Asymmetric Reduction

Asymmetric reduction of ketophosphonates is one of the most convenient methods for the synthesis of chiral hydroxyphosphonates [187–199]. Asymmetric reduction can occur (i) under the control of a chiral reductant, (ii) under the control of asymmetric catalysis (organometallic or organocatalyst), (iii) via the transfer of chirality from chirogenic phosphorus, or (iv) via the chiral center in a side chain. For example, acylphosphonates were directly reduced by sodium borohydride to the corresponding hydroxyphosphonates and dihydroxyalkanebisphosphonates (Scheme 3.89).

The reduction of [(*N*-*p*-toluenesulfonyl) amino]-β-ketophosphonates **123** with different borohydrides gave [(*N*-*p*-toluenesulfonyl)amino]-β-hydroxyphosphonates **124** and **125** in good chemical yields and moderate diastereoselectivity (Table 3.3). The best

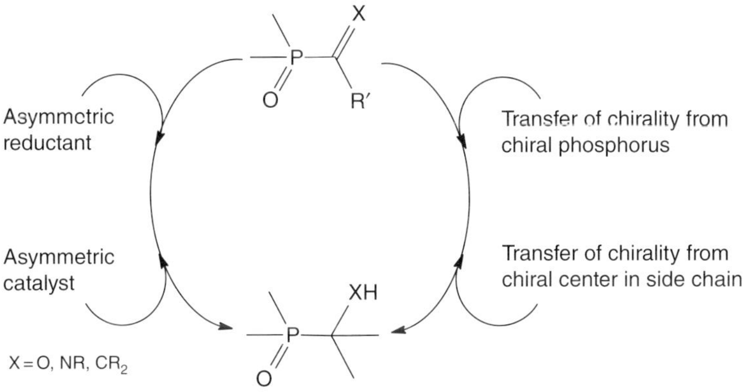

Scheme 3.89 Version of asymmetric reduction.

Table 3.3 Reduction of β-ketophosphonates with different borohydrides.

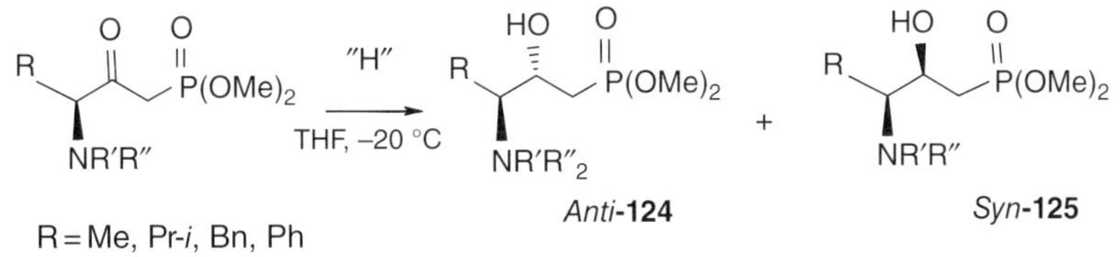

R	R′	R″	"H" = Hydride	Yield (%)	*dr*	References
i-Pr	H	*p*-Ts	$LiBH_4$/THF	98	53 : 47	[200]
i-Pr	H	*p*-Ts	$NaBH_4$/MeOH	99	81 : 19	[200]
i-Bu	H	*p*-Ts	$NaBH_4$/MeOH	98	29 : 71	[200]
Ph	H	*p*-Ts	$NaBH_4$/MeOH	97	63 : 37	[200]
Ph	H	Bn	$NaBH_4$/MeOH	75	63 : 37	[187]
i-Pr	Bn	Bn	$NaBH_4$/THF	44	85 : 15	[188]
i-Pr	H	*p*-Ts	$Zn(BH_4)_2$/THF	91	77 : 23	[200]
Ph	H	Bn	$Zn(BH_4)_2$/THF	80	88 : 12	[187]
i-Pr	H	Bn	$Zn(BH_4)_2$/THF	85	96 : 4	[187]

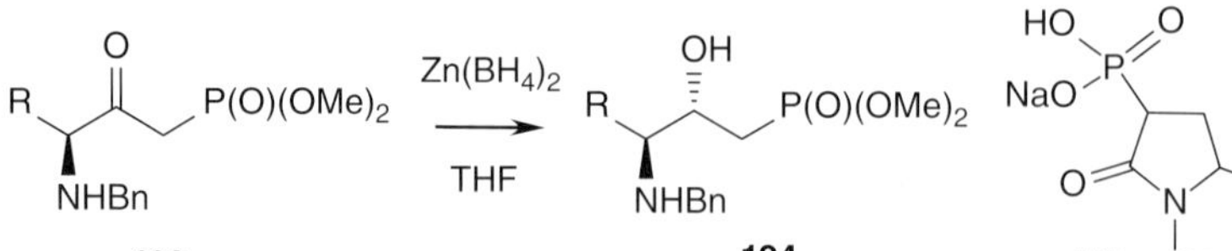

123 124 125

Scheme 3.90 Reduction of γ-*N*-benzylamino β-keto phosphonates with zinc borohydride.

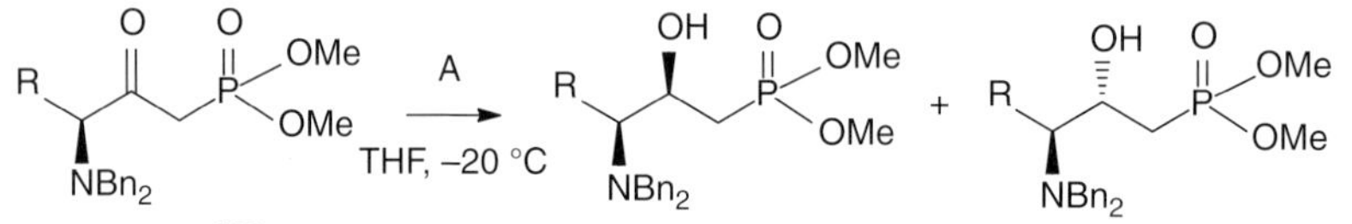

123

R = Me, *i*-Pr, Bn, Ph

Syn-**126** *Anti*-**126**

A	Yield (%)	85:86
DIBAL-H	50	82:18
$NaBH_4$	44	85:15
Catecholborane	69	>98:2
Catecholborane	85	>98:2
Catecholborane	89	>98:2
Catecholborane	82	90:10

Scheme 3.91 Reduction of ketophosphonates.

stereoselectivity was attained with zinc borohydride reduction of **123** resulted in the formation of *anti*-β-hydroxy-α-aminophosphonates **124** with good diastereoselectivity [188]. This methodology was used to prepare 1,5-dihydroxy-2-oxopyrrolidine-3-phosphonic acid **125** (SF-2312), which is active against gram-positive and gram-negative bacteria (Schemes 3.90 and 3.91) [187, 188]. The reduction of the γ-*N*-benzylamino-β-ketophosphonate derived from readily available (*S*)-tribenzylated amino acids was achieved with catecholborane to afford γ-amino-β-hydroxyphosphonates in high diastereoselectivity [200]. The reduction of β-keto-γ-*N*,*N*-dibenzylaminophosphonates **123** with catecholborane resulted in the formation of the β-hydroxy-γ-aminophosphonates *syn*- and *anti*-**126** in good yields and with high diastereoselectivity [189, 190]. The results summarized in Table 3.3 show that the diastereoselectivity of the reduction of *N*,*N*-dibenzylamino-β-ketophosphon4 and **126** were separated and their configuration was confirmed by X-ray analysis and NMR spectroscopy (Scheme 3.91) [191].

The reduction of dimethyl 3-*N*,*N*-di(α-methylbenzyl)amino-2-ketophosphonates **127** with catecholborane at 78 °C in presence of $LiClO_4$, gave γ-amino-β-hydroxy phosphonates **128** in good yield and with excellent diastereoselectivity. The hydrolysis and hydrogenation of **128** afforded the (*R*)-α-amino-β-hydroxypropylphosphonic acid **129** (GABOBP). The reduction of β-ketophosphonates bearing two chiral α-methylbenzyl auxiliaries proceeded with higher diastereoselectivity than the reduction of β-ketophosphonates with one α-methylbenzyl group [191] (Scheme 3.92).

The reduction of γ-*N*-benzylamino-β-ketophosphonate **130**, which was prepared from L-proline, can be carried out in high diastereoselectivity with catecholborane (CB) in THF at −78 °C to produce the γ-*N*-benzylamino-β-hydroxyphosphonates *syn*-**131**

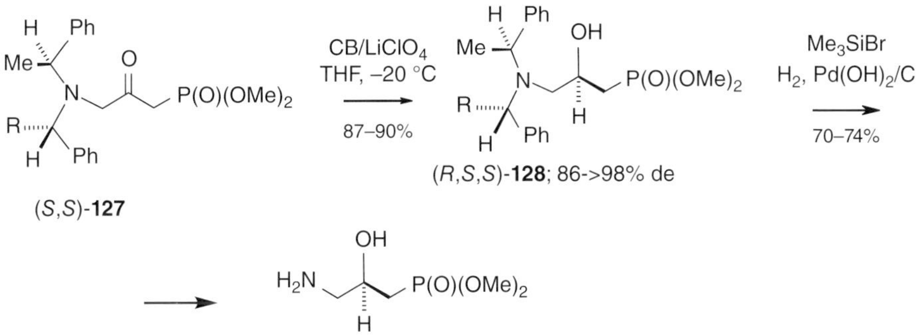

(*S*,*S*)-**127**

(*R*,*S*,*S*)-**128**; 86->98% de

(*R*)-**129**; 86->98% de

Entry	R	**127**	Hydride	Conditions	Yield (%)	de (%)
1	Me	(*S*)	$NaBH_4$	MeOH,0 °C	96	70
2	Me	(*S*)	$NaBH_4$	THF,0 °C	95	60
3	Me	(*S*)	$LiBH_4$	THF, 78 °C	93	62
4	Me	(*S*)	$Zn(BH_4)_2$	THF, 78 °C	95	54
5	Me	(*S*)	DIBAL-H	THF, 78 °C	98	14
6	Me	(*S*)	CB	THF, 78 °C	91	86
7	Me	(*S*)	$CB/LiClO_4$	THF, 78 °C	89	>98
8	Me	(*R*)	CB	THF, 78 °C	91	86
9	Me	(*R*)	$CB/LiClO_4$	THF, 78 °C	87	>98
10	H	(*R*)	CB THF	78 °C	95	16
11	H	(*R*)	$CB/LiClO_4$,THF	78 °C	91	24

Scheme 3.92 Reduction of **127** with various reducing agents.

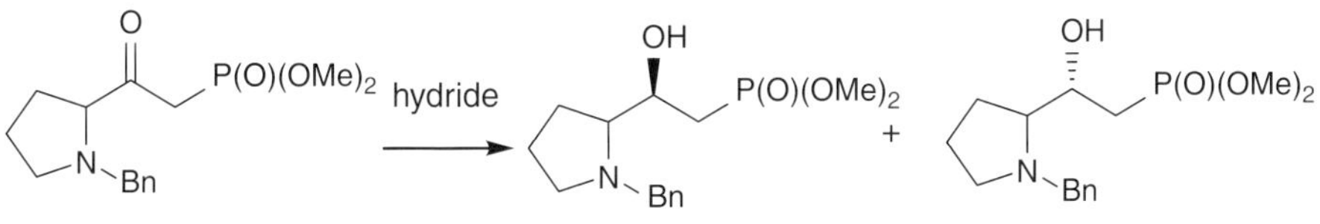

130 *Syn*-**131** *Anti*-**131**

Hydride	Conditions	Yield,%	Syn:Anti
$NaBH_4$	MeOH, 25 °C	70	69:31
$LiBH_4$	THF, –78 °C	69	75:25
DIBAL-H	THF, –78 °C	69	79:21
CB	THF, –78 °C	78	>96:4

Scheme 3.93

and *anti*-**131** in 96 : 4 ratio (Scheme 3.93) [192]. Proline-like 2,4-dialkyl-5-phosphonylpyrrolidines **132** were obtained stereoselectively by reduction of the corresponding β-iminophosphonates with $NaBH_4$ (Scheme 3.94) [193].

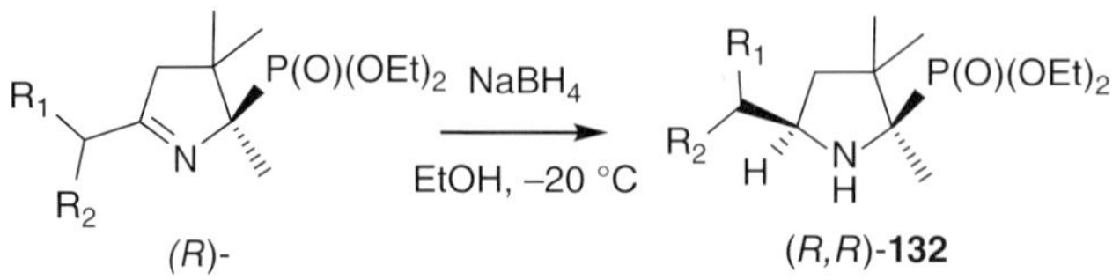

Scheme 3.94 Synthesis of 2,4-dialkyl-5-phosphonylpyrrolidines.

The formation of the γ-amino-β-hydroxyphosphonates *syn*-**131** as a major diastereoisomer in the reduction of the β-ketophosphonates **130** with catecholborane indicates that the reduction proceeds under the non-chelation or Felkin-Anh model control, and the bulkiness of the *N*-benzylamino- and *N*,*N*-dibenzylamino-groups in the β-ketophosphonates are sufficient to limit the rotamer populations blocking the *Re* face of the carbonyl group that leads to the addition of hydride by the *Si* face (Figure 3.3) [191].

Palacios has reported an efficient method for the preparation of quaternary α-aminophosphonic acid derivatives **135**. The nucleophilic addition of organometallic reagents to α-ketiminophosphonates **134** led to the formation of quaternary α-aminophosphonates. The diastereoselective synthesis of a quaternary α-aminophosphonate using a chiral tartaric-acid-derived phosphonate was also described [153] (Scheme 3.95).

Barco *et al.* [190] reported the synthesis of β-amino-α-hydroxyphosphonates **138** by reduction of β-phthalimido-α-keto phosphonates **136** with borane-dimethylsulfide complex and oxazaborolidine as catalyst in good yields and high diastereoselectivity. The phosphonates **137** were deprotected with hydrazine to furnish diethyl 2-amino-1-hydroxy phosphonates **138** in quantitative yields. Oshikawa and Yamashita [194] reported that the reduction of β-phthalimido-α-ketophosphonates **136** (R = Me, *i*-Bu) with sodium cyanoborohydride, resulted in the formation of β-amino-α-hydroxyphosphonates **137** in almost quantitative yield but in low diastereoselectivity (ratio of 2 : 1 to 3 : 1) (Scheme 3.96).

The reduction of α-ketophosphonates **139** with (−)-chlorodiisopinocampheylboranes (Ipc_2B-Cl) yielded (*S*)-α-hydroxyphosphonates with 65% ee [195]. This method allowed to obtain α-hydroxy-β-aminophosphonates and β-hydroxy-γ-iminophosphonates

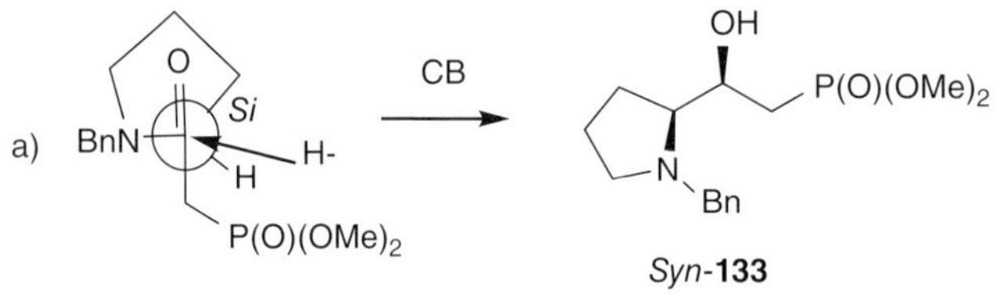

Figure 3.3 Preparation of γ-amino-β-hydroxyphosphonates **131**.

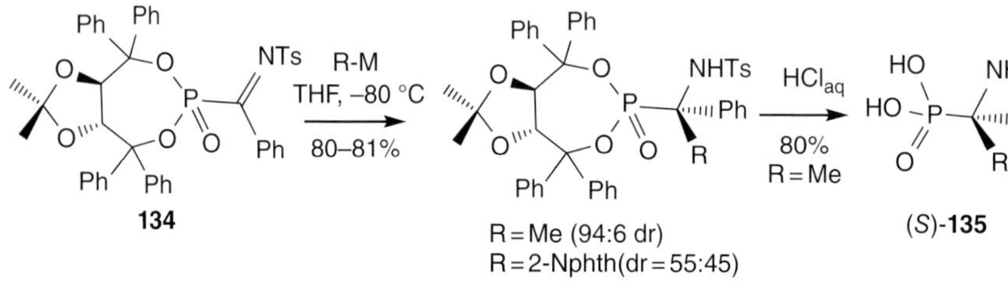

Scheme 3.95

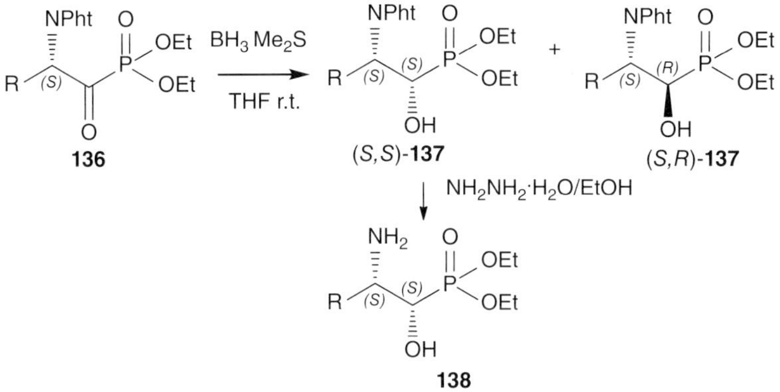

R	Yield (%)	Ratio 4/5
Bn	66	8:1
p-$BnOC_6H_4CH_2$	60	8:1
i-Bu	66	9:1
Me	62	10:1

Scheme 3.96 Reductions of keto phosphonates **136** with the borane-dimethylsulfide complex.

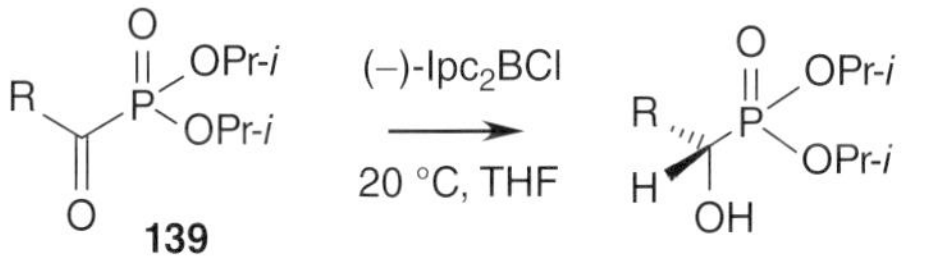

Scheme 3.97 Asymmetric reduction of ketophosphonates **139** with (–)-Ipc_2BCl.

(Scheme 3.97). Maier used catecholborane and oxazaborolidine as catalyst for the asymmetric reduction of ketophosphonates [196, 197].

The reduction of ketophosphonates **140** with sodium borohydride in THF proceeded with low stereoselectivity (30–35% de) [187] which was increased via the formation of a chiral complex of sodium borohydride with natural (*R*,*R*)-tartaric acid [198]. The reduction of ketophosphonates **140** with this chiral complex afforded the diethyl (1*S*)-α- and β-hydroxyphosphonates with 60–85% ee and the dimenthyl (1*S*)-α-hydroxyphosphonates with diastereomeric purities of 80–93%. This methodology was applied to the synthesis of (*S*)- and (*R*)-phosphocarnitines **142** and **143** with both enantiomers being prepared using the (*R*,*R*)- and (*S*,*S*)-tartaric acid (Scheme 3.98) [198]. The phosphocarnitines **142** and **143** were isolated as colorless pure solids.

Diethyl (R)-(D)-2-hydroxy-3-chloropropylphosphonate. (*R*,*R*)-(+)-Tartaric acid (1.5 g, 10 mmol) was added to a suspension of sodium borohydride (0.36 g, 10 mmol) in 50 ml of THF and then the reaction mixture was refluxed for 4 h. After that, a solution of ketophosphonate (2.5 mmol) in 10 ml of THF was added at 30 °C and the reaction mixture was stirred at this temperature for 24 h. Then 20 ml of ethyl acetate and 30 ml of 1 N hydrochloric acid was added dropwise to the reaction mixture. The organic layer was separated, the aqueous phase was saturated with NaCl, and extracted two times with ethyl acetate (15 ml). The organic extracts were washed with a saturated solution of Na_2CO_3 (320 ml) and dried with Na_2SO_4. The solvent

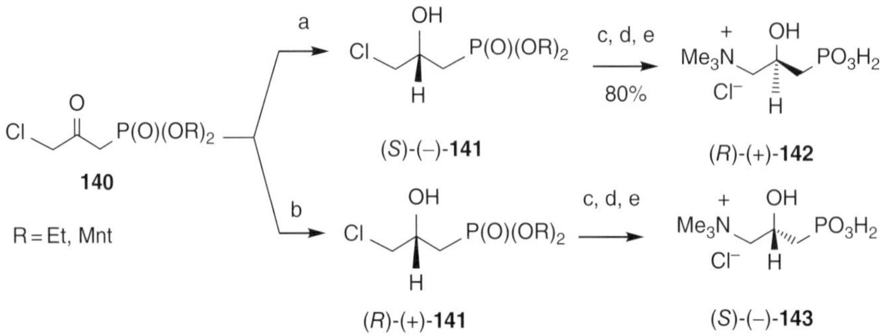

a = $NaBH_4$/(R,R)-TA; b = $NaBH_4$/(S,S)-TA; c = H_3O^+; d = Me_3N/H_2O; e = $Me_3NH^+Cl^-$

Scheme 3.98 Synthesis of (S)- and (R)-phosphocarnitines.

was removed under vacuum and the residue was crystallized from acetonitrile (yield: 85%, yellow oil $[\alpha]_D^{20}$ +12.4 (c = 3.2, $CHCl_3$) δ_P (121.4 MHz, $CDCl_3$) 29.4).

3.6 Asymmetric Oxidation

The asymmetric oxidation of carbon–carbon bonds or carbanions is a useful route to a variety of hydroxyl and dihydroxyphosphonates. For example, the chiral α-hydroxyphosphonates **144** of high ee (96–98% ee) were prepared by the stereoselective oxaziridine-mediated hydroxylation of dialkyl benzylphosphonates **143**. α Hydroxyphosphonates were converted into corresponding free phosphonic acids and retained a high degree of stereochemical purity (90–98% ee) (Scheme 3.99) [199, 201].

The Sharpless asymmetric dihydroxylation is the chemical reaction of an alkene with chiral osmium catalysts (Os_2O_3 with quinine ligand) to form a vicinal diol.

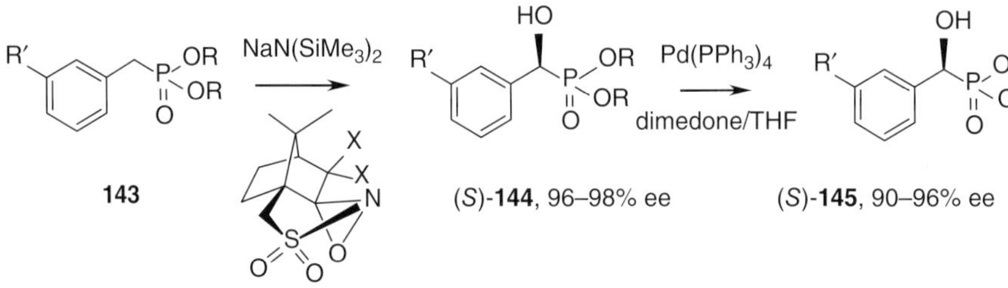

R	R′	X	ee (%)
Me	H	H	69
Me	H	Cl	93
Me	NO_2	Cl	80
Me	Cl	Cl	87
Me	MeO	Cl	81
Et	CF_3	Cl	98
All	H	Cl	96
All	PhO	Cl	98

Scheme 3.99

The Sharpless dihydroxylation was used for the enantioselective preparation of 1,2-dihydroxyphosphonates from alkenphosphonates. This procedure was performed with an osmium catalyst ($(DHQD)_2$PHAL, $(DHQ)_2$PHAL or their derivatives) and a stoichiometric oxidant (e.g., $K_3Fe(CN)_6$ or *N*-methylmorpholine oxide (NMO)); it was carried out in a buffered solution to ensure a stable pH because the reaction proceeds too rapidly under basic conditions [202–206]. The dihydroxyalkylphosphonates **147** were prepared by asymmetric dihydroxylation of alkenphosphonates **146**. The dihydroxy derivatives **147** were converted into 2-amino-1-hydroxy- and 3-amino-2-hydroxyalkylphosphonates through formation of azidoderivatives and the catalytic reduction of the azide group [202] (Scheme 3.100).

Asymmetric aminohydroxylation of diethyl (*E*)-alkenylphosphonates **146** with potassium osmate(VI) dihydrate, toluene sulfonchloramide T, and $(DHQD)_2$PHAL as a chiral ligand led to the *threo*-1-hydroxy-2-aminophosphonic acids **148** in good yields (55–75%) and enantioselectivities from 45% to 92% ee [203, 204] (Scheme 3.101). Asymmetric dihydroxylation of (*E*)-alkenylphosphonates **146** with AD-mix-α or AD-mix-β reagents led to the formation of optically active *threo*-α,β-dihydro-xyphosphonates **151** (Scheme 3.102) [206–208].

The highest level of enantioselectivity (>88% ee) was observed in the oxidation of (*E*)-alkenylphosphonates with conjugated aromatic substituents. Enantioselectivities and yields were significantly improved when the dihydroxylation reaction was carried out with the dimethyl phosphonate instead of diethyl phosphonate.

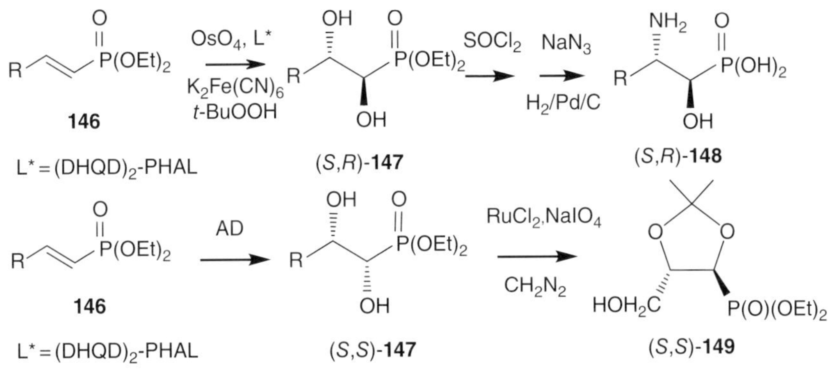

AD	R	Yield (%)	ee (%)
AD-mix-β	Ph	42	91
AD-mix-α	4-MeOC_6H_4	71	95
AD-mix-β	4-MeOC_6H_4	69	98
AD-mix-α	4-ClC_6H_4	65	98
AD-mix-α	1-Naphthyl	80	93
AD-mix-β	*n*-C_7H_{13}	63	84
AD-mix-α	3-MeOC_6H_4	67	96
AD-mix-α	2-Furyl	17	88

Scheme 3.100

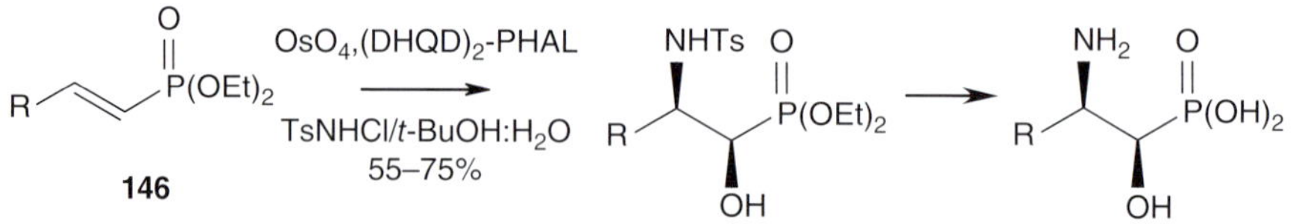

146 (*R*,*R*)-**149**, 45–75% ee (*R*,*R*)-**148**

R = Ph, 4-BrC_6H_4, 4-$O_2NC_6H_4$, 4-$MeOC_6H_4$, H

Scheme 3.101

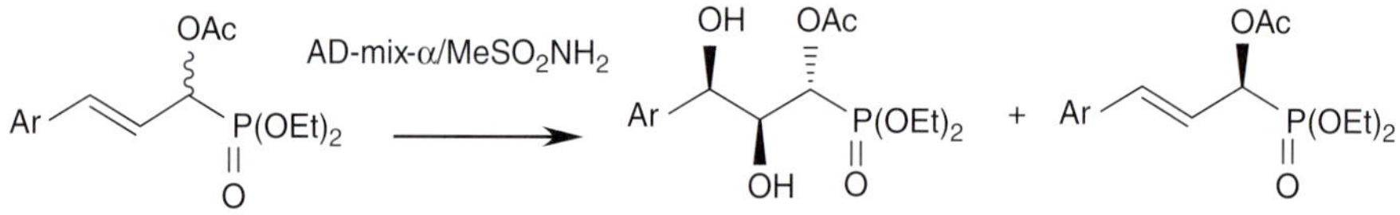

(*S*/*R*)-**150** (*R*,*R*)-**151** (*S*)-**150**, 40–99% ee

Scheme 3.102

α,β-Dihydroxyphosphonates **147** (R = *p*-An) via a two-step synthesis (oxidization with $RuCl_3/NaIO_4$ and reduction with $NaBH_4$) were transformed into dioxolane **149**, which is a chiron, for the asymmetric synthesis of α-heteroatom-substituted phosphonates [197]. The racemic mixture of (*S*/*R*)-1-acetoxy-2-(*E*)-alkenylphosphonates **150** was resolved by kinetically controlled dihydroxylation with AD-mix-α or AD-mix-β [205]. The stereochemistry of dihydroxylation depends on the configuration of the α-carbon atom and the nature of substituents at the double bond. For example, AD-mix-β oxidizes mainly the (*R*)-enantiomer of alkenphosphonate **150**; hence, in addition to the optically active (1*R*,2*S*,3*R*)-1-acetoxy-2,3-dihydroxyphosphonate **151** formed, the reaction mixture contains nonconsumed (*S*)-acyloxyphosphonate **150**, which was isolated with a good optical purity (Scheme 3.102) [207].

Cristau, Kafarski *et al.* have developed a diastereoselective way of synthesis of 2-amino-1-hydroxy-2-arylethylphosphonic esters **153** by opening of *trans*-1,2-epoxy-2-arylethylphosphonic esters with 28% NH_{3aq} in methanol [209, 210]. Diastereomeric diethyl (1*R*,2*R*)- and (1*S*,2*R*)-2,3-epoxy-1-benzyloxypropylphosphonates **152** were obtained from the 2,3-*O*-cyclohexylidene-1-hydroxypropylphosphonates via the following sequence of reactions: benzylation, acetal hydrolysis, and transformation of the terminal diol into the epoxide using the Sharpless protocol. These epoxides were regioselectively opened with dibenzylamine to afford the compounds **153** after acetylation and hydrogenolysis [211] (Scheme 3.103).

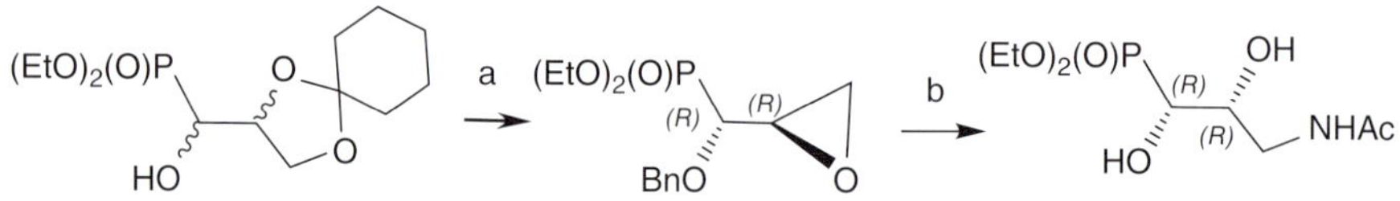

(1*R*,2*R*)-**152** (1*R*,2*R*)-**153**

(a) BnBr, Ag_2O, HCl, dioxane, $MeC(OMe)_3$, PPTS, AcBr, K_2CO_3, MeOH;
(b) $HNBn_2$, Ac_2O, NEt_3, H_2-Pd(Î)$_2$/C

Scheme 3.103

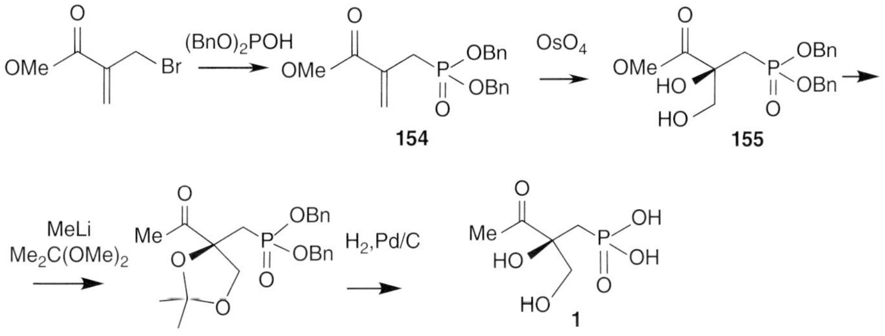

Scheme 3.104 First synthesis of phosphonothrixin **1**.

Asymmetric epoxidation of allylphosphonates was extensively used in the asymmetric synthesis of naturally occurring hydroxyphosphonates. In particular, Nakamura *et al.* developed several methods of syntheses of PTX **1**. The first method was developed starting from bromomethylallylketone, which, via a Michaelis–Becker reaction, was converted into allylphosphinate **154**. This was oxidized by dihydroxylation with osmium tetroxide to give the dihydroxyphosphonate **155**, which was converted into PTX **1** (Scheme 3.104) [211].

The second synthesis of PTX **1** by the same authors employed the dienyl-alcohol as the starting reactant. Catalytic asymmetric epoxidation of dienyl-alcohol using D-DETA furnished the chiral (*R*)-epoxy alcohol **157** (92% ee, 57% yield). The C–P bond was formed by adding the chloromagnesium of dibenzyl phosphite. Ozonolysis and debenzylation gave (*S*)-PTX in 79% yield and 92% ee. The (*R*)-enantiomer of PTX **1** was also prepared using this methodology. On the basis of the specific rotation and biological activity, the natural product was determined to have (*S*)-configuration (Scheme 3.105)[211, 212].

The third synthesis of PTX **1** was accomplished in six steps with 24% overall yield from the commercially available methyl 3-hydroxy-2-methylene butyrate **159**, which was phosphorylated with diethyl chlorophosphite in the presence of triethylamine to give the *E*-allylphosphonate **160** (60% yield) forming the key C–P bond via an intramolecular Arbusov rearrangement. The vicinal dihydroxylation to diol **161** followed by the oxidation led to the formation of protected PTX in 80% yield, which was deprotected by excess TMSI in CH_2C1_2 and aqueous HF in MeCN (Scheme 3.106) [212].

Natural 1-HO-AEP **162**, which was isolated from the membrane of the soil amoeba *Acanthamoeba castellanii*, was synthesized by the addition of sodium dimethyl phosphite to *N*-(2-oxoethyl)phthalimide followed by removal of the phthaloyl group with hydrazine and acid hydrolysis of the ester groups [213, 214] (Scheme 3.107).

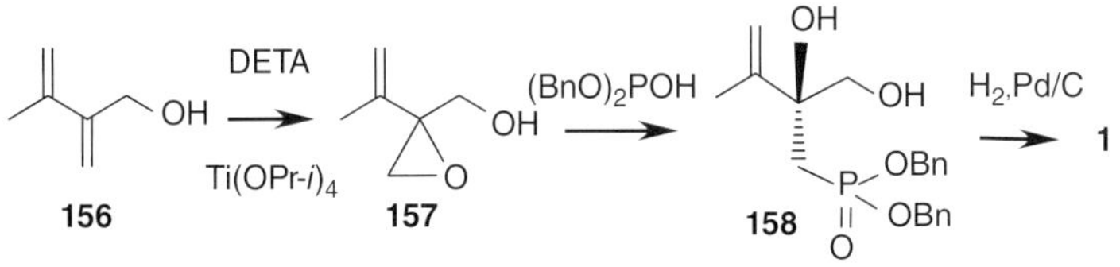

Scheme 3.105 Second synthesis of phosphonothrixine **1**.

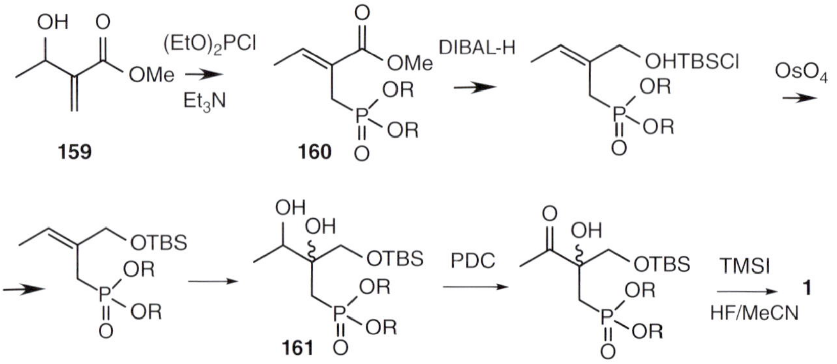

Scheme 3.106 Third synthesis of phosphonothrixine **1**.

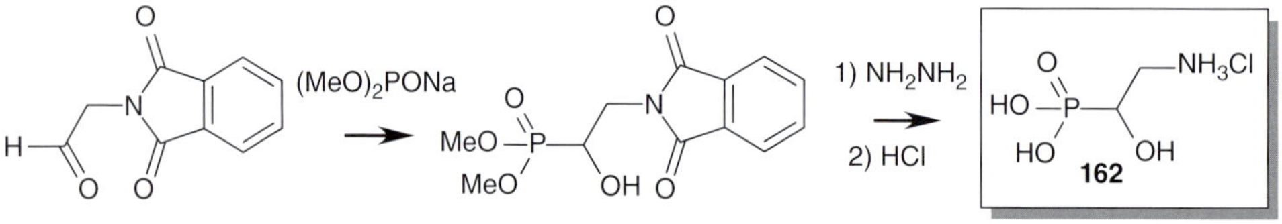

Scheme 3.107 Synthesis of 1-hydroxy-2-aminoethylphosphonic acid.

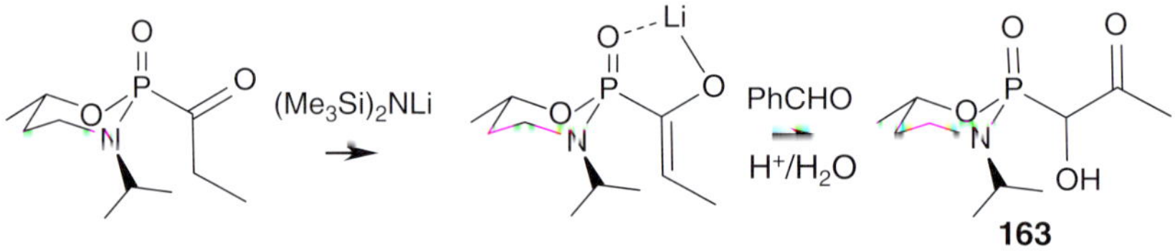

R = EtO, R^2 = *i*-PrN$(CH_2)_3$O

Scheme 3.108

3.7 C-Modification

The addition of the lithium derivative of (*S*,*S*)-2-(α-hydroxypropenyl)-2-oxo-l,3,2-oxazaphosphorinane to benzaldehyde results in a diastereomer mixture of α-hydroxy-2-oxo-l,3,2-oxaza-phosphorinanes **163** in 3 : 1 ratio (Scheme 3.108) [215, 216]. β-Hydroxyphosphonates **164** were obtained by reaction of metal phosphonates with carbonyl compounds in THF or DME-TMEDA (Scheme 3.109) [217]. β-Hydroxyalkylphosphonates were prepared with moderate diastereoselectivity by the reaction of phosphonate carbanions with carbonyl compounds in THF. The lithium derivative of diethyl isothiocyanomethylphosphonate was reacted with aldehydes to afford a mixture of *cis*- and *trans*-(2-thioxo-oxazolidine-4-yl)-phosphonates **165**, which were separated by column chromatography and converted into the *N*-Boc-1-amino-2-hydroxyalkylphosphonates **166** (Scheme 3.110)[217, 218].

The reaction of achiral *N*,*O*-acetals with triphenylphosphite in the presence of $TiCl_4$ led to the formation of *N*-acetylated α-aminophosphonates **167** in low diastereoselectivity. Pure diastereomers **167** were obtained by fractional

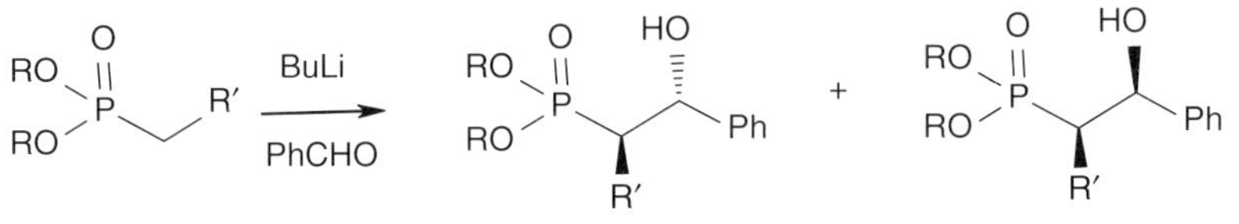

(R,R)-**164** (R,S)-**164**

R	R′	Solvent	T (°C)	de (%)
Me	Me	THF	−95	85
Me	Me	THF–TMEDA	−95	81
Me	Me	Et_2O	−78	42
Me	Me	DME-TMEDA	−78	61

Scheme 3.109

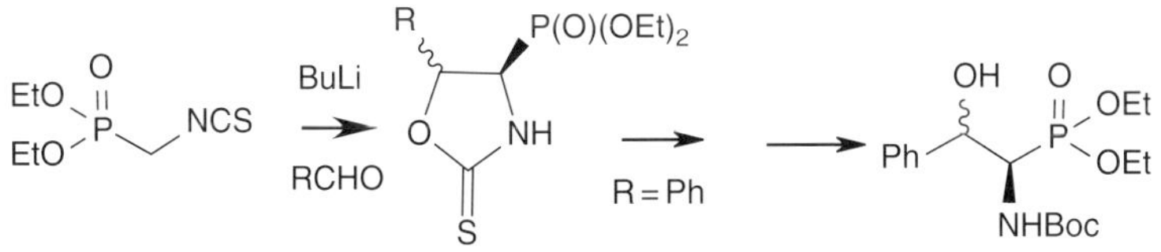

165 **166**

Scheme 3.110

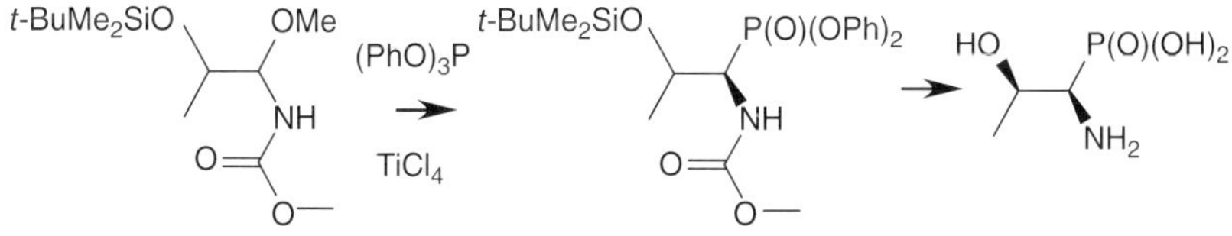

(R,R)/(S,S)-**167**, dr=2:1

Scheme 3.111

recrystallization. Diastereomerically enriched hydroxyphosphonates prepared by a phospha-aldol reaction were additionally purified by preparative column chromatography (Scheme 3.111) [219].

3.8 Asymmetric Cycloaddition

Cycloaddition is a process in which two or more reactants combine to form a stable cyclic molecule and during which no small fragments are eliminated and bonds are formed but not broken. Cycloadditions are conveniently classified by the number of ring atoms contributed by each component to the new ring, or the number of electrons (usually π electrons) involved in each component. In this section, asymmetric induction in the Diels-Alder reaction and stereoselective 1,3-dipolar cycloadditions are discussed. For example, [3+2]-cycloadditions of alkene-phosphine oxides to nitrones are represented by 1,3-dipolar cycloadditions. In the addition of alkenes to nitrones, either partner may be chiral. In the examples shown in Scheme 3.112, the tertiary alken-phosphine or nitrone is chiral.

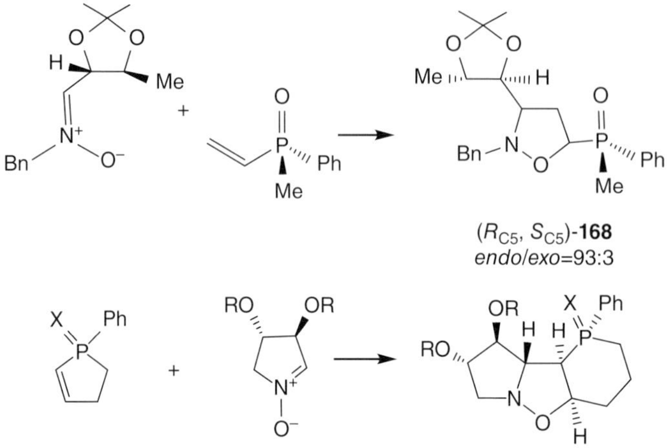

(R_{C5}, S_{C5})-**168**
endo/exo=93:3

169

Scheme 3.112 Asymmetric induction in [3 + 2]-cycloaddition.

Brandi and Pietrusiewicz demonstrated examples of 1,3-dipolar cycloadditions of prochiral divinylphosphine derivatives to a five-membered **168, 169** with predictable stereochemistry at phosphorus [220–222]. Enantiomerically pure five-membered ring nitrones derived from L-tartaric acid via C_2-symmetric *O,O′*-protected 3,4-dihydroxypyrrolidines undergo highly regio- and stereoselective cycloaddition reactions with racemic 2,3-dihydro-1-phenyl-1*H*-phosphole 1-oxide and 1-sulfide. In all cases, the formation of only two diastereomeric cycloadducts was observed and their ratio (up to 10 : 1) was dependent on the size of the protecting group at the nitrone and on the extent of conversion (Scheme 3.112). The 1,3-dipolar cycloaddition of nitrones to diphenylvinylphosphine oxides, sulfides, and selenides proceeded stereoselectively to form cycloadducts **170** under conditions that avoid cycloreversion. The selectivity decreases with an increase in the electron-withdrawing ability of the substituent according to the sequence: $Ph_2P > PhMeP(O) > Ph_2P(S) \geq Ph_2P(O) > (EtO)_2P(O)$ [222] (Scheme 3.113).

The asymmetric 1,3-dipolar cycloaddition of the P-stereogenic dipolarophile **171** to *C,N*-diphenylnitrone led to P-stereogenic isoxazolinyl diphosphine dioxides (R_P,S_P)- and (R_P,S_P)-**172**, with 1.5 : 1 dr, which were separated by column chromatography. The reaction time was reduced by means of microwave irradiation from 48 h to 40 min. The stereospecific reduction of isoxazolinyl diphosphine dioxides with $Ti(OiPr)_4$/PMHS resulted in enantio- and diastereomerically pure diphosphines (R_P,S_P)-**173** in high yield with retention of configuration at the phosphorus atoms. The treatment of the crude reduction mixtures with BH_3/THF resulted in the formation of diastereomerically

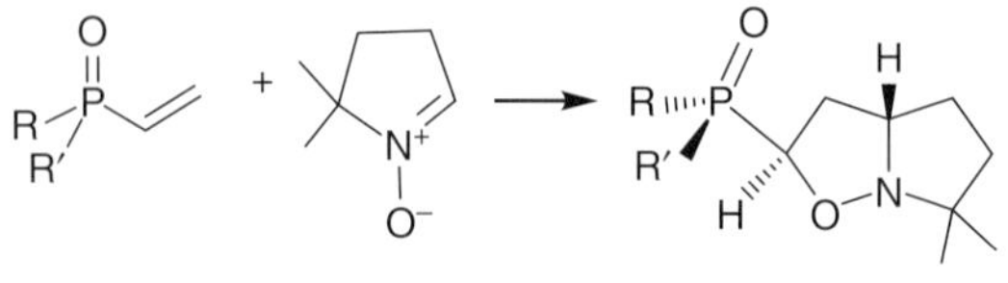

170

Scheme 3.113

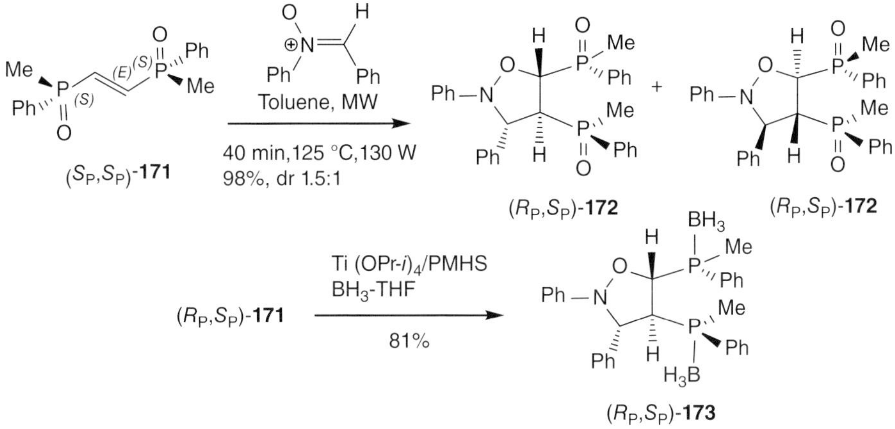

Scheme 3.114 The Diels–Alder cycloaddition of the di-P-stereogenic dienophile.

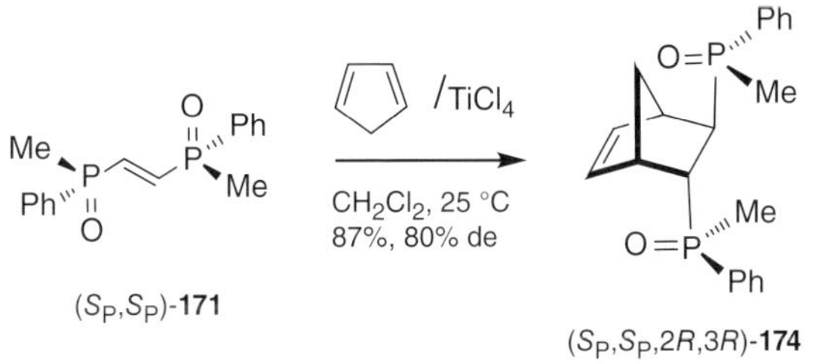

Scheme 3.115 The Diels–Alder cycloaddition of the di-P-stereogenic dienophile (S_P,S_P)-**171**.

pure diphosphine diboranes (S_P,R_P)-**173** in 81–84% yield (Scheme 3.114) [223, 224]. The Diels–Alder cycloaddition of the di-P-stereogenic dienophile (S_P,S_P)-**171** to cyclopentadiene led to di-P-stereogenic norbornene derivatives **174** and proceeded with moderate diastereoselectivity in the absence of a catalyst. However in the presence of $TiCl_4$, the diastereoselectivity was raised to 9 : 1. The separation of the diastereomeric cycloadducts was performed by fractional crystallization in the presence of (−)-*O*,*O*-dibenzoyltartaric acid monohydrate. The X-ray analysis of the structure confirmed that the (S_P)-methyl(phenyl)phosphinoyl groups are in *exo*- and *endo*-positions, *trans*- to one another, and that the absolute configurations at the C2, C3 carbon atoms are (*R*) [224] (Scheme 3.115).

3.9 Multiple Stereoselectivity

One of the methods to increase the stereoselectivity of reactions is multiple stereoselectivity (multiple stereodifferentiation and multiple asymmetric induction), when the stereochemical process proceeds under the control of more than one chiral auxiliary [225]. In order to obtain the highest stereoselectivity, it is necessary to introduce two or several chiral asymmetric centers into the reaction system. In this case, we obtain double stereoselectivity or multistereoselectivity if we have several chiral asymmetric centers taking part in the asymmetric synthesis. It is obvious that every additional chiral auxiliary in a reaction system affects asymmetric induction and changes the difference between the activated diastereomeric forms. The individual stereochemical properties of chiral auxiliaries present in a reacting system can, as a rule, reinforce one another (*matched asymmetric synthesis*) or, on the contrary, counteract each other (*mismatched asymmetric synthesis*) [225, 226]. Stereochemical control of the formed stereogenic centers of synthesized compounds is attained by selection of (*R*)- or (*S*)-chiral reagents, and also their stereoselectivities There is particular interest in the cooperative catalysis, consisting in introduction of several chiral centers into a ligand or cooperation of two different catalysts in one reaction process. The reaction of chiral di- and trialkylphosphites, derivatives of (1*R*,2*S*,5*R*)-menthol, *endo*-bomeol, or di-*O*-isopropylidene-1,2:5,6-α-D-glucofuranose with aldehydes under condition of phospha-aldol reaction took place with the transfer of chirality from phosphorus to the α-carbon atom in alkylphosphonates. Stereoselectivity of the reaction depended on structure of initial reagents from reaction conditions. For example, as shown in the Scheme 3.116, the (*S*) prolinal reacted with diethylphosphite with low stereoselectivity to give the diastereomeric mixture of (*S*,*R*) and (*S*,*S*)-hydroxyphosphonates in 2 : 1 ratio (single asymmetric induction); however, the stereoselectivity of reaction was raised when the chiral (*S*)-prolinal reacted with chiral trimenthylphosphite or dimenthylphosphite (double asymmetric induction) [227]. The chiral $(R^*O)_2POH$ and (R*O)[3]P essentially allow to increase the stereoselectivity of phospha-aldol reactions catalyzed by quinine or cinchonidine. For example, the quinine catalyzes the enantioselective phospha-aldol reaction of dialkyl phosphites with *ortho*-nitrobenzaldehyde leading to the formation of the *a*-hydroxyphosphonates **175** with moderate enantioselectivities [228]. However the stereoselectivity of the reaction was increased when dimenthyl phosphite or dibornylphosphite were reacted with aldehydes in the presence of quinine or cinchonidine, owing to double asymmetric induction [229]. The double as well as the triple asymmetric induction was applied to increase the

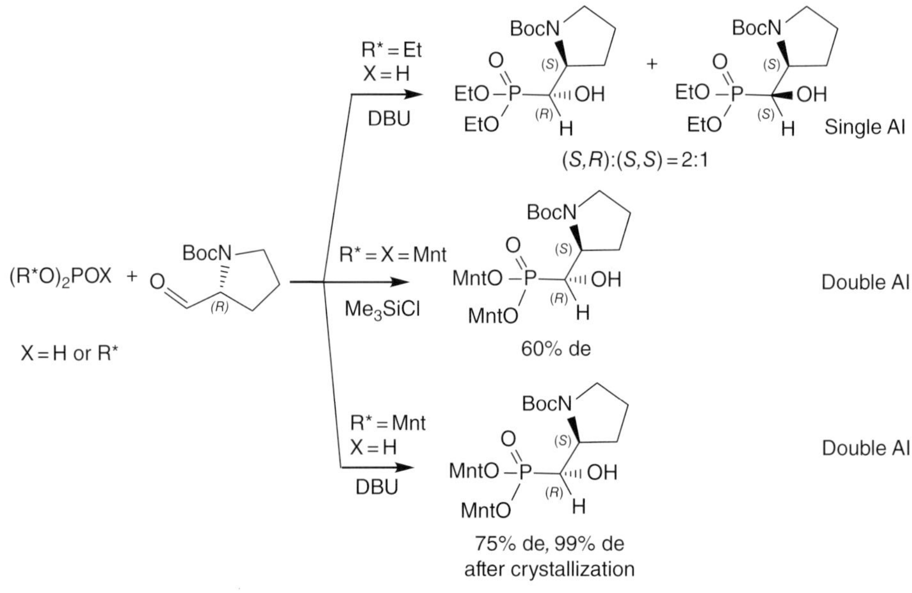

Scheme 3.116 Examples of single and double asymmetric induction (AI) of phospha-aldol reaction.

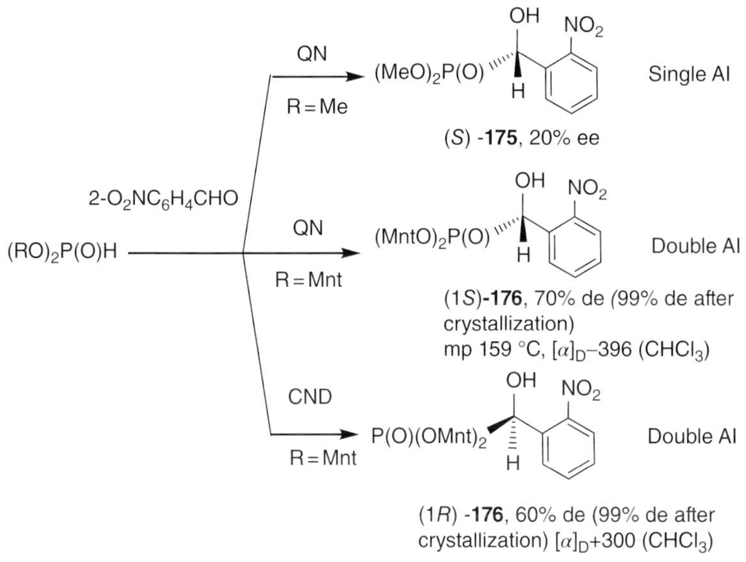

Scheme 3.117 Single and double asymmetric induction (AI) of phospha-aldol reaction catalyzed by cinchona alkaloids.

stereoselectivity of phospha-aldol reactions [230–232]. Double stereoselectivity was attained in the reaction between chiral di(l*R*,2*S*,5*R*)-menthylphosphite and chiral 2,3-*O*-isopropylidene-D-glyceraldehyde. The reaction of chiral phosphite with chiral aldehyde catalyzed by chiral (*S*)-ALB involving three chiral auxiliaries proceeded with highest stereoselectivity (85% de) (Scheme 3.117).

At the same time, the (*R*)-ALB catalyst did not increase the stereoselectivity (55% de). Hence, the highest stereoselectivity in the phospha-aldol reaction was achieved in the reaction involving three chiral auxiliaries, which reinforced one another in matched asymmetric induction as in the case of (*R*)-glyceraldehyde/(l*R*,2*S*,5*R*)-menthyl/(*S*)-BINOL. The stereoselectivity did not increase, if the absolute configurations of chiral auxiliaries were mismatched as in the case of (*R*)-glyceraldehyde/(l*R*,2*S*,5*R*)-menthyl/(*S*)-BINOL [233]. Introduction of two or three asymmetric inductors in the reacting system additively increased the stereoselectivity of the reaction of (*R*)-glyceraldehyde acetonide phosphorous diesters if the absolute configurations of chiral groups in the initial compounds (e.g., (*R*)-glyceraldehyde/(1*R*,2*S*,5*R*)-menthyl/(*S*)-BINOL) acted in one direction to increase the diastereofacial selectivity of the reagents (Scheme 3.118) [99].

Feng *et al.* proposed the bifunctional chiral Al(III)-BINOL complex bearing two chiral centers for the effective enantioselective hydrophosphonylation of aldehydes [234]. Al(III) complexes of BINOL **179e** and BINOL derivatives **179a–d** which contained two *tert*-amine moieties on the BINOL showed high reactivity, and the aluminum complex of (*R*,*S*)-**179a** gave the product with the best result (85% yield and 70% ee) In contrast, ligand (*S*,*S*)-**179b** with an (*S*)-BINOL fragment displayed low enantioselectivity and favored the chiral α-hydroxy phosphonate with an opposite configuration as that produced by (*R*,*S*)-**179**. These results suggested that the axial chirality of the BINOL moieties had more influence in determining the absolute configuration of the hydrophosphonylation product, and the matched stereogenic elements are (*R*)-BINOL and (*S*)-l-phenylethanamine components. As shown in Scheme 3.119, the Al(III) center

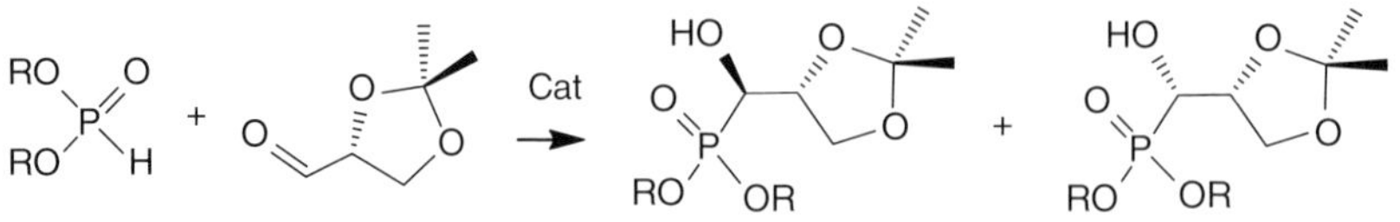

(1*R*,2*R*)-**177** (1*S*,2*R*)-**178**

R	Cat	(1*R*,2*R*)-**177**/(1*S*,2*R*)-**178**	Stereoselectivity
Et	DBU*b*	45 : 55	Single
Mnt	DBU	20 : 80	Double
Mnt	DBU	28 : 72	Double
Mnt	(*S*,*S*)-ALB*c*	5 : 95	Triple matched
Mnt	(*R*,*R*)-ALB	22 : 78	Triple mismatched

a without solvent; *b* DBU, *c* ALB = Al-Li-bis(binaphthoxid)

Scheme 3.118 Double and triple asymmetric induction of phospha-aldol reaction.

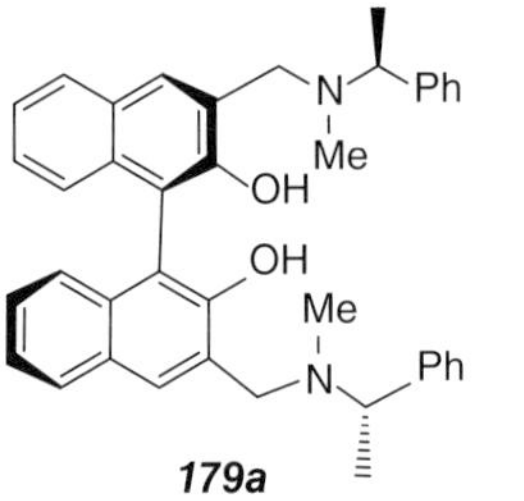

179a

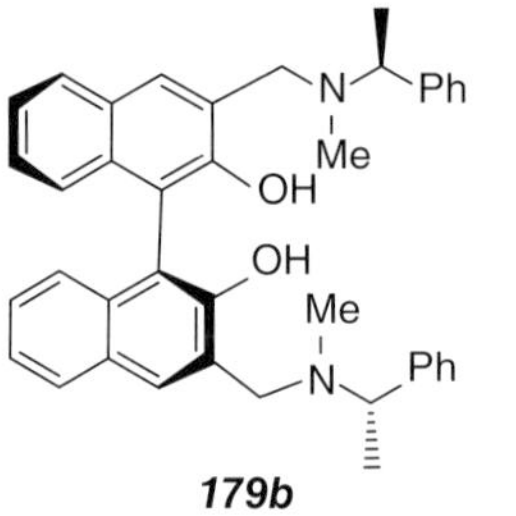

179b

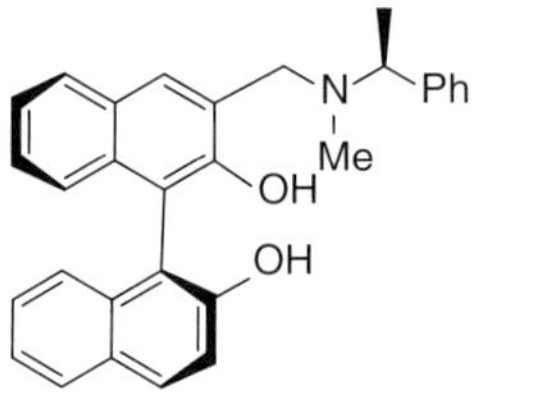

179c

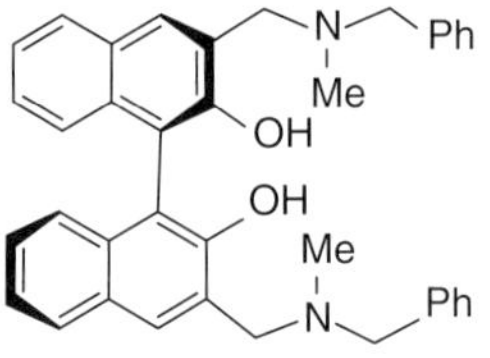

179d

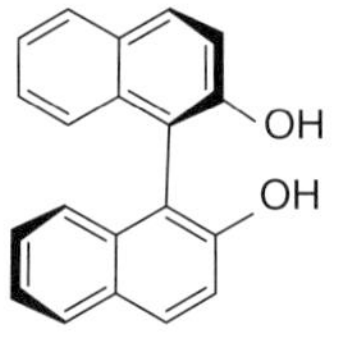

179e

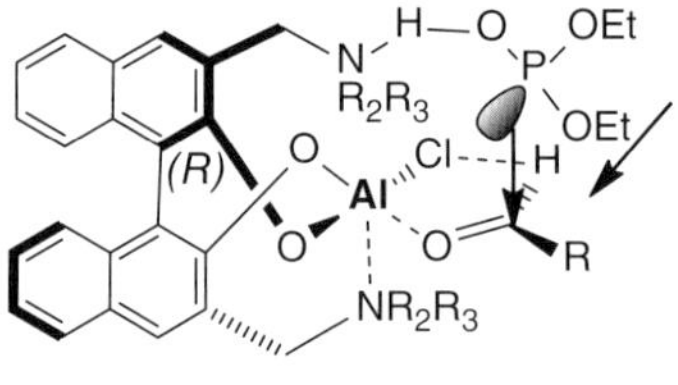

***Si** Favored* **A**

Ligand **7**	Time (h)	Yield (%)	ee (%)/Configuration	Al
a	48	85	70/(*S*)	Double matched
b	24	49	27/(*R*)	Double mismatched
c	24	<19	0	Double mismatched
d	48	81	65/(S)	Single
e	48	—	7/(S)	Single

Scheme 3.119 Reaction intermediate in catalytic cycle for the asymmetric hydrophosphonylation of aldehydes.

of the catalyst acted as a Lewis acid and activated the aldehyde. The aldehyde was fixed on the catalyst and activated by two kinds of interactions: the strong one involved the coordination between Al(III) and the oxygen of the carbonyl group and the weak one was the hydrogen bond effect between the proton of the aldehyde and the chlorine atom.

The carbonyl preferred to approach the catalyst along a vector *trans* to oxygen (the less hindered way, as shown in Scheme 3.119).

The combination of the organometallic BINOL catalysts and cinchona organocatalysts in combination with $Ti(OiPr)_4$ for the asymmetric hydrophosphonylation of aldehydes allowed to raise the stereoselectivity of the phospha-aldol reaction. The chiral Lewis base moiety (cinchona alkaloid) in these self-assembled bifunctional catalysts spontaneously coordinates to the central metal of the chiral Lewis acidic moiety (BINOL-Ti complex) to form the metal-organic assemblies. The bifunctional catalyst was generated from the metal-organic self-assembly of substituted BINOLs **180a–e** and **181** for the highly efficient asymmetric hydrophosphonylation of aldehydes [235]. The chiral Lewis base (cinchona alkaloid) of the bifunctional catalyst coordinates to the central metal of a chiral Lewis acid (BINOL-Ti complex), with the formation of an organometallic complex **B**. In comparison with the usual bifunctional catalysts in which Lewis acid/Lewis base exists in one molecule through covalent bonds, in this case there is a coordination of asymmetric induction of the chiral ligand, a metal ion, and a substrate that provides the high efficiency of the catalyst (Scheme 3.120) (Table 3.4).

Synthesis of organophosphorus analogs of C-13 lateral chains of taxoides were carried out. Scheme 3.121 shows the full synthesis of both stereoisomers of these compounds, starting from optically pure natural amino acids. Reaction of achiral diethylphosphite with aminoaldehydes proceeded with low stereoselectivity to give a mixture of stereoisomers. At the same time, the reaction of chiral dimenthylphosphites and dibornylphosphites with leucinal and phenylalaninal under conditions of double asymmetric induction gave chiral aminohydroxyphosphonates **180**, **183** with good stereoselectivity. The reaction of aldehydes with dibornylphosphite gave the (1*S*,2*R*)-stereoisomer, whereas dimenthylphosphite gave the (1*R*,2*R*)-stereoisomer. The aminohydroxyphosphonates **182**, **183** were obtained in stereochemically pure state after purification by column chromatography and crystallization (Scheme 3.121)

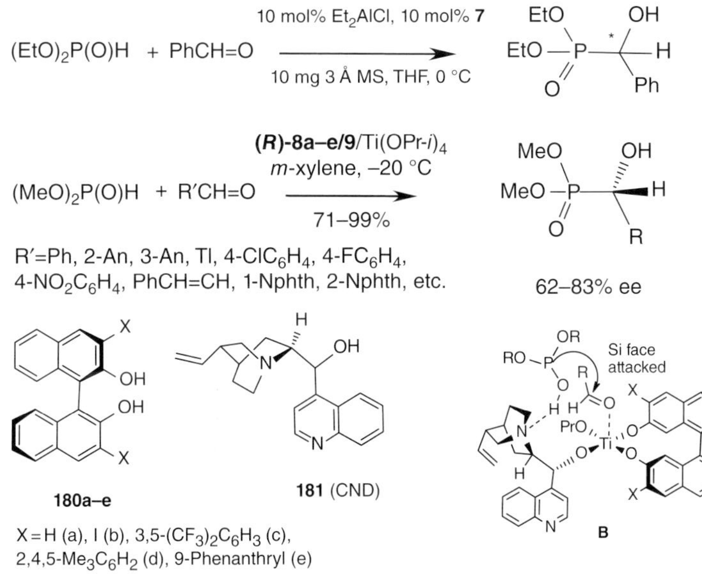

Scheme 3.120 Catalysts for phospha-aldol reactions (Table 3.4) and transition complex.

Table 3.4 Double asymmetric catalysis of phospha-aldol reaction, with Cat = (*R*)-**180c**/**181**/Ti(O*i*-Pr)$_4$.

$$(MeO)_2P(O)H + R'CH=O \xrightarrow{cat} (MeO)_2P(O)CH(OH)R'$$

Entry	R′	Yield (%)	ee (%)
1	Ph	92	94
2	3-MeC_6H_4	90	95
3	4-MeC_6H_4	92	94
4	2-MeOC_6H_4	95	96
5	2-ClC_6H_4	99	90
6	4-ClC_6H_4	87	90
7	4-$NO_2C_6H_4$	99	90
8	1-Naphthyl	97	>99
9	2-Naphthyl	97	>99
10	PhCH_2CH_2	96	92
11	Cyclohexyl	93	92
12	*n*-Oct	98	94
13	*i*-Pr	90	94

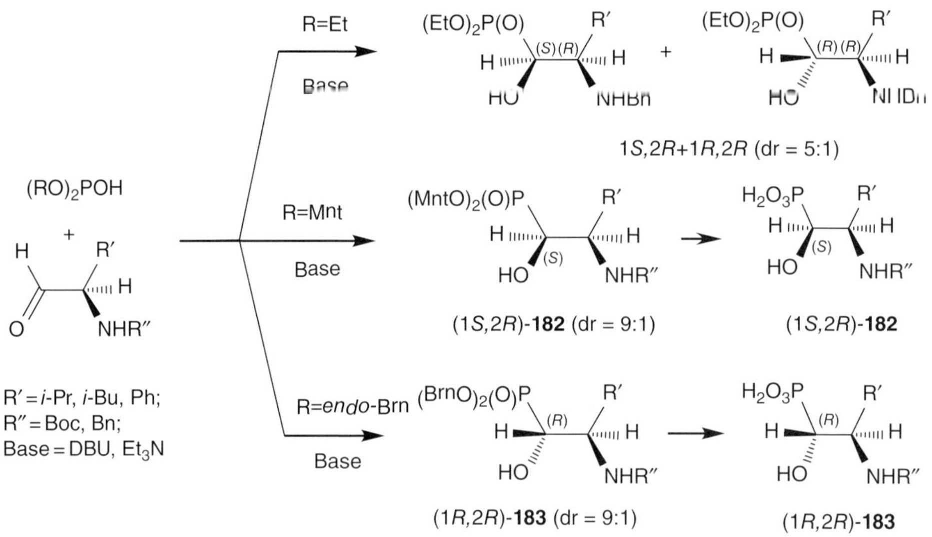

Scheme 3.121 Full synthesis of enantiomers of phosphonate analogs of lateral C-13 chains of taxoids.

[231, 236, 237]. Scheme 3.121 shows the phosphonate analog synthesis of a C-13 taxoid lateral chain [107] which was developed on a basis of trimenthylphosphite and tris(trimethylsilyl)phosphite with the use of double asymmetric synthesis methodology [231, 236, 237]. The chiral dialkylphosphites reacted with chiral aldehydes under the control of two chiral auxiliaries which reinforced one another to yield the diastereomers **182**, **183** with very good de.

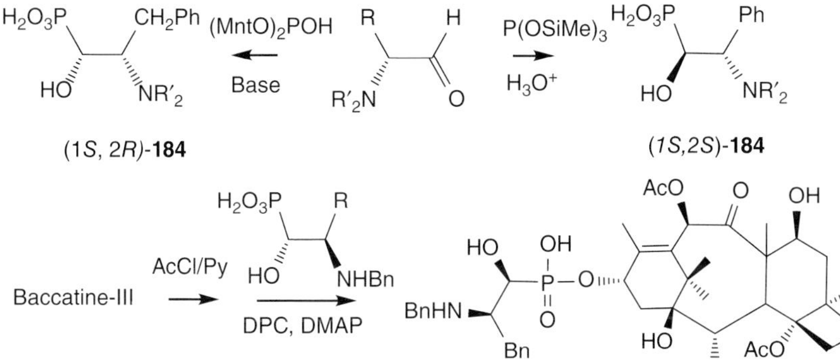

R = Boc, R′ = H (Docetaxel); R = Bz, R′ = Ac (Paclitaxel)

Scheme 3.122 Asymmetric synthesis of phosphonate analogs of lateral C-13 chain of taxoids.

The reaction of dibenzylphenylalanynal with tris(trimethylsilyl)phosphite afforded the (1*R*,2*S*)-1-hydroxy-2-aminoalkylphosphonic acid **184**, which was purified by recrystallization. The stereoselectivity of the reaction depended on the solvent, the nature of the bases, and temperature. The chiral 1-hydroxy-2-aminophosphonic acids were used for modification of baccatin III **185** in the synthesis of new taxoids-potential anticancer drugs (Scheme 3.122).

The chiral β-amino-α-hydroxy-*H*-phosphinates **187** were obtained by hydrophosphonylation of *N*,*N*-dibenzyl-α-aminoaldehydes **186** catalyzed by (*S*)- or (*R*)-ALB complexes [238–240]. The hydrophosphinylation using ethyl phosphinate afforded both *syn*- and *anti*-β-amino-α-hydroxy-*H*-phosphinates **187**, **188** with high diastereoselectivities by tuning the chirality of ALB. In these cases, diastereofacial selectivity was controlled predominantly by the chirality of the asymmetric catalyst rather than that of the α-amino aldehydes. Furthermore, the diastereoselectivity of the hydrophosphinylation with (*S*)-ALB was generally higher than that with (*R*)-ALB (Scheme 3.123).

The reactions of achiral dialkyl phosphites with chiral aldimines as well as that of chiral di-(1*R*,2*S*,5*R*)-menthyl phosphite with achiral aldimines resulted in low diastereomeric enrichment of the addition compound [145, 146, 241–244]. Thus, the reaction of chiral di(1*R*,2*S*,5*R*)-menthyl phosphite with achiral benzylbenzaldimine as well as that of achiral diethyl phosphite with chiral (*S*)-α-methylbenzylbenzaldimine controlled

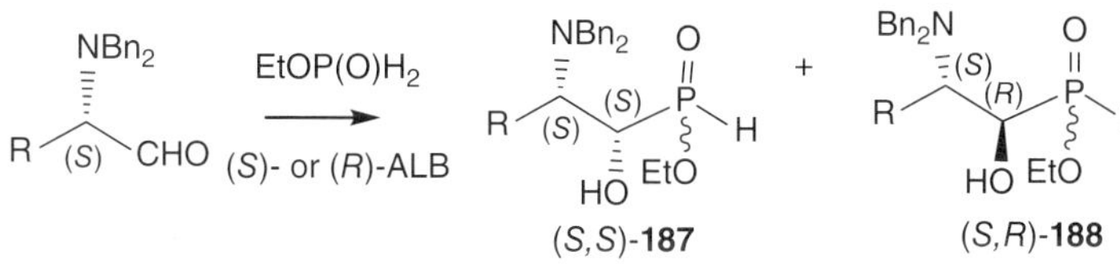

Entry	R	ALB	Syn/anti	AS
1	Bn	(*S*)	6:94	Matched
2	Bn	(*R*)	87:13	Mismatched
3	i-Bu	(*S*)	2:98	Matched
4	i-Bu	(*R*)	94:6	Mismatched

Scheme 3.123 Hydrophosphonylation of *N*,*N*-dibenzyl-α-aminoaldehydes catalyzed by (*S*)- or (*R*)-ALB complexes.

stereochemically by one asymmetric inductor resulted in 30–45% diastereomeric enrichment of aminophosphonic acid diesters **189, 190**. However, the stereoselectivity of reaction was increased when additional chiral inductors were introduced into the reaction system. For example, chiral di(1*R*,2*S*,5*R*)-menthyl phosphite reacted with chiral (*S*)-α-methylbenzylbenzaldimine on heating to 80 °C to form practically single (1*R*,2*S*)-diastereomer of *N*-substituted aminophosphonic acid diester **189**. The stereoselectivity of the reaction depended on the nature of the substituents in the aromatic ring of the Schiff bases. As seen from the results presented in Table 3.5, electron-acceptor substituents in the aromatic ring reduced reaction stereoselectivity, while electron donors increased it. For example, the highest diastereoselectivity was observed in the case of compounds containing the substituents 4-MeO, 4-Me_2N, and H in the aromatic ring. However, the compounds **189**, containing F, Br, and NO_2 in the aromatic ring, were obtained with low stereoselectivity [146] (Table 3.5).

Lyzwa [245] has reported interesting results on the double asymmetric induction observed in the addition reaction of lithium salts of enantiomeric dimenthyl phosphites to both enantiomers of *N*(*p*-tolylsulfinyl) benzaldimine **191** proceeding with formation of α-aminophosphonates **192**. It involves the nucleophilic addition of anions of enantiomeric dimenthyl phosphites to both (+)-(*S*)-**191** and (−)-(*R*)-**191** enantiomers of *N*(*p*-tolylsulfinyl)benzaldimine and subsequent acidic hydrolysis of the adducts to form the enantiomers of (−)-**193** and (+)-**193** (entries 1 and 4, matched pairs of isomers).The

Table 3.5 Stereoselective addition of dialkylphosphites to the Schiff bases.

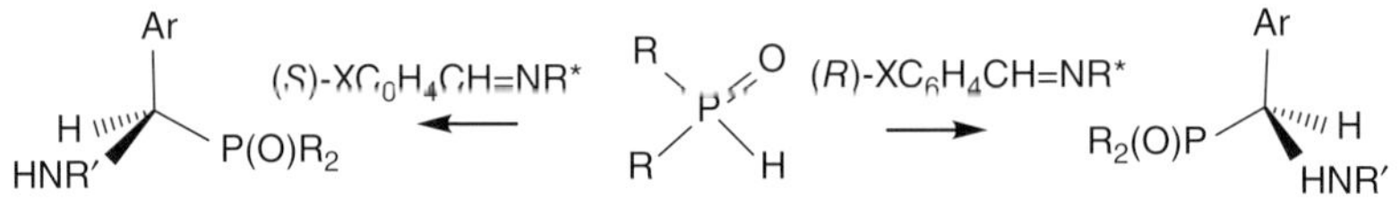

(R_S)-**189**

Ar = C_6H_4X; R* = CH(Me)Ph

(S_R)-**190**

R_2	Configuration *R**	X	Configuration	*dr*	Diastereoselectivity	References
$(MntO)_2$	(*R*)	H	(*SR*)	75 : 25	Mismatched	[242]
$(MntO)_2$	(*S*)	4-MeO	(*RS*)	94 : 6	Matched	[242]
$(MntO)_2$	(*R*)	4-MeO	(*SR*)	78 : 22	Mismatched	[242]
$(MntO)_2$	(*S*)	4-NMe_2	(*RS*)	93 : 7	Matched	[242]
$(MntO)_2$	(*R*)	4-NMe_2	(*SR*)	75 : 25	Mismatched	[242]
$(MntO)_2$	(*S*)	4-NO_2	(*RS*)	85 : 15	Matched	[242]
$(MntO)_2$	(*R*)	4-NO_2	(*SR*)	70 : 30	Mismatched	[242]
$(MntO)_2$	(*S*)	4-F	(*RS*)	88 : 12	Matched	[242]
$(MntO)_2$	(*R*)	4-F	(*SR*)	75 : 25	Mismatched	[242]
$(MntO)_2$	(*S*)	4-Br	(*RS*)	86 : 14	Matched	[242]
$(MntO)_2$	(*R*)	4-Br	(*SR*)	74 : 26	Mismatched	[242]
$(MntO)_2$	(*S*)	Ph	—	93 : 7	Matched	[242]
$(MntO)_2$	(*R*)	Ph	—	80 : 20	Mismatched	[242]
(*R*)-Ph(Mnt)	(*S*)	Ph	(*S*)	72 : 28	Matched	[243]
(*R*)-Ph(Mnt)	(*R*)	Ph	(*R*)	91 : 9	Mismatched	[244]

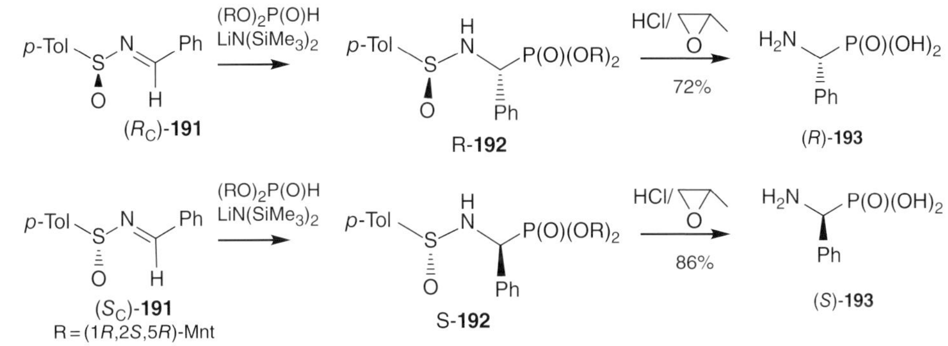

Entry	R	191	192	dr	Yield (%)	Configuration. of 193
1	(1*R*,2*S*,5*R*)-Mnt	(+)(*S*)	(S_SR_C)-:(S_SS_C)-	100:0	88	(+)(*R*)
2	(1*R*,2*S*,5*R*)-Mnt	(−)(*R*)	(R_SS_C)-:(R_SR_C)-	92:8	77	(−)(*S*)
3	(1*S*,2*R*,5*S*)-Mnt	(+)(*S*)	(S_SR_C)-:(S_SS_C)-	91:9	75	(+)(*R*)
4	(1*S*,2*R*,5*S*)-Mnt	(−)(*R*)	(R_SS_C)-:(R_SR_C)-	100:0	86	(−)(*S*)

Scheme 3.124 Asymmetric addition of dimenthylphosphites to *N*-(*p*-toluenesulfinyl)benzaldimine.

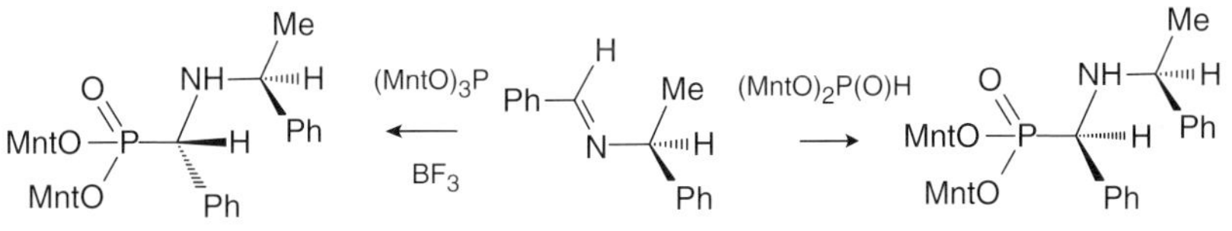

(*S*,*S*)-**194** (*S*) (*R*,*S*)-**195**

Scheme 3.125 Stereochemical course of reaction of chiral aldimines with di- and trialkylphosphites.

single diastereomers of the adducts **192** formed have opposite absolute configurations at the newly formed stereogenic center of the α-carbon atom. On the other hand, the stereomers **192** and **193** (entries 2 and 3) were formed by mismatched pairs of starting chiral reactants and the dr of the major adducts is above 10 : 1 (Scheme 3.124).

Tris(1*R*,2*S*,5*R*)-menthyl phosphite reacted with C=N compounds in the presence of boron trifluoride etherate to form aminophosphonic acid derivatives with the absolute configuration opposite to that appearing in the reaction of di(1*R*,2*S*,5*R*)-menthyl phosphite with the same C=N compounds (Scheme 3.126). When two chiral inductors were involved in the reaction of trialkyl phosphites with C=N compounds, stereoselectivity increased. While the reaction of [tri(1*R*,2*S*,5*R*)-menthyl] phosphite with (*S*)-α-methylbenzylbenzaldimine resulted predominantly in the formation of (*N*-1-phenylethyl)-aminobenzylphosphonate (1*S*,2*S*)-**194**, dimenthyl phosphite reacted with (*S*)-α-methylbenzylbenzaldimine to form the *N*-(1-phenylethyl)aminobenzylphosphonate (1*R*,2*S*)-**195** (Scheme 3.125) [146].

Difference in steric results of the reaction of imino compounds with chiral phosphorous di- and triesters can be explained by the formation of complexes **196** of these compounds with boron trifluoride (Scheme 3.126). Stereochemical analysis of reaction of the imino compounds with phosphorous esters by means of theoretical

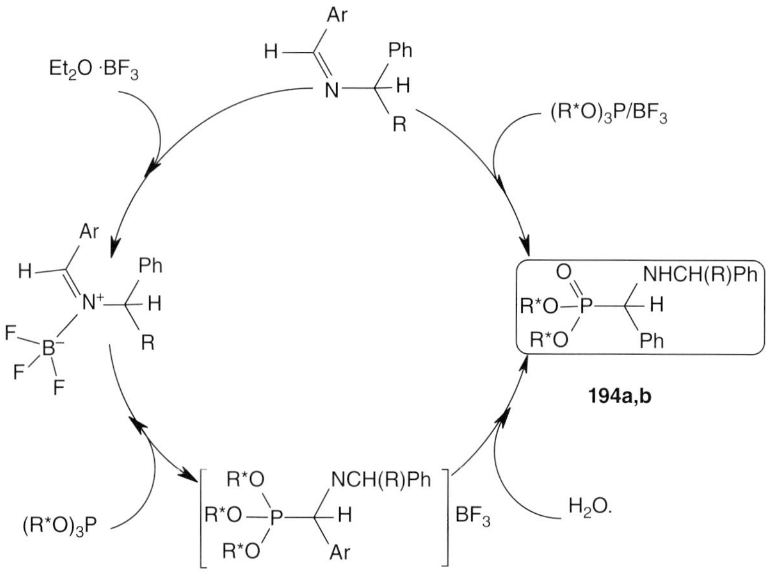

194a,b

196a,b

R = Me, H; R* = Mnt (a), Et (b)

Scheme 3.126 Mechanism of reaction of trialkylphosphites with Schiff's bases.

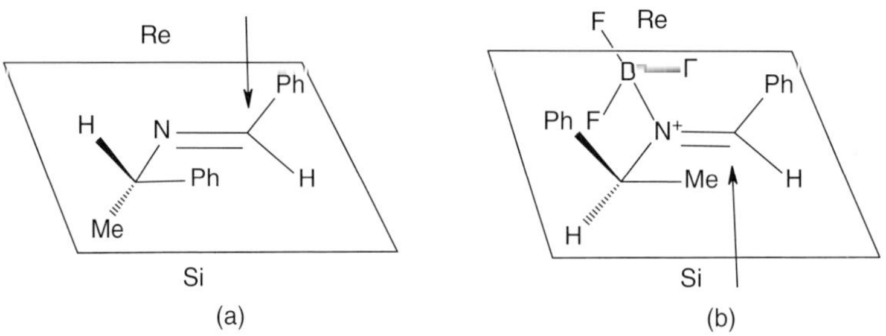

Figure 3.4 Stereochemical models of reaction: (a) dimenthyl phosphite with (*S*)-methylbenzylbenzaldimine and (b) trimenthyl phosphite with the boron trifluoride (*S*)-methylbenzylbenzaldimine complex.

calculations showed that the conformation of imino compound bearing phenyl groups in diastereo-zero plane (Figure 3.4) corresponds to the energy minimum. Therefore, the carbon atom of the C=N group is attacked at the diastereofacial side *Re* where diastereogenic center possess a hydrogen atom, the lesser substituent by size, resulting in the (1*R*,2*S*)-configuration of aminophosphonate. The presence of the charged BF_3 group at the iminium carbon atom in the complex of imino compound with boron trifluoride makes preferable the configuration with methyl group in diastereo-zero plane, resulting in attack at the C=N carbon atom by the reagent at the *Si*-diastereofacial side. This leads to formation of (1*S*,2*S*)-aminophosphonate **194** (Figure 3.4) [246].

The reduction of ketophosphonates **197a,b** with the chiral complex of $NaBH_4$/(*R*)-tartaric acid proceeded with the highest stereoselectivity to give the diethyl (1*S*)-α-hydroxybenzylphosphonates **198a** with 60% ee and the dimenthyl

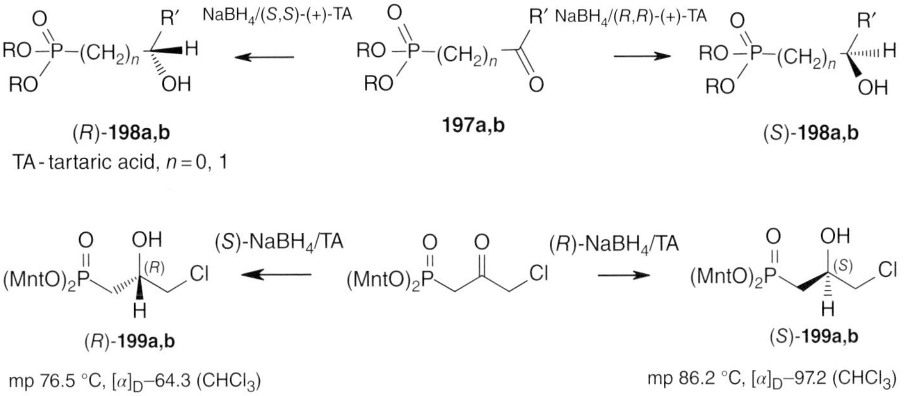

R	R′	Configuration TA	ee	Configuration of **199**	Stereoselectivity
Et	Ph	*R,R*	60	*S*	Single
Et	Ph	*S,S*	60	*R*	Single
Mnt	Ph	*R,R*	92.5	*S*	Matched double
Mnt	Ph	*S,S*	46	*R*	Mismatched double

Scheme 3.127 Double stereoselective reduction of β-ketophosphonates **197**.

(1*S*)-α-hydroxyarylmethylphosphonates **198b** with 80–93% de. Higher stereoselectivity of reduction of dimenthyl arylketophosphonates was explained by the effect of double asymmetric induction [247–252]. Evidently in this case, the asymmetric inductions of chiral (l*R*,2*S*,5*R*)-menthyl groups and (*R*,*R*)-tartaric acid act in one direction, increasing the diastereofacial selectivity of the reagents, whereas asymmetric inductions of (l*R*,2*S*,5*R*)-menthyl groups and (*S*,*S*)-tartaric acid are mismatched and act in opposite directions decreasing the resulting stereoselectivity (Scheme 3.127) [249].

3.10 Summary

This chapter has introduced the stereoselective addition of phosphoric nucleophiles to C=X bonds: phospha-aldol, phospha-Mannich, and phospha-Michael reactions. The phospha-aldol and related reactions discussed in this chapter are very important reactions in organophosphorus chemistry because these reaction products constitute the backbone of many important antibiotics, anticancer drugs, and other bioactive molecules. Indeed, studies of the phospha-aldol, phospha-Mannich, and phospha-Michael reactions are actively pursued in order to improve reaction conditions, enhance stereoselectivity, and widen the scope of applicability of this type of reaction. Multiple stereoselectivity creates two or more chiral centers in one step and is regarded as one of the most efficient synthetic strategies in organic synthesis. The further development of this methodology should involve studies of the stereochemistry of chiral multifunctional catalysis, for example, in the case of the cooperative catalysis.

References

1 Hildebrand, R.L. (1983) *The Role of Phosphonates in Living System*, CRC Press, Boca Raton, FL.

2 Kafarski, P. and Lejczak, B. (1991) Biological activity of aminophosphonic acids. *Phosphorus, Sulfur Silicon Relat. Elem.*, **63** (1), 193–215.

3 Horiguchi, M. and Kandatsu, M. (1959) Isolation of 2-aminoethane. Phosphonic acid from rumen protozoa. *Nature*, **184**, 901–902.

4 Kolodiazhnyi, O.I. (2015) in *Topics in Current Chemistry*, vol. **361** (ed. J.-L. Montchamp), Springer International Publishing, Switzerland, pp. 161–236.

5 Hughes, A.B. (ed.) (2009) *Amino Acids, Peptides and Proteins in Organic Chemistry*, Modified Amino Acids, Organocatalysis and Enzymes, vol. **2**, Wiley-VCH Verlag GmbH, Weinheim, 705 pp.

6 Kudzin, Z.H., Kudzin, M.H., Drabowicz, J., and Stevens, C.V. (2011) Aminophosphonic acids–phosphorus analogues of natural amino acids. Part 1: syntheses of α-aminophosphonic acids. *Curr. Org. Chem.*, **15** (12), 2015–2071.

7 Guin, J., Wang, Q., van Gemmeren, M., and List, B. (2015) The catalytic asymmetricAbramov reaction. *Angew. Chem. Int. Ed.*, **54** (1), 355–358.

8 Orsini, F., Sello, G., and Sisti, M. (2010) Aminophosphonic acids and derivatives. Synthesis and biological applications. *Curr. Med. Chem.*, **17** (3), 264–289.

9 Kolodiazhnyi, O.I. (2011) Enzymatic synthesis of organophosphorus compounds. *Russ. Chem. Rev.*, **80** (9), 883–910.

10 Blazis, V.J., Koeller, K.J., and Spilling, S.D. (1995) Reactions of chiral phosphorous acid diamides: the asymmetric synthesis of chiral alpha,-hydroxy phosphonamides, phosphonates, and phosphonic acids. *J. Org. Chem.*, **60** (4), 931–940.

11 Denmark, S.E. and Amburgey, J. (1993) A new, general, and stereoselective method for the synthesis of trisubstituted alkenes. *J. Am. Chem. Soc.*, **115** (22), 10386–10387.

12 le Bec, C. and Wickstrom, E. (1994) Stereospecific grignard activated coupling of a deoxynucleoside : methylphosphonate on a polyethylene glycol support. *Tetrahedron Lett.*, **35** (51), 9525–9528.

13 Bennani, Y.L. and Hanessian, S. (1990) A versatile asymmetric synthesis of α-amino α-alkyl-phosphonic acids of high enantiomeric purity. *Tetrahedron Lett.*, **31** (45), 6465–6468.

14 Hanessian, S., Bennani, Y.L., and Delorme, D. (1990) The asymmetric synthesis of α-chloro α-alkyl and α-methyl α-alkyl phosphonic acids of high enantiomeric purity. *Tetrahedron Lett.*, **31** (45), 6461–6464.

15 Cramer, C.J., Denmark, S.E., Miller, P.C., Dorow, R.L., Swiss, K.A., and Wilson, S.R. (1994) Structure and dynamics of phosphorus(V)-stabilized carbanions: a comparison of theoretical, crystallographic, and solution structures. *J. Am. Chem. Soc.*, **116** (6), 2437–2447.

16 Kranz, M., Denmark, S.E., Swiss, K.A., and Wilson, S.R. (1996) An ab initio study of the P-C bond rotation in phosphoryl- and thiophosphoryl-stabilized carbanions: five- and six-membered heterocycles. *J. Org. Chem.*, **61** (24), 8551–8563.

17 Leyssens, T. and Peeters, D. (2008) Negative hyperconjugation in phosphorus stabilized carbanions. *J. Org. Chem.*, **73** (7), 2725–2730.

18 Denmark, S.E. and Cramer, C. (1990) The theoretical structures of neutral, anionic, and lithiated P-allylphosphonic diamide. *J. Org. Chem.*, **55** (6), 1806–1813.

19 Denmark, S.E., Miller, P.C., and Wilson, S.R. (1991) Configuration, conformation, and colligative properties of a phosphorus-stabilized anion. *J. Am. Chem. Soc.*, **113** (4), 1468–1470.

20 Denmark, S.E., Swiss, K.A., and Wilson, S.R. (1996) Solution and solid-state structure of a lithiated phosphine oxide. *Angew. Chem. Int. Ed. Engl.*, **35**, 2515–2517.

21 Denmark, E.S., Swiss, K.A., and Wilson, S.R. (1993) Solution- and solid-state structure and dynamics of thiophosphonamide anions: electronic tuning of rotational barriers. *J. Am. Chem. Soc.*, **115** (9), 3826–3827.

22 Aggarval, V.K. (1994) Enantioselective transformations and racemization studies of heteroatom substituted organolithium compounds. *Angew. Chem. Int. Ed. Engl.*, **33** (2), 175–177.

23 Rein, T. and Pedersen, T.M. (2002) Asymmetric Wittig type reactions. *Synthesis*, **2002** (5), 579–594.

24 Hanessian, S., Gomtsyan, A., and Malek, N. (2000) Asymmetric conjugate additions of chiral phosphonamide anions to α,β-unsaturated carbonyl compounds. A versatile method for vicinally substituted chirons. *J. Org. Chem.*, **65** (18), 5623–5631.

25 Denmark, S.E. and Kim, J.H. (2000) Diastereoselective alkylations of chiral, phosphorus-stabilized carbanions: *N*-alkyl substituent effects in P-alkyl-1,3,2-diazaphosphorinane 2-oxides. *Can. J. Chem.*, **78** (6), 673–688.

26 Toru, T. and Nakamura, S. (2003) Enantioselective synthesis by lithiation adjacent to sulfur, selenium or phosphorus, or without an adjacent activating heteroatom. *Top. Organomet. Chem.*, **5**, 177–216.

27 Denmark, S.E., Chatani, N., and Pansare, S.V.T. (1992) Asymmetric electrophilic amination of chiral phosphorus-stabilized anions. *Tetrahedron*, **48** (11), 2191–2208.

28 Denmark, S.E. and Chen, C.T. (1994) Alkylation of chiral, phosphorus-stabilized carbanions: substituent effects on the alkylation selectivity. *J. Org. Chem.*, **59** (11), 2922–2924.

29 Pagliarin, R., Papeo, G., Sello, G., Sisti, M., and Paleari, I. (1996) Design of β-amino alcohols as chiral auxiliaries in the electrophilic amination of 1,3,2-oxazaphospholanes. *Tetrahedron*, **52** (43), 13783–13794.

30 Baumgartner, T. (2014) Insights on the design and electron-acceptor properties of conjugated organophosphorus materials. *Acc. Chem. Res.*, **47** (5), 1613–1622.

31 Hanessian, S., Bennani, Y.L., and Herve, Y. (1993) A novel asymmetric synthesis of α- and β-amino aryl phosphonic acids. *Synlett*, **1993** (1), 35–36.

32 Hanessian, S., Gomtsyan, A., Payne, A., Herve, Y., and Beaudoin, S. (1993) Asymmetric conjugate additions of chiral allyl- and crotylphosphonamide anions to. alpha, beta, unsaturated carbonyl compounds: highly stereocontrolled access to vicinally substituted carbon centers and chemically asymmetrized chirons. *J. Org. Chem.*, **58** (19), 5032–5034.

33 Hanessian, S. and Gomstyan, A. (1994) Highly stereocontrolled sequential asymmetric Michael addition reactions with cinnamate esters – generation of three and four contiguous stereogenic centers on seven-carbon acyclic motifs. *Tetrahedron Lett.*, **35** (41), 7509–7512.

34 Focken, T. and Hanessian, S. (2014) Application of cyclic phosphonamide reagents in the total synthesis of natural products and biologically active molecules. *Beilstein J. Org. Chem.*, **10** (8), 1848–1877.

35 Paquette, L.A., Wang, T.Z., and Pinard, E. (1995) Total synthesis of natural (+)-acetoxycrenulide. *J. Am. Chem. Soc.*, **117** (4), 1455–1456.

36 Wang, T.Z., Pinard, E., and Paquette, L.A. (1996) Asymmetric synthesis of the diterpenoid marine toxin (+)-acetoxycrenulide. *J. Am. Chem. Soc.*, **118** (6), 1309–1318.

37 Wu, X., Zhou, J., and Snider, B.B. (2009) Synthesis of (-)-berkelic acid. *Angew. Chem. Int. Ed.*, **48** (7), 1283–1286.

38 Foucher, V., Guizzardi, B., Groen, M.B., Light, M., and Linclau, B. (2010) A novel, versatile D → BCD steroid construction strategy, illustrated by the enantioselective total synthesis of estrone. *Org. Lett.*, **12** (4), 680–683.

39 Hailes, H.C., Isaac, B., and Javaid, M.H.T. (2001) 1,4-addition of chiral 2-propenylphosphonamide anions to substituted cyclopentenones: use in enantioselective syntheses of methyl dihydrojasmonates and methyl jasmonates. *Tetrahedron Lett.*, **42** (41), 7325–7328.

40 Marinozzi, M. and Pellicciari, R. (2000) Novel enantioselective synthesis of (2*S*,2′*R*,3′*R*)-2-(2′,3′-dicarboxycyclopropyl)glycine (DCG-IV). *Tetrahedron Lett.*, **41** (47), 9125–9128.

41 Clive, D.L.J. and Liu, D. (2008) Synthesis of the potent anticancer agents ottelione A and ottelione B in both racemic and natural optically pure forms. *J. Org. Chem.*, **73**, 3078–3087.

42 Hanessian, S., Cantin, L.D., Roy, S., Andreotti, D., and Gomstyan, A. (1997) The synthesis of enantiomerically pure, symmetrically substituted cyclopropane phosphonic acids. A constrained analog of the GABA antagonist Phaclophen. *Tetrahedron Lett.*, **38** (7), 1103–1106.

43 Reissig, H.-U. (1996) Recent developments in the enantioselective syntheses of cyclopropanes. *Angew. Chem. Int. Ed. Engl.*, **35** (9), 971–973.

44 Hanessian, S., Andreotti, D., and Gomstyan, A. (1995) Asymmetric synthesis of enantiomerically pure and diversely functionalized cyclopropanes. *J. Am. Chem. Soc.*, **117** (41), 10393–10394.

45 Haynes, R.K., Stokes, J.P., and Hambley, T.W. (1991) The preparation of (*R*)- and (*S*)-(E)-but-2-enyl-t-butylphenylphosphine oxides and their enantiospecific conversion into enantiomeric hydrindenones related to vitamin D. *J. Chem. Soc., Chem. Commun.*, (1), 58–60.

46 Haynes, R.K., Freeman, R.N., Mitchell, C.R., and Vonwiller, S.C. (1994) Preparation of enantiomerically pure tertiary phosphine oxides from, and assay of enantiomeric purity with, (*R*p)- and (*S*p)-tert-butylphenylphosphinothioic acids. *J. Org. Chem.*, **59** (11), 2919–2921.

47 Denmark, S.E., Marlin, J.E., and Rajendra, G. (2013) Carbanion-accelerated claisen rearrangements: asymmetric induction with chiral phosphorus-stabilized anions. *J. Org. Chem.*, **78** (1), 66–82.

48 Tan, J., Cheon, C.-H., and Yamamoto, H. (2012) Catalytic asymmetric Claisen rearrangement of enolphosphonates: construction of vicinal tertiary and all-carbon quaternary centers. *Angew. Chem. Int. Ed.*, **51** (33), 8264–8267.

49 Denmark, S.E. and Miller, P.C. (1995) Asymmetric [2,3]-wittig rearrangements with chiral, phosphorus anion-stabilizing groups. *Tetrahedron Lett.*, **36** (37), 6631–6634.

50 Gulea-Purcarescu, M., About-Jaudet, E., and Collignon, N. (1995) Sigmatropic (2,3]-Wittig rearrangement of α-allylic-heterosubstituted methylphosphonates. Part 31: high diastereoselectivity in the rearrangement of anion of dimenthyl allyloxymethylphosphonate. *Tetrahedron Lett.*, **36** (37), 6635–6638.

51 Bestmann, H.J., Heid, E., Ryschka, W., and Lienert, J. (1974) Bestimmung der absoluten Konfiguration des (+)-4-Benzylidencyclohexancarbonsaure-athylesters. *Liebigs Ann. Chem.*, **1974** (10), 1684–1687.

52 Tanaka, K., Furuta, T., and Fuji, K. (2004) in *Modern Carbonyl Olefination: Methods and Applications*, Chapter 7 (ed. T. Takeda), Wiley-VCH Verlag GmbH, Weinheim, 365 pp.

53 Harmat, N.J.S. and Warren, S. (1990) Chiral synthesis of Z-2-butylidenecyclohexan-1-OL and -1-YL phenylsulphide from optically active phosphine oxides. *Tetrahedron Lett.*, **31**, 2743–2746.

54 Hanessian, S., Giroux, S., and Merner, B.L. (2013) *Design and Strategy in Organic Synthesis: From the Chiron Approach to Catalysis*, Wiley-VCH Verlag GmbH, Weinheim, 848 pp.

55 Zhang, Y. and Schuster, G.B. (1994) A search for photoresolvable mesogens: synthesis and properties of a series of liquid crystalline, axially chiral 1-benzylidene-4-[4′-[(p-alkylphenyl) ethynyl] phenyl]cyclohexanes. *J. Org. Chem.*, **59** (7), 1855–1862.

56 Suarez, M. and Schuster, G.B. (1995) Photoresolution of an axially chiral bicyclo[3.3.0]octan-3-one: phototriggers for a liquid-crystal-based optical switch. *J. Am. Chem. Soc.*, **117** (25), 6732–6738.

57 Denmark, S.E. and Chen, C.T. (1992) Electrophilic activation of the Homer-Wadsworth-Emmons-Wittig reaction: highly selective synthesis of dissymmetric olefins. *J. Am. Chem. Soc.*, **114** (26), 10674–10676.

58 Hanessian, S. and Fu, J.-m. (2001) Total synthesis of polyoximic acid. *Can. J. Chem.*, **79** (11), 1812–1826.

59 Hanessian, S., Focken, T., and Oza, R. (2010) Total synthesis of Jerangolid. *Org. Lett.*, **12** (14), 3172–3175.

60 Hanessian, S., Focken, T., Mi, X., Oza, R., Chen, B., Ritson, D., and Beaudegnies, R. (2010) Total synthesis of (+)-ambruticin S: probing the pharmacophoric subunit. *J. Org. Chem.*, **75** (16), 5601–5618.

61 Headley, C.E. and Marsden, S.P. (2007) Synthesis and application of P-stereogenic phosphines as superior reagents in the asymmetric Aza-Wittig reaction. *J. Org. Chem.*, **72** (19), 7185.

62 Bennani, Y.L. and Hanessian, S. (1997) trans-1,2-Diaminocyclohexane derivatives as chiral reagents, scaffolds, and ligands for catalysis: applications in asymmetric synthesis. *Chem. Rev.*, **97** (8), 3161–3196.

63 Kawasaki, T., Ogawa, A., Takashima, Y., and Sakamoto, M. (2003) Enantioselective total synthesis of (−)-pseudophrynaminol through tandem olefination, isomerization and asymmetric Claisen rearrangement. *Tetrahedron Lett.*, **44** (8), 1591–1593.

64 Kawasaki, T., Shinada, M., Kamimura, D., Ohzono, M., and Ogawa, A. (2006) Enantioselective total synthesis of (−)-flustramines A, B and (−)-flustramides A, B via domino olefination/isomerization/Claisen rearrangement sequence. *J. Chem. Soc., Chem. Commun.*, (4), 420–422.

65 Kramp, G.J., Kim, M., Gais, H.J., and Vermeeren, C. (2005) Fully stereocontrolled total syntheses of the prostacyclin analogues 16S-iloprost and 16S-3-oxa-iloprost by a common route, using alkenylcopper-azoalkene conjugate addition, asymmetric olefination, and allylic alkylation. *J. Am. Chem. Soc.*, **27** (50), 17910–17920.

66 Marinozzi, M., Serpi, M., and Amori, L. (2007) Synthesis and preliminary pharmacological evaluation of the four stereoisomers of

(2S)-2-(20-phosphono-30-phenylcyclopropyl)glycine. *Bioorg. Med. Chem.*, **15**, 3161–3170.
67 Amori, L., Serpi, M., Marinozzi, M., Costantino, G., Diaz, M.G., Hermit, M.B., Thomsen, C., and Pellicciari, R. (2006) Synthesis and preliminary biological evaluation of (2S,1′R,2′S)- and (2S,1′S,2′R)-2-(2′-phosphonocyclopropyl)glycines, two novel conformationally constrained l-AP4 analogues. *Bioorg. Med. Chem. Lett.*, **16** (1), 196–199.
68 Hanessian, S. (1984) *Total Synthesis of Natural Products: The "Chiron Approach"*, Pergamon, Oxford, 310 pp.
69 Hanessian, S. and Beaudoin, S. (1992) Studies in asymmetric olefinations–the synthesis of enantiomerically pure allylidene, alkylidene, and benzylidene cyclohexanes. *Tetrahedron Lett.*, **33** (50), 7655–7658.
70 Kolodiazhnyi, O. (1999) *Phosphorus Ylides. Chemistry and Application in Organic Synthesis*, Wiley-VCH Verlag GmbH, Weinheim, New York, Chichester, pp. 1–565.
71 Li, C.-Y., Zhu, B.-H., Ye, L.-W., Jing, Q., Sun, X.-L., Tang, Y., and Shen, Q. (2007) Olefination of ketenes for the enantioselective synthesis of allenes via an ylide route. *Tetrahedron*, **63** (33), 8046–8053.
72 Kolodiazhnyi, O.I. (2005) Asymmetric synthesis of hydroxyphosphonates. *Tetrahedron: Asymmetry*, **16** (20), 3295–3340.
73 Abramov, V.S. (1950) *Dokl. Akad. Nauk SSSR*, **73**, 487–489; *Chem. Abstr.*, **45** (1951) 2855.
74 Mahrwald, R. (ed.) (2004) *Modern Aldol Reactions*, Wiley-VCH Verlag GmbH, Weinheim, p. 679.
75 Nixon, T.D., Dalgarno, S., Ward, C.V., Jiang, M., Halcrow, M.A., Kilner, C., Thornton-Pett, M., and Kee, T.P. (2004) Stereocontrol in asymmetric phospho-aldol catalysis. Chirality relaying in action. *C.R. Chim.*, **7** (8), 809–821.
76 Keglevich, G. and Bálint, E. (2012) The Kabachnik–fields reaction: mechanism and synthetic use. *Molecules*, **17**, 12821–12835.
77 Fu, X., Loh, W.-T., Zhang, Y., Chen, T., Ma, T., Liu, H., Wang, J., and Tan, C.-H. (2009) Chiral guanidinium salt catalyzed enantioselective phospha-Mannich reactions. *Angew. Chem. Int. Ed.*, **121** (40), 7523–7526.
78 Enders, D., Saint-Dizier, A., Lannou, M.-I., and Lenzen, A. (2006) The phospha-Michael addition in organic synthesis. *Eur. J. Org. Chem.*, **2006** (1), 29–49.
79 Lenker, H.K., Richard, M.E., Reese, K.P., Carter, A.F., Zawisky, J.D., Winter, E.F., Bergeron, T.W., Guydon, K.S., and Stockland, R.A. Jr., (2012) Phospha-Michael additions to activated internal alkenes: steric and electronic effects. *J. Org. Chem.*, **77** (3), 1378–1385.
80 Blazis, V.J., Koeller, K.J., and Spilling, S.D. (1994) Asymmetric synthesis of α-hydroxyphosphonamides, phosphonates and phosphonic acids. *Tetrahedron: Asymmetry*, **5** (3), 499–502.
81 Pudovik, A. and Arbuzov, A. (1950) Addition of dialkyl phosphites to unsaturated ketones, nitriles and esters. *Dokl. Akad. Nauk SSSR, Ser. Khim.*, **73**, 327–355.
82 Pudovik, A.N. (1954) Thiophosphinic and phosphinic acids, new method for the synthesis of their esters. *Usp. Khim.(Russ. Chem. Rev.)*, **23** (5), 547–580.
83 Sum, V. and Kee, T.P. (1993) Synthetic, stereochemical and mechanistic studies on the asymmetric phosphonylation of aldehydes via 2-triorganosiloxy-1,3,2-oxazaphospholidines. *J. Chem. Soc., Perkin Trans. 1*, (22), 2701–2712.

84 Kolodiazhnyi, O.I. and Kolodiazhna, O.O. (2012) New catalyst for phosphonylation of C = X electrophiles. *Synth. Commun.*, **42** (11), 1637–1649.

85 Olszewski, T.K. (2015) Asymmetric synthesis of α-hydroxymethylphosphonates and phosphonic acids via hydrophosphonylation of aldehydes with chiral H-phosphonate. *Tetrahedron: Asymmetry*, **26** (7), 393–399.

86 Gancarz, R. (1995) Nucleophilic addition to carbonyl compounds. Competition between hard (amine) and soft (phosphite) nucleophile. *Tetrahedron*, **51** (38), 10627–10632.

87 Wroblewski, A.E. and Konieczko, V.T. (1984) Stereochemistry of 1,2-oxaphospholanes. III. Evidence for the retro-Abramov pathway in methoxide-catalysed equilibration of substituted 2-methoxy-2-oxo-1,2-oxaphospholan-3-ols. *Monatsh. Chem.*, **115** (6–7), 785–791.

88 Wroblewski, A.E. (1986) Synthesis, stereochemistry, and ring opening reactions of diastereomeric 3-hydroxy-2-methoxy-5-methyl-l,2-oxaphospholan-2-ones. *Liebigs Ann. Chem.*, (8), 1448–1455.

89 Kolodiazhnyi, O.I., Gryshkun, E.V., Kachkovskyi, G.O., Kolodiazhna, A.O., Kolodiazhna, O.O., and Sheiko, S.Y. (2011) New methods for the synthesis of phosphonic analogues of natural compounds. *Phosphorus, Sulfur Silicon Relat. Elem.*, **186** (4), 644–651.

90 Kolodyazhnaya, O.O. and Kolodyazhnyi, O.I. (2011) Chiral N-Moc-pyrrolidine bisphosphonate. *Russ. J. Gen. Chem.*, **81** (1), 145–146.

91 Moreno, G.E., Quintero, L., Berne's, S., and Anaya de Parrodi, C. (2004) Diastereoselective carbonyl phosphonylation using chiral N,N'-bis-[(S)-a-phenylethyl]-bicyclic phosphorous acid diamides. *Tetrahedron Lett.*, **45** (22), 4245–4248.

92 Kolodiazhnyi, O.I., Grishkun, E.V., Sheiko, S., Demchuk, O., Thoennessen, H., Jones, P.G., and Schmutzler, R. (1998) Chiral symmetric phosphoric acid esters as sources of optically active organophosphorus compounds. *Tetrahedron: Asymmetry*, **9** (9), 1645–1649.

93 Kolodiazhnyi, O.I., Grishkun, E.V., Sheiko, S., Demchuk, O., Thoennessen, H., Jones, P.G., and Schmutzler, R. (1999) Asymmetric synthesis of α–substituted alkylphosphonates on the base of symmetric dialkylphosphites. *Russ. Chem. Bull. (Izvestia AN)*, **48** (8), 1588–1593.

94 Kolodiazhnyi, O.I., Sheiko, S.Y., and Grishkun, E.V. (2000) C3-symmetric trialkyl phosphites as starting compounds of asymmetric synthesis. *Heteroat. Chem.*, **11** (2), 138–143.

95 Cain, M.J., Cawley, A., Sum, V., Brown, D., Thornton-Pett, M., and Kee, T.P. (2002) Stereoselective redox transformations of phosphorus heterocycles: stereocontrol in the asymmetric phospho-Mukaiyama aldol reaction. *Inorg. Chim. Acta*, **345**, 154–172.

96 Devitt, P.G. and Kee, T.P. (1995) Diastereoselective phosphonylation of aldehydes using chiral diazaphospholidine reagents. *Tetrahedron*, **51** (40), 10987–10996.

97 Wroblewski, A.E. and Piotrowska, D.G. (2001) Stereochemistry of the addition of dialkyl phosphites to (S)-NN-dibenzylphenylglycinal. *Tetrahedron: Asymmetry*, **12** (21), 2977–2984.

98 Wroblewski, A.E. and Balcerzak, K.B. (1998) Synthesis of enantiomeric diethyl (1R,2R)- and (1S,2R)-1,2,3-trihydroxypropylphosphonates. *Tetrahedron*, **54** (24), 6833–6840.

99 Kolodiazhnaia, A.O., Kukhar, V.P., Chernega, A.N., and Kolodiazhnyi, O.I. (2004) Double and triple stereoselectivity in phosphaaldol reaction. *Tetrahedron: Asymmetry*, **15** (13), 1961–1963.

100 Kolodiazhnaia, A.O., Kukhar, V.P., and Kolodiazhnyi, O.I. (2004) Asymmetric synthesis of phosphathreoninic acid. *Russ. J. Gen. Chem.*, **74** (12), 1945–1946.

101 Kolodiazhnaia, A.O., Kukhar, V.P., and Kolodiazhnyi, O.I. (2004) Asymmetric synthesis of (1S,2R)-(1,2,3-trihydroxypropyl)diphenylphosphine oxide. *Russ. J. Gen. Chem.*, **74** (6), 965–966.

102 Wroblewski, A.E. and Piotrowska, D.G. (1999) Enantiomeric phosphonate analogs of the paclitaxel C-13 side chain. *Tetrahedron: Asymmetry*, **10** (10), 2037–2043.

103 Wroblewski, A.E. and Piotrowska, D.G. (2002) Enantiomerically pure N-Boc- and N-benzoyl-(S)-phenylglycinals. *Tetrahedron: Asymmetry*, **13** (22), 2509–2512.

104 Yokomatsu, T., Yamagishi, T., and Shibuya, S. (1993) Stereodivergent synthesis of β-amino-α-hydroxyphosphonic acid derivatives by lewis acid mediated stereoselective hydrophosphonylation of α amino aldehydes. *Tetrahedron: Asymmetry*, **4** (7), 1401–1404.

105 Pousset, C. and Larcheveque, M. (2002) An efficient synthesis of α- and β-aminophosphonic esters from α-amino acids. *Tetrahedron Lett.*, **43** (30), 5257–5260.

106 Wroblewski, A.E. and Balcerzak, K.B. (2001) Synthesis of 2-amino-1,3-dihydroxypropyl-phosphonates from Garner aldehyde. *Tetrahedron: Asymmetry*, **12** (3), 427–432.

107 Wroblewski, A.E. and Piotrowska, D.G. (2000) Enantiomeric phosphonate analogs of the docetaxel C-13 side chain. *Tetrahedron: Asymmetry*, **11** (12), 2615–2624.

108 Piotrowska, D.G., Halajewska Wosik, A., and Wroblewski, A.E. (2000) Racemization-free recovery of α-hydroxyphosphonates from their carboxylic esters. *Synth. Commun.*, **30** (21), 3935–3940.

109 Wroblewski, A.E. and Piotrowska, D.G. (1998) Phosphonate analogs of *N*-benzoyl- and *N*-Boc-3-phenylisoserine, the taxol C-13 side chain. *Tetrahedron*, **54** (28), 8123–8132.

110 Patel, D.V., Rielly-Gauvin, K., Ryono, D.E., Free, C.A., Rogers, W.L., Smith, S.A., DeForrest, J., Oehl, R.S., and Petrillo, E.W. Jr., (1995) α-hydroxy phosphinyl-based inhibitors of human renin. *J. Med. Chem.*, **38** (22), 4557–4569.

111 Stowasser, B., Budt, K.H., Jian-Qi, L., Peyman, A., and Ruppert, D. (1992) New hybrid transition state analog inhibitors of HIV protease with peripheric C_2-symmetry. *Tetrahedron Lett.*, **33** (44), 6625–6628.

112 Peyman, A., Budt, K.H., Spanig, J., Stowasser, B., and Ruppert, D.C. (1992) 2-symmetric phosphinic acid inhibitors of HIV protease. *Tetrahedron Lett.*, **33** (32), 4549–4552.

113 (a) Gulyaiko, I.V. and Kolodyazhnyi, O.I. (2009) Chiral fluorophosphono-methylbenzaldehydes. *Russ. J. Gen. Chem.*, **79** (1), 146–147; (b) Yokomatsu, T., Yamagishi, T., Matsumoto, K., and Shibuya, S. (1996) Reconfirmation of high enantiomeric excesses in the enantioselective transformation of meso-epoxides. *Tetrahedron*, **52** (43), 11725–11738.

114 Kabat, M.M. (1993) A novel method of highly enantioselective synthesis of γ-hydroxy-β-keto phosphonates via allene oxides. *Tetrahedron Lett.*, **34** (52), 8543–8544.

115 Yamagishi, T., Kusano, T., Kaboudin, B., Yokomatsu, T., Sakuma, C., and Shibuya, S. (2003) Diastereoselective synthesis of β-substituted α-hydroxyphosphinates through hydrophosphinylation of α-heteroatom-substituted aldehydes. *Tetrahedron*, **59** (6), 767–772.

116 Bongini, A., Panunzio, M., Bandini, E., Martelli, G., and Spunta, G. (1995) Stereoselective synthesis of α,β-disilyloxy phosphonates by stereoselective phosphonylation of α-silyloxyaldehydes. *Synlett*, **1995** (5), 461–462.

117 Bongini, A., Camerini, R., Panunzio, M., Bandini, E., Martelli, G., and Spunta, G. (1996) Stereochemical aspects of the asymmetric synthesis of chiral α,β-dihydroxy phosphonates. Synthesis of α,β-dihydroxy phosphonic acids. *Tetrahedron: Asymmetry*, **7** (12), 3485–3504.

118 Bandini, E., Martelli, G., Spunta, G., and Panunzio, M. (1995) Silicon directed asymmetric synthesis of (1R, 2S)-(−)-(1,2-Epoxypropyl)phosphonic acid (fosfomycin) from (S)-lactaldehyde. *Tetrahedron: Asymmetry*, **6** (9), 2127–2130.

119 Bongini, A., Camerini, R., and Panunzio, M. (1996) Efficient synthesis of the four diastereomers of phosphothreonine from lactalhehyde. *Tetrahedron: Asymmetry*, **7** (5), 1467–1476.

120 Wroblewski, A.E. and Glowacka, I.E. (2002) Synthesis of (1R,2S)- and (1S,2S)-3-azido-1,2-dihydroxypropylphosphonates. *Tetrahedron: Asymmetry*, **13** (9), 989–994.

121 Yokomatsu, T., Yoshida, Y., and Shibuya, S. (1994) Stereoselective synthesis of.beta.−oxygenated alpha.−hydroxyphosphonates by Lewis acid-mediated stereoselective hydrophosphonylation of alpha.−benzyloxy aldehydes. An application to the synthesis of phosphonic acid analogs of oxyamino acids. *J. Org. Chem.*, **59** (25), 7930–7933.

122 Hanaya, T., Miyoshi, A., Noguchi, A., Kawamoto, H., Armour, M.A., Hogg, A.M., and Yamamoto, H. (1990) A convenient synthesis of (2R)-1-amino-1-deoxy-l-phosphinylglycerols. *Bull. Chem. Soc. Jpn.*, **63** (12), 3590–3594.

123 Yokomatsu, T. and Shibuya, S. (1992) Enantioselective synthesis of α-amino phosphonic acids by an application of stereoselective opening of homochiral dioxane acetals with triethyl phosphite. *Tetrahedron: Asymmetry*, **3** (3), 377–378.

124 Wroblewski, A.E. (1984) Synthesis of carbohydrates having phosphorus in the anomeric position. *Carbohydr. Res.*, **125** (1), C1–C4.

125 Wroblewski, A.E. (1984) Synthesis of 1,3-di-O-benzyl-D-glycero-tetrulose. *Carbohydr. Res.*, **131** (2), 325–330.

126 Wroblewski, A.E. (1986) Notizen analogues of methyl D-ribo and D-arabino-furanosides having phosphorus in the anomeric position. *Z. Naturforsch., B*, **41** (6), 791–792.

127 Hanaya, T. and Yamamoto, H. (1989) Synthesis and structural analysis of 4-deoxy-4-(hydroxyphosphinyl and phenylphosphinyl)-D-ribofuranoses. *Bull. Chem. Soc. Jpn.*, **62** (7), 2320–2327.

128 Harvey, T.C., Simiand, C., Weiler, L., and Withers, S.G. (1997) Synthesis of cyclic phosphonate analogs of ribose and arabinose. *J. Org. Chem.*, **62** (20), 6722–6725.

129 Yamamoto, H., Hanaya, T., Kawamoto, H., Inokawa, S., Yamashita, M., Armour, M.A., and Nakashima, T.T. (1985) Synthesis and structural analysis of 5-deoxy-5-C-(hydroxyphosphinyl)-D-xylo- and -glucopyranoses. *J. Org. Chem.*, **50** (19), 3516–3521.

130 Thiem, J. and Guenter, M. (1984) Abramow-reaktion von kohlenhydratderivaten mit freien anomeren. *Phosphorus, Sulfur Silicon Relat. Elem.*, **20** (1), 67–79.

131 Thiem, J., Guenter, M., Paulsen, H., and Kopf, J. (1977) Phosphorhaltige Kohlenhydrate, XVI. Ringerweiterung von Furanose-Ringen zu 1,2λ5-Oxaphosphorinanen. *Chem. Ber.*, **110** (9), 3190–3200.

132 Wroblewski, A.E. (1986) Analogues of branched-chain tetrafuranosides having phosphorus in the anomeric position. *Tetrahedron*, **42** (13), 3595–3606.

133 Wroblewski, A.E. (1986) Synthesis of enantiomeric dimethyl (1,2,3,4-tetrahydroxybutyl) phosphonates. *Liebigs Ann. Chem.*, **1986** (11), 1854–1862.

134 Kachkovskyi, G.O. and Kolodiazhnyi, O.I. (2010) Synthesis of the phosphonoanalog of benzo[c]pyroglutamic acid. *Phosphorus, Sulfur Silicon Relat. Elem.*, **185** (11), 2238–2242.

135 Guliaiko, I.V. and Kolodiazhnyi, O.I. (2004) Highly stereoselective addition of diphenyl(trimethylsilyl) phosphine to a chiral aldehyde. *Russ. J. Gen. Chem.*, **74** (10), 1623–1624.

136 Kolodiazhna, A.O., Guliaiko, I.V., and Kolodiazhnyi, O.I. (2005) Diasteroselective addition of mono and bis-silylphosphines to chiral aldehydes. *Phosphorus, Sulfur Silicon Relat. Elem.*, **180** (10), 2335–2346.

137 Kolodiazhnyi, O.I., Guliaiko, I.V., and Kolodiazhna, A.O. (2004) Highly stereoselective addition of silylphosphines to chiral aldehydes. *Tetrahedron Lett.*, **45** (37), 6955–6957.

138 Gordon, N.J. and Evans, S.A. Jr., (1993) Diastereoselective condensation of oxazaphosphites with aliphatic and aromatic aldehydes. *J. Org. Chem.*, **58** (20), 5293–5294.

139 Merino, P., Marqués-López, E., and Herrera, R.P. (2008) Catalytic enantioselective hydrophosphonylation of aldehydes and imines. *Adv. Synth. Catal.*, **350**, 1195–1208.

140 Zefirov, N.S. and Matveeva, E.D. (2008) Catalytic Kabachnik-Fields reaction: new horizons for old reaction. *ARKIVOC*, (i), 1–17.

141 Gröger, H. and Hammer, B. (2000) Catalytic concepts for the enantioselective synthesis of α-amino and α-hydroxy phosphonates. *Chem. Eur. J.*, **6**, 943–948.

142 Bera, K., Nadkarni, D., and Namboothiri, I.N.N. (2013) Asymmetric synthesis of γ-aminophosphonates: the bio-isosteric analogs of γ-aminobutyric acid. *J. Chem. Sci.*, **125** (3), 443–465.

143 Gilmore, W.F. and McBride, H.A. (1972) Synthesis of an optically active α-aminophosphonic acid. *J. Am. Chem. Soc.*, **94** (12), 4361.

144 Yuan, C. and Cui, S. (1991) Studies on organophosphorus compounds XL VIII structural effect on the induced asymmetric addition of dialkyl phosphite to chiral aldimine derivatives. *Phosphorus, Sulfur Silicon Relat. Elem.*, **55** (1–4), 159–164.

145 Sheiko, S., Guliaiko, I., Grishkun, E., and Kolodiazhnyi, O.I. (2002) Double asymmetric induction during addition of chiral phosphites to C=N bond. *Phosphorus, Sulfur Silicon Relat. Elem.*, **177** (8-9), 2269–2270.

146 Kachkovskii, G.A., Andrushko, N.V., Sheiko, S.Y., and Kolodiazhnyi, O.I. (2005) Dual stereochemical control in reaction of diand trialkyl phosphites with aldimines. *Russ. J. Gen. Chem.*, **75** (11), 1735–1743.

147 Lewkowski, J., Tokarz, P., Lis, T., and Slepokura, K. (2014) Stereoselective addition of dialkyl phosphites to di-salicylaldimines bearing the (R,R)-1,2-diaminocyclohexane moiety. *Tetrahedron*, **70** (4), 810–816.

148 Gowiak, T., Sawka-Dobrowolska, W., Kowalik, J., Mastalerz, P., Soroka, M., and Zon, J. (1977) Palladium catalyzed thienylation of allylic alcohols with 3-bromothiophene. *Tetrahedron Lett.*, **45** (18), 3965–3968.

149 Lewkowski, J. and Karpowicz, R. (2010) Diastereoselective addition of Di-(trimethylsilyl)phosphite to chiral *N*-(R)-α-methylbenzyl and *N*-(1-methoxycarbonyl-iso-pentyl) Schiff bases of various aldehydes. *Heteroat. Chem*, **21** (5), 326–331.

150 Lewkowski, J., Karpowicz, R., and Skowronski, R. (2009) First synthesis of ferrocenyl-*N*-alkylaminophosphonic acids. *Phosphorus, Sulfur Silicon Relat. Elem.*, **184** (4), 815–819.

151 Cottier, L., Descotes, G., Lewkowski, J., and Skowronski, R. (1996) Synthesis and its stereochemistry of aminophosphonic acids derived from 5-hydroxymethylfurfural. *Phosphorus, Sulfur Silicon Relat. Elem.*, **116** (1), 93–100.

152 Yager, K.M., Taylor, C.M., and Smith, A.B. (1994) Asymmetric synthesis of α-aminophosphonates via diastereoselective addition of lithium diethyl phosphite to chelating imines. *J. Am. Chem. Soc.*, **116** (20), 9377–9378.

153 Vicario, J., Ortiz, P., and Palacios, F. (2013) Synthesis of tetrasubstituted α-aminophosphonic acid derivatives from trisubstituted α-aminophosphonates. *Eur. J. Org. Chem.*, **2013** (31), 7095–7100.

154 Boratynski, P.J., Skarzewski, J., and Sidorowicz, L. (2012) Stereochemistry of hydrophosphonylation of 9-aminoquinine Schiff bases. *Arkivoc*, (IV), 204–215.

155 Lefebvre, I.M. and Evans, S.A. (1997) Studies toward the asymmetric synthesis of α-amino phosphonic acids via the addition of phosphites to enantiopure sulfinimines. *J. Org. Chem.*, **62** (22), 7532–7533.

156 Davis, F.A., Lee, S., Yan, H., and Titus, D.D. (2001) Asymmetric synthesis of quaternary amino phosphonates using sulfinimines. *Org. Lett.*, **3** (11), 1757–1760.

157 Lyzwa, P. and Mikołajczyk, M. (2010) Asymmetric synthesis of aminophosphonic acids mediated by chiral sulfinyl auxiliary: recent advances. *Pure Appl. Chem.*, **82** (3), 577–582.

158 Chen, Q., Li, J., and Yuan, C. (2008) Sulfinimine-mediated asymmetric synthesis of acyclic and cyclic α-aminophosphonates. *Synthesis*, **2008** (18), 2986–2990.

159 Lyzwa, P., Blaszczyk, J., Sieron, L., and Mikołajczyk, M. (2013) Asymmetric synthesis of structurally diverse aminophosphonic acids by using enantiopure *N*-(p-tolylsulfinyl)cinnamaldimines as reagents. *Eur. J. Org. Chem.*, **2013** (11), 2106–2115.

160 Roschenthaler, G.-V., Kukhar, V.P., Kulik, I.B., Belik, M.Y., Sorochinsky, A.E., Rusanov, E.B., and Soloshonok, V.A. (2012) Asymmetric synthesis of phosphonotrifluoroalanine and its derivatives using *N*-tert-butanesulfinyl imine derived from fluoral. *Tetrahedron Lett.*, **53** (5), 539–542.

161 Yao, Q. and Yuan, C. (2013) Enantioselective synthesis of H-phosphinic acids bearing natural amino acid residues. *J. Org. Chem.*, **78** (14), 6962–6974.

162 Khan, H.A. and Elmman, J. (2013) Asymmetric synthesis of α-aminophosphonate esters by the addition of dialkyl phosphites to tert-butanesulfinyl imines. *Synthesis*, **45** (22), 3147–3150.

163 Mikołajczyk, M., Łyzwa, P., and Drabowicz, J. (2002) A new efficient procedure for asymmetric synthesis of α-aminophosphonic acids via addition of lithiated bis(diethylamino)phosphine borane complex to enantiopure sulfinimines. *Tetrahedron: Asymmetry,* **13** (23), 2571–2576.
164 Mikołajczyk, M., Łyzwa, P., and Drabowicz, J. (1997) Asymmetric addition of dialkyl phosphite and diamido phosphite anions to chiral, enantiopure sulfinimines: a new, convenient route to enantiomeric α-aminophosphonic acids. *Tetrahedron: Asymmetry,* **8** (24), 3991–3994.
165 Chen, Q. and Yuan, C. (2007) A new and convenient asymmetric synthesis of α–amino- and α–alkylaaminophosphonic acids using *N*-tert-butylsulfinyl imines as chiral auxiliaries. *Synthesis,* (24), 3779–3786.
166 Roschenthaler, G.V., Kukhar, V., Barten, J., Gvozdovsk, N., Belik, M., and Sorochinsky, A. (2004) Asymmetric synthesis of α,α-difluoro-β-amino phosphonic acids using sulfinimines. *Tetrahedron Lett.,* **45** (35), 6665–6667.
167 Roschenthaler, G.-V., Kukhar, V.P., Belik, M.Y., Mazurenko, K.A.I., and Sorochinsky, A.E. (2006) Diastereoselective addition of diethyl difluoromethylphosphonate to enantiopure sulfinimines: synthesis of α,α-difluoro-β-aminophosphonates, phosphonic acids, and phosphonamidic acids. *Tetrahedron,* **62** (42), 9902–9910.
168 Cherkasov, R.A. and Galkin, V.I. (1998) The Kabachnik–Fields reaction: synthetic potential and the problem of the mechanism. *Russ. Chem. Rev.,* **67** (10), 857–882.
169 Fang, Z., Yang, H., Miao, Z., and Chen, R. (2011) Asymmetric Mannich-type synthesis of *N*-phosphinyl α-aminophosphonic acid monoesters. *Helv. Chim. Acta,* **94** (9), 1586–1593.
170 Cui, Z., Zhang, J., Wang, F., Wang, Y., Miao, Z., and Chen, R. (2008) The diastereoselective synthesis of methyl 5-deoxy-5-(dialkylphosphono)-5-(dialkylphosphorylamido)-2,3-O-isopropylidene-beta-D-ribofuranosides. *Carbohydr. Res.,* **343** (15), 2530–2534.
171 Tibhe, G.D., Reyes-González, M.A., Cativiela, C., and Ordóñez, M. (2012) Microwave-assisted high diastereoselective synthesis of α-aminophosphonates under solvent and catalyst free-conditions. *J. Mex. Chem. Soc.,* **56** (2), 183–187.
172 Palacios, F., Aparicio, D., de Retana, O.M.A., de los Santos, J.M., Gil, J.I., and Lopez de Munain, R. (2003) Asymmetric synthesis of 2H-aziridine phosphonates, and α- or β-aminophosphonates from enantiomerically enriched 2H-azirines. *Tetrahedron: Asymmetry,* **14** (6), 689–700.
173 Kolodiazhnyi, O.I., Kukhar, V.P., and Kolodiazhna, A.O. (2014) Asymmetric catalysis as a method for the synthesis of chiral organophosphorus compounds. *Tetrahedron: Asymmetry,* **25** (12), 865–922.
174 Haynes, R.K., Lam, W.W.L., and Yeung, L.L. (1996) Stereoselective preparation of functionalized tertiary P-chiral phosphine oxides by nucleophilic addition of litbiated tert-butylpbenylphospbine oxide to carbonyl compounds. *Tetrahedron Lett.,* **37** (29), 4729–4732.
175 Tedeschi, L. and Enders, D. (2001) Asymmetric synthesis of α-phosphono malonates via Fe_2O_3-mediated phospha-Michael addition to Knoevenagel acceptors. *Org. Lett.,* **3** (22), 3515–3517.
176 Jaklińska, M., Cordier, M., and Stankevič, M. (2016) Enantioselective organocatalytic phospha-Michael reaction of α,β-unsaturated aldehydes. *J. Org. Chem.,* **81** (4), 1378–1390.

177 Hanaya, T., Yamamoto, H., and Yamamoto, H. (1992) Orientation of the addition of dimethyl phosphonate to 5,6-dideoxy-6-nitro-n-hex-5-enofuranoses. *Bull. Chem. Soc. Jpn.*, **65** (4), 1154–1156.

178 Castelot-Deliencourt, G., Pannecoucke, X., and Quirion, J.C. (2001) Diastereoselective synthesis of α-substituted β-amidophosphonates. *Tetrahedron Lett.*, **42** (6), 1025–1028.

179 Castelot-Deliencourt, G., Roger, E., Pannecoucke, X., and Quirion, J.C. (2001) Diastereoselective synthesis of chiral amidophosphonates by 1,5-asymmetric induction. *Eur. J. Org. Chem.*, **2001** (16), 3031–3038.

180 Liu, X., Hu, X.E., Tian, X., Mazur, A., and Ebetino, F.H. (2002) Enantioselective synthesis of phosphinyl peptidomimetics via an asymmetric Michael reaction of phosphinic acids with acrylate derivatives. *J. Organomet. Chem.*, **646** (1–2), 212–222.

181 Brunner, H. and Net, G. (1995) Enantioselective catalysis, part 93: 1 optically active expanded phosphanes derived from 1,2-bisphosphanobenzene and amides and esters of acrylic acid. *Synthesis*, **1995** (4), 423–426.

182 Knühl, G., Sennhenn, P., and Helmchen, G. (1995) New chiral β-phosphinocarboxylic acids and their application in palladium-catalysed asymmetric allylic alkylations. *J. Chem. Soc., Chem. Commun.*, (18), 1845–1846.

183 Jansen, J.F.G.A. and Feringa, B.L. (1990) Michael addition of lithiodiphenylphosphine to menthyloxy-2[5H]-furanone:enantioselective synthesis of SS-CHIRAPOS. *Tetrahedron: Asymmetry*, **1** (10), 719–720.

184 Gourdel, Y., Pellon, P., Toupet, L., and LeCorre, M. (1994) Stereoselective synthesis of new functionalized bisphosphines. *Tetrahedron Lett.*, **35** (8), 1197–1200.

185 Hanaya, T., Ohmori, K., Yamamoto, H., Armour, M.A., and Hogg, A.M. (1990) Synthesis and structural analysis of 5-deoxy-5-[(R)- and (S)-methylphosphinyl]-α,β-D-manno- and –L-gulopyranoses. *Bull. Chem. Soc. Jpn.*, **63** (4), 1174–1179.

186 (a) Yamashita, M., Sugiura, M., Tamada, Y., Oshikawa, T., and Clardy, J. (1987) First X-ray study on orientation of addition of phosphorus compounds to 3-O-alkyl-5,6-dideoxy-1,2-O-isopropylidene-6-C-nitro-α–D-xylohexo-5-(Z)–enofuranoses. *Chem. Lett.*, (7), 1407–1408; (b) Zhang, H., Sun, Y.M., Zhao, Y., Zhou, Z.-Y., Wang, J.-P., Xin, N., Nie, S.-Z., Zhao, C.-Q., and Han, L.-B. (2015) One-pot process that efficiently generates single stereoisomers of 1,3-bisphosphinylpropanes having five chiral centers. *Org. Lett.*, **17** (1), 142–145.

187 Ordonez, M., de la Cruz-Cordero, R., Quinones, C., and Gonzalez-Morales, A. (2004) Highly diastereoselective synthesis of anti–γ-*N*-benzylamino-β-hydroxyphosphonates. *J. Chem. Soc., Chem. Commun.*, (06), 672–673.

188 Ordonez, M., de la Cruz, R., Fernandez-Zertuche, M.M., and Munoz-Hernandez, M.A. (2002) Diastereoselective reduction of α-ketophosphonates derived from amino acids. A new entry to enantiopure-hydroxy--aminophosphonate derivatives. *Tetrahedron: Asymmetry*, **13** (02), 559–562.

189 Burke, T.R. Jr., Smyth, M.S., Nomizu, M., Otaka, A., and Roller, P.P. (1993) Preparation of fluoro- and hydroxy-4-(phosphonomethyl)-D,L-phenylalanin suitably protected for solid-phase synthesis of peptides containing hydrolytically stable analogues of O-phosphotyrosinel. *J. Org. Chem.*, **58** (06), 1336–1340.

190 Barco, A., Benetti, S., Bergamini, P., De Risi, C., Marchetti, P., Pollini, G.P., and Zanirato, V. (1999) Diastereoselective synthesis of β-amino−α−hydroxy phosphonates via oxazaborolidine catalyzed reduction of β−phthalimido−α−ketophosphonates. *Tetrahedron Lett.*, **40** (43), 7705–7708.

191 Ordonez, M., Gonzalez-Morales, A., and Salazar-Fernandez, H. (2004) Highly diastereoselective reduction of β−ketophosphonates bearing homochiral bis(α−methylbenzyl)amine: preparation of both enantiomers of phosphogabob (GABOBP). *Tetrahedron: Asymmetry*, **15** (17), 2719–2725.

192 Ordóñez, M., Lagunas-Rivera, S., Hernández-Núñez, E., and Labastida-Galván, V. (2010) Synthesis of syn-γ-amino-β-hydroxyphosphonates by reduction of β-ketophosphonates derived from L-proline and L-serine. *Molecules*, **15** (3), 1291–1301.

193 Amedjkouh, M. and Grimaldi, J. (2002) Stereoselective reduction and reductive dephosphonylation of β-iminophosphonates. *Tetrahedron Lett.*, **43** (20), 3761–3764.

194 Oshikawa, T. and Yamashita, M. (1990) Preparation of optically active (S)-2-aminoalkylphosphonic acids from (S)-amino acids without racemization. *Bull. Chem. Soc. Jpn.*, **63** (9), 2728–2730.

195 Meier, C. and Laux, W.H.G. (1996) Asymmetric synthesis of chiral, nonracemic dialkyi-hydroxyalkylphosphonates via a (-)-chlorodiisopinocampheylborane (Ipc_2B-CI) reduction. *Tetrahedron: Asymmetry*, **7** (1), 89–94.

196 Meier, C. and Laux, W.H.G. (1995) Asymmetric synthesis of chiral, nonracemic dialkyl-α−, β−, and γ−hydroxyalkylphosphonates via a catalyzed enantioselective catecholborane reduction. *Tetrahedron: Asymmetry*, **6** (5), 1089–1092.

197 Meier, C. and Laux, W.H.G. (1996) Enantioselective synthesis of diisopropyl α−, β−, and γ−hydroxyarylalkylphdosphonates from ketophosphonates: a study on the effect of the phosphonyl group. *Tetrahedron*, **52** (02), 589–598.

198 Nesterov, V. and Kolodiazhnyi, O.I. (2005) Enantioselective synthesis of α−hydroxyalkyl-phosphonates. *Russ. J. Gen. Chem.*, **75** (7), 1161–1162.

199 Pogatchnik, D.M. and Wiemer, D.F. (1997) Enantioselective synthesis of α−hydroxy phosphonates via oxidation with (Camphorsulfonyl)oxaziridines. *Tetrahedron Lett.*, **38** (20), 3495–3498.

200 Ordonez, M., De la Cruz-Cordero, R., Fernandez-Zertuche, M., Munoz-Hernandez, M.A., and Garcıa-Barradas, O. (2004) Diastereoselective reduction of dimethyl c-[(*N*-p-toluenesulfonyl)amino. β−ketophosphonates derived from amino acids. *Tetrahedron: Asymmetry*, **15** (19), 3035–3043.

201 Skropeta, D. and Schmidt, R.R. (2003) Chiral, non−racemic α−hydroxyphosphonates and phosphonic acids via stereoselective hydroxylation of diallyl benzylphosphonates. *Tetrahedron: Asymmetry*, **14** (2), 265–273.

202 Yamagishi, T., Fujii, K., Shibuya, S., and Yokomatsu, T. (2004) Synthesis of chiral γ-amino-β-hydroxyphosphonate derivatives from unsaturated phosphonates. *Synlett*, **2004** (14), 2505–2508.

203 Cravotto, G., Giovenzana, G.B., Pagliarin, R., Palmisano, G., and Sisti, M. (1998) A straightforward entry into enantiomerically enriched β-amino−α−hydroxyphosphonic acid derivatives. *Tetrahedron: Asymmetry*, **9** (5), 745–748.

204 Li, G., Chang, H.T., and Sharpless, K.B. (1966) Katalytische asymmetrische Aminohydroxylierung (AA) von Olefinen. *Angew. Chem.*, **108** (4), 449–452.

205 Yokomatsu, T., Yamagishi, T., Sada, T., Suemune, K., and Shibuya, S. (1998) Asymmetric dihydroxylation of 1-acyloxy-2(E)-alkenylphosphonates with AD-mix

reagents. Effects of 1-acyloxy functional groups on the asymmetric dihydroxylation. *Tetrahedron*, **54** (5-6), 781–790.

206 Yokomatsu, T., Suemune, K., Yamagishi, T., and Shibuya, S. (1995) Highly regioselective silylation of α,β-dihydroxyphosphonates: an application to stereoselective synthesis of α-amino-β-hydroxyphosphonic acid derivatives. *Synlett*, **1995** (8), 847–849.

207 Yokomatsu, T., Yamagishi, T., Suemune, K., Yoshida, Y., and Shibuya, S. (1998) Enantioselective synthesis of threo-α,β-dihydroxyphosphonates by asymmetric dihydroxylation of 1(E)-alkenylphosphonates with AD-mix reagents. *Tetrahedron*, **54** (5-6), 767–780.

208 Yokomatsu, T., Yoshida, Y., Suemune, K., Yamagishi, T., and Shibuya, S. (1995) Enantioselective synthesis of threo-α,β-dihydroxyphosphonates by asymmetric dihydroxylation of vinylphosphonates. An application to the stereocontrolled synthesis of (4S,5S)-4-diethylphosphono-5-hydroxymethyl-2,2-dimethyl-1,3-dioxolane. *Tetrahedron: Asymmetry*, **6** (2), 365–368.

209 Drag, M., Latajka, R., Gancarz, R., Kafarski, P., Pirat, J.L., and Cristau, H.J. (2002) Regio- and stereoselective synthesis, solution conformations of 2-amino-1-hydroxy-2-arylethylphosphonic esters and acids. *Phosphorus, Sulfur Silicon Relat. Elem.*, **177**, 2191–2192.

210 Cristau, H.J., Pirat, J.L., Drag, M., and Kafarski, P. (2000) Regio- and stereoselective synthesis of 2-amino-1-hydroxy-2-aryl ethylphosphonic esters. *Tetrahedron Lett.*, **41** (50), 9781–9785.

211 Nakamura, K., Kimura, T., Kanno, H., and Takagashi, E. (1995) Total synthesis of (+/-)-phosphonothrixin, a novel herbicidal antibiotic containing C-P bond. *J. Antibiot.*, **48** (10), 1134–1137.

212 Nakamura, K. and Yamamura, S.T. (1997) Enantioselective synthesis of phosphonothrixin and its absolute stereochemistry. *Tetrahedron Lett.*, **38** (3), 437–438.

213 Field, S.C. (1998) Total synthesis of (±)-phosphonothrixin. *Tetrahedron Lett.*, **39** (37), 6621–6624.

214 Tone, T., Okamoto, Y., and Sakurai, H. (1978) Preparation of 1-hydroxy-2-aminoethylphosphonic acid. *Chem. Lett.*, (12), 1349–1350.

215 Gordon, N.J. and Evans, S.A. Jr., (1993) Synthesis and enantioselective aldol reaction of a chiral 2-oxo-2-propionyl-1,3,2-oxazaphosphorinane. *J. Org. Chem.*, **58** (20), 5295–5297.

216 Hammerschmidt, F. and Wuggenig, F. (1998) Enzymes in organic chemistry, 8.[11] resolution of in a biphasic system protease-catalyzed kinetic α-chloroacetoxyphosphonates. *Phosphorus, Sulfur Silicon Relat. Elem.*, **141** (1), 231–238.

217 Kawashima, T., Ishii, T., Inamoto, N., Tokitoh, N., and Okazaki, R. (1998) The olefin synthesis from P-hydroxyalkylphosphonates induced by fluorides or relatively weak bases. *Bull. Chem. Soc. Jpn.*, **71** (1), 209–219.

218 Blazewska, K., Sikora, D., and Gajda, T. (2003) Con synthesis of protected diethyl 1-amino-2- hydroxyalkylphosphonates. *Tetrahedron Lett.*, **44** (25), 4747.

219 Wroblewski, A.E. and Halajewska-Wosik, A. (2003) An efficient synthesis of an enantiomerically pure phosphonate analogue of l-GABOB. *Tetrahedron: Asymmetry*, **14** (21), 3359–3363.

220 Goti, A., Cichi, S., Brandi, A., and Pietruchiewicz, K.M. (1991) Nitrone cycloadditions to 2,3-dihydro-1-phenyl-1H-phosphole 1-oxide. Double asymmetric

induction and kinetic resolution by a chiral nitrone. *Tetrahedron: Asymmetry*, **2** (12), 1371–1378.
221 Brandi, A., Cichi, S., Goti, A., Koprowski, M., and Pietrusiewicz, K.M. (1994) Kinetic resolution in 1,3-dipolar cycloaddition of tartaric acid-derived nitrones to 2,3-dihydro-1-phenyl-1H-phospholes. An enantioselective approach to the 2,2'-coupled pyrrolidine-phospholane ring system. *J. Org. Chem.*, **59** (6), 1315–1318.
222 Brandi, A., Cichi, S., Goti, A., Pietrusiewicz, K.M., and Wisniewski, W. (1990) The regioselectivity of nitrone cycloadditions to vinyl phosphorus compounds. *Tetrahedron*, **46** (20), 7093–7104.
223 Vinokurov, N., Pietrusiewicz, K.M., Frynas, S., Wiebckec, M., and Butenschoen, H. (2008) Asymmetric 1,3-dipolar cycloaddition with a P-stereogenic dipolarophile: an efficient approach to novel P-stereogenic 1,2-diphosphine systems. *Chem. Commun.*, (8), 5408–5410.
224 Vinokurov, N., Pietrusiewicz, K.M., and Butenschoen, H. (2009) Asymmetric Diels-Alder cycloaddition of a Di-P-stereogenic dienophile with cyclopentadiene. *Tetrahedron: Asymmetry*, **20** (9), 1081–1085.
225 Izumi, A. and Tai, S. (1977) *Stereodifferentiating Reactions*, Kodansha Ltd., Tokyo, Academic Press, New York, 334 pp.
226 Kolodiazhnyi, O.I. (2003) Multiple stereoselectivity and its application in organic synthesis. *Tetrahedron*, **59** (32), 5953–6018.
227 Kolodiazhnyi, O.I. (2002) Double asymmetric induction as method for the synthesis of chiral organophosphorus compounds. *Phosphorus, Sulfur Silicon Relat. Elem.*, **177** (8–9), 2111–2115.
228 Wynberg, H. and Smaardijk, A.A. (1983) Asymmetric catalysis in carbon-phosphorus bond formation. *Tetrahedron Lett.*, **24** (52), 5899–5900.
229 Kolodyazhna, A.O., Kukhar, V.P., and Kolodyazhnyi, O.I. (2008) Organic catalysis of phospha-aldol condensation. *Russ. J. Gen. Chem.*, **78** (11), 2043–2051.
230 Kolodyazhnaya, O.O., Kolodyazhnaya, A.O., and Kolodyazhnyi, O.I. (2014) Synthesis of phosphonic analog of (S)-homoproline. *Russ. J. Gen. Chem.*, **84** (1), 169–170.
231 Guliaiko, I.V. and Kolodiazhnyi, O.I. (2008) Asymmetric syntheses of new phosphonotaxoids. *Phosphorus, Sulfur Silicon Relat. Elem.*, **183** (2-3), 677–678.
232 Kolodiazhnyi, O.I., Guliayko, I.V., Gryshkun, E.V., Kolodiazhna, A.O., Nesterov, V.V., and Kachkovskyi, G.O. (2008) New methods, and strategies for asymmetric synthesis of organophosphorus compounds. *Phosphorus, Sulfur Silicon Relat. Elem.*, **183** (2-3), 393–398.
233 Matsunaga, S. and Shibasaki, M. (2014) Recent advances in cooperative bimetallic asymmetric catalysis: dinuclear Schiff base complexes. *Chem. Commun.*, **50** (9), 1044–1057.
234 Gou, S., Zhou, X., Wang, J., Liu, X., and Feng, X. (2008) Asymmetric hydrophosphonylation of aldehydes catalyzed by bifunctional chiral Al(III) complexes. *Tetrahedron*, **64** (12), 2864–2870.
235 Yang, F., Zhao, D., Lan, J., Xi, P., Yang, L., Xiang, S., and You, J. (2008) Self-assembled bifunctional catalysis induced by metal coordination interactions: an exceptionally efficient approach to enantioselective hydrophosphonylation. *Angew. Chem. Int. Ed.*, **47** (20), 5646–5649.
236 Gulyaiko, I.V. and Kolodyazhnyi, O.I. (2005) Asymmetric synthesis of l-hydroxy-2-alkylphosphonic acids. *Russ. J. Gen. Chem.*, **75** (11), 1848–1849.

237 Gulyaiko, I.V. and Kolodyazhnyi, O.I. (2008) Synthesis of a phosphorus analog of iso-statine. *Russ. J. Gen. Chem.*, **78** (8), 1626–1627.
238 Yamagishi, T., Suemune, K., Yokomatsu, T., and Shibuya, S. (2001) Diastereoselective synthesis of chiral 3-amino-α-hydroxy-H-phosphinates through hydrophosphinylation of α-amino aldehydes. *Tetrahedron Lett.*, **42** (30), 5033–5036.
239 Yamagishi, T., Suemune, K., Yokomatsu, T., and Shibuya, S. (2002) Asymmetric synthesis of P-amino-α–hydroxyphosphinic acid derivatives through hydrophosphinylation of α-amino aldehydes. *Tetrahedron*, **58** (13), 2577–2583.
240 Yamagishi, T., Suemune, K., Yokomatsu, T., and Shibuya, S. (2002) Stereoselective synthesis of [3- Amino–α–hydroxy(allyl)phosphinates and an application to the synthesis of a building block for phosphinyl peptides. *Synlett*, **2002** (9), 1471–1474.
241 Kolodyazhnii, O.I. and Kolodyazhnaya, A.O. (2015) A new approach towards synthesis of phosphorylated alkenes. *Russ. J. Gen. Chem.*, **85** (2), 359–365.
242 Kolodyazhnyi, O.I. and Sheiko, S.Y. (2001) Double asymmetric induction in the addition reaction of chiral phosphites to C=N compounds. *Russ. J. Gen. Chem.*, **71** (6), 977–978.
243 Zhou, Z.-Y., Zhang, H., Yao, L., Wen, J.-H., Nie, S.-Z., and Zhao, C.-Q. (2016) Double asymmetric induction during the addition of (R_P)-menthyl phenyl phosphine oxide to chiral aldimines. *Chirality*, **28** (2), 132–135.
244 Kolodyazhnyi, O.I. and Sheiko, S. (2001) Enantioselective addition of dimenthyl and dibornyl phosphites to schiff bases. *Russ. J. Gen. Chem.*, **71** (7), 1155–1156.
245 Lyzwa, P. (2014) Double asymmetric induction in the synthesis of enantiomeric α–aminophosphonic acids mediated by sulfinimines. *Heteroat. Chem.*, **25** (1), 15–19.
246 Matsumoto, K., Sawayama, J., Hirao, S., Nishiwaki, N.i., Sugimoto, R., and Saigo, K. (2014) Enantiopure O-ethyl phenylphosphonothioic acid: a solvating agent for the determination of enantiomeric excesses. *Chirality*, **26** (10), 614–619.
247 Nesterov, V.V. and Kolodyazhnyi, O.I. (2006) Enantioselective reduction of ketophosphonates using chiral acid adducts with sodium borohydride. *Russ. J. Gen. Chem.*, **76** (7), 1022–1030.
248 Nesterov, V.V. and Kolodiazhnyi, O.I. (2007) Efficient method for the asymmetric reduction of α– and β–ketophosphonates. *Tetrahedron*, **63** (29), 6720–6731.
249 Nesterov, V.V. and Kolodiazhnyi, O.I. (2007) Di(lR,2S,5R)-menthyl 2-hydroxy-3-chloropropylphosphonate as useful chiron for the synthesis of α– and β– hydroxyphosphonates. *Synlett*, **2007** (15), 2400–2404.
250 Gryshkun, E.V., Nesterov, V.V., and Kolodyazhnyi, O.I. (2012) Enantioselective reduction of ketophosphonates using adducts of chiral natural acids with sodium borohydride. *Arkivoc*, **IV**, 100–117.
251 Guliaiko, I., Nesterov, V., Sheiko, S., Kolodiazhnyi, O.I., Freytag, M., Jones, P.G., and Schmutzler, R. (2008) Synthesis of optically active hydroxyphosphonates. *Heteroat. Chem*, **19** (2), 133–139.
252 Kolodiazhnyi, O.I. and Kolodiazhna, A.O. (2016) Multiple stereoselectivity in organophosphorus chemistry. *Phosphorus, Sulfur Silicon Relat. Elem.*, **191** (3), 444–458.

4

Asymmetric Catalysis with Metal Complexes

Chapters 4–6 are devoted to methods of asymmetric catalysis that are most often used and studied in modern organic chemistry. The methods of asymmetric metal complex catalysis, oraganocatalysis, and enzymatic biocatalysis are discussed first. These methods attract the attention of academic chemists who are interested in the development of fundamental organic and theoretical chemistry and also that of the experts, including chemists, working in fine organic synthesis, pharmaceutical chemistry, and agrochemistry [1–10].

4.1 Introduction

Chiral phosphorus compounds play an important role in many areas of science, including biologically active pharmaceuticals, agrochemicals, and ligands for transition metal complexes [1, 2]. Many methods are used to prepare enantiomerically pure organophosphorus compounds, including classical resolution via diastereoisomers, chemical kinetic resolution, enzymatic resolution, chromatographic resolution, and asymmetric catalysis. Over the last few years, great success has been achieved in the areas of asymmetric catalytic synthesis of organophosphorus compounds and catalyzed asymmetric hydrogenation reactions, and many articles devoted to the synthesis of chiral organophosphorus compounds have been published. An example of a significant commercial-scale enantioselective catalytic process (>10 000 tons annually) is the production of the (*S*)-metolachlor that is widely used as a herbicide. Metolachlor is an atropoisomeric compound forming four stereoisomers of which two (*S*)-diastereomers are biologically active. Metolachlor is produced from 2-ethyl-6-methylaniline (MEA) via condensation with methoxyacetone. The resulting imine is hydrogenated to give primarily the (*S*)-stereoisomeric amine. The key step in the production of metolachlor is the asymmetric hydrogenation of the imine with the chiral phosphine ligand Josiphos, accompanied by chloroacetylation of (*S*)-N-substituted aniline [10, 11] (Scheme 4.1).

Asymmetric Synthesis in Organophosphorus Chemistry: Synthetic Methods, Catalysis and Applications, First Edition. Oleg I. Kolodiazhnyi.

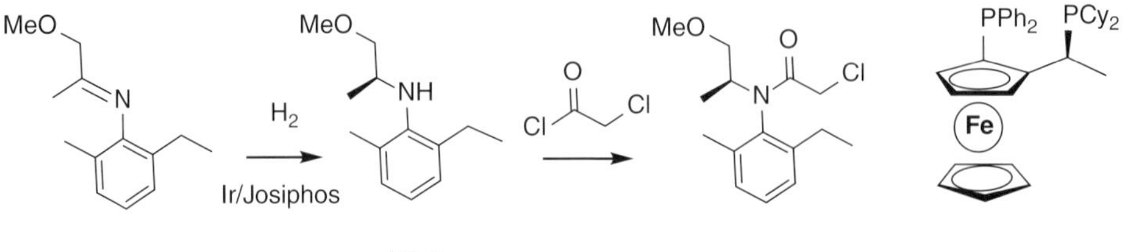

(*S*)-**1** (*S*)-Metolachlor **2**, Josiphos

Scheme 4.1 An industrial method for production of (*S*)-metolachlor, catalyzed by Josiphos.

4.2 Asymmetric Catalytic Hydrogenation and Other Reactions of Reduction

Asymmetric catalytic hydrogenation of unsaturated phosphonates is widely used in the synthesis of aminophosphonates and aminophosphonic acids which are of biological interest. Homogeneous asymmetric hydrogenation with chiral complexes of transition metals is one of the most important industrial methods for the preparation of enantiomerically pure organic molecules. Asymmetric hydrogenation of prochiral aminophosphonates, ketophosphonates, and keto iminophosphonates is an effective, practical, and economic synthetic method for the preparation of chiral organophosphorus compounds. Various chiral complexes of transition metals bearing chiral phosphine ligands have been used as catalysts for the asymmetric hydrogenation of unsaturated phosphorus compounds (Scheme 4.2) [1–3]. Examples of the most commonly used ligands for hydrogenation of unsaturated phosphonates are shown in Scheme 4.3.

4.2.1 Hydrogenation of C=C Phosphorus Compounds

The first work devoted to the asymmetric synthesis of aminophosphonates by catalytic hydrogenation of unsaturated phosphonates was published about 30 years ago. In 1985, Schollkopf *et al.* [12] reported the asymmetric hydrogenation of *N*-[1-(dimethoxyphosphoryl)-ethenyl]formamide **15** using a rhodium catalyst with (+)-DIOP 10 chiral ligand to afford the (1-aminoethyl) phosphonate **L-16** in good yields and enantioselectivity of 76% enantiomeric excess (ee). The initially formed formamide **L-15** was hydrolyzed with concentrated hydrochloric acid to give the aminophosphonic acid **L-16**. Crystallization from water/methanol increased the enantiomeric purity of **L-16** up to 93% ee (Scheme 4.4). The hydrogenation of α-enamidophosphonates has attracted the interest of several groups as a method for the synthesis of chiral aminophosphonates and a number of articles have been published on this subject [12–15]. Oehme *et al.* [13] reported that chiral Rh(I) complexes with BPPM ligands 6 or PROPRAPHOS 7 are active catalysts for the asymmetric hydrogenation of

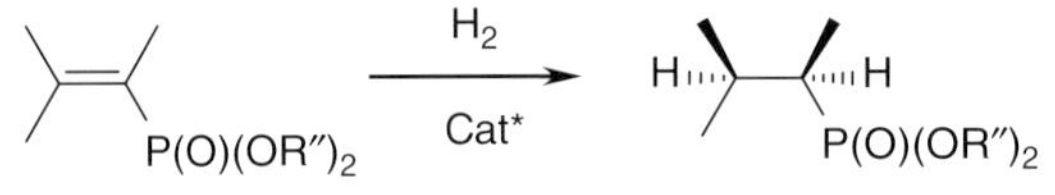

Cat* = chiral catalyst

Scheme 4.2 Asymmetric catalytic hydrogenation of vinylphosphonates.

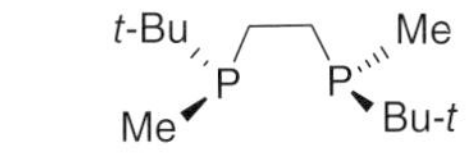

3, (*S,S*)-BisP

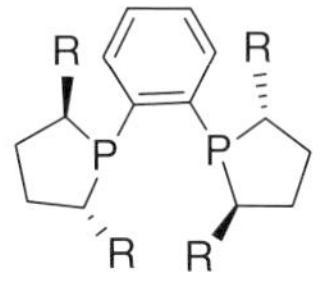

4a, (*S,S*)-MeDuPhos, R = Me,
4b, (*S,S*)-EtDuPhos, R = Et,
4c, (*S,S*)-PrDuPhos, R = Pr

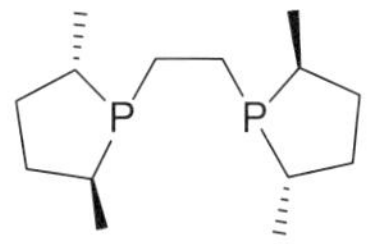

5a, (*S,S*)-Me-BPE,
5b, (*S,S*)-Et-BPE

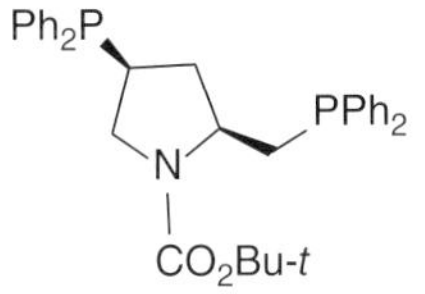

6, (*S,S*)-BPPM

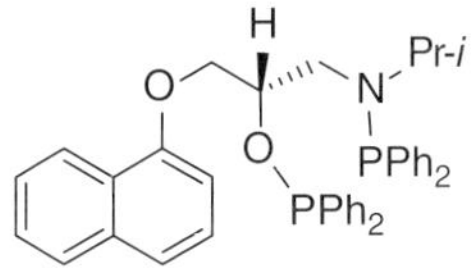

7, (*S*)-PROPRAPHOS

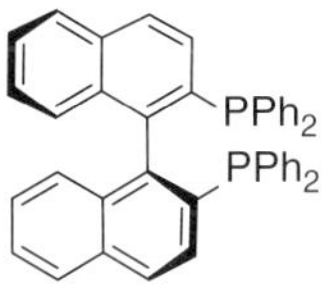

8, (*S*)-BINAP

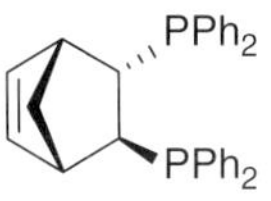

9, (*S,S*)-NORPHOS

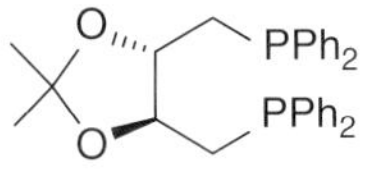

10, (*S,S*)-DIOP

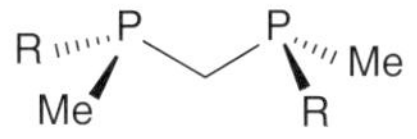

11, (*S,S*)-MiniPhos

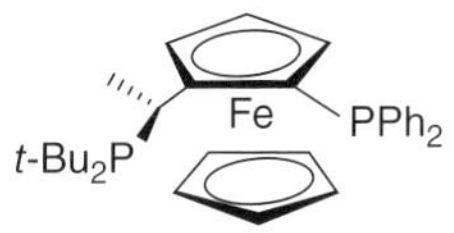

12

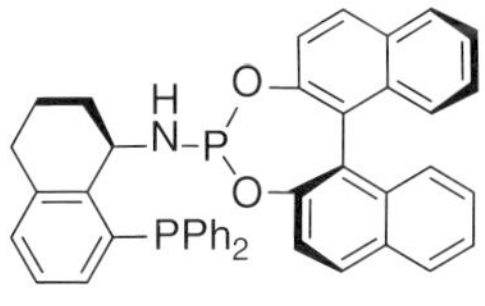

13, (*R,R*)-THNAPhos

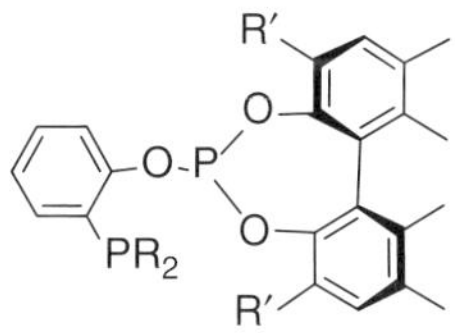

14a–c, P-OP
R = Ph, R′ = H (a); R = *i* - Pr, R′ = H (b);
R = *i*-Pr, R′ = *t*-Bu (c)

Scheme 4.3 Examples of chiral ligands used in asymmetric catalysis.

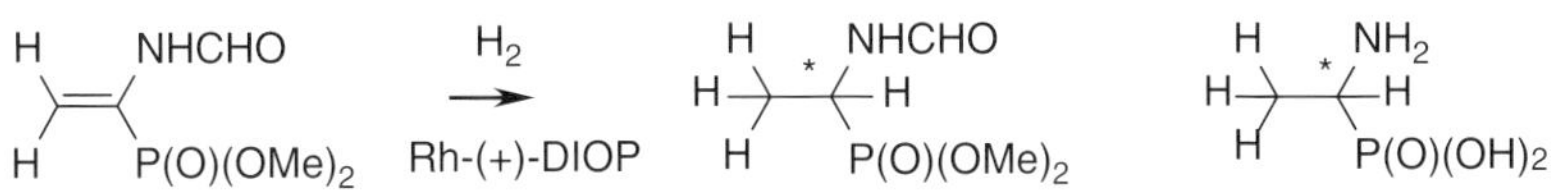

L-**15**, 76% ee
$[\alpha]_D = -12.9$

L-**16**, 93% ee
$[\alpha]_D = -15.6$ (1N NaOH)

Scheme 4.4 Hydrogenation of *N*-[1-(dimethoxyphosphoryl)ethenyl]formamide in the presence of a chiral rhodium/(+)-DIOP catalyst.

(*E*)-phenylenamidophosphonates displaying high reaction rates and relatively high stereoselectivities. For example, the hydrogenation of amido vinylphosphonates **17a,b** catalyzed by the BPPM(**6**)/Rh complex afforded α-aminophosphonates **18a,b** with 96% ee, and as PROPRAPHOS **7** is available in both configurations, (*R*)- and (*S*)-α-aminophosphonic acid esters **18c** could be obtained with enantioselectivities of 88–96% ee (Scheme 4.5).

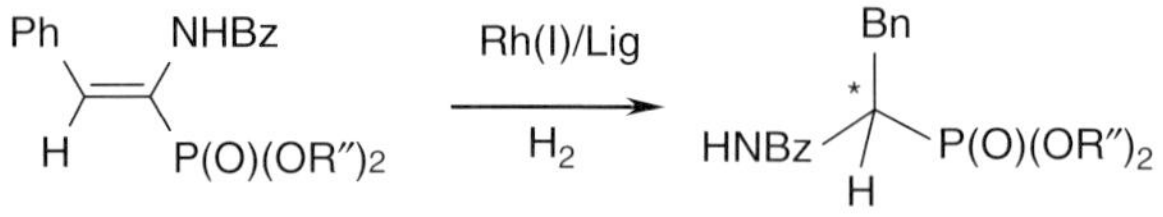

17a,b 18

R" = Me (a), Et (b); Lig = **6,9**

Lig/ee/Conf =(2*S*,4*S*)-BPPM, ee = 96%, (*S*),(R)-PROPRAPHOS, ee = 89%, (*R*), (*S*)-PROPRAPHOS, 92%, (*S*), (R)-Ph-β-GlupOH, 91%, (4*R*,5*R*)-NORPHOS, ee = 63%, (*S*), (2*S*,3*S*)-DIOP, 83%, (*S*), (*S*)-PP Cyclopent, ee = 91% (*S*)

Scheme 4.5 Asymmetric hydrogenation of vinylphosphonates **17** in the presence of rhodium catalysts (25 °C, 0.1 MPa H_2, 1 mmol substrate, 0.01 mmol catalyst, 15 Ml of MeOH; catalyst prepared *in situ* using Rh(COD)$_2$]BF$_4$.

Table 4.1 Enantioselective hydrogenation of enamidophosphonates, catalyzed by Rh complexes bearing DuPhos **4a,b**, BPE **5a,b**, and BisP **3** ligands.

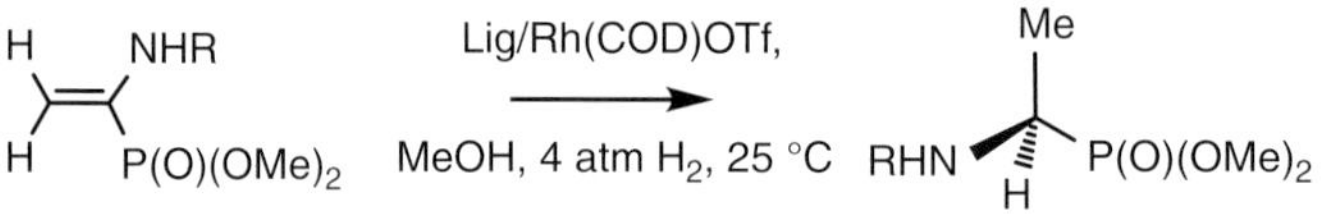

R	Ligand	Conversion (%)	ee (%)	Configuration
Ac	(*R,R*)-**3**	100	90	(*R*)-(−)
Ac	(*S,S*)-**4a**	100	93	(*R*)-(−)
Ac	(*S,S*)-**4b**	100	95	(*R*)-(−)
Ac	(*S,S*)-**4c**	90	68	(*R*)-(−)
Cbz	(*S,S*)-**5a**	88	88	(*R*)-(−)
Cbz	(*S,S*)-**4a**	72	90	(*R*)-(−)
Cbz	(*S,S*)-**4b**	100	94	(*R*)-(−)
Cbz	(*S,S*)-**5b**	100	81	(*R*)-(−)

Burk *et al.* [15] proposed cationic rhodium complexes of C_2 symmetric DuPHOS **4a,b** and BPE **5a,b** ligands as effective catalysts for the asymmetric hydrogenation of *N*-aryl and *N*-benzyloxycarbonylenamido phosphonates **17** (Scheme 4.5). The catalyst Et-DuPHOS/Rh(COD) provided good enantioselectivity for both types of substrates **17** (95% and 94% ee, respectively).The reaction was completed in 12 h in methanol at room temperature and a pressure of 4 atm to give the aminophosphonates **18** with enantioselectivities of up to 95% ee [14]. Similar results were also obtained by Beletskaya, Gridnev, and others with the Rh/(*R,R*)-*t*-Bu-BisP and Rh(P-OP) catalysts (Table 4.1) [16–18].

Wang *et al.* [19] applied readily available and inexpensive chiral phosphine–aminophosphine ligands **20** for the enantioselective hydrogenation of various α-enol ester phosphonates and α-enamidophosphonates. The phosphine–aminophosphine ligands **20** exhibited superior enantioselectivities to those obtained with BoPhos analogs **21**. Very good enantioselectivities (93–97% ee) were achieved in the hydrogenation of various substrates catalyzed by an (*S*)-**20**/[Rh(COD)]BF$_4$ complex, thus

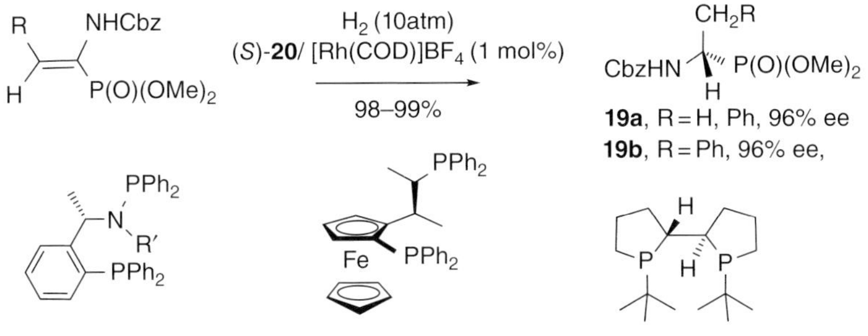

(*S*)-**20a,b**
R′ = H (a), Me (b)

21a, R = H, H-BoPhos
21b, R = Me, Me-Me-BoPhos

22, TangPhos

Scheme 4.6 Hydrogenation of α-enamidophosphonates catalyzed by Rh/(*S*)-**26** complex.

demonstrating the high potential of these ligands **20** in the preparation of optically active α-aminophosphonates **19** (Scheme 4.6).

Asymmetric hydrogenation of β-enamidophosphonates catalyzed by iridium complex followed by desulfidation afforded 2-amino-1-phosphinoalkanes **24**, offering a new approach to chiral N,P-ligands that can serve as ligands in asymmetric reactions. Oshima *et al.* developed a convenient method for the synthesis of chiral phosphine sulfides **23** [20]. Chiral iridium complexes containing chiral ferrocene ligand **12** catalyzed the enantioselective hydrogenation of amino-1-thiophosphinyl-1-alkenes with the formation of optically active (*E*)-2-amino-1-thiophosphinylalkanes **23** in high yields and with high ee; however, the absolute configuration of the resultant compounds was not determined. The subsequent desulfidation of phosphine sulfides **23** led to the formation of 2-amino-1-phosphinoalkanes **24**, including the optically active phosphines **25** and **26** (Scheme 4.7).

Boerner *et al.* showed that the Rh-catalyzed asymmetric hydrogenation of prochiral β-*N*-acetylamino-vinylphosphonates led to the formation of chiral β-*N*-acetylamino-phosphonates with excellent yields (up to 100%) and with high enantioselectivities (89–92% ee) [21]. The reaction was dependent on the chiral bidentate phosphorus ligand and the solvent employed. In some cases, an inversion of the induced chirality was observed on using the corresponding *E*- or *Z*-isomeric substrates. Catalysts were generated *in situ* by mixing $[Rh(COD)_2]BF_4$ with equimolecular amounts of bidentate phosphorous ligand. The phosphines **2**, **29–30** were the most effective among 240 chiral ligands. The highest enantioselectivity was achieved

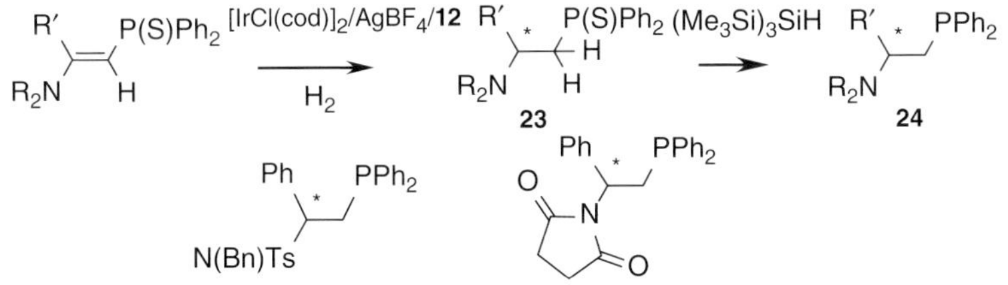

25, 93% yield, 97% ee
26, 99% yield, 89% ee

Scheme 4.7 Enantioselective hydrogenation of vinylphosphonates **22** with iridium complexes.

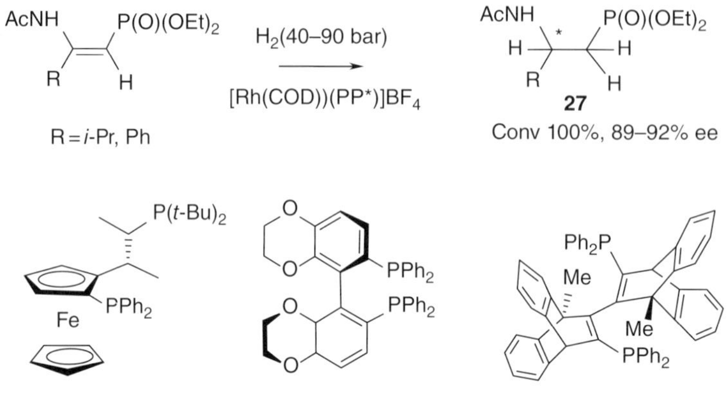

28, (*R*,*S*)-DPPF-*t*-BP **29**, (*R*)-SYNPHOS **30**, (*R*)-Me-CATPHOS

Scheme 4.8 Asymmetric hydrogenation of β-*N*,*N*-vinylphosphonates with Rh-complexes.

in dichloromethane or tetrahydrofuran (THF), at room temperature and a hydrogen pressure of 4 bar. The enantiomeric purity of the hydrogenation products was determined by high performance liquid chromatography (HPLC); however, the absolute configurations of the products were not reported (Scheme 4.8).

Doherty *et al.* [22] have reported that rhodium complexes with (*R*,*S*)-JOSIPHOS **2** or (*R*)-Me-CATPHOS **30** ligands are effective catalysts for the asymmetric hydrogenation of (*E*)- and (*Z*)-β-aryl-β-(enamido)phosphonates, but well-known ligands such as TangPhos **22**, PHANEPHOS, and DuPhos **4** are ineffective. These complexes, Rh/JOSIPHOS **2** and Rh/Me-CATPHOS **30**, form a complementary pair of catalysts for the efficient asymmetric hydrogenation of (*E*)- and (*Z*)-β-aryl-β-(enamido)phosphonates, respectively. In the majority of cases, hydrogenation with these catalysts afforded phosphonates in good yields (72–97%) and with very good enantioselectivity (99% ee). The authors reported the specific rotations of the products; however, the absolute configurations were not determined (Table 4.2).

Phosphine–phosphinites and phosphine–phosphites are examples of nonsymmetric ligands that differ in the electronic and the steric properties of their respective binding groups [18]. The P-OP ligands studied in asymmetric hydrogenation encompass diverse carbon backbones and stereogenic elements between the two phosphorus functionalities, many of which provide a highly stereodifferentiating environment around the catalytic rhodium metal center. The hydrogenation of various functionalized alkenes catalyzed by Rh/P-OP complexes led to the formation of products with high enantioselectivities even at low loadings of the catalyst. For example, Pizzano *et al.* [18] studied the hydrogenation of β-(acylamino) vinylphosphonates **31** with Rh/P-OP catalysts **14** leading to the formation of β-acyliminophosphonates **32** with enantioselectivities of up to 99% ee (Scheme 4.9).

Analysis of these results showed that catalysts containing electron-donating groups such as P(*i*-Pr)$_2$, were more active and enantioselective than the PPh$_2$-substituted catalysts.

NMR studies on the interaction between vinylphosphonates and the catalyst indicated the formation of chelates with the olefin cis to the phosphite group of the Rh(P-OP)$^+$ fragment. In all cases, the catalyst containing (*S*)-P-OP ligands afforded (*R*)-enantiomers while those with (*R*)-P-OP ligands led to the formation of (*S*)-hydrogenated products.

Table 4.2 Enantioselective hydrogenation of (*E*)-/(*Z*)-β-aryl-β-(enamido) phosphonates with Rh complexes containing the ligands (*R,S*)-JOSIPHOS **2** and (*R*)-Me-CATPHOS **30**.

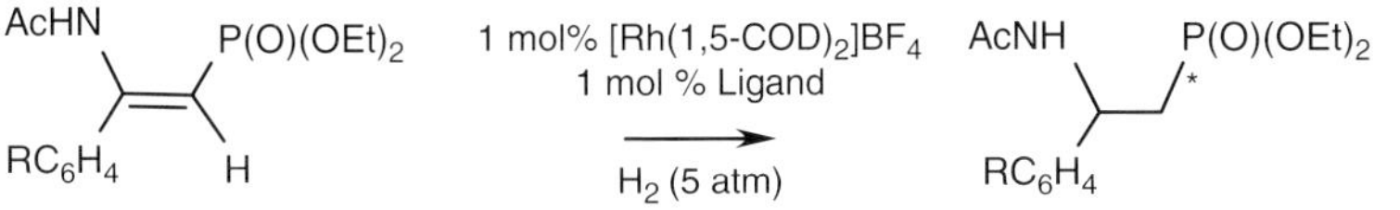

Entry	Ligand	*E/Z*	R	Yield (%)	ee (%)
1	(*R,S*)-JOSIPHOS	*Z*	4-Me	77	99 (+)
2	(*R,S*)-JOSIPHOS	*Z*	H	72	97 (+)
3	(*R,S*)-JOSIPHOS	*Z*	4-F	66	96 (+)
4	(*R,S*)-JOSIPHOS	*Z*	4-Cl	81	94 (+)
5	(*R,S*)-JOSIPHOS	*Z*	4-Br	79	>99 (+)
6	(*R,S*)-JOSIPHOS	*Z*	4-MeO	82	>99 (+)
7	(*R*)-Me-CATPHOS	*E*	4-Me	94	>99 (+)
8	(*R*)-Me-CATPHOS	*E*	H	97	99 (+)
9	(*R*)-Me-CATPHOS	*E*	4-F	78	>99 (+)
10	(*R*)-Me-CATPHOS	*E*	4-Cl	80	99 (+)
11	(*R*)-Me-CATPHOS	*E*	4-Br	87	99 (+)
12	(*R*)-Me-CATPHOS	*E*	4-MeO	79	99 (+)

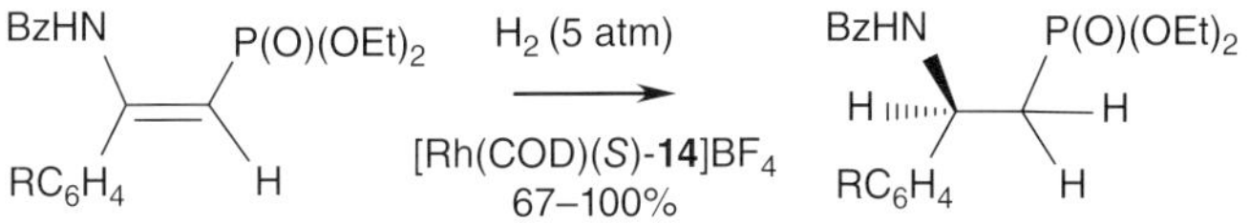

31

32, 94–99% ee

R = Ph, *p*-Tl, *p*-An, 4-BrC_6H_4, 2-ClC_6H_4

Scheme 4.9 Catalytic hydrogenation of β-(acylamino)vinylphosphonates **31**.

Significant attention has been paid to the asymmetric hydrogenation of α- and β-enolphosphonates as a method for the synthesis of chiral hydroxyphosphonates which, as well as aminophosphonates, exhibit diverse and interesting biological and biochemical properties. Chiral phosphinic ligands 1,2-bis(alkylmethylphosphino) ethanes (BisP) **3** and bis(alkylmethylphosphino)methanes (MiniPHOS) **11** display high enantioselectivity in the hydrogenation of diethyl benzoyloxyvinylphosphonate **33** catalyzed by rhodium complexes [16, 18, 23, 24]. An important feature of these ligands is that a bulky and a small alkyl group (methyl group) are bound to each phosphorus atom. These ligands form five- or four-membered C_2-symmetric chelates, and therefore the imposed asymmetric environment ensures high enantioselectivity in catalytic asymmetric reactions. The asymmetric catalytic hydrogenation of **33** was performed in methanol at 4 bar pressure of H_2 to give (*S*)-α-benzoyloxyethylphosphonates **34** with 93% ee (Scheme 4.10) [23].

The hydrogenation of enolphosphonates **33** catalyzed by rhodium complexes with chiral P-OP ligands **14** also proceeded with good enantioselectivity. Complexes

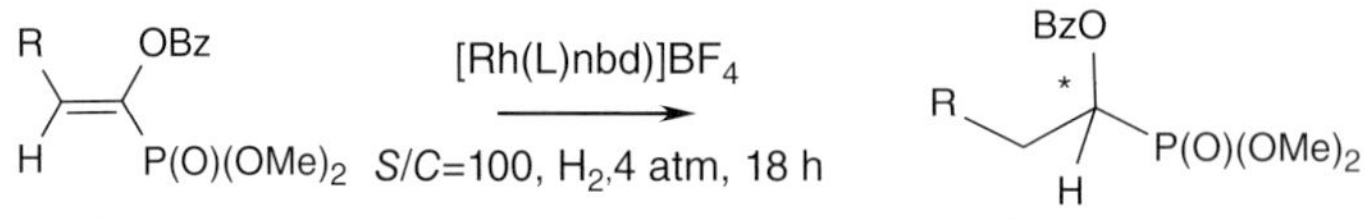

33 (*S*)- or (*R*)-**34**

R = H	L = (*S*,*S*)-**3**	88% ee
R = H	L = (*R*,*R*)-**11**	96% ee
R = Me	L = (*R*,*R*)-**3**	86% ee
R = Me	L = (*R*,*R*)-**11**	89% ee
R = Et	L = (*R*,*R*)-**3**	87% ee
R = Et	L = (*R*,*R*)-**11**	93% ee

Scheme 4.10 Asymmetric hydrogenation of α-benzoyloxyvinylphosphonates.

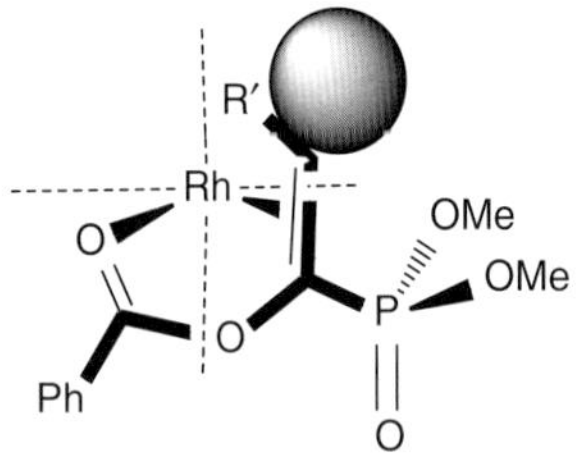

Figure 4.1 Preferable organization of the intermediate Rh-olefin complex for the stereochemical sense of the asymmetric hydrogenation of **35**.

[Rh(COD)(**14**)]BF_4 show fluxional behavior in solution, consistent with backbone oscillation around the coordination plane (Figure 4.1).

Depending upon the steric characteristics of the ligands and substrate, the hydrogenations catalyzed with these complexes provide products **34** with enantioselectivity of greater than 90% ee as shown in Scheme 4.11. Detailed studies of

33 + H_2 —[Rh(COD)(L = **14a,b**)]BF_4→ (*R*)-**34**

R=H	L = **14a**	100%	85% ee
R=H	L = **14b**	100%	91% ee
R=Et	L = **14a**	100%	89% ee
R=Et	L = **14b**	100%	95% ee
R=Et	L = **14b**	100%	96% ee
R=*i*Pr	L = **14b**	100%	98% ee
R=Bu	L = **14a**	100%	91% ee
R=Bu	L = **14b**	100%	96% ee
R=Ph	L = **14a**	100%	82% ee
R=Ph	L = **14b**	100%	92% ee
R=*p*-Tl	L = **14a**	100%	87% ee
R=*p*-Tl	L = **14b**	80%	83% ee
R=3,4-$(MeO)_2C_6H_3$	L = **14a**	100%	82% ee
R=3,4-$(MeO)_2C_6H_3$	L = **14b**	60%	86% ee

Scheme 4.11 Hydrogenation of β-arylalkenphosphonates catalyzed by the [Rh(cod)(**14**)]BF_4 complex.

the phosphane–phosphite ligands **14a–c** (Scheme 4.3) demonstrated the influence of the steric characteristics on the enantioselectivities; 98% ee was thus obtained with substrates bearing an alkyl substituent at the β-position, while for their aryl counterparts values of up to 92% ee were achieved. Rh/P-OP complexes are excellent catalysts for enantioselective hydrogenations of β-(acyloxy)vinylphosphonates.

For example, the hydrogenation of substrates **35** catalyzed by the complex [Rh(COD)[(*S*)-**14a,b**]]BF_4 led to the formation of chiral phosphonates **36** with good yields and enantioselectivities of 95–99% ee [18]. It was noticed that the β-alkyl substrates were more reactive than β-aryl ones, and reacted for a short time with full conversion. Phosphonates **36** were transformed without racemization into the corresponding alcohols. For example, the deprotection of a phosphonate **36** gives easy access to β-hydroxy-γ-aminophosphonic acids **37**, which are precursors of biologically active phosphono-GABOB. The NMR data and the magnitude of coupling constants $^1J_{Rh,P}$ of the ^{31}P nuclei of P-OP ligands indicated a *cis*-olefin coordination to the phosphite: $\delta = 13.6$ (dd, $J_{P,Rh} = 140$ Hz, $J_{P,P} = 63$ Hz, PC), 14.9 [s, $P(O)(OMe)_2$], 132.0 (dd, $J_{P,Rh} = 262$ Hz, $J_{P,P} = 62$ Hz, PO) (Schemes 4.12 and 4.13) [18].

The development of a method for the enantioselective synthesis of hydroxyphosphonates by catalytic hydrogenation of β-aryl-, β-alcoxy-, and β-alkyl-substituted enolphosphonates has attracted considerable attention. Thus, Wang *et al.* [23] developed an enantioselective method for the synthesis of α-benzyloxyphosphonates by hydrogenation of enolphosphonates **35**, including β-aryl-, β-alkoxy-, and β-alkyl-substituted substrates, in the presence of rhodium complexes containing unsymmetrical phosphine–phosphoramidite ligands THNAPhos **13**.

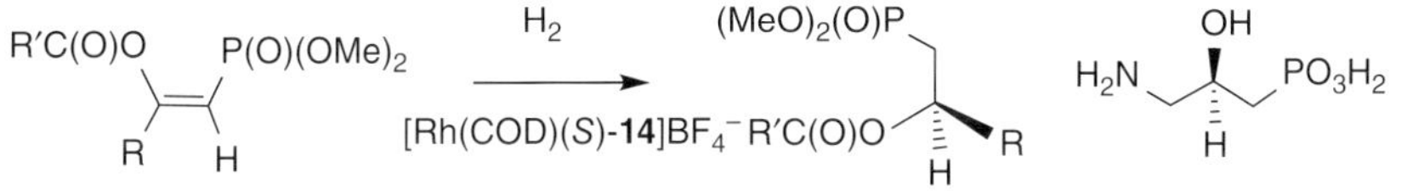

35a–h (*R*)-**36a–i**, upto 99% ee **37**

R = Me (a), *i*-Pr (b), Bu (c), 4-Tl (d), 4-An (e), 4-BrC_6H_4 (f), 2-Nphth (g), CH_2NHBoc (h); R′ = Ph

Scheme 4.12 Enantioselective hydrogenation of β-(acyloxy)vinylphosphonates **35** catalyzed by rhodium complexes of chiral P-OP ligands **14**.

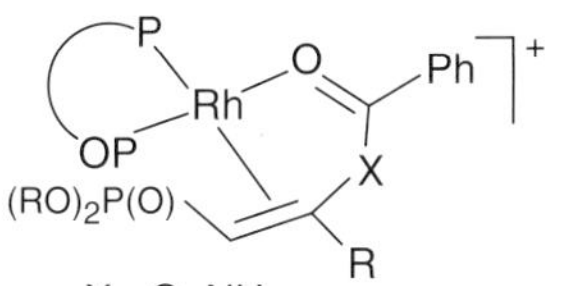

X = O, NH

38

Scheme 4.13 Coordination mode of vinylphosphonates for the complex {Rh(**35**)[(*S*)-**14b**]}BF_4.

(R)-2-Hydroxy-1-dimethoxyphosphoryl-propane. In a reactor, (100 ml) was charged with the enolphosphonate **35a** (0.22 mmol) and catalyst precursor [Rh(COD)[(*S*)-**14**]]BF_4 (0.002 mmol) in CH_2Cl_2 (5 ml). The vessel was pressurized with H_2 to 4 atm. The reaction mixture was continuously stirred for 20–24 h. Then the reactor was depressurized, the mixture was evaporated to dryness, and the residue was analyzed by ^{31}P NMR spectroscopy. Purification of the residue was performed by column chromatography and the enantiomeric excess of (*R*)-2-benzoyloxy-1-dimethoxyphosphorylpropane was determined by chiral HPLC (yield 88%, $[\alpha]_D^{20}$ −3.5 ($c = 1.0$, THF), $\delta_P = 28.5$).Sodium carbonate (1.4 mmol) was added to a solution of (*R*)-2-benzoyloxy-1-dimethoxyphosphorylpropane (0.34 mmol) in MeOH (5 ml). The reaction mixture was stirred for 14–16 h and the solvent was evaporated. EtOAc (10 ml) was added to the resulting mixture, which was then washed with saturated solutions of $NaHCO_3$ (10 ml) and NaCl (10 ml). The organic phase was dried over $MgSO_4$, filtered, and solvent evaporated. The residue was then purified by column chromatography (silicaGel, AcOEt/MeOH 9 : 1) affording the hydroxyphosphonate as a colorless oil (yield: 60%, 99% ee, $[\alpha]_D^{20}$ −11.8 ($c = 1.2$, THF)). The optical purity was analyzed by ^{31}P NMR in the presence of quinine as CSA [18].

After asymmetric hydrogenation, the phosphonates **39** were prepared with very high enantioselectivities of up to 99.9% ee. Rhodium complexes with phosphine–aminophosphine ligands (*R,R*)-**4b** also displayed a good enantioselectivity of 97% ee in the asymmetric hydrogenation of dimethyl α-benzoyl vinylphosphonates **40**, containing α-aryl, α-alkyl, and α-alkoxy substituents. The enantioselectivity of phosphine–aminophosphine ligands **24a** was higher than that with well-known BoPhoz and DuPhos ligands [23, 25] (Scheme 4.14). Hydrogenation of enolphosphonates **35** in the presence of cationic rhodium catalysts Lig/Rh(COD)OTf containing C_2-symmetric ligands (Lig) BPE **5a,b** or DuPHOS **4a–d** occurred with good ee at room temperature and a low pressure of hydrogen. The highest stereoselectivity for the unsubstituted enolphosphonates **35** was obtained with the Et-DuPHOS-Rh catalyst. Alkyl-substituted enolbenzoate substrates **35** were reduced with optimum enantioselectivity using the less bulky Me-DuPHOS-Rh as shown in Scheme 4.14 [15]. Chiral phosphine–phosphoramidite ligand (*S*)-HYPhos, which was prepared by a simple two-step method from 1-naphthylamine and 2,20-dihydroxyl-1,10-binaphthyl (BINOL)-phosphite, was successfully applied in the Rh-catalyzed asymmetric hydrogenation of functionalized olefins, including α-(acetamido)cinnamates, enamides, and enolphosphonates, with 98–99% ee (Scheme 4.15). The asymmetric hydrogenation of

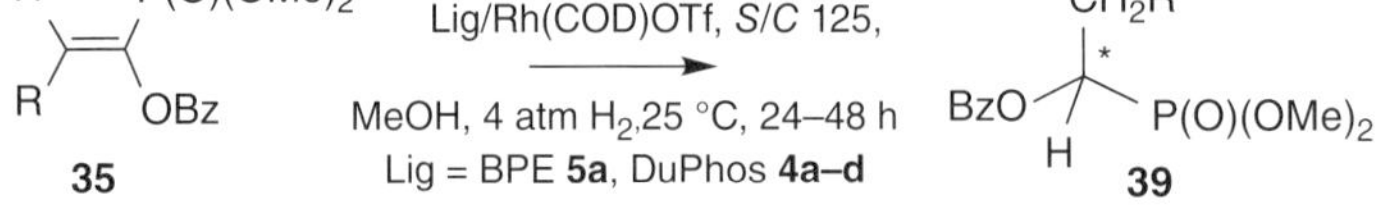

R = H	(*R,R*)-**5a**	Conv = 27%	64%	(S)-(−)
R = H	(*S,S*)-**4a**	Conv = 100%	84%	(R)-(+)
R = H	(*R,R*)-**4b**	Conv = 100%	96%	(S)-(−)
R = H	(*R,R*)-**4c**	Conv = 100%	92%	(S)-(−)
R = H	(*R,R*)-**4d**	Conv = 40%	85%	(R)-(+)

Scheme 4.14 Asymmetric hydrogenation of β-alkoxy and β-aryl substrates.

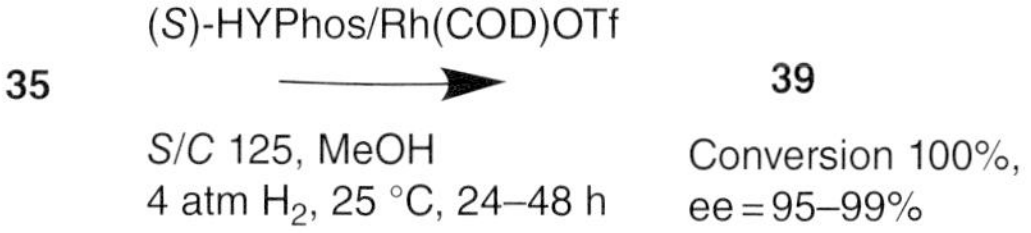

R = H	ee = 99%	(*S*)
R = H	ee = 99%	(*R*)
R = Me	ee = 99%	(*R*)
R = Et	ee = 98%	(*R*)
R = $CH_2(CH_2)_8CH_3$	ee = 98%	–
R = Ph	ee = 98%	(*R*)
R = 4-FC_6H_4	ee = 98%	–
R = 4-ClC_6H_4	ee = 99%	–
R = 4-$NO_2C_6H_4$	ee = 98%	–
R = 4-$MeOC_6H_4$	ee = 95%	–
R = 3-ClC_6H_4	ee = >99%	–
R = OEt	ee = 96%	–
R = OPr-*i*	ee = 98%	–

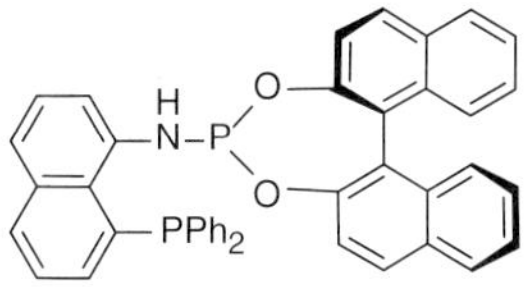

(*S*)-HY-Phos

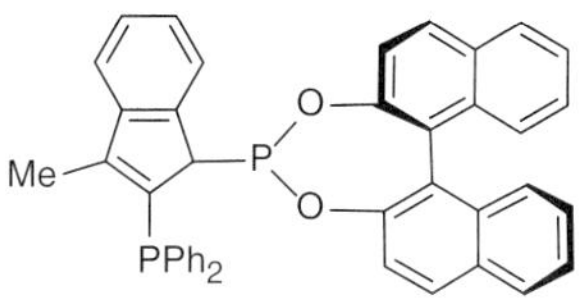

IndolPhos

Scheme 4.15 Hydrogenation of enolphosphonates in the presence of cationic rhodium catalysts.

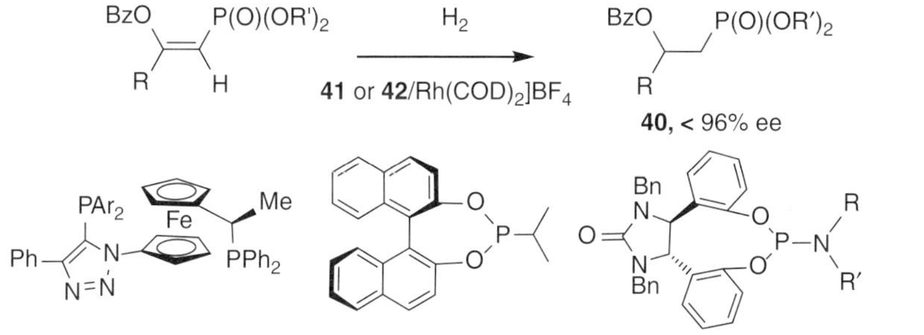

41 | 42

Ar=3,5-Xyl (ClickFerrophos II) (*R*)-MonoPhos (DpenPhos) R = H, Bn; R′ = Me, Pr-*i*

Scheme 4.16 Hydrogenation of unsaturated phosphonates catalyzed by rhodium complexes.

vinylphosphonates **35** catalyzed by rhodium complexes containing (*R,R*)-TADDOL or (*S*)-BINOL-phosphite derivatives of indole (IndolPhos ligands) were also tested and gave enantiomerically enriched (*S*)-phosphonates **39** with 32–87% ee.

Rhodium complexes containing ClickFerrophos II and (*R*)-MonoPhos ligands were used as catalysts for the hydrogenation of various unsaturated phosphonates. The hydrogenation of α,β-unsaturated phosphonates including β-alkyl-, β-aryl-, and β-dialkylphosphonates, (*Z*)-β-enolphosphonates, and α-phenylethylphosphonates, allowed the preparation of corresponding chiral phosphonates in good yields and with high enantioselectivities (up to 96% ee) [24, 26–29] (Scheme 4.16). Zhang *et al.* [27] reported that Rh(I) complexes of monodentate phosphoramides bearing primary amines (DpenPhos) **42** effectively catalyze the asymmetric hydrogenation of α- or β-acyloxy α,β-unsaturated phosphonates, providing the corresponding biologically important chiral α- or β-hydroxyphosphonates **40** with very good enantioselectivities (93–96% ee). The asymmetric hydrogenation of alkenephosphonates represents an attractive method for the preparation of chiral alkylphosphonates and chiral tertiary

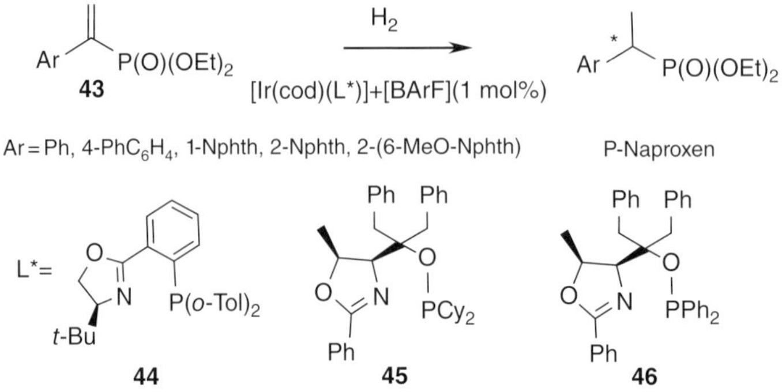

Scheme 4.17 Enantioselective catalytic hydrogenation of vinylphosphonates **43**.

phosphine oxides that can be used as new drugs or new chiral ligands. Genet and Beletzskaya reported the enantioselective hydrogenation of vinylphosphonates **43** catalyzed by complexes of iridium containing phenyloxazoline ligands **44–46** [16, 17]. The effectiveness of chiral iridium catalysts (70–94% ee) was proved on a number of substrates. For example, the optically active phosphorous analogs of Naproxen (Ar = 2-(6-MeO-Nphth)), were synthesized with 92–95% ee by the hydrogenation of vinylphosphonates in methylene chloride, at room temperature or by gentle heating and an H_2 pressure of 5–60 bar (Scheme 4.17).

A number of chiral alkylphosphonates **47** bearing a β-stereogenic center, were synthesized by catalytic hydrogenation of the corresponding β-substituted α,β-unsaturated phosphonates using rhodium complexes with ferrocene-based monophosphoramidite ligands **48** (Scheme 4.18). Under mild conditions, the hydrogenation proceeded with 100% conversion to give products with high enantioselectivities: 99.5% ee in case of the *E*-substrates, and 98.0% ee in case of the *Z*-substrates. Hydrogenation with the Rh/(R_C,S_C)-FAPhos catalyst led to the formation of compounds **49** with the (*R*)-configuration, although in other cases the absolute configurations of the products were not defined. Asymmetric hydrogenation of β,γ-unsaturated phosphonates

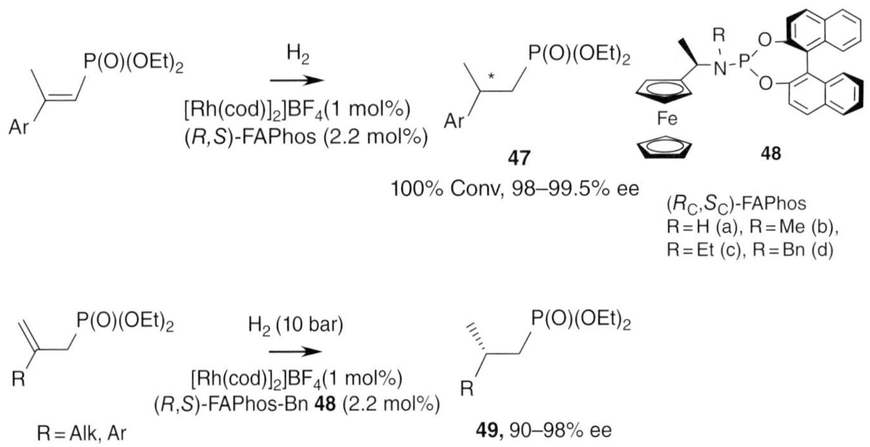

Scheme 4.18 Hydrogenation of α,β-unsaturated phosphonates catalyzed by Rh/(R_C,S_C)-FAPhos.

Table 4.3 Rh-catalyzed asymmetric hydrogenation of enolphosphonates.

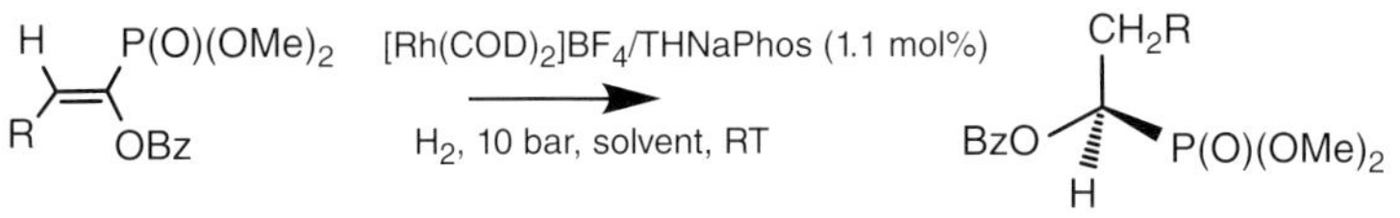

Entry	L	Substrate (R)	Yield (%)	ee (%) (configuration)
1	Me-BoPhoz	Ph	98	5
3	H-BoPhoz	Ph	99	89
4	**L-1**	Ph	98	96 (*S*)
5	Me-DuPHOS	Ph	96	92
6	**L-1**	*p*-FC_6H_4	99	96 (+)
7	**L-1**	*p*-ClC_6H_4	98	94 (+)
8	**L-1**	*p*-BrC_6H_4	95	97 (+)
9	**L-1**	*p*-$NO_2C_6H_4$	99	95 (+)
10	**L-1**	*p*-$MeOC_6H_4$	99	95 (*S*)
11	**L-1**	*m*-$MeOC_6H_4$	98	96 (+)
12	**L-1**	*o*-ClC_6H_4	98	94 (+)
13	**L-1**	1-Naphthyl	98	95 (+)
14	**L-1**	2-Thienyl	98	95 (+)
15	**L-1**	H	94	93 (*S*)
16	**L-1**	Me	99	96 (*S*)
17	**L-1**	Et	99	96 (*S*)
18	**L-1**	$(CH_2)_9CH_3$	97	96 (*S*)
19	**L-1**	OMe	99	94 (S)
20	**L-1**	OEt	99	93 (+)

catalyzed by an Rh/(R_C,S_C)-FAPhos–Bn complex led to the formation of chiral β-substituted alkanephosphonates **49** with 98% ee (Table 4.3) [26, 28, 30].

Chiral 1-aryl- or 1-alkyl-substituted ethylphosphonates **51** were synthesized with enantioselectivities of 92–98% ee and very good yields by asymmetric hydrogenation of the corresponding 1-aryl- or 1-alkylethenylphosphonates **50** in the presence of rhodium complexes containing P-chiral aminophosphine-phosphine BoPhoz-type ligands **52**. The authors reported that this catalyst is especially effective for the asymmetric hydrogenation of 1-arylethenylphosphonates. A number of substrates were hydrogenated with this catalyst with enantioselectivities of up to 98% ee. Hydrogenation proceeded under mild conditions (room temperature under 10 bar H_2, and 0.2 mol% of catalyst) to provide chiral 1-aryl- or 1-alkyl-substituted ethylphosphonates **52** (Scheme 4.19) [23, 28].

Diphenylvinylphosphine oxides and di- and trisubstituted vinylphosphonates were employed as substrates for asymmetric hydrogenations catalyzed by iridium complex **54**. Complete conversion and excellent enantioselectivities (up to and above 99% ee) were observed for a range of substrates with both aromatic and aliphatic groups at the prochiral carbon atom. A large number of compounds **53** with high enantioselectivities

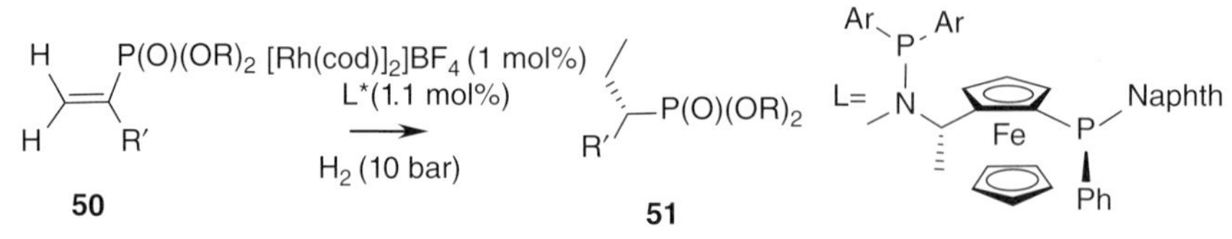

R = Me, Et, *i*-Pr, R′ = Et, Ph, *p*-Tl, *o*-, *m*-, *p*-An, *p*-XC$_6$H$_4$,
X = F, Cl, Br, 1-naphthyl, 6-methoxy-2-naphthyl,

52, Ar = 4-CF$_3$C$_6$H$_4$

Scheme 4.19 Asymmetric hydrogenation of alkenphosphonates **50** with catalyst Rh/(S_C,R_{FC},R_P)-**52**.

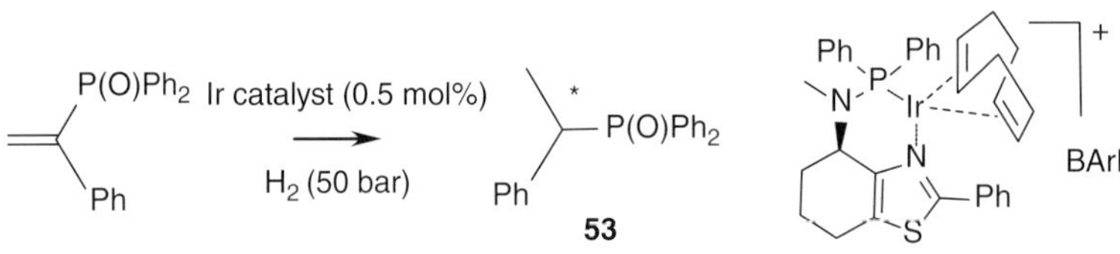

54, 99% conv, >99% ee

Scheme 4.20 Asymmetric hydrogenation of carboxyethylvinylphosphonates.

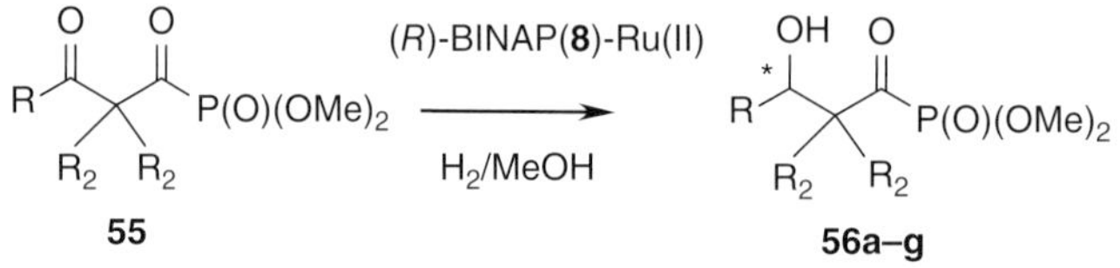

R^1 = Me, R^2 = H (a); R^1 = Me, R^2 = Et (b); R^1 = R^2 = Me (c); R^1 = C$_5$H$_{11}$, R^2 = Me (d),
R^1 = *i*-Pr, R^2 = Me (e); R^1 = C$_5$H$_5$, R^2 = Me (f); (R^1 = Me, R^2 = Br (g)

(*R*)-**55a**	99%	ee = 98%	(*R*)-**56a**
(*R*)-**55a**	97%	ee = 91%	(*R*)-**56a**
(*R*)-**55a**	96%	ee = 98%	(*R*)-**56a**
(*R*)-**55d**	98%	ee = 96%	(*R*)-**56b**
(*R*)-**55c**	97%	ee = 98%	(*R*)-**56c**
(*S*)-**55d**	98%	ee = 94%	(*S*)-**56d**
(*S*)-**55e**	96%	ee = 96%	(*S*)-**56e**
(*R*)-**55f**	96%	ee = 95%	(*R*)-**56f**
(*S*)-**55g**	95'%	ee = 98%	(1*R*,2*S*)-**56g**

Scheme 4.21 Catalytic asymmetric hydrogenation of ketophosphonates **55**.

were described. The hydrogenation of electron-deficient carboxyethylvinylphosphonates was also carried out with stereoselectivities of up to 99% ee (Schemes 4.20 and 4.21) [31].

4.2.2 Hydrogenation of C=O Phosphorus Compounds

Chiral complexes of transition metals catalyze hydrogenation and hydroxylation of prochiral ketones. From the practical point of view, catalytic asymmetric hydrogenation and hydroxylation of ketophosphonates is one of the convenient methods for the synthesis of chiral hydroxyphosphonates. The described synthetic approaches are based on the asymmetric hydrogenation using various catalysts, in particular, Ru(II)-2,20-bis(diphenylphosphino)-1,10-binaphthyl (BINAP) complexes. Noyori *et al.* [32, 33] in 1995–1996 disclosed that Ru(II)-BINAP (1 mol% [$RuCl_2$(*R*)-BINAP]$(dmf)_n$) complexes catalyze the enantioselective hydrogenation of β-ketophosphonates in methanol under low-pressure hydrogen and at 30 °C with formation of the corresponding β-hydroxyphosphonates in very high yields and with 97% ee. The hydrogenation with an (*S*)-BINAP-Ru(II) catalyst afforded predominantly the (*R*)-products, while the (*R*)-BINAP complexes formed the (*S*)-enriched compounds. Phosphaalanine, phosphaethylglycine, and phosphaphenylalanine with high enantiomeric purity were obtained by this method. It was noticed that (*E*)-alkenes are more reactive than their (*Z*)-isomers (Scheme 4.20). Racemic α-acetamido-β-ketophosphonates **57** in the presence of (*R*)-BINAP-Ru catalyst were hydrogenated to (1*R*,2*R*)-hydroxyphosphonates **58** with high diastereoselectivity (*syn:anti* = 97 : 3) and with enantioselectivity of 98% ee (Scheme 4.22). Then the product (1*R*,2*R*)-**58** was successfully transformed into enantiomerically pure (1*R*,2*R*)-phosphothreonine **59** in 92% yield (Scheme 4.22). The hydrogenation of racemic, configurationally labile *rac*-**57** led to the formation of four diastereomers (1*R*,2*R*)-, (1*R*,2*S*)-, (1*S*,2*R*)-, and (1*S*,2*S*)-**58**. The optimization of reaction conditions and high stereoselectivity of the (*R*)-BINAP ligand allowed to obtain predominantly (1*R*,2*R*)-α-amido hydroxyphosphonates **58** with high enantio- and diastereoselectivity (98 : 2 dr and 95% ee for the *syn*-isomer). Noyori developed a method for the synthesis of the antibiotic fosfomycin **62** using asymmetric hydrogenation catalyzed by the BINAP-Ru complex. Fosfomycin **62** was obtained in 84% yield with 98% ee and with *syn:anti* ratio – 90 : 10, starting from racemic β-keto-α-bromphosphonate *rac*-**60** (Scheme 4.23) [31, 32]. Asymmetric hydrogenation of α-ketophosphonates catalyzed by Rh-complexes with chiral phosphine–phosphoramidite ligands proceeded at 30 bar H_2 to afford α-hydroxyphosphonates [34].

Genet reported the asymmetric hydrogenation of ketophosphonates catalyzed by chiral (*R*)- and (*S*)-BINAP-Ru **63** and (*R*)-MeO-BIPHEP **65c** complexes and obtained hydroxyphosphonates **64** with enantioselectivities up to 99% ee (Scheme 4.24) [34, 35]. The authors hydrogenated a large variety of β-ketophosphonates and β-ketothiophosphonates, including heterocycles, catalyzed by chiral Ru(II) complexes, with high enantioselectivities. For example, asymmetric hydrogenation of diethyl 2-oxopropylphosphonate at 1 bar and 50 °C with (*S*)-Binap/Ru(II) led to the formation of β-hydroxyphosphonates **64** with 99% ee [35].

Asymmetric hydrogenation of α-amido-β-ketophosphonates **57**, catalyzed by atropoisomeric ruthenium complexes **65**, **66** with SunPhos ligands via dynamic kinetic resolution (dynamic kinetic resolution (DKR)) led to the formation of the corresponding

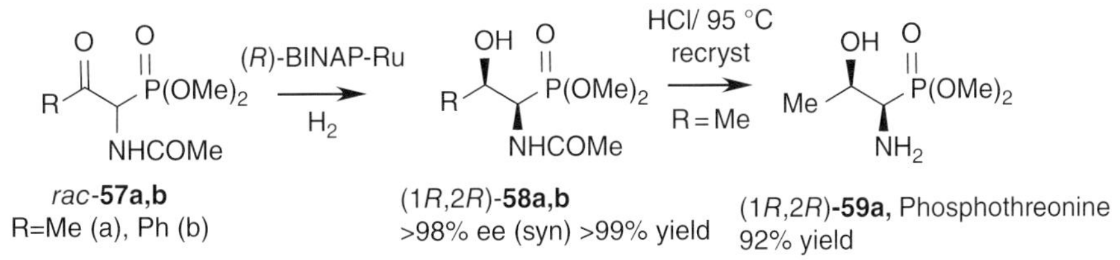

Scheme 4.22 Enantioselective synthesis of phosphothreonine **59a**.

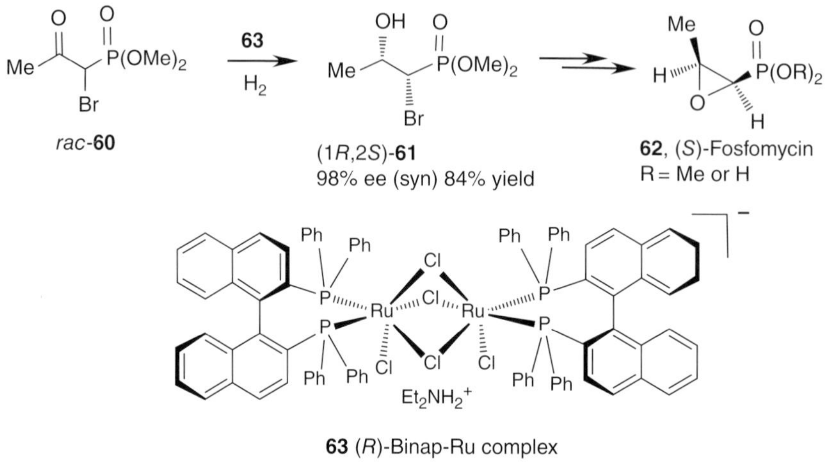

Scheme 4.23 Enantioselective synthesis of phosphomycin **62**.

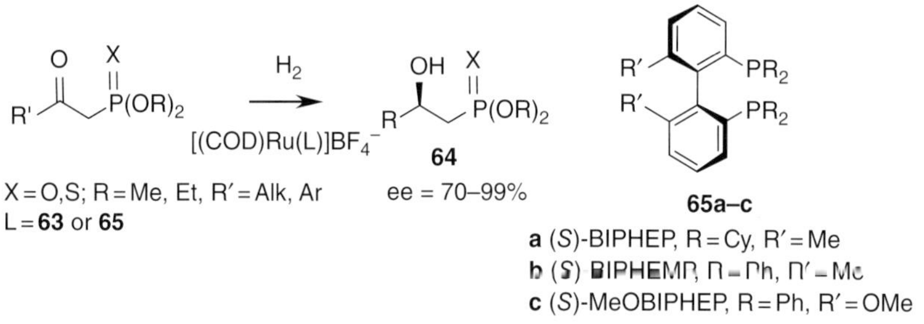

Scheme 4.24 BINAP-Ru(II)-catalyzed asymmetric hydrogenation of β-ketophosphonates.

β-hydroxy-α-amidophosphonates **58** with high diastereoselectivities (up to 99 : 1 dr) and enantioselectivities (up to 99.8% ee). The catalysts were prepared using atropoisomeric (*S*)-SunPhos ligands and $[RuCl_2(benzol)]_2$. Hydrogenation was carried out under 10 bar of hydrogen pressure, 50° C in methanol to result in chiral phosphonates **58** with 98.0% ee and in 97 : 3 *syn:anti* selectivity (Scheme 4.25) [36, 37].

Hydrogenation of β-ketophosphonates **67** catalyzed by Ru-(*S*)-SunPhos afforded the corresponding β-hydroxy-α-amidophosphonates **68** with enantioselectivity greater than 98% ee. Hydrogenation of methyl, ethyl, and isopropyl (2-*oxo*-2-phenylethyl) phosphonates led to the formation of alcohol **68** with 99.7%, 95.5%, and 90.0% ee, correspondingly. It was found that additives increased diastereo- and enantioselectivities of the reaction. For example, the addition of catalytic amounts of $CeCl_3{\cdot}7H_2O$ raised the stereoselectivity of the reaction to 99 : 1 dr and 99.8% ee [37]. Electron-donating groups in the *para*-position of the phenyl group of ketophosphonates increased yields and ee values of the products, while electron-withdrawing groups reduced the chemical yields. A wide range of α-keto phosphonates were hydrogenated with a new chiral phosphine–phosphoramidite ligand to afford the corresponding (*R*)-α-hydroxy phosphonates with good enantioselectivities (up to 87% ee) (Scheme 4.26) [37].

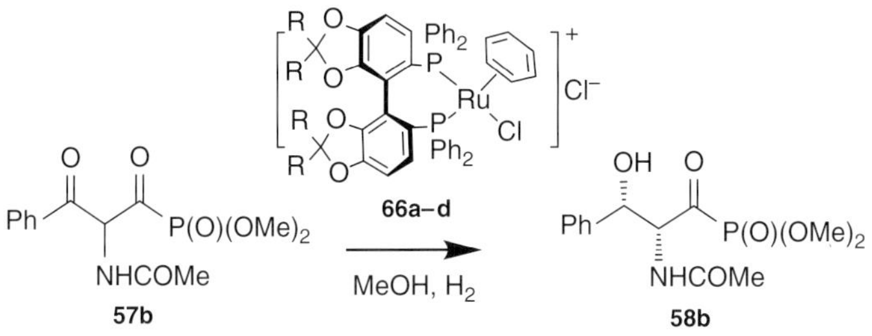

a, Ar = Ph, R = Me, (*S*)-SunPhos; **b**, Ar = 4-MeC_6H_4, R = Me (*R*)-Tol-SunPhos; **c**, Ar = Ph, R = H; d, **65c**, (*S*)-MeOBIPHEP

L = **65a**	dr = 97:3	ee = 98.0%
L = **66b**	dr = 94:6	ee = 99.9%
L = **65c**	dr = 89:11	ee = 95.7%
L = **65d**	dr = 89:11	ee = 96.7%

Scheme 4.25 Dynamic kinetic resolution of α-amido-β-ketophosphonates **57** via asymmetric hydrogenation catalyzed by ruthenium complexes **65, 66**.

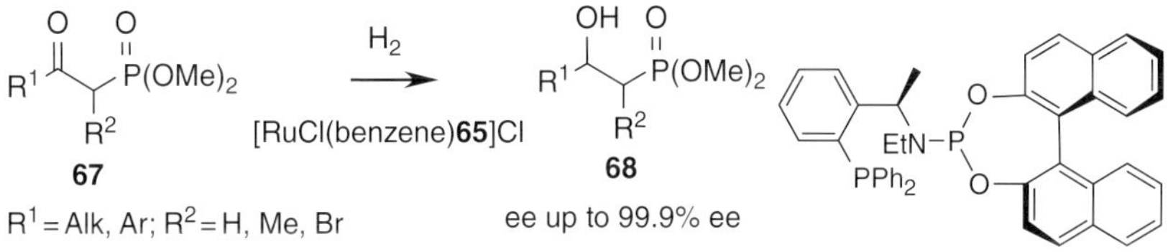

Scheme 4.26 Hydrogenation of β-ketophosphonates with Ru-(*S*)-SunPhos catalyst.

4.3 Asymmetric Reduction and Oxidation

Enantioselective reduction of prochiral ketophosphonates is one of important methods for the preparation of enantioenriched hydroxyphosphonates that are important biologically active compounds and initial reagents for the synthesis of many enantiopure products, including natural compounds. Various methods for the enantioselective reduction of keto- and ketiminophosphonates were developed, including reduction with chirally modified boranes, complex metal hydrides, biocatalytic reduction, and others. A number of enantioenriched hydroxyphosphonates and aminophosphonates were prepared by this method. Biocatalytic reduction is an especially convenient method which was developed over recent years. Baker's yeast and other microorganisms were effectively used for the reduction of ketophosphonates to be promising reagents for practical application (see Chapter 6). One of the best methods for asymmetric catalytic reduction of ketophosphonates is reduction with chiral-modified borohydrides. Besides the borohydride anion (BH_4^-), which can be modified by a chiral counterion, the hydrides can be substituted by chiral alcohol, carboxylic acids, alcoholic acids, and hydroxy amines. Chiral-modified borohydrides immobilized on a polymer were also available for reuse. Asymmetric CBS (Cory-Bakshi-Shibata) catalytic reduction (enantioselective reduction using borane and a chiral oxazaborolidine as CBS catalyst) represents an effective method for the preparation of various chiral alcohols [38, 39].

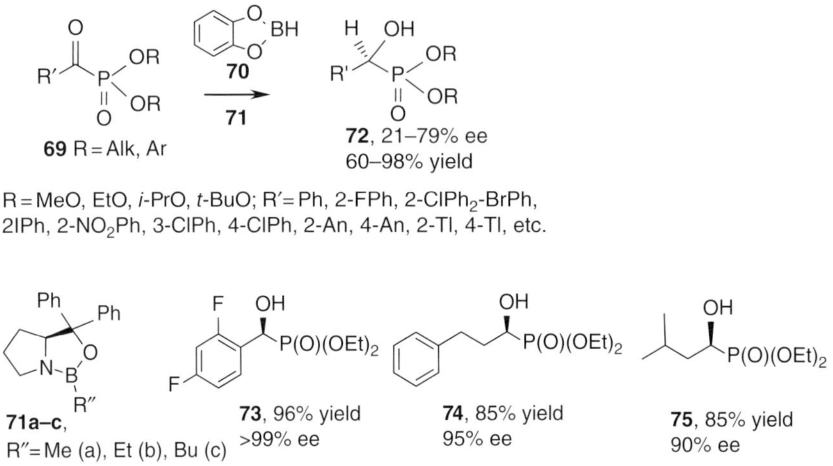

Scheme 4.27 Enantioselective reduction of ketophosphonates **69** by catecholborane **70**.

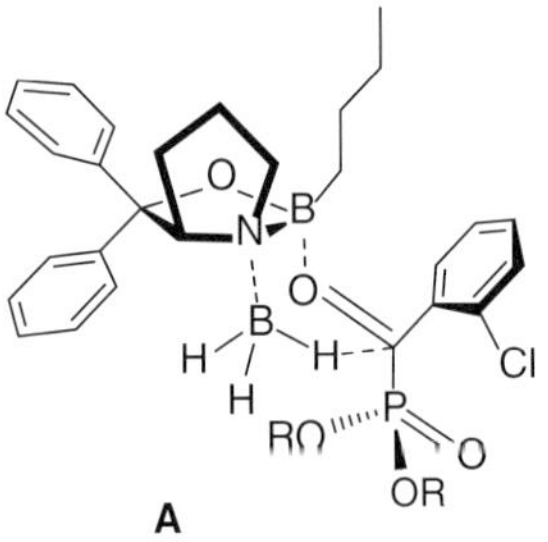

Figure 4.2 Model of the reaction complex of **69**, (*S*)-**71c**, and borane.

4.3.1 Reduction of C=O, C=N, and C=C bonds

Using the reduction of α-ketophosphonates **69** with catecholborane (catBH) as reducing reagent and oxazaborolidine **71** as catalyst, enantiomerically enriched hydroxyphosphonates **72** were synthesized with good enantioselectivities (Scheme 4.27) [40–43]. Enantioselective reduction of α-ketophosphonates led to the formation of α-hydroxyarylmethylphosphonates **73–75** with enantioselectivities from moderate to good (up to 80% ee). The mechanism of catalytic reduction was studied by *ab initio* MO calculations (Figure 4.2) [40]. According to Corey's model, the carbonyl group of the α-ketophosphonates **69** is complexed to the boron atom of the (*S*)-2-*n*-butyloxazaborolidine **71c** in such a way that the hydride from the borane complexed to the nitrogen atom attacks the carbon atom from the *Re* face. This leads to the differentiation of the two residues flanking the carbonyl group: the phosphoryl group is the "large" substituent, whereas the aromatic system is the "small" group. The co-ordination of the borohydride with an oxazaborolidine nitrogen atom increases the acidity of the intracyclic boron atom to facilitate the reduction of ketones **69** (Scheme 4.27).

Enantiopure carboxylic acids of natural origin were applied for chiral modifications of borohydrides [44–55]. In particular, the chiral reductant $NaBH_4$-Pro, obtained from

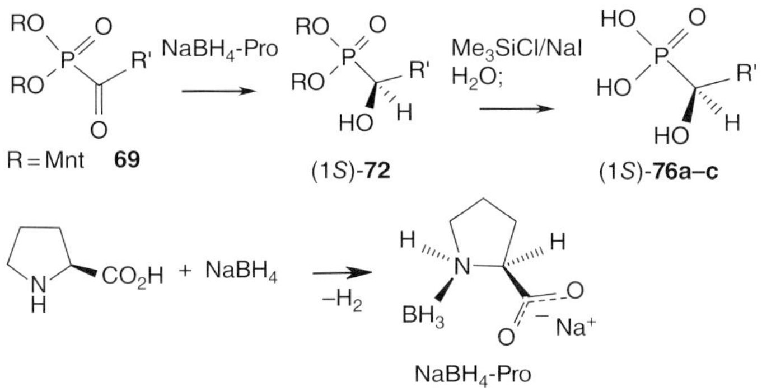

Scheme 4.28 Reduction of ketophophonates with $NaBH_4$-Pro.

$NaBH_4$ and (*S*)-proline, reduced ketophosphonates **69** with 50–70% ee. This reductant was applied for the synthesis of a number of hydroxyphosphonates **72** (Scheme 4.28) [44].

(S)-Diisopropyl hydroxyphenylmethylphosphonate **72**. A THF solution of 1.1 ml of catBH **70** (1 M; 1.1 equiv.) was added at −80 °C to a solution of 1.00 mmol of the ketophosphonate **69** and (*S*)-5,5-diphenyl-2-butyl-3,4-l.3.2-oxazaborolidine **71a** (0.12 mmol, 0.12 equiv.) in 3.0 ml toluene. The reaction mixture was stored for 5 h at −20 °C. After that time, the mixture was diluted at room temperature by the addition of 20 ml Et_2O, extracted with 4 × 5 ml each of a saturated $NaHCO_3$ solution, dried (Na_2SO_4), and concentrated *in vacuo*. The residue was purified by column chromatography (eluent $CH_3OH + CH_2Cl_2$), yield of (*S*)-**72** 65%, 90% ee, $[\alpha]_D$ −18.5 ($c = 1$, $CHCl_3$)) [40].

The enantioselective reduction of ketophosphonates by borohydrides in the presence of tartaric acid is another interesting method [45–47]. Natural (*R,R*)-(+)-tartaric acid and borohydride form a chiral complex that is a convenient stereoselective reagent for reduction of ketophosphonates [45, 46]. Reduction of ketophosphonates **78** with this complex was performed on cooling to 30 °C in THF. Reduction of diethyl α-ketophosphonates **78b** with the $NaBH_4$/(*R,R*)-TA chiral complex yielded diethyl (1*S*)-α-hydroxybenzylphosphonates **79b** with optical purity of 60%, while reduction of dimenthyl ketophosphonates **78c** led to formation of (1*S*)-α-hydroxybenzylphosphonates **77** with purity of up to 80–93% de (see Section 3.7). The stereoselectivity of reduction of ketophosphonates **78** containing chiral menthyl groups at the phosphorus atom with $NaBH_4$/(*R,R*)-TA was higher than in the case of ketophosphonates containing achiral methyl or ethyl groups at phosphorus (Table 4.4 and Scheme 4.29).

Stereoisomers of dimenthyl 2-hydroxy-3-chloropropylphosphonate (*S*)- and (*R*)-**80b** were prepared with optical purity of 96% ee [50].These compounds represent useful chiral synthons (chirons) for the synthesis of enantiomerically pure β-hydroxyphosphonates. Biologically important chiral β-hydroxyphosphonic acids: phosphono-carnitin and phosphono-GABOB, were obtained in multigram quantities by means of these chirons [54].

Table 4.4 Asymmetric reduction of ketophosphonates to hydroxyphosphonates (Scheme 4.29).

Entry	R	R′	n	Yield (%)	A	Configuration	ee (%)
1	Ph	Mnt	0	90	L-Pro	*S*	52.6
2	2-F-C_6H_4	Mnt	0	90	L-Pro	*S*	79.2
3	2-An	Mnt	0	90	L-Pro	*S*	60.6
3	Ph	Mnt	0	95	L-TA	*R*	92.4
4	Ph	Mnt	0	98	D-TA	*S*	46
5	2-F-C_6H_4	Mnt	0	97	L-TA	*S*	80.5
6	2-An	Mnt	0	96	L-TA	*S*	74
7	Piperonyl	Mnt	0	97	L-TA	*S*	96
8	*i*-Pr	Mnt	0	97.6	L-TA	*S*	68
9	Ph	Et	0	95	L-TA	*S*	60
10	Ph	Et	0	94	D-TA	*R*	60
11	CH_2Cl	Et	1	86	L-TA	*S*	80
12	CH_2Cl	Et	1	82	D-TA	*R*	80
13	CH_2Cl	Mnt	1	94	L-TA	*S*	96
14	CH_2Cl	Mnt	1	80	D-TA	*R*	82
15	Ph	Et	1	95	D-TA	*S*	44

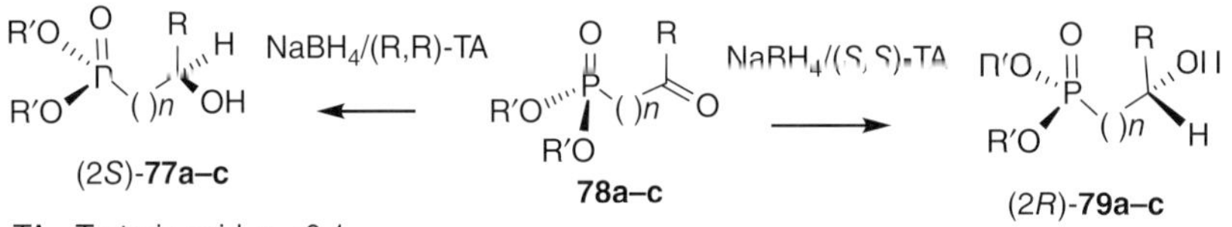

TA - Tartaric acid, $n = 0,1$
R=Me (a), Et (b), (1*S*,2*R*,5*R*)-Mnt (c); R′ = Alk, Ar, CH_2Cl, Pyperonyl

Scheme 4.29 The reduction of ketophosphonates **78** with $NaBH_4$/(*R,R*)-or (*S,S*)-TA.

Di(1R,2S,5R)-menthyl (R)-hydroxy(phenyl)methylphosphonate. (*R,R*)-(+)-tartaric acid (10 mmol) was added to a suspension of sodium borohydride (10 mmol) in 50 ml of THF and the reaction mixture was refluxed for 4 h. Asolution of ketophosphonate (2.5 mmol) in 10 ml of THF was then added at −30 °C and the reaction mixture was stirred at this temperature for 24 h. To the reaction mixture, 20 ml of ethyl acetate and 30 ml of 1 N hydrochloric acid was added dropwise. The organic layer was separated and the aqueous phase was saturated with NaCl and extracted two times with ethyl acetate (15 ml). The organic extracts were washed with a saturated solution of Na_2CO_3 (3 × 20 ml) and dried with Na_2SO_4. The solvent was removed under vacuum and the residue was crystallized from acetonitrile (yield: 95%, white solid, mp 139 °C (hexane), $[\alpha]_D^{20}$ −70 ($c = 1.0$, $CHCl_3$)) [45].

Corbett and Johnson [55] recently described a method for selective DKR of α-aryl acyl phosphonates **80**, providing β-stereogenic-α-hydroxy phosphonic acid

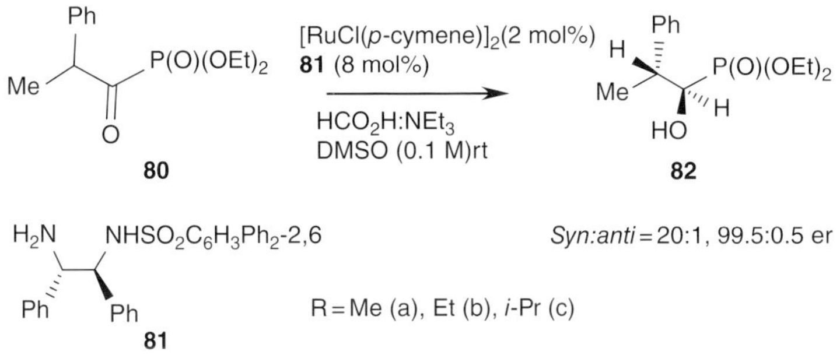

Scheme 4.30

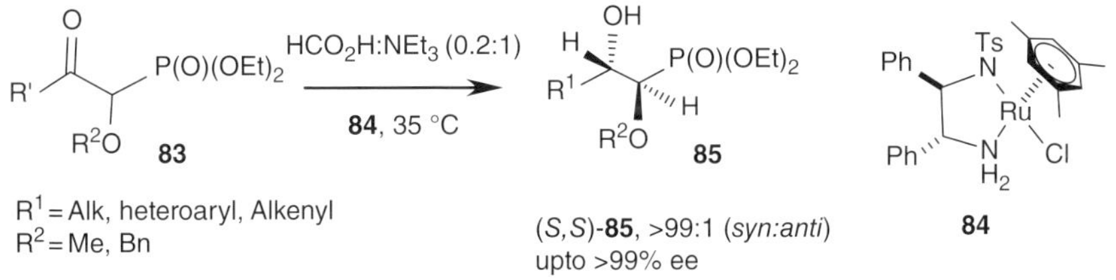

Scheme 4.31 Reduction of acyl phosphonates with formic acid and triethylamine, catalyzed by Ru-complex.

derivatives. The reduction of acyl phosphonates **80** with formic acid and triethylamine, catalyzed by RuCl[(*S,S*)-TsDPEN] (*p*-cymene) complex containing chiral aminosulfonamid ligand **81**, led to the formation of (*R*)-hydroxyphosphonates **82** with high diastereo- and enantioselectivities up to 99% ee. The absolute configurations of the products was established as (1*R*,2*R*) via X-ray crystallographic analysis, confirming the *anti*-orientation of the OH and Ar groups (Scheme 4.30). Son and Lee [56] applied the DKR-based asymmetric transfer to the hydrogenation of a wide range of 2-substituted α-alkoxy-β-ketophosphonates **83** and obtained corresponding 2-substituted α-alkoxy-β-hydroxyphosphonates **85** with excellent levels of diastereo- and enantioselectivity (Scheme 4.31).

Barco *et al.* [57] described the diastereoselective borohydride reduction of β-phthalimido-α-ketophosphonates **86a–d** catalyzed by chiral oxazaborolidines leading to the formation of β-amino-α-hydroxyphosphonates **87** [57]. The reduction of **86** with borohydride-dimethyl sulfide complex in THF led to the formation of an (*S,S*)- and (*S,R*)-diastereomeric mixture (*S,S*)-**87** and (*S,R*)-**88** (dr = 8 : 1 – 10 : 1); at the same time, the reduction of ketophosphonates **86** with catBH **70** and oxazaborolidine **71**a (12 mol%) in toluene at −60 °C provided only a single (*S,S*)-diastereomer **87a–d** in good yield (Scheme 4.32).

Only a few examples of the enantioselective reduction of phosphorus compounds bearing C=N or C=C bonds have been described [58]. An interesting example of the enantioselective reduction of iminophosphonates **90** was reported by Onys'ko and Mikolajczyk [59]. The CBS catalytic reduction of 1-imino-2,2,2-trifluoroethylphosphonates **91** led successfully to the formation of aminophosphonates **92** (yield 65–98% and 30–72% ee). The reduction of **91** with catBH catalyzed by methyloxazaborolidine **71a** formed aminophosphonates **92** and

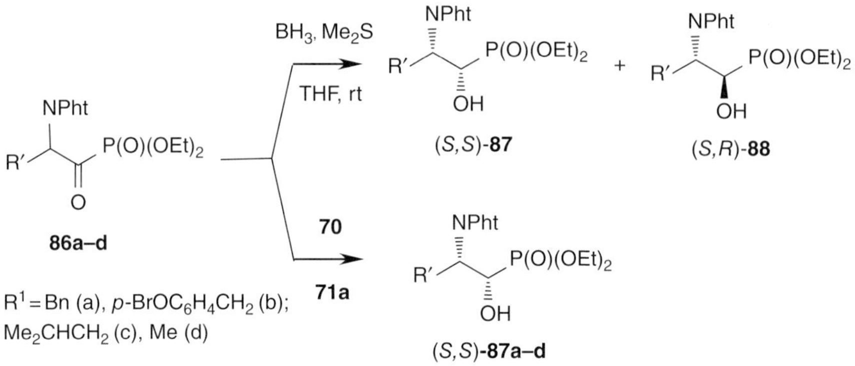

Scheme 4.32 Reduction of ketophosphonates with catecholborane and oxazaborolidine.

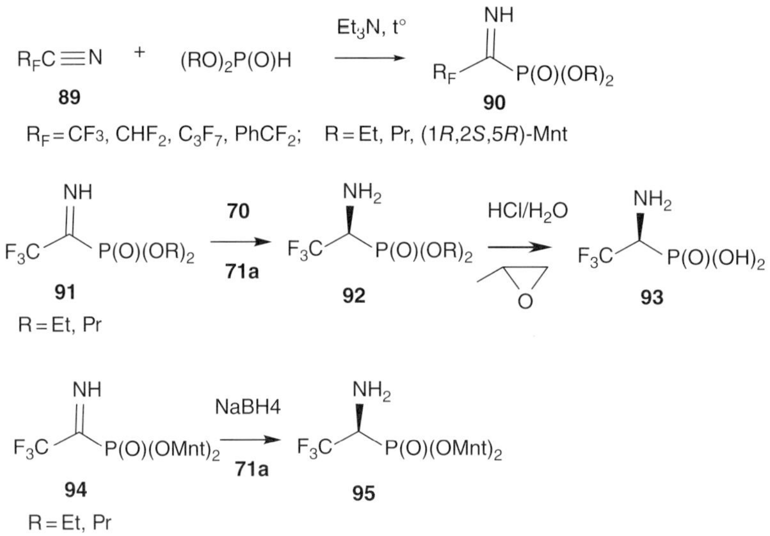

Scheme 4.33 Enantioselective catalytic reduction of 1-imino-2,2,2-trifluoroethylphosphonates.

aminophosphonic acid **93** in 98% yields and with 72% ee. Evidently, the starting iminophosphonates **91** activated with the electron-withdrawing CF_3 group are able to coordinate the reagent/catalyst (Scheme 4.33).

Dimenthyl α-trifluoromethyl-a-aminomethylphosphonate **95**. A solution of (*R*)-1-methyl-3,3-diphenylpyrrolidinooxazaborolidine **71a** (0.045 ml, 1 M solution in toluene, 0.045 mmol) was dissolved in THF (2 ml), cooled to −15 °C, and catBH **70** (1.35 ml, 1 M solution in THF, 1.35 mmol) was added. A solution of α-aminophosphonate **93** (0.9 mmol) in THF (3 ml) was added dropwise over a period of 3 h. After the addition was completed, the reaction mixture was stirred at −15 °C for 2 h. The reaction mixture was quenched with aqueous 1 N HCl (3 ml) and allowed to warm to room temperature. The mixture was extracted with diethyl ether (3 × 5 ml) and the layers were separated. The aqueous layer was basified with saturated aqueous $NaHCO_3$ (2 ml) and was extracted with ethyl acetate (3 × 10 ml). The combined extracts were dried over anhydrous magnesium sulfate and evaporated under reduced pressure to leave a colorless oil, which was purified on silica gel using diethyl ether as eluent to give analytically pure aminophosphonate (−)-**95** (98%, 72% ee) as a colorless oil ($[\alpha]_D^{25}$ −2.26 (c = 2.0, $CHCl_3$)).

A wide range of 3-aryl-4-phosphonobutenoates **96** were reduced with PMHS (polymethylhydrosiloxane) in the presence of a Lig/$Cu(OAc)_2$ H_2O catalyst, where Lig = (S)-SegPhos **98**, (S,R)-t-Bu-JosiPhos **99**, (S,R)-XyliPhos, and (S)-TolBINAP (1–5 mol%) in *tert*-butanol with good enantioselectivities of up to 94% ee as shown in Scheme 4.34 [60]. Various silanes were screened (PMHS, $PhSiH_3$, 1,1,3,3-tetramethyldisiloxane) with similar results, although PMHS was superior with respect to enantioselectivity. The reduction was influenced by the steric and electronic effects of the substrates. The substrates with an electron-withdrawing group at the *para*-position of the phenyl ring showed greater reduction in ee value than those with an electron-donating group. A similar reduction of enamine phosphonates was also reported.

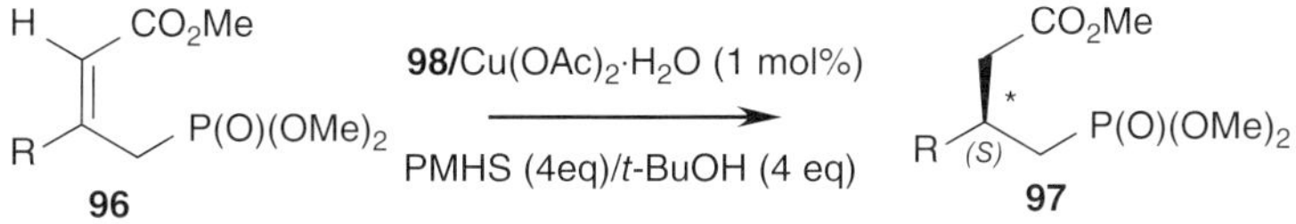

96 **97**

R = Ph	95%	94% ee	(–)
R = 2-An	77%	68% ee	(–)
R = 3-An	90%	93% ee	(–)
R = 4-An	95%	83% ee	(–)
R = 4-BrC_6H_4	93%	92% ee	(–)
R = 4-ClC_6H_4	94%	92% ee	(S)
R = 4-FC_6H_4	86%	91% ee	(–)
R = 4-$NO_2C_6H_4$	95%	90% ee	(–)
R = 3-ClC_6H_4	95%	72% ee	(–)
R = 2-nphth	94%	90% ee	(–)
R = 2-thienyl	85%	91% ee	(–)

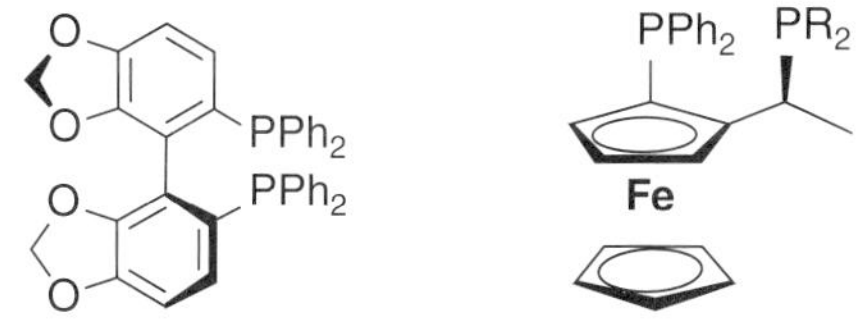

99

98, (S)-SegPhos
72–94% ee

R = t-Bu, t-Bu-JosiPhos, 90% ee
R = Mes, XyliPhos, 50% ee

Scheme 4.34 Reduction of phosphonobutenoates **96** with PMHS catalyzed by (S)-Segphos/$Cu(OAc)_2 \cdot H_2O$.

4.3.2 Asymmetric Oxidation

There are only a few examples of asymmetric oxidation of organophosphorus compounds. Thomas and Sharpless [61], and Sisti [62] applied the catalytic asymmetric reactions of dihydroxylation and aminohydroxylation of vinylphosphonates for preparation of dihydroxyphosphonates **100,** phosphonoepoxides **101**, aminohydroxyphosphonates **102**, and phosphonoaziridines **103** (Scheme 4.35). The catalytic asymmetric synthesis of β-amino-α-hydroxyphosphonates **102** was attained via aminohydroxylation of α,β-unsaturated phosphonates with potassium osmium(VI) complexes bearing $(DHQ)_2$PHAL ligands. The *syn*-β-amino-α-bromination of unsaturated phosphonates was performed under typical Sharpless reaction conditions with excess *N*-bromoacetamide [63]. One of the most interesting examples of asymmetric oxidation of vinylphosphonates is the synthesis of fosfomycin **106** and fosfadecin **107**, which are well-known as *antibiotics* used against Gram-negative and Gram-positive bacteria. For their preparation, Kobayashi *et al.* [64] used the Sharpless asymmetric dihydroxylation of *trans*-propenylphosphonates (Scheme 4.36) [65].

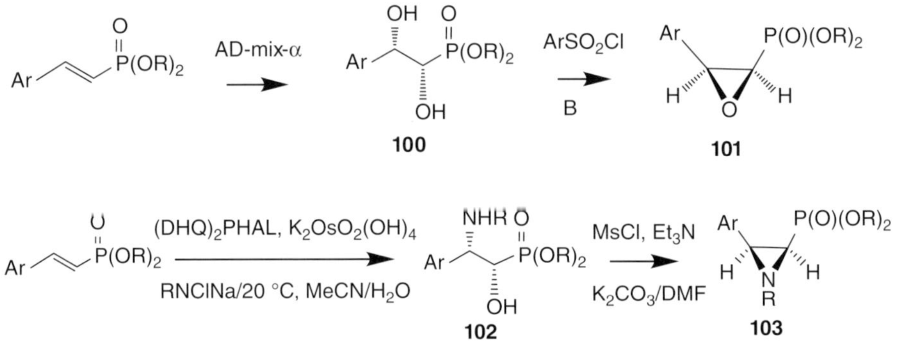

Scheme 4.35 Asymmetric dihydroxylation and aminohydroxylation of alkenphosphonates.

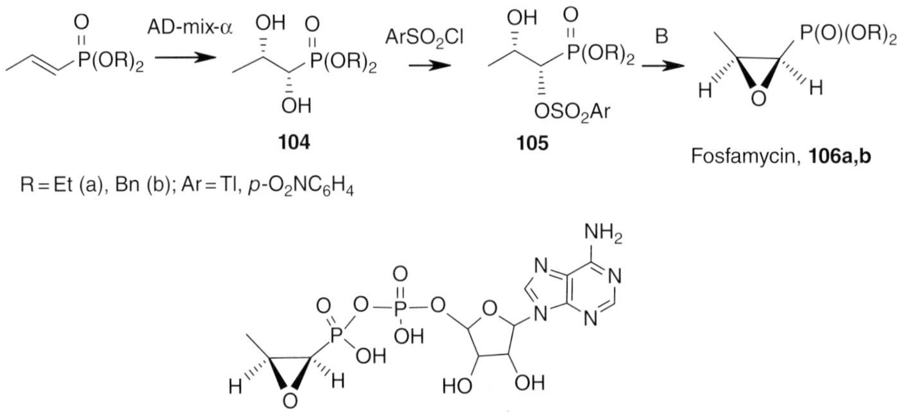

Scheme 4.36 The synthesis of fosfamycin and fosfadecin.

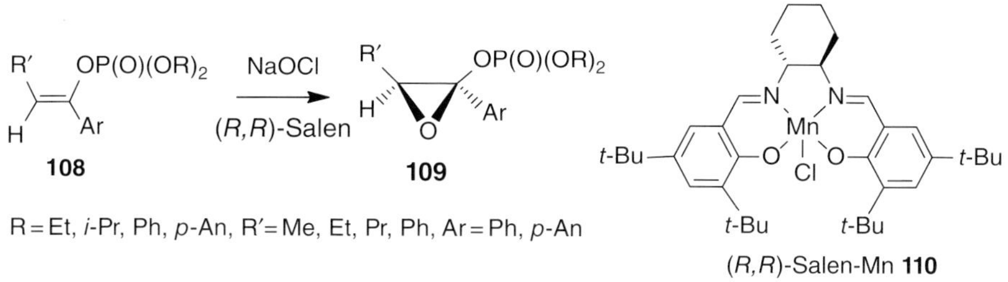

Scheme 4.37 Asymmetric oxidation of enol phosphates with the NaOCl/Mn(III)(salen).

The oxidation of alkene with AD-Mix-α (asymmetric dihydroxylation) led to the formation of diol **104** in 65% yield and with greater than 99% ee after crystallization from a mixture of hexane/ethyl acetate (yield 95% and 78% ee before crystallization). Subsequent monosulfonylation of the resulting diol **104** and treatment with K_2CO_3 in acetone afforded the dibenzyl epoxide **105a** (R = Bn) and then the fosfomycin **106b** (R = H) (Scheme 4.36). Asymmetric dihydroxylation of olefin **105b** with AD-mix-α and $MeSO_2NH_2$ in *t*-BuOH resulted in the formation of diol **106** in 85% yield and with 96% ee. After recrystallization, the pure diol was obtained and converted into epoxide **107**.

Krawczyk *et al.* [66, 67, 68] synthesized the optically active epoxides **109** using the asymmetric oxidation of enol phosphates **108** with NaOCl in the presence of Mn(III)(salen)complex **110**. The hydrolysis of **109** led to the formation of chiral hydroxy α-ketones in good yields and with enantioselectivity of 68–96% ee (Scheme 4.37). Chen *et al.* [69] have described the kinetic resolution of α-hydroxyphosphonates **111**, catalyzed by chiral vanadyl(V)methoxide complex **114** bearing *N*-salicylidene-α-aminocarboxylates, effecting highly enantioselective and chemoselective aerobic oxidations at ambient temperature. Excellent reaction rates and selectivity factors ($k_{rel} > 99$) were observed in the cases of the most electron-withdrawing 4-nitro and 4-carbomethoxy substrates. The reactions were completed at 50% conversion, leading to recovery of the enantiomers (*S*)-**113** with 99% ee. This method works well with various α-aryl- and α-heteroaryl-α-hydroxyphosphonates. It was found that the more sterically encumbered diastereomeric adduct **B** was faster reacting for the subsequent α-proton elimination process leading to α-ketophosphonate **112** with concomitant reduction of the vanadyl(V) species **114** into the corresponding vanadium(III)OH (Scheme 4.38, Table 4.5).

4.4 Electrophilic Asymmetric Catalysis

Nucleophilic and electrophilic catalyses have been well-known for many years. There are many examples of asymmetric nucleophilic or electrophilic activations of organophosphorus compounds that have attracted the attention of many chemists. Usually, the electrophilic catalysts used are Lewis acids and nucleophilic catalysts employ organic bases. Lewis acids catalyze reactions through electrophilic activation of organic groups. The addition of a Lewis acid to the substrate containing a free electronic pair is accompanied by an increase of reactivity of the complex generated. Typical examples of electrophilic asymmetric activation of organophosphorus compounds by chiral Lewis acids are catalytic alkylation reactions, arylation, halogenations, catalysis of chemical reactions by enzymes, and others.

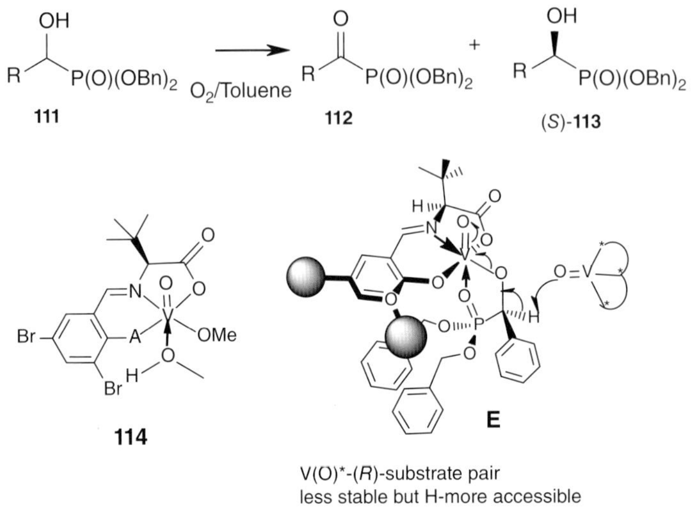

Scheme 4.38 Effect of substituents on the asymmetric aerobiotic oxidation of racemic α-hydroxyphosphonates.

Table 4.5 Asymmetric aerobiotic oxidation of racemic α-hydroxyphosphonates (Scheme 4.38).

R	% conversion	Yield (%)	% ee	K_{rel}*
Ph	51	47	99	>99
4-Tl	49	46	96	>99
4-An	50	49	99	>99
2-An	50	49	99	>99
4-$Me_2NC_6H_4$	50	50	97	>99
4-ClC_6H_4	49	49	96	>99
4-$NO_2C_6H_4$	50	49	99	>99
4-CNC_6H_4	51	47	95	81
4-MeOC(O)C_6H_4	50	49	>99	>99
2-Furanyl	49	47	90	95
2-Thiophenyl	50	49	99	>99
trans-CH_3CH=CH	49	49	96	>99

*Selectivity factor: $K_{rel} = \ln[(1 - C)(1 - ee)]/\ln[(1 - C)(1 + ee)]$, C = conversion, ee = enantiomeric excess.

4.4.1 Catalytic Electrophilic Substitution at the Phosphorus Atom

Over the last few years, catalytic asymmetric synthesis of tertiary phosphines has attracted the attention of many chemists [70–78]. Interesting results have been published in articles and reviews [71–77]. One route leading to α-stereogenic phosphines is electrophilic substitution at the phosphorus atom of secondary phosphines, as a result of asymmetric catalysis in which the catalyst activates a phosphorous nucleophile or a carbon electrophile, creating an asymmetric environment, that is, creating a

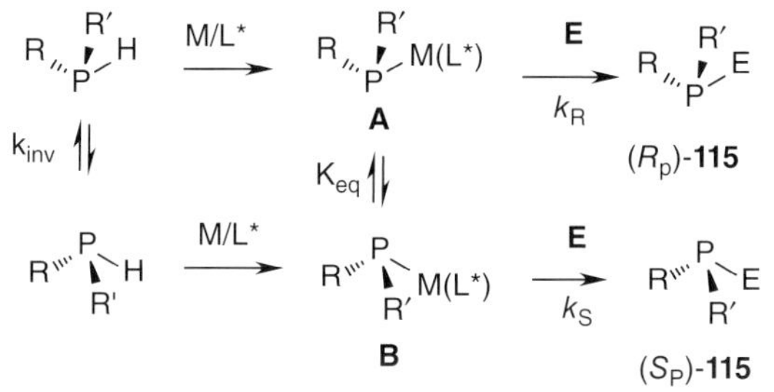

Scheme 4.39 Mechanism of catalytic electrophilic substitution at trivalent phosphorus.

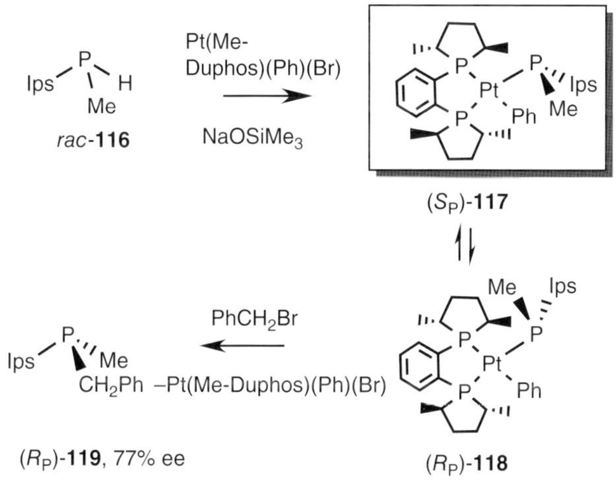

Scheme 4.40 Asymmetric alkylation of secondary phosphines catalyzed by Pt(II) complexes.

preference for one of the *Si* or the *Re* faces at the reactive center [71–74]. Upon reaction with a chiral metal complex, racemic secondary phosphines are converted into diastereomeric metal-phosphide complexes A or B, which interconvert rapidly by P-inversion. If the equilibrium A B is faster than the reaction of A or B with the electrophile E, then P-stereogenic phosphines **115**, in which pyramidal inversion is slow, can be formed enantioselectively. The product ratio in this dynamic kinetic asymmetric transformation depends both on K_{eq} and on the rate constants k_S and k_R (Scheme 4.39).

4.4.1.1 Alkylation and Arylation of P(III) Compounds

The asymmetric arylation or alkylation of racemic secondary phosphines catalyzed by chiral Lewis acids led in many cases to the formation of enantiomerically enriched tertiary phosphines [75]. For example, Glueck *et al.* [75–84] found that the racemic secondary phosphines **116** form with a platinum complex Pt(Me-Duphos)(Ph)(Br) and $NaOSiMe_3$ in toluene adduct **117**, which interconvert rapidly by P-inversion (S_P)-**117**⇌(R_P)-**118** (Scheme 4.40).

Adduct **117** was isolated and studied by low-temperature NMR and X-ray monocrystal analysis. The crystal structure of the adduct showed that the major enantiomer of **117** had an (R_P)-absolute configuration [79]. The treatment of adduct **118** with benzyl bromide led to the formation of tertiary phosphine (R_P)-**119** with 77% ee and to initial catalyst Pt(Me-Duphos)(Ph)(Br), which confirms the proposed mechanism. Substitution at the tricoordinated phosphorus atom of the secondary phosphine **116** proceeded with retention of absolute configuration at phosphorus, according to classical representations. On the basis of these results, the authors came to a conclusion that the enantioselectivity was determined mainly by the thermodynamic preference for one of the interconverting diastereomers of (S_P)-**117**⇌(R_P)-**118**, although their relative rates of alkylation were also important (Curtin–Hammett kinetics) (Scheme 4.40).

The asymmetric arylation or alkylation of racemic secondary phosphines catalyzed by chiral Lewis acids led in many cases to the formation of enantiomerically enriched tertiary phosphines [79–82]. Chiral complexes of ruthenium, platinum, and palladium were commonly used. For example, the chiral complex Pt(Me-Duphos)(Ph)Br catalyzed the asymmetric alkylation of secondary phosphines with various RCH_2X (X = Cl, Br, I) compounds with formation of tertiary phosphines (or their boranes **121**) with 50–93% ee [75, 76, 79]. The enantioselective alkylation of secondary phosphines with benzyl halides catalyzed by complexes $[RuH(i\text{-Pr-PHOX})_2]^+$ led to the formation of tertiary phosphines **121** with 57–95% ee [79, 80]. Catalyst [(*R*)-difluorphos(dmpe)Ru(H)][BPh_4] was effective in the asymmetric alkylation of secondary phosphines with benzyl bromides, whereas (*R*)-MeOBiPHEP/dmpe was more effective in case of benzyl chlorides (Scheme 4.41) [80, 81] The arylation of secondary phosphines with aryl halides, catalyzed by chiral complexes of platinum [75–77, 83], ruthenium [80, 81], and palladium [84–86], in many cases proceeded with good enantioselectivity to give enantiomerically enriched tertiary phosphines. For example, the reaction of aryl iodides with secondary arylphosphines **120**, catalyzed by the chiral complex Pd((*R*,*R*)-Me-Duphos) (*trans*-stilbene), furnished tertiary phosphines with enantioselectivities of up to 88% ee [82–84]. The arylation of secondary phosphines **120** with *ortho*-aryl iodides catalyzed by the *in situ* generated complex $Pd_2(dba)_3 \times CHCl_3$, containing chiral ligand Et,Et-FerroTANE **124** and LiBr, led to the formation of the corresponding tertiary phosphines with an enantioselectivity of 90% ee. [84–86]. The palladium complex **126** also showed high enantioselectivity in the arylation of secondary phosphines [85]. Some examples of the arylation reaction of secondary phosphines with low ee have

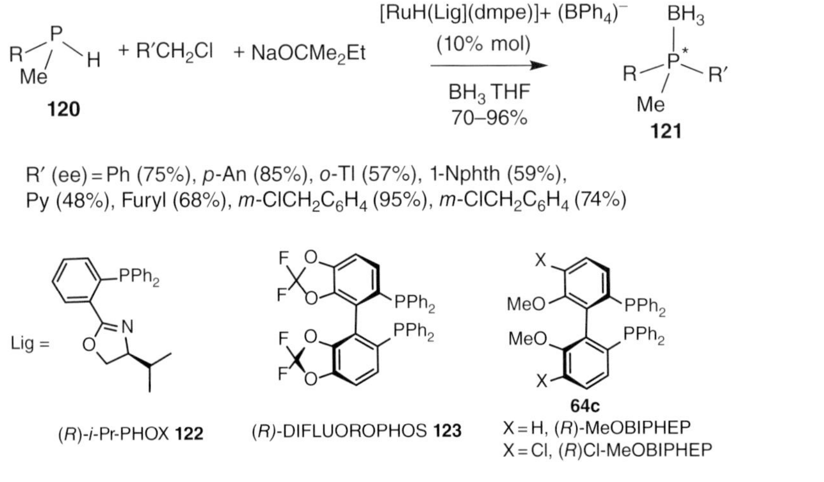

Scheme 4.41 Asymmetric alkylation of *sec*-phosphines catalyzed by chiral Ru complexes.

been described. The asymmetric arylation of phosphine boranes with anisyl iodide, catalyzed by the chiral complex of oxazoline phosphine **125** led to the formation of enantiomerically enriched tertiary phosphines with 45% ee [85]. The complex of (*R,S*)-*t*-Bu-JOSIPHOS catalyzed the arylation of PH(Me)(Ph)(BH_3) by *o*-anisyl iodide with the formation of PAMP-BH_3 with 10% ee (Table 4.6) [73].

The reaction of secondary phosphine boranes **127** with anisyl iodide, catalyzed by a chiral Pd complex with (*S,S*)-Chiraphos, proceeded with retention of absolute configuration at the phosphorus atom [78]. The addition of Pd((*S,S*)-Chiraphos)(*o*-An) to enantioenriched secondary phosphine **127** in the presence of $NaOSiMe_3$ led to the formation of the stable complex **129**. Heating this complex to +50 °C in excess diphenylacetylene converted it into (R_P)-**130** in 70% yield and with an enantiomeric purity of 98% ee (Scheme 4.42).

Table 4.6 The arylation of secondary phosphines catalyzed by palladium complexes.

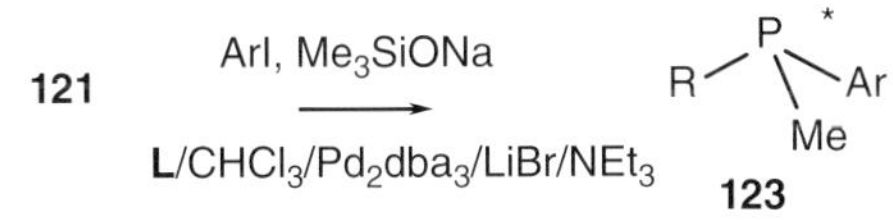

R	ArI	L	Yield (%)	ee (%) Config.
2-PhC_6H_4	2-*t*-BuOCOC_6H_4I	**124**	76	90 (*S*)
2-An	2-*t*-BuOCOC_6H_4I	**124**	43	86 (*S*)
2-$CF_3C_6H_4$	2-*t*-BuOCOC_6H_4I	**124**	39	93 (*R*)
2-PhC_6H_4	2-MeOCOC_6H_4I	**124**	69	85 (*S*)
t-Bu	3-AnI	**125**	—	45
2,4,6-(*i*-Pr)$_3C_6H_2$	PhI	**126**	84	78 (*S*)
2,4,6-(*i*-Pr)$_3C_6H_2$	*p*-PhOC_6H_4I	**126**	89	88 (*S*)

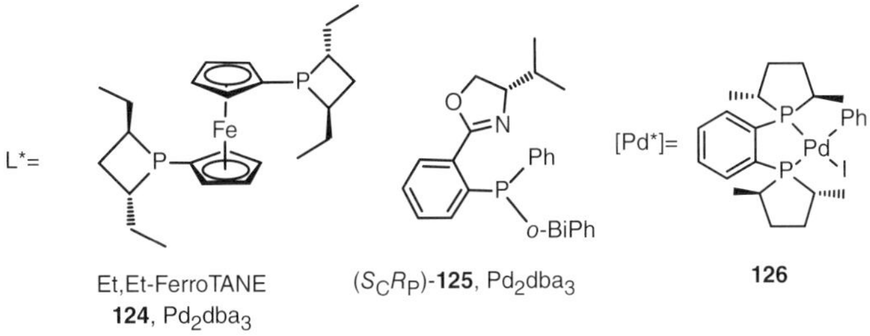

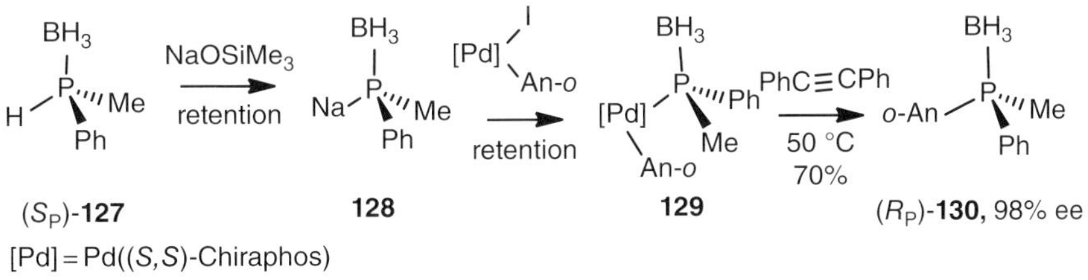

Scheme 4.42 Reaction of **126** with anisyl iodide catalyzed by chiral Pd(*S,S*)-Chiraphos complex.

Alkylation of silylated alkylarylphosphines **131** instead of P–H phosphines for the preparation of chiral tertiary phosphines led in some cases to an appreciable increase in the enantioselectivity. For example, Toste and Bergman [85] reported the reaction of aryl-substituted iodides **131** with silylphosphines catalyzed by Pd(Et-FerroTANE)Cl_2 in the presence of *N,N′*-dimethyl-*N,N,N*-propyleneurea (DMPU) leading to the formation of P-chiral tertiary phosphine sulfides **132** with 98% ee (Scheme 4.43).

Enantioselective intramolecular cyclization of secondary phosphines **133** or their boranes, catalyzed by chiral palladium (diphosphine) complexes, afforded P-stereogenic benzophospholanes **134** with moderate stereoselectivity (<70% ee). This reaction provided chiral phospholanes which are valuable ligands in asymmetric catalysis (Scheme 4.44) [86].

Examples of electrophilic additions of secondary phosphines to alkenes or alkynes have been described. Glueck [70] reported the platinum-catalyzed enantioselective alkylated/arylation of primary phosphines proceeding with the formation of chiral phosphaacenaphthenes.

The reaction was catalyzed by chiral platinum or palladium complexes, bearing an (*R,R*)-MeDUPHOS **4** ligand and proceeded with the formation of P-chirogenic phosphines **136** but with low ee (Scheme 4.45) [77]. Attempts to apply chiral ammonium salts as phase-transfer catalysts in the asymmetric alkylation of racemic secondary phosphines were undertaken. However, the alkylation of phenyl phosphine borane with methyl iodide in the presence of a chiral quaternary cinchonine ammonium salt proceeded also with low enantioselectivity (17% ee) (Ref. [5]).

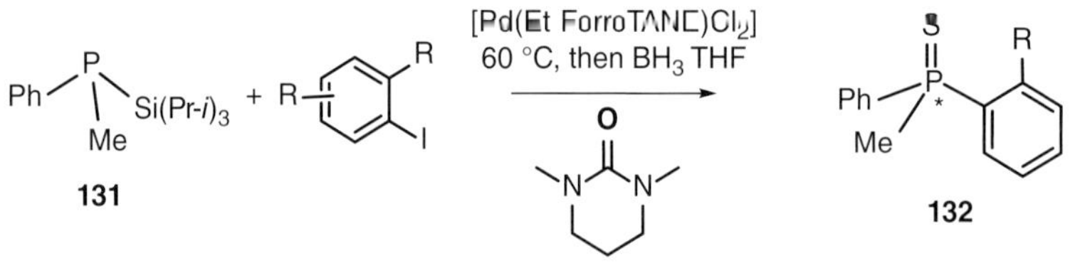

131 **132**

R = OMe	74%	ee = 55%	(*R*)
R = SMe	73%	ee = 63%	(*R*)
R = COOBu-*t*	83%	ee = 78%	–
R = $COOC_6H_4Me_2$-2,6	76%	ee = 82%	–
R = $COONEt_2$	63%	ee = 79%	–
R = COONPr-i_2	53%	ee = 98%	–

Scheme 4.43 Reaction of aryl-substituted iodides with silylphosphines catalyzed by [Pd(Et-FerroTANE)Cl_2].

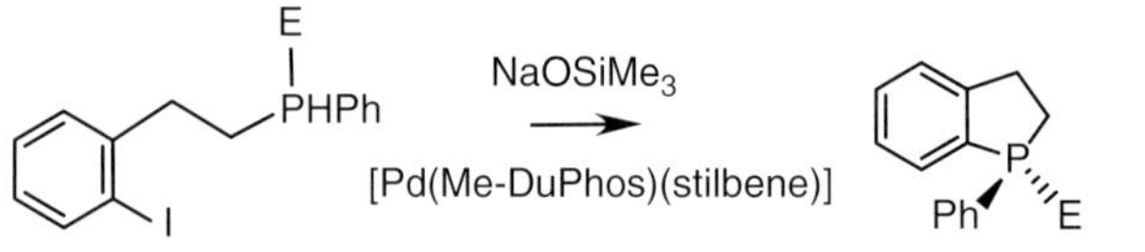

133 E = BH_3 S **134**, 63–70% ee

Scheme 4.44 Enantioselective intramolecular cyclization of secondary phosphines **133**.

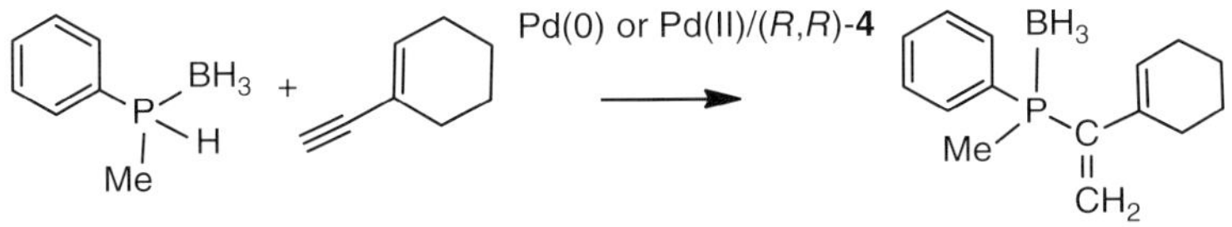

135 **4**-Me-DuPhos **136**, 39% ee

Scheme 4.45 Example of electrophilic additions of secondary phosphines to alkenes.

4.4.2 Catalytic Electrophilic Substitution in a Side Chain

4.4.2.1 Alkylation

A number of successful syntheses of chiral alkylated ketophosphonates were realized by enantioselective catalysis with chiral bis(oxazoline)-copper complexes. For example, Shibata *et al.* [87] recently reported on the enantioselective alkylation of β-ketophosphonates **138** with aromatic alcohols as electrophiles, catalyzed by Cu(II)(OSO_2CF_3)$_2$/**140a–c** leading to the formation of products **138** in good yields and with relatively high enantioselectivity. The asymmetric vinylogous aldol reaction between α-keto phosphonates **137** and 2-(trimethylsilyloxy)furan was also realized by using bis(oxazoline)-copper catalyst **140d** (Scheme 4.46) [88]. Enantioselective alkylation of β-ketophosphonates **142** with propargylic alcohols **141** in the presence of thiolate diruthenium complex **144** and a chiral complex of copper bearing a bis(4,5-diphenyl-4,5-dihydrooxazoline) ligand **140** as cocatalyst led to the formation of alkylated propargyl products **143** in good yields and with diastereoselectivity of 16 : 1 to 20 : 1 and enantioselectivities up to 97% ee [87]. In THF at ambient temperature for 40 h the reaction resulted in 2-oxocyclopentylphosphonate **143** with 63% ee of two diastereoisomer mixtures (*anti*-**143**:*syn*-**143** = 15 : 1) with 89% ee of *anti*-**143** (Scheme 4.47) [87a].

The Friedel–Crafts enantioselective alkylation of indoles with α,β-unsaturated ketophosphonates, catalyzed with various chiral metallocomplexes and organocatalysts has attracted significant attention [89, 90]. Evans reported that the pybox

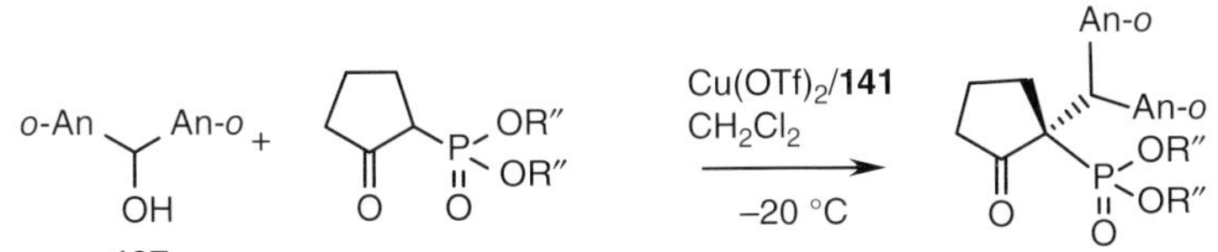

137 **138** **139** (80–97% ee)

Lig = **140a**	95%	ee = 84%
Lig = **140a**	68%	ee = 86%
Lig = **140a**	29%	ee = 74%
Lig = **140b**	46%	ee = 80%
Lig = **140c**	14%	ee = 48%
Lig = **140a**	61%	ee = 85%
Lig = **140a**	9%	ee = 87%
Lig = **140a**	57%	ee = 86%

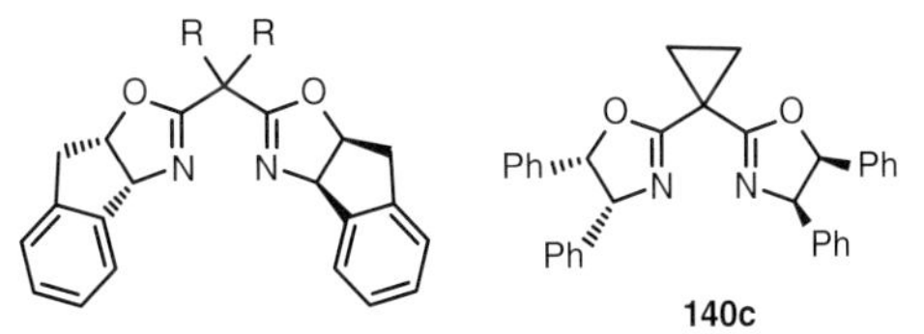

140a ($R + R = CH_2CH_2$), **140b** ($R_2 = Me_2$) **140c**

Scheme 4.46 Enantioselective catalytic alkylation of β-ketophosphonates with aromatic alcohols.

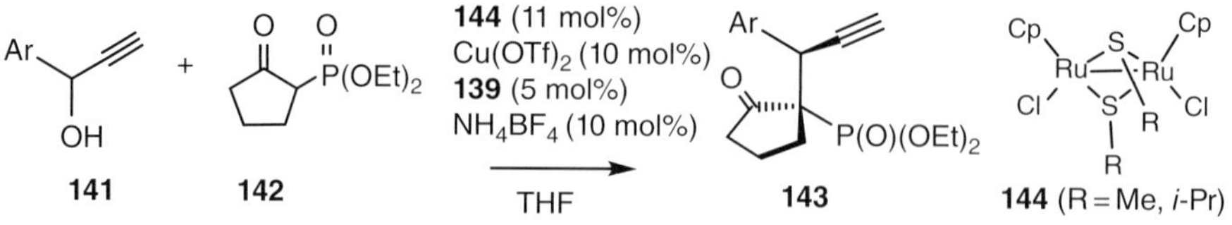

R = Et	89%	*anti/syn* = >20/1	ee = 91%
R = Me	97%	*anti/syn* = 19/1	ee = 90%
R = Pr	92%	*anti/syn* = >20/1	ee = 92%
R = Bu	78%	*anti/syn* = 16/1	ee = 91%
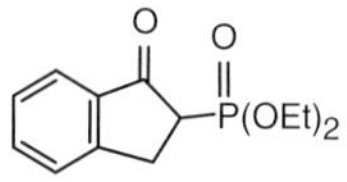	92%	*anti/syn* = >20/1	ee = 82%

Scheme 4.47 Enantioselective alkylation of β-ketophosphonates **142** by propargylic alcohols.

scandium(III) triflate complex **148** catalyzes the conjugated addition of indole to α,β-unsaturated acyl phosphonates **145** to give acyl phosphonates **146** with high yields (51–83%) and enantioselectivities greater than 99 ee [89]. Yamamoto used chiral aluminum complexes **149** for asymmetric phosphonylation of indoles and achieved enantioselectivities of 98% ee and yields of 85% [89]. Jørgensen *et al.* [90] have performed a variety of stereoselective conjugate additions of carbon-based nucleophiles (oxazolones, indoles, and 1,3-dicarbonyl compounds) to α,β-unsaturated acyl phosphonates catalyzed by chiral thioureas and obtained products in satisfactory yields and with enantioselectivities of 72–90% ee. The stereoselectivity of the 1,3-dicarbonyl addition to acyl phosphonates can be explained by bifunctional coordination of the nucleophilic and electrophilic reaction partners to the quinine-derived catalyst. It was proposed that the acyl phosphonate is hydrogen bonded to the squaramide motif, placing the alkene side chain in the sterically less-demanding area away from the C-9 center of the catalyst, while the 1,3-dicarbonyl compound is deprotonated and directed for nucleophilic attack by the tertiary nitrogen atom of the catalyst. The R-group of the nucleophile is oriented away from the reaction site and the subsequent conjugate addition approaches the *Si*-face of the C=C bond, accounting for both the enantio- and diastereoselectivities of the reaction. Subsequent treatment of the reaction mixture with methanol and 1,8-diazobicyclo[5.4.0]undec-7-ene (DBU) led to the formation of methyl 3-(indol-3-yl)-propanoates **147** in good yields (65–82%) and with high enantioselectivities (up to 99% ee). The high selectivity of the reaction was explained by coordination of the β,γ-unsaturated α-ketophosphonate **145** with palladium catalyst **151** by a two-centered coordinate bond in a bidentate fashion as the indole attacks the double bond, as shown in the formula. Since the *Re*-face of the double bond of the phosphonate was blocked preferentially by one of the phenyl groups of the (*R*)-BINAP, the addition of indole proceeded from the *Si*-face in a highly enantioselective manner. Bachu and Akiyama [92] used a BINOL-phosphoric acid **151** as a catalyst for this reaction. A variety of indoles underwent enantioselective Friedel–Crafts alkylation with α,β-unsaturated ketophosphonates **145** in the presence of 10 mol% of chiral BINOL-based phosphoric acid (Scheme 4.48). A catalytic asymmetric vinylogous Mukaiyama aldol reaction of ketophosphonates with furan derivatives catalyzed with chiral copper-sulfoximine catalyst provided phosphonic

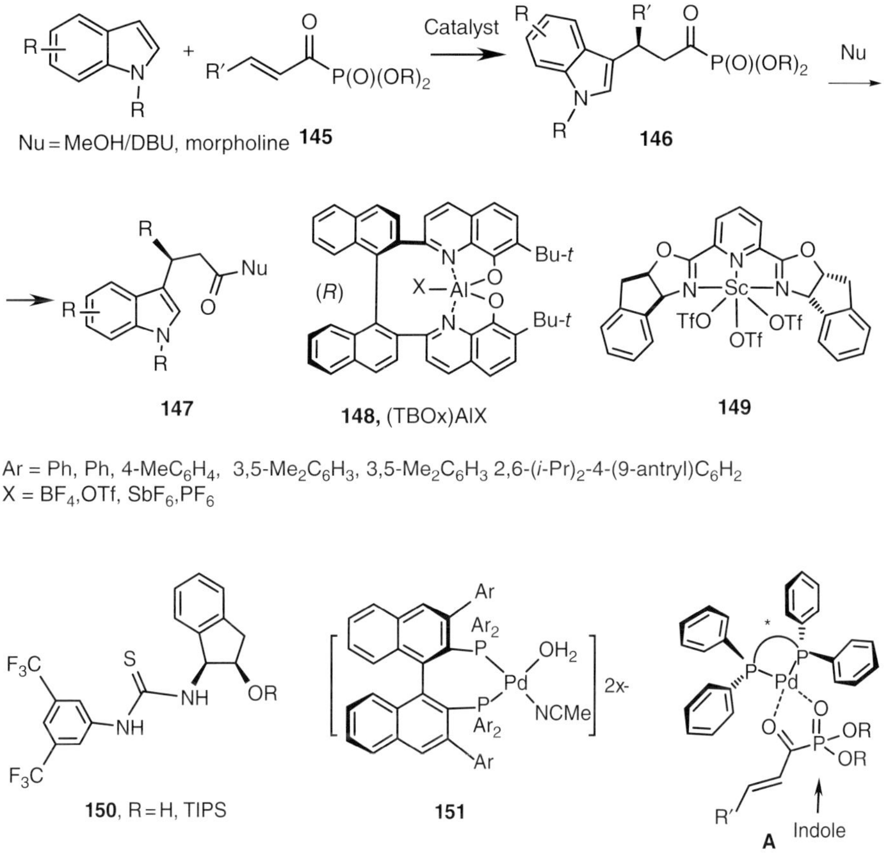

Scheme 4.48 Enantioselective Friedel–Crafts alkylation with α,β-unsaturated ketophosphonates.

γ-(hydroxyalkyl)butenolides bearing adjacent quaternary stereogenic centers with high diastereo- and enantioselectivity [88b].

4.4.2.2 Halogenation

Bernardi and Jørgensen [91] have developed a catalytic enantioselective chlorination and fluorination of β-ketophosphonates, using *N*-chlorosuccinimide (NCS) and *N*-fluorobenzenesulfonimide (NFSI). The reaction proceeded smoothly for both acyclic and cyclic β-keto phosphonates **152** giving the corresponding optically active α-chloro and α-fluoro-β-keto phosphonates **153** in high yields and enantioselectivities using an (*R*,*R*)-**154**/Zn(II) catalyst. The acyclic β-ketophosphonates with both aromatic and alkyl substituents at the β-position were converted into the corresponding optically active α-chloro-β-ketophosphonates **153** in yields of 80–98% and with enantioselectivities of 78–94% ee (Scheme 4.49).

The catalytic fluorination of the β-ketophosphonates by NFSI using a chiral **154**/Zn(II) catalyst also proceeds readily. For the fluorination reaction, a more easily prepared catalyst formed by a combination of $Zn(ClO_4)_2{\cdot}6H_2O$ and Ph-DBFOX **154** in the presence of 4 Å molecular sieves (MS) was used with comparable results. The introduction of two stereogenic centers in ligand **155** improved the enantioselectivity

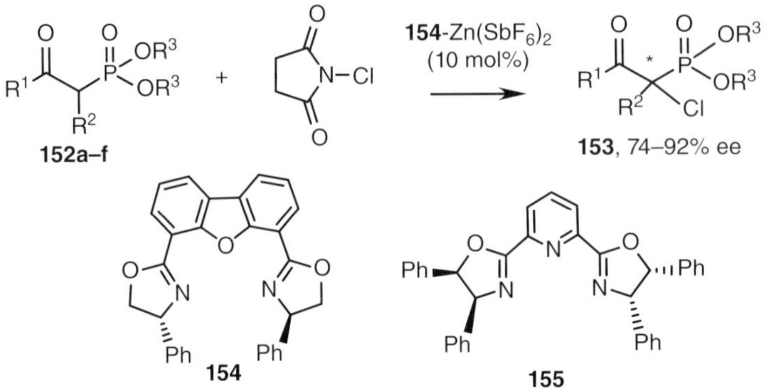

a	R^1 = Ph	R^2 = Me	R^3 = Et	98%	92% ee
b	R^1 = Ph	R^2 = Me	R^3 = Me	97%	78% ee
c	R^1 = Ph	R^2 = Allyl	R^3 = Et	93%	92% ee
d	R^1 = 2-Np	R^2 = Me	R^3 = Et	97%	93% ee
e	R^1 = Me	R^2 = Me	R^3 = Et	80%	94% ee
f	$R^1 + R^2 = (CH_2)_3$		R^3 = Et	40%	80% ee

Scheme 4.49 Catalytic chlorination of the β-ketophosphonates by NFSI.

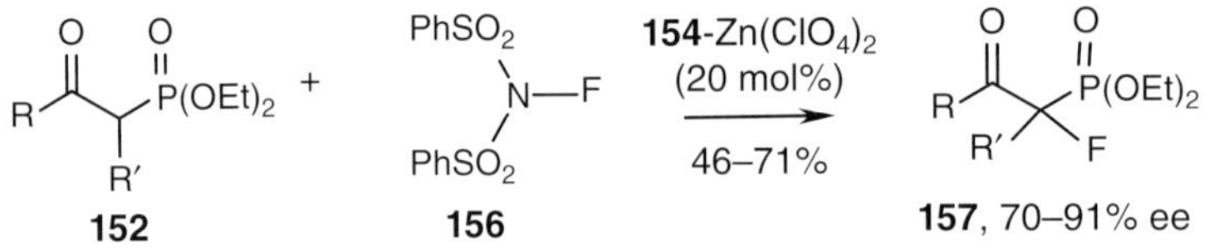

R = Ph, 2-Np, Me, R′ = Me, Allyl, R + R′ = $(CH_2)_2$

Scheme 4.50 Enantioselective fluorination of β-ketophosphonates by *N*-fluorobenzenesulfonimide **154**.

to 91% ee. Enantioselective fluorination of β-ketophosphonates **152** performed by the action of NFSI **156** catalyzed by zinc complexes with **154**, resulted in formation of the optically active α-fluorine- β-ketophosphonates **157** with yields from moderate to good and enantioselectivities of up to 91% ee (Scheme 4.50) [92]. Sodeoka *et al.* [93, 94] have developed an efficient catalytic enantioselective fluorination of cyclic and acyclic β-ketophosphonates by the action of NFSI in the presence of chiral palladium complexes **161**, **162** (1–10 mol%), containing (*R*)-BINAP ligands. The fluorination proceeded under mild conditions (in alcohol, acetone, or THF, at room temperature) and led to the formation of chiral fluorinated ketophosphonates **167** in yields of 57–97% and enantioselectivities of 94–96% ee [95, 96]. α-Fluorinated phosphonates were then converted into phosphonic acids (Scheme 4.51).

The absolute configuration predominating in the reaction indicates that the fluorinating reagent reacts from the less hindered side of the enolates, because the two ethoxy groups of the β-ketophosphonates **152** are positioned to cause the minimum amount of steric repulsion with the aryl group on the phosphine as shown in Figure 4.3.

The palladium catalysts **163** bearing BINAP type ligands have been utilized for the synthesis of various fluorinated cyclic and acyclic ketophosphonates [95, 96]. The best

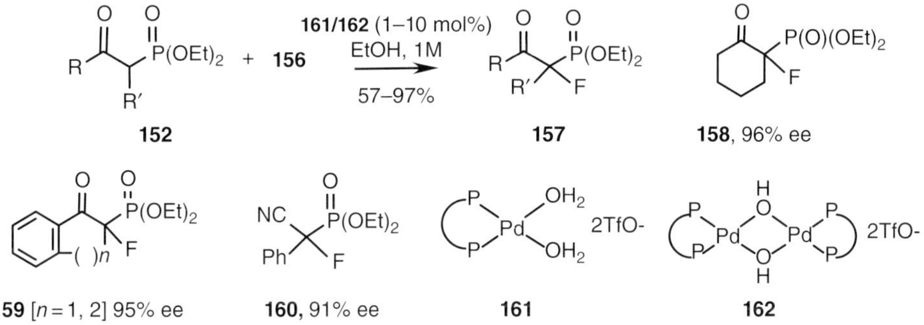

Scheme 4.51 Enantioselective fluorination of β-ketophosphonates by NFSI catalyzed by palladium complexes **161**, **162**.

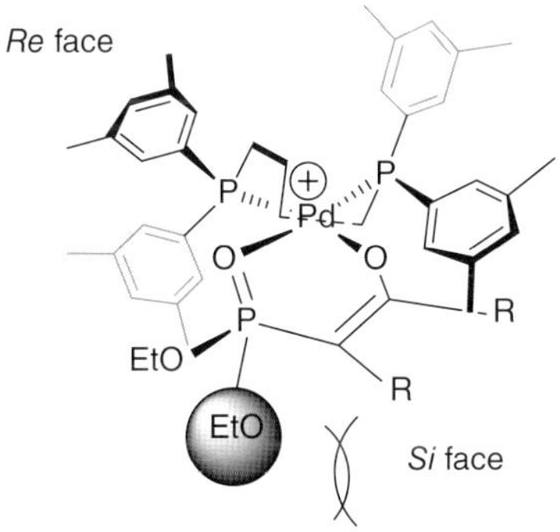

Figure 4.3 Plausible transition state model of catalytic fluorination.

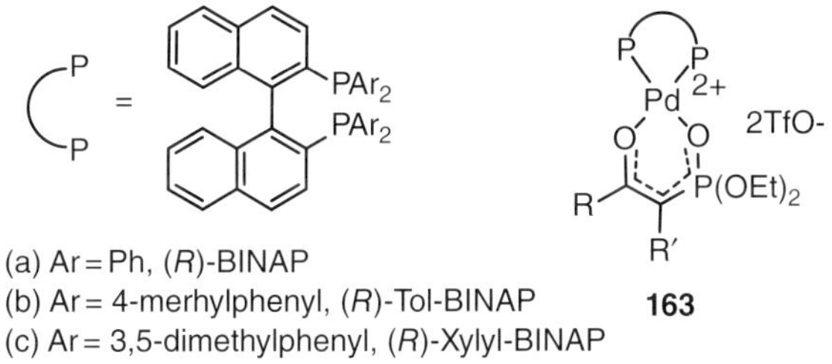

Scheme 4.52 Catalytic enantioselective chlorination and fluorination of β-ketophosphonates.

results were obtained with NSFI, which fluorinated β-ketophosphonates to afford the α-fluorinated β-ketophosphonates **157**, **159**, and **160** in yields of 50–93% and with an enantioselectivity of 87–97% ee. The complexes bearing ligands (*R*)-DM-BINAP or (*R*)-DM-SEGPHOS, which form chiral enolate complexes with β-ketophosphonates were the most effective for a broad range of palladium catalysts. Other catalysts provided low yields and moderate enantioselectivities (Scheme 4.52).

4.4.2.3 Amination

Kim *et al.* applied chiral palladium complexes **161** and **162** containing BINAP ligands for the catalysis of enantioselective amination of β-ketophosphonates **152**. The treatment of β-ketophosphonates with diethyl azodicarboxylate (DEAD) under mild reaction

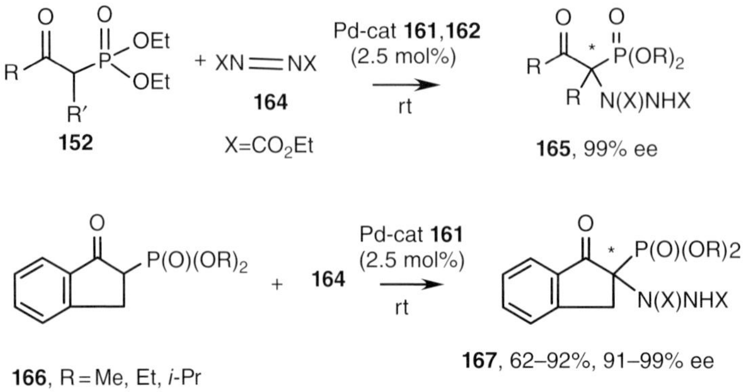

Scheme 4.53 Enantioselective amination of β-ketophosphonates, catalyzed by palladium complexes **165, 167**.

Table 4.7 The reaction of β-ketophosphonates with azodicarboxylate, catalyzed by (*S*)-**170**.

Entry	α-Keto phosphonate			Yield (%)	ee (%)
	R	R′	R″		
1	Ph	Me	Et	85	92
2	2-Np	Me	Et	93	92
3	Bn	Me	Et	60	95
4	Me	Me	Et	75	85
5	Ph	Allyl	Et	85	98
6	Ph	Me	Me	97	94
7	$(CH_2)_3$		Et	98	95
8	$(CH_2)_4$		Et	98	94

conditions, afforded substituted β-ketophosphonates **165** and **167** in satisfactory yields and with very good ee (Scheme 4.53) [95, 96].

Jørgensen reported enantioselective addition of β-ketophosphonates to imines or azodicarboxylates catalyzed by chiral complexes of copper or zinc with bis-oxazoline ligands **170–172** leading to the formation of chiral amination products. The reaction was performed in solvent (ether, methylene chloride, or ethylene dichloride) at room temperature. After deprotection, the corresponding optically active α-amino-β-hydroxyphosphonic acid derivatives were obtained in high yields and with enantioselectivities greater than 90% ee (Table 4.7) [97]. The (*R,R*) absolute configuration of optically active aminophosphonates were defined by X-ray monocrystal analysis (Scheme 4.54).

The asymmetric alkylation of phosphonates and phosphine oxides **173** by Grignard reagents catalyzed by copper complexes with chiral ligands TaniaPhos **176** or **177** leading to the formation of chiral phosphorus synthons **174, 175** has been reported (Scheme 4.55) [98].

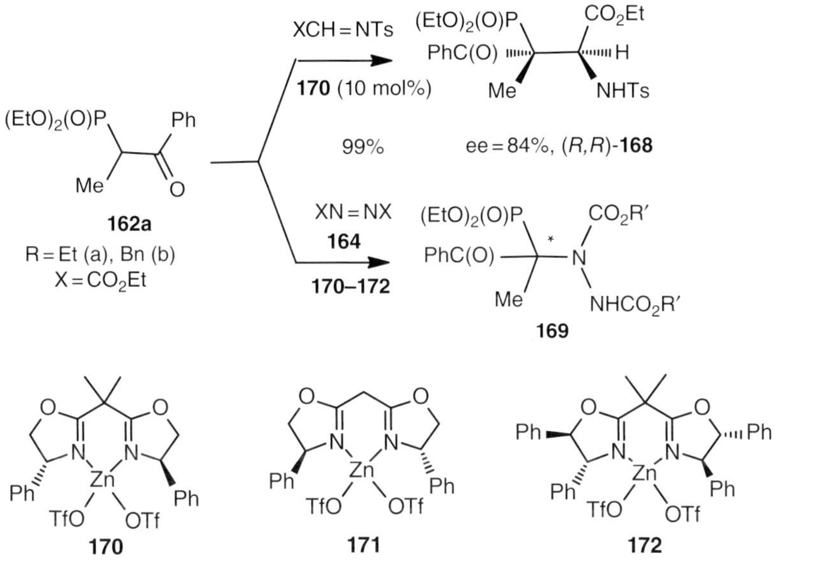

Scheme 4.54 Enantioselective addition of β-ketophosphonates to imines or azodicarboxylates.

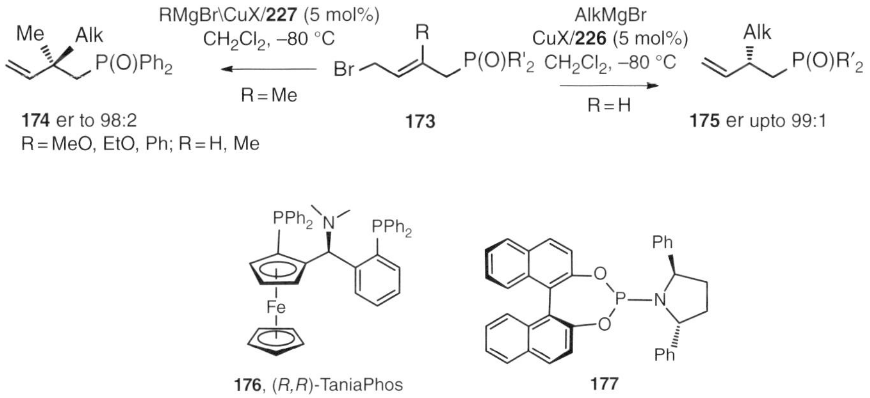

Scheme 4.55 The asymmetric alkylation of phosphine oxides **173** by Grignard reagents.

4.5 Nucleophilic Asymmetric Catalysis

4.5.1 Asymmetric Addition of Phosphorus Nucleophiles to Multiple Bonds

The fast development of the chemistry and biology of phosphonic acid derivatives over the last decade has been determined by the development of highly effective methods for their preparation. Chiral phosphonic acids can be prepared by various methods. The main method for the synthesis of phosphonates is the phosphonylation of carbonyl compounds, mainly via the phospha-aldol reaction, phospha-Mannich reaction, or phospha-Michael reaction [99].

4.5.1.1 Phospha-Aldol Reaction

Two types of phospho-aldol reaction are possible: (i) the reaction of dialkylphosphites with carbonyl reagents proceeding in the presence of a base catalyst, which shifts the P(O)H⇌P–OH tautomeric equilibrium toward the H(O)-form and (ii) the addition reaction of phosphoric acid triesters to carbonyl compounds proceeding in the presence of proton-donating reagents (phenol, carboxylic acids, hydrochlorides of aniline, etc.) or Lewis acids. The addition reaction of phosphoric acid esters with carbonyl compounds (the Abramov reaction) [99–103] involves two steps: the first step is the formation of a P–C bond, while the second step is the cleavage of the ester function with the formation of a phosphonyl group [99, 100]. In the case of chiral catalysts, the asymmetric version of the Abramov reaction is possible [101, 104, 105]. The asymmetric phospho-aldol reaction has been studied intensively because α-hydroxyphospohonates are important components of enzyme inhibitors. The preparation of hydroxyphosphonates also used catalytic methods including metallocomplex catalysis, organo-, and biocatalysis, leading to the formation of functionalized molecules with high enantiomeric purity and, therefore having high potential in synthetic chemistry (Scheme 4.56). Shibasaki described the first enantioselective hydroxyphosphonylation of aldehydes catalyzed by the heterobimetallic complexes ALB [Al,Li(binaphthoxide)$_2$] and Ln-Li-is(binaphthoxide) (LLB) [102], Shibuya used the Sharpless catalyst [103]. Spilling has tested complexes of chiral diols with Ti(OPr-*i*)$_4$ [104]. The best results were obtained in the reaction of dimethylphosphite with cinnamaldehyde, catalyzed by a complex of titanium isopropoxide with (*S*,*S*)-cyclohexanediol. Qian obtained analogous results (35–74% ee) in the Abramov reaction catalyzed by lanthanum complexes bearing BINOL ligands [106]. These results have already been reviewed (Scheme 4.56).

Double and triple asymmetric induction that use two or three chiral centers in reacting system allows to increase the stereoselectivity of phospha-aldol reactions (see Section 3.8). For example, the reaction of chiral di(1*R*,2*S*,5*R*)-menthyl phosphite with chiral 2,3-D-isopropylidene-(*R*)-glyceraldehyde catalyzed with chiral (*R*)-ALB proceeds under the stereochemical control of three chiral inductors and therefore results in the hydroxyphosphonates with 95% ee [107]. The bifunctional chiral Al(III)-BINOL **185** complex bearing two chiral centers provides the effective enantioselective hydrophosphonylation of aldehydes. A number of aromatic, heteroaromatic, α,β-unsaturated, and aliphatic aldehydes were phosphonylated in the presence of this catalyst with the formation of α-hydroxyphosphonates in yields of up to 99% and with enantioselectivity of up to 87% ee [108]. The bifunctional catalysts generated from chiral BINOL derivative in combination with Ti(O*i*-Pr)$_4$ and cinchona alkaloid provided the efficient catalyst for the asymmetric hydrophosphonylation of aldehydes with 91–99% ee (Schemes 4.57 and 4.58) [109, 110].

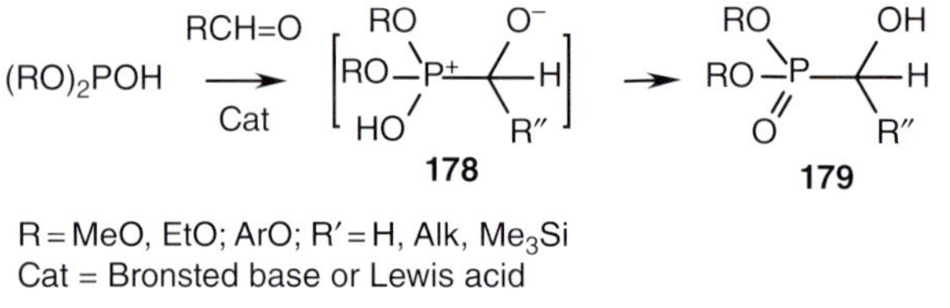

Scheme 4.56

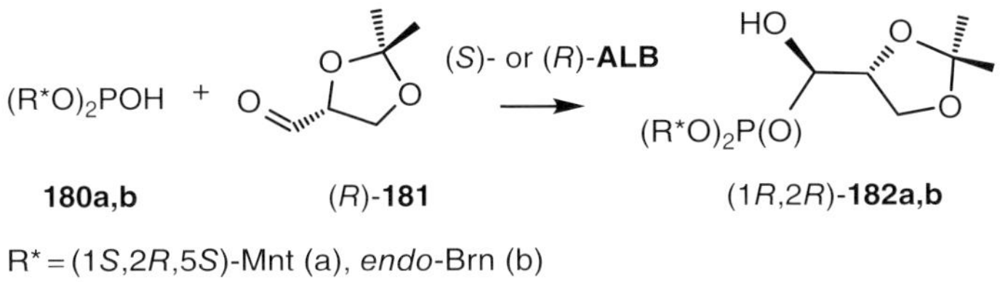

Scheme 4.57 Double and triple asymmetric induction of phospha-aldol reactions.

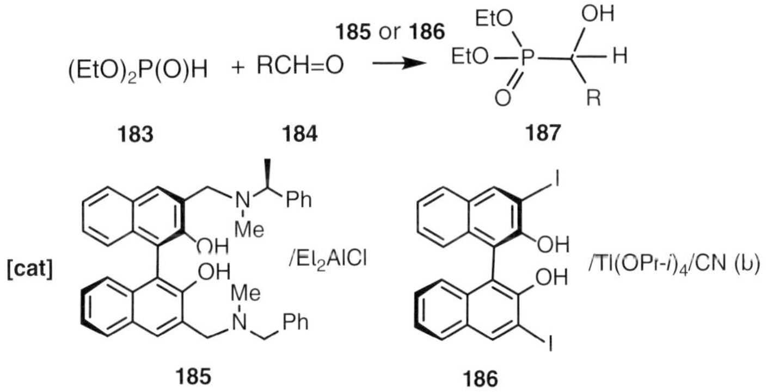

Scheme 4.58 Enantioselective co-operative catalysis of the Abramov reaction.

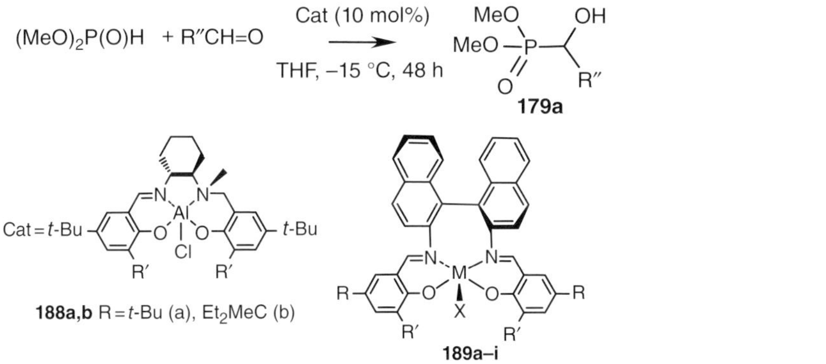

a: M = Co, R = R = F, X = I; **b**: M = Co, R = *t*-Bu, R′ = NO_2, X = I; **c**: M = Al, R = R′ = F, X = Cl; **d**: M = Al, R = *t*-Bu, R′ = NO_2, X = Cl; **e**: M = Al, R = R′ = *t*-Bu, X = Cl; **f**: M = Al, R = *t*-Bu, R′ = H, X = Cl; **g**: M = Al, R = *t*-Bu, R′ = Br, X = Cl; **h**: M = Al, R = *t*-Bu, R′ = I, X = Cl; **i**: M = Al, R = *t*-Bu, R′ = OMe, X = Cl

Scheme 4.59 Asymmetric hydrophosphonylation of aldehydes catalyzed by Al(salalen)-complexes.

Optically active aluminum-salalen complexes **188** enantioselectively catalyze the hydrophosphonylation of aldehydes yielding the corresponding α-hydroxyphosphonates [110] (Scheme 4.59 and Table 4.8). Kee *et al.* [111] have reported that α-hydrophosphonylation of aromatic aldehydes Al(salen) and Al(salan) complexes containing cyclohexandiamine substituents, display moderate enantioselective effect to 61% ee, X-ray analysis of Al(salan) complexes showed that they possess di-*m*-hydroxo structure

Table 4.8 Asymmetric hydrophosphonylation of aldehydes with catalyst **188a**.

Entry	R	R′	Yield (%)	ee (%)	Configuration	Reference
1	*p*-$O_2NC_6H_4$	*t*-Bu	95	94	*S*	[109]
2	*p*-ClC_6H_4	*t*-Bu	88	88	*S*	[109]
3	*p*-$MeOC_6H_4$	*t*-Bu	87	81	*S*	[109]
4	*o*-ClC_6H_4	*t*-Bu	96	91	—	[109]
5	(*E*)-PhCH=CH	*t*-Bu	77	83	*S*	[109]
6	$PhCH_2CH_2$	*t*-Bu	94	91	—	[109]
7	$(CH_3)_2CH$	*t*-Bu	89	89	—	[109]
8	CH_3CH_2	*t*-Bu	61	89	*S*	[109]
9	*p*-$O_2NC_6H_4$	Et_2MeC	98	98	—	[110]
10	*p*-ClC_6H_4	Et_2MeC	95	98	—	[110]
11	*o*-$ClCC_6H_4$	Et_2MeC	94	97	—	[110]
12	(*E*)-PhCH=CH	Et_2MeC	97	95	—	[110]
13	$PhCH_2CH_2$	Et_2MeC	93	97	—	[110]

and the salan ligand occupies the *cis*-β-conformation. These results proved that the chiral complex possessing two coordination centers in *cis*-position can be catalyst of asymmetric hydrophosphonylation [111, 112].

Katsuki *et al* found that the addition of potassium carbonate significantly enhanced the reaction rate of the Al(salalen)-catalyzed asymmetric hydrophosphonylation of aldehydes with dimethyl phosphonate. The enantioselectivity of hydrophosphonylation increases to 93–98% ee even if the catalyst loadings reduces by several times. Chemists have studied the mechanism of the hydrophosphonylation reaction between dimethylphosphite and benzaldehyde catalyzed by the Al(salalen)complex, using density functional theory (DFT) and ONIOM methods. They came to the conclusion that the stereochemistry of the reaction catalyzed by the chiral Al(salalen) complex is controlled by the steric repulsion between the ortho *t*-Bu groups of the ligand and dimethylphosphite, as well as the coordination mode of dimethylphosphite to the catalyst. The calculations reproduce the predominant product in (*S*)-configuration with high ee, which accounts well for the experimental observations. The high enantioselectivity of the complex is explained by its unique structure, which has a deformed trigonal-bipyramidal configuration that allows salalen ligand to occupy the cisoid position in which the chiral amino group is located close to the aluminum atom [112, 113]. The replacement of the achiral cyclohexyl group in complex **188** on chiral (*R*)-binaphthyl, increasing the molecular asymmetry of catalyst **189**, promoted an increase in enantioselectivities of the reaction [114]. On the whole, the phosphonylation of aromatic aldehydes is carried out more enantioselectively than that of the aliphatic aldehydes (Table 4.8). Katsuki *et al.* [109, 110] studied in detail the chiral, trigonal-pyramidal metal (salalen) complexes **189a–i**. Complexes **189a,b**, containing cobalt were inactive. At the same time, the aluminum complexes **189c–i** catalyzed enantioselective hydrophosphonylations of dimethylphosphite with various aldehydes [109, 110, 114]. The complexes **189** possessing *tert*-butyl group in C3 and C3′ positions catalyzed the reaction in good yields and with good enantioselectivity. Substituents in

Table 4.9 Asymmetric hydrophosphonylation of aldehydes with catalysts **189c–i**.

Entry	189c–i	R″	Time (d)	Yield (%)	ee (%)	Configuration
1	**c**	*p*-ClC_6H_4	4	29	11	*R*
2	**d**	*p*-ClC_6H_4	4	86	76	*R*
3	**e**	*p*-ClC_6H_4	3	100	65	*R*
4	**f**	*p*-ClC_6H_4	5	83	68	*R*
5	**g**	*p*-ClC_6H_4	3	78	84	*R*
6	**h**	*p*-ClC_6H_4	3	69	78	*R*
7	**i**	*p*-ClC_6H_4	3	68	83	*R*
8	**g**	Ph	5	62	79	*R*
9	**g**	*p*-FC_6H_4	5	69	82	—
10	**g**	*p*-An	2	55	79	*R*
11	**g**	*p*-Tl	2	82	80	*R*
12	**g**	*o*-FC_6H_4	2	69	80	—
13	**g**	*o*-Tl	2	79	75	—
14	**g**	$PhCH_2CH_2$	1	71	83	—
15	**g**	*c*-Hex	1	86	86	—
16	**g**	*n*-Hex	2	79	86	—
17	**g**	(*E*)-PhCH=CH	2	82	64	*R*

C5 and C5′ positions also effected good yields and enantioselectivities, although to a lesser degree. Complexes **189** containing Br at C3, were the most effective catalysts from the viewpoint of enantioselectivity (84% ee). Complexes **189** containing the electron-donating MeO group also initiated high enantioselectivity, although in low chemical yield. Electron-donating groups in *para*-position to OH group and also increase in volume of the *ortho*-substituents R′ of aromatic substituent salalen complex (*t*-Bu, Ad, Et_2MeC) increased the enantioselectivity of the reaction (Scheme 4.59, Table 4.9).

Chiral complexes of Al(III) with tridentate Schiff bases **192a–e** catalyzed the asymmetric hydrophosphonylation of aldehydes and trifluoromethylketone without any side reactions. Iron/camphor-based tridentate Schiff base complexes $[FeCl(SBAIB\text{-}d)]_2$ produces as well α-hydroxyphosphonates in high yields and excellent enantioselectivities (up to 99%) [115] (Scheme 4.60).

Dimethyl 4-chlorophenyl-1-hydroxymethylphosphonate **179**. The Al(salalen)-complex **188a** (0.02 mol) and dimethyl phosphite (0.21 mmol) were dissolved in THF (1.0 ml) and the solution was cooled to −15 °C. Then the solution of *p*-ClC_6H_4CHO (0.20 mmol) in THF was added and the reaction mixture was stirred for 48 h at −15 °C followed by further stirring for 1 h at room temperature. The reaction mixture was quenched with a water solution of NH_4Cl and extracted with ethyl acetate. The organic extract was filtered and dried with Na_2SO_4. After evaporation of the solvent under reduced pressure, the residue was chromatographed on silica gel with hexane-ethylacetate as eluent to give the corresponding α-hydroxyphosphonate (yield 72%, 80% ee) [110].

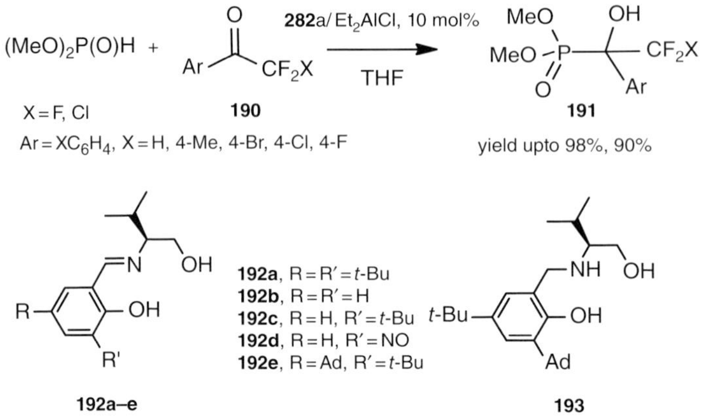

Scheme 4.60 The asymmetric hydrophosphonylation of aldehydes catalyzed by chiral complexes **282a–e**.

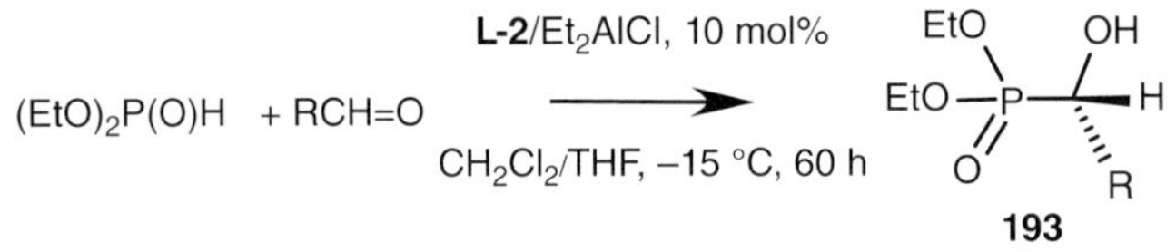

R = Ph	96%	95% ee	(*S*)
R = Tl	89%	97% ee	(*S*)
R = 4-An	94%	97% ee	(*S*)
R = 4-ClC$_6$H$_4$	82%	95% ee	(*S*)
R = 2-thienyl	90%	93% ee	(*S*)
R = 2-furyl	89%	94% ee	(*S*)

Scheme 4.61 Abramov reaction catalyzed by complexes **193**/Et$_2$AlCl.

The chiral ligand influenced the enantioselectivity and the derivatives of L-valinol catalyzed the reaction of dialkylphosphites with various alkylaldehydes, arylaldehydes, and trifluoromethylketones with higher enantioselectivity, than other ligands [115–118]. Ligands with large *ortho*-substituents on the phenol group, such as adamantyl, also raised enantioselectivities (to 85% ee). The reaction with acetophenone proceeded with low enantioselectivity, but in high yields. Counterions of complexes **192**/Et$_2$AlX where X = Cl, Et, *i*-PrO also influenced the reaction enantioselectivity; the highest enantioselectivities were attained with Et$_2$AlCl (Scheme 4.61).

Yamamoto and Abell [119] synthesized α-hydroxy- and α-aminophosphonates in high yields and with high enantioselectivity using the bis(8-chinolinato) (TBO$_x$)Al complex **195** (0.5–1 mol%). Under the optimized reaction conditions, the electron-rich aldehydes were more reactive and more selective than electron-deficient aromatic aldehydes. Aliphatic aldehydes also reacted as well with satisfactory selectivity. All reactions proceeded very rapidly, in high yields and enantioselectivities. Reducing the catalyst loading to 0.5 mol% did not influence the enantioselectivity or yield. The best results were obtained with phosphites containing trifluoroethyl groups (Scheme 4.62).

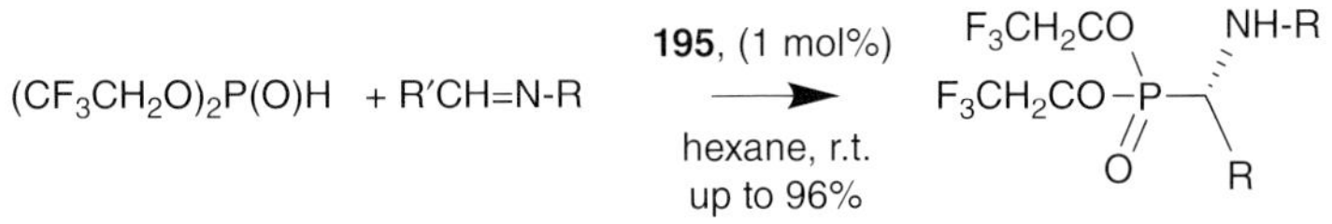

R′=Aryl, Heteroaryl

194, upto 97% ee

R′ = Ph	95%	ee = 96%
R′ = 2-naphthyl	98%	ee = 95%
R′ = 4-ClC$_6$H$_4$	94%	ee = 95%
R′ = 4-BrC$_6$H$_4$	96%	ee = 95%
R′ = 4-NO2C$_6$H$_4$	93%	ee = 92%
R′ = 4-MeOC$_6$H$_4$	93%	ee = 97%
R′ = 4-MeC$_6$H$_4$	94%	ee = 94%
R′ = 3-MeOC$_6$H$_4$	93%	ee = 95%
R′ = 2-MeOC$_6$H$_4$	93%	ee = 93%
R′ = 2-MeC$_6$H$_4$	95%	ee = 95%
R′ = *c*-hexyl	95%	ee = 82%

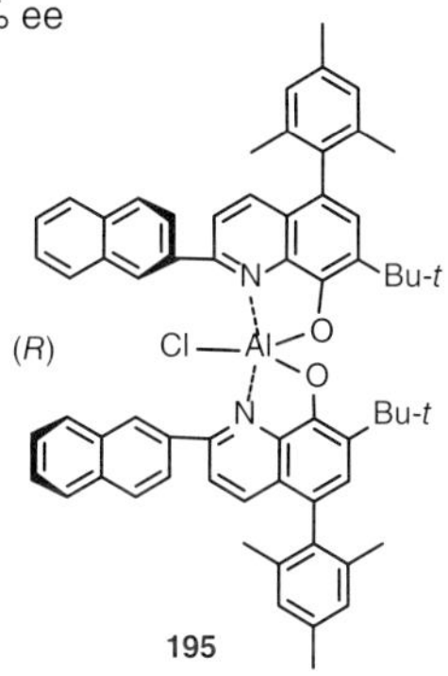

195

Scheme 4.62 Synthesis of α-hydroxy and α-aminophosphonates using the bis(8-chinolinato) (TBO$_x$)Al complex **195**.

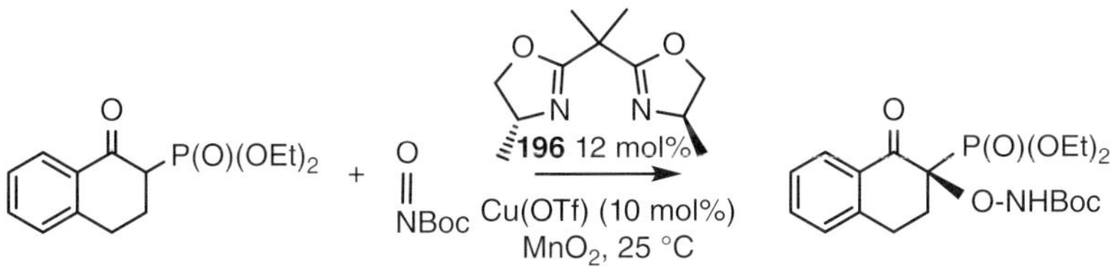

Scheme 4.63 Enantioselective copper-catalyzed hydrophosphonylation of aldehydes.

A highly enantioselective copper-catalyzed hydrophosphonylation of aldehydes in the presence of bis(oxazoline) ligand proceeded smoothly under mild conditions and resulted in α-hydroxyphosphonates in high yields and enantioselectivities up to 98% ee [120]. The copper-catalyzed α-oxidation of β-ketophosphonates using *in situ* generated nitrosocarbonyl compounds as an electrophilic oxygen source leads to the formation of the tertiary α-hydroxy phosphonic acid derivatives in high yields (up to 95%) and enantioselectivities (up to >99% ee). This method was also applied for the synthesis of α,β-dihydroxyphosphonates and β-amino-α-hydroxyphosphonates (Scheme 4.63) [121].

4.5.1.2 Phospha-Mannich Reaction

The synthesis of enantioenriched α-aminophosphonates and aminophosphonic acids by means of asymmetric catalytic hydrophosphonylation of imines or by an asymmetric reaction has attracted constant interest [122–124]. A special case of the phospha-Mannich is the Kabachnik–Fields reaction representing a one-pot, three-component procedure, which includes carbonyl compound, amine, and dialkyl phosphite [122]. For asymmetric aminophosphonylation, catalysts such as chiral metal complexes, LLB, chiral thioureas, BINOL phosphoric acid, and quinine [123, 125] have been used successfully. Shibasaki *et al.* reported examples of asymmetric Mannich

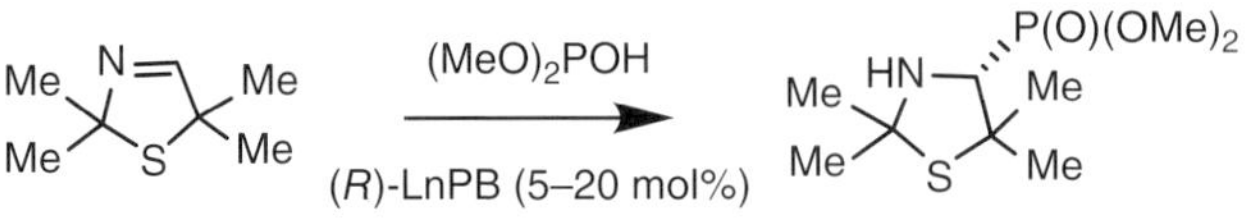

(S)-**197**

Scheme 4.64 Asymmetric addition of dimethylphosphite to thiazolines catalyzed by LnPB.

reactions. The asymmetric addition of dimethyl phosphite to imines catalyzed by LLB, LPB, and LSB bimetallic complexes resulted in the formation of aminophosphonates with 50–96% ee.

Martens *et al.* [124] have described the diastereoselective hydrophosphonylation of cyclic imines using heterobimetallic lanthanum BINOL complexes LLB as chiral inductors with a satisfactory diastereoselectivity (dr 95 : 5) (Scheme 4.64).

The BF_3-activated addition of binaphthol-phosphite to 3-thiazoline yielded almost exclusively 4-thiazolidinylphosphonates the structures of which were proven by X-ray crystal analysis. Theoretical studies of the mechanism of the enantioselective reaction by a two-layer ONIOM (B3LYP/6-31G (d)/AM1) method showed that the reaction proceeds in two-steps, involving proton transfer and nucleophilic addition, which is the stereo-controlling step. The energy differences between *Si*-facial attack and *Re*-facial attack are significant for the hydrophosphonylation of the aldimine. Feng *et al.* [126, 127] reported that the *N,N′*-dioxide (**198**)/Sc(III) complex catalyzes three-component Kabachnik–Fields reactions, yielding the corresponding α-aminophosphonates with good yields and enantioselectivities of up to 87% ee (Scheme 4.65). A one-pot Kabachnik–Fields reaction catalyzed by the optically active aluminum salen complex **188** also proceeded with good enantioselectivities to afford the corresponding α-aminophosphonates **189** and **201** with high enantioselectivity [128]. The highest enantioselectivities were attained with alkynyl- or alkenylaldimines prepared from phenylpropargylaldehyde and diphenylmethylamine with aldimines and containing an R = 4-methoxy-3-methylphenyl group at nitrogen. The high catalytic activity of the complex was attributed to its unique structure: the distorted trigonal bipyramidal configuration allowed the salen ligand to take a *cis*-like structure wherein the chiral amino group was located close to the metal center (Schemes 4.66 and 4.67).

The sterically constrained 8-aluminum bis-(TBO_x) complex **195** displayed high enantioselectivity in the reaction of dialkyl phosphites with aldimines [119]. The hydrophosphonylation of aldimines, substituted with various groups and catalyzed by

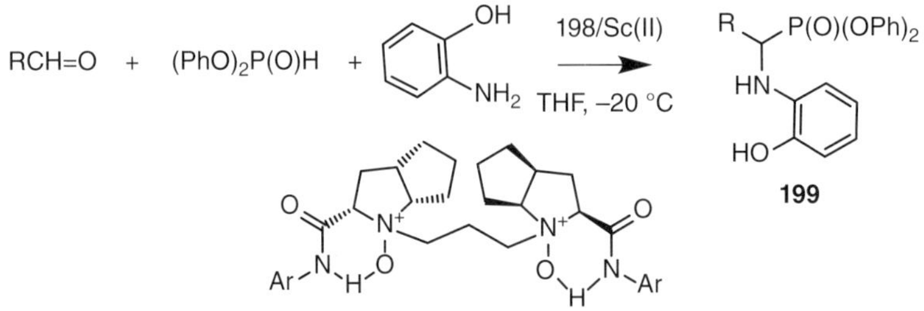

198, Ar = 2,6-*i*-$Pr_3C_6H_3$

Scheme 4.65 Asymmetric Kabachnik–Fields reactions catalyzed by complex (**198**)/Sc(III).

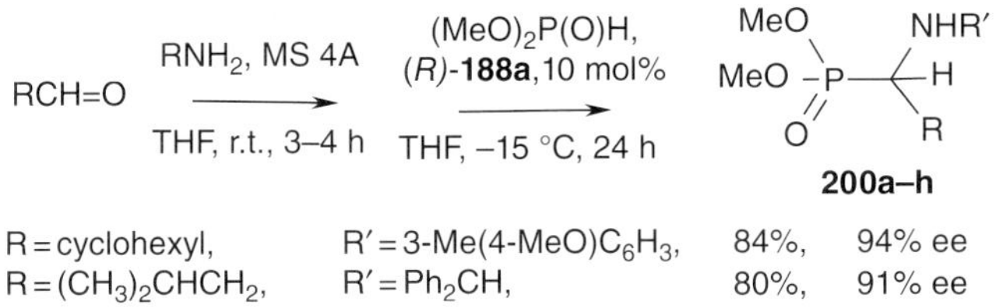

Scheme 4.66

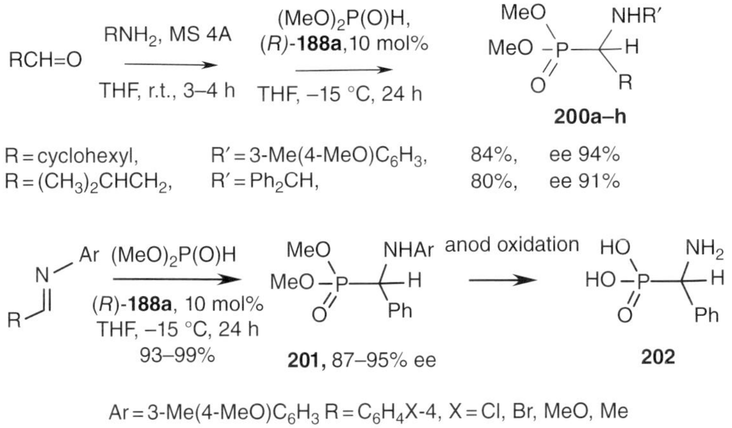

Scheme 4.67 One-pot Kabachnik–Fields hydrophosphonylation of aldimines.

complex **195**, proceeded with high enantioselectivity even when the catalyst loading was decreased to 0.5–1 mol%. The enantioselectivity of the reaction depended on the size of the substituent at the nitrogen atom. Electron-rich aldimines showed higher activity while the reaction of electron-deficient aldimines proceeded with low enantioselectivity, although with high yields of aminophosphonates. Cyclic (*R*)-BINOL phosphoric acids used as chiral Brønsted acids (10 mol%) catalyzed the hydrophosphonylation of aldimines with diisopropyl phosphite at room temperature to give aminophosphonates with good to high enantioselectivities (Scheme 4.68).

The addition of phosphoric nucleophile to imines on the Mannich-type reaction represents a useful method for the preparation of enantioenriched α-aminophosphonates [129, 130]. Complexes $Cu(OTf)_2$ with diamine **208** are effective catalysts for this reaction, yielding aminophosphonates **209** in high yields and with good enantioselectivity. The addition of 3-Å MS and slow addition of the substrates were found to improve the yield, selectivity, and reproducibility (71% yield, 86% ee). The adducts **209** were transformed into α-aminophosphonate **210**, an intermediate for the synthesis of inhibitors of endothelin-converting enzymes as shown in Scheme 4.69. Endothelin-converting enzymes are involved in the proteolytic processing of endothelin-1 (EDN1), endothelin-2 (EDN2), and endothelin-3 (EDN3) to biologically active peptides (Scheme 4.69).

Zhao and Dodda have developed an enantioselective method for the synthesis of enantioenriched α-aminopropargylphosphonates **211**. High yields and good enantioselectivities (60–81% ee) were achieved using a complex of monovalent copper and pybox

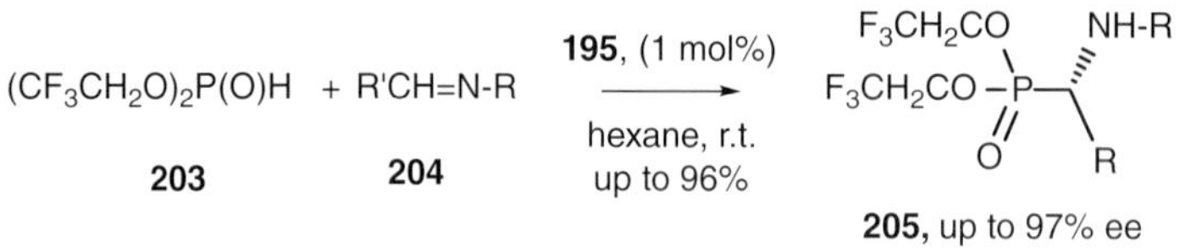

R'=Aryl, Heteroaryl

R=Ph	98%	ee = 96%
R=3-An	93%	ee = 98%
R=4-An	91%	ee = 90%
R=4-ClC_6H_4	85%	ee = 90%
R=4-BrC_6H_4	88%	ee = 92%
R=4-$NO_2C_6H_4$	90%	ee = 88%
R=4-Tl	92%	ee = 96%
R=2-Tl	96%	ee = 92%
R=2-furyl	89%	ee = 91%
R=2-thienyl	93%	ee = 94%

Scheme 4.68 Hydrophosphonylation of aldimines, catalyzed by Al bis-(TBO_x) complex.

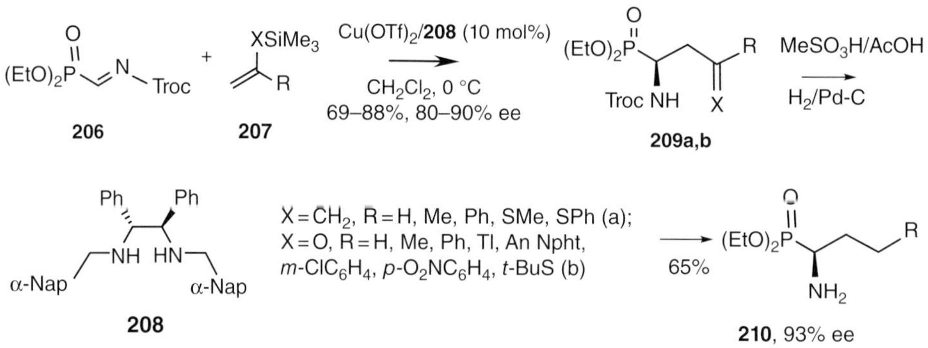

Scheme 4.69 The Mannich addition reaction of alkenes to iminophosphonates.

ligand **212** as the catalyst. A loading of 2 mol% of catalyst **212** was sufficient enough to give a reasonable conversion. Under the optimized reaction conditions (2 mol% of loading catalyst, $CHCl_3$ as solvent, room temperature), the reaction was successful with various terminal alkynes (Scheme 4.70) [131, 132, 58].

4.5.1.3 Phospha-Michael Reaction

The phospha-Michael reaction is a nucleophilic addition of phosphoric anion to an activated multiple bond and is one of the most useful methods for the formation of P–C bonds [133]. The asymmetric phospha-Michael reaction can be accomplished by two main methods: organometallic and organocatalytic. Organometallic catalysis is the most convenient for primary and secondary phosphine additions to activated C=C bond with the formation of chiral tertiary phosphines. The addition of primary or secondary phosphines to electron-deficient alkenes is a convenient method for the preparation of chiral tertiary phosphines of interest as ligands for complexes with transition metals [134]. For example, Gluck described the addition of secondary phosphines to activated

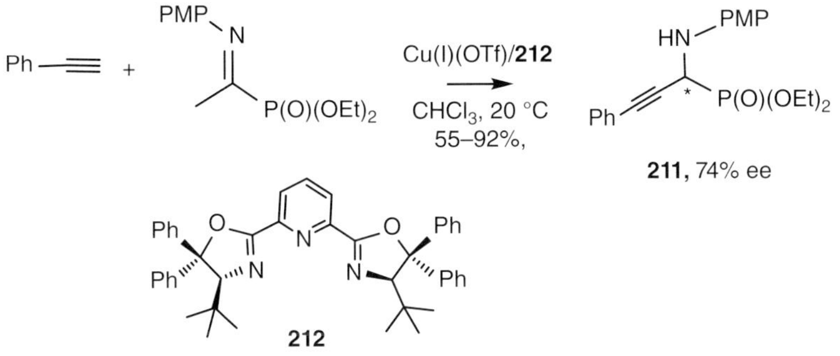

Scheme 4.70 Enantioselective synthesis of enantioenriched α-aminopropargylphosphonates.

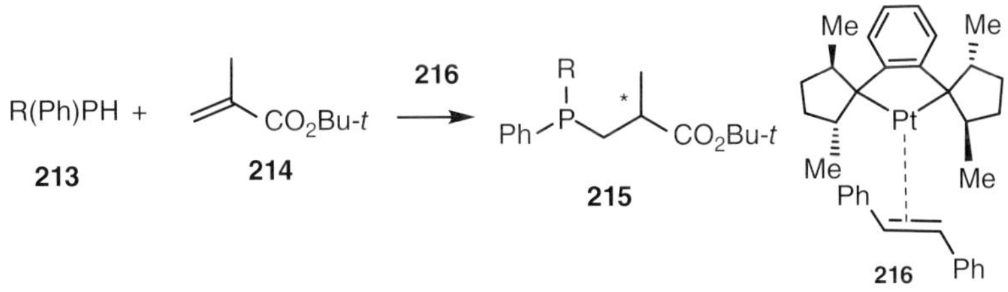

Scheme 4.71 Asymmetric hydrophosphination of olefins catalyzed by Pt complex **216**

olefins, catalyzed by a platinum complex bearing the chiral ligand Pt(*R,R*)-Me-Duphos) *trans*-stilbene), which resulted in chiral phosphines **215** in good yields and moderate enantioselectivity [134] (Scheme 4.71). Later Togni *et al.* [136, 137] reported on an asymmetric hydrophosphination reaction of methacrylonitrile with secondary phosphines, catalyzed by the dicationic nickel complex [Ni(Pigiphos)(THF)]$(ClO_4)_2$. The reaction led to the formation of chiral 2-cyanopropylphosphines **217** in good yield and ee's up to 94%. The absolute configuration of the obtained tertiary phosphine was defined as (*S*) (Scheme 4.72). The authors proposed a mechanism involving coordination of methacrylonitrile to the dicationic nickel catalyst followed by a 1,4-addition of the phosphine and then rate-determining proton transfer: A = methacrylonitrile ligand, B = Ni ketenimine intermediate, and C = Ni-coordinated hydrophosphination product. The mechanism was supported by an experimentally determined rate law, a large primary deuterium isotope effect kH/kD 4.6(1) for the addition of t-$Bu_2PH(D)$, the isolation of the species $[Ni(k3\text{-}Pigiphos)(kN\text{-}methacrylonitrile)]^{2+}$, and DFT calculations of model compounds.

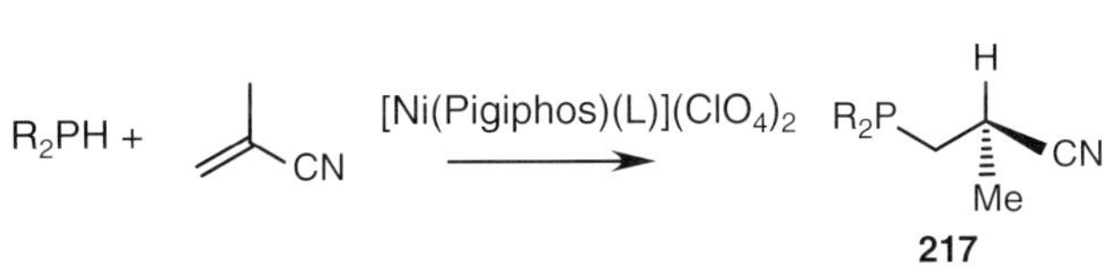

R = Cy, 71%, 70% ee; R = Ph,10%, 32% ee; R = *i*-Pr, -%, 70% ee; R = *t*-Bu, 87%, 89% ee;
R = 1-Ad, 95%, 94% ee

Scheme 4.72 Enantioselective phospha-Michael addition of secondary phosphines, catalyzed by organonickel complex.

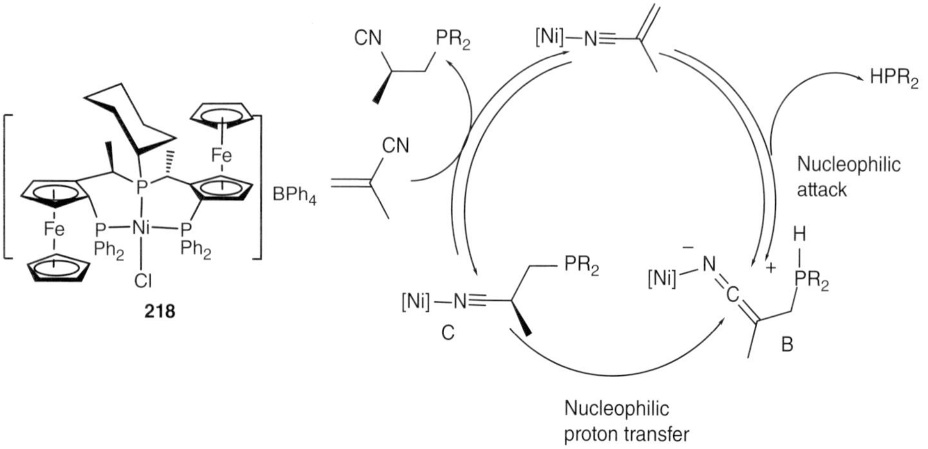

Scheme 4.73 (a) Pigiphos-nickel(II) complex **335** and (b) catalytic cycle for the hydrophosphination of methylacrylonitrile by the complex [Ni (3-Pigiphos) ($NCMeCCH_2$)].

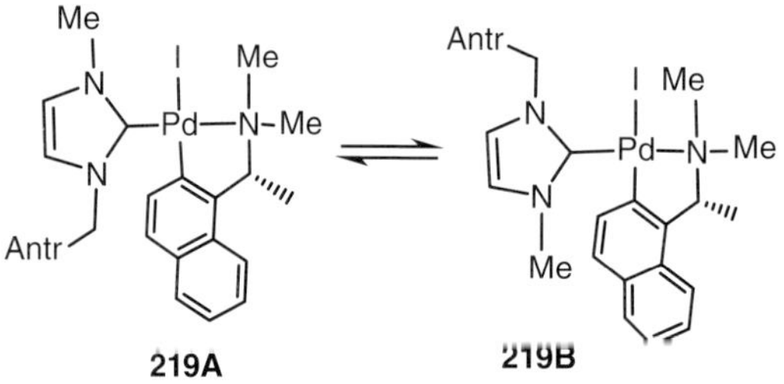

Scheme 4.74 Atropoisomerism **219A⇌219B** of complexes.

This mechanism assumes stereospecific transfer of a proton, reversible bonding P–C, and also the formation of an unusual Ni ketenimine intermediate ([137] (Scheme 4.73)).

Sabater developed interesting chiral palladacycles with *N*-heterocyclic carbene ligands starting from commercially available enantiomerically pure benzylamines. These complexes exist in the form of two atropisomers **219A⇌219B** that were isolated using column chromatography and characterized by NMR spectroscopy (Scheme 4.74) [138, 139]. The epimerization of the isolated complexes slowly occurs in solution. The authors used complexes **219** as catalysts in the 1,4-additions of diarylphosphines to α,β-unsaturated ketones, which resulted in the formation of tertiary phosphine oxides **221** in good yields and with good enantioselectivities (63–93% yields, 90–99% ee) (Scheme 4.75).

Cyclic palladium complexes **224** catalyzed diastereo- and enantioselective addition of primary and secondary phosphines to enones and enamines [140, 141]. The reaction of bis(enones) with phenyl phosphine allows to carry out intermolecular construction of chiral tertiary phosphoric heterocycles **223** by the one-pot method in high yields and with high enantioselectivities as shown in Scheme 4.76. This highly active, chemo- and enantioselective reaction was used for the synthesis of a number of chiral tertiary enaminophosphites **225** [142] (Scheme 4.77).

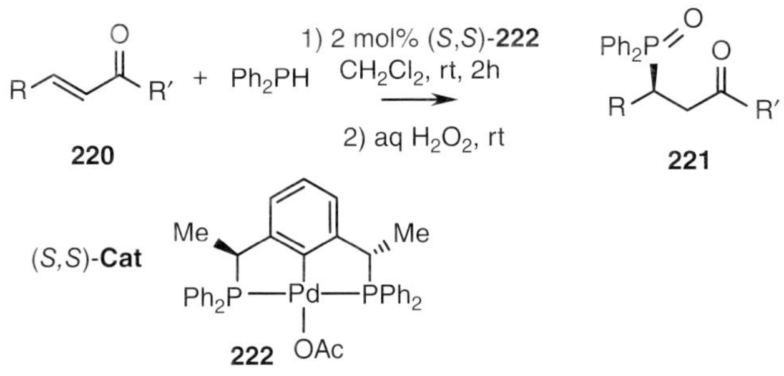

219	R = Ph	R′ = Ph	91%	ee = 56%
222	R = Ph	R′ = Ph	93%	ee = 99%
222	R = *p*-BrC_6H_4	R′ = Ph	89%	ee = 99%
222	R = *p*-$MeOC_6H_4$	R′ = Ph	75%	ee = 98%
222	R = *m*-BrC_6H_4	R′ = Ph	93%	ee = 97%
222	R = *p*-$O_2NC_6H_4$	R′ = Ph	78%	ee = 95%
222	R = Ph	R′ = p-BrC_6H_4	90%	ee = 98%
222	R = Ph	R′ = p-$O_2NC_6H_4$	88%	ee = 99%
222	R = Ph	R′ = m-BrC_6H_4	90%	ee = 99%
222	R = Ph	R′ = o-$MeOC_6H_4$	69%	ee = 90%
222	R = Ph	R′ = p-MeC_6H_4	63%	ee = 90%
222	R = Me	R′ = p-BrC_6H_4	71%	ee = 96%

Scheme 4.75 Asymmetric addition of diarylphosphines to enones catalyzed by palladium complexes (*S,S*)-**219** or (*S,S*)-**222**.

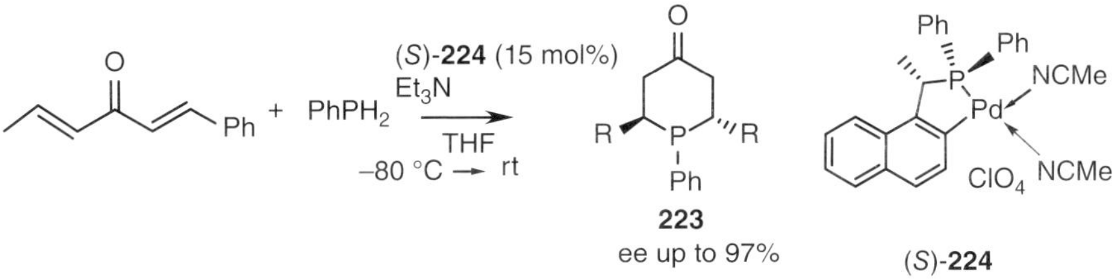

Scheme 4.76 The synthesis of chiral tertiary phosphoric heterocycles **223**.

Addition of $R_2P(O)H$ compounds to the C=C bond proceeded more easily than the addition of secondary phosphines and furnished usually products with high yields and enantioselectivities [135, 143–145]. For example, Ishihara reported recently an enantioselective 1,4-hydrophosphinylation of α,β-unsaturated esters with diaryl phosphine oxides and an enantioselective 1,2-hydrophosphonylation of α,β-unsaturated ketones with dialkyl phosphites by the use of chiral agnesium(II) binaphtholate aqua complexes as cooperative Brønsted/Lewis acid–base catalysts [143]. Wang *et al.* [135, 144, 145] reported the asymmetric 1,4-addition reaction of diethyl phosphite to simple enones catalyzed by a dinuclear zinc complex **226**, leading to the formation of ketophosphonates **227** in high yields and with enantioselectivities up to 99% ee (Scheme 4.78). This catalytic phospha-Michael reaction was screened for a number of β-aryl- or

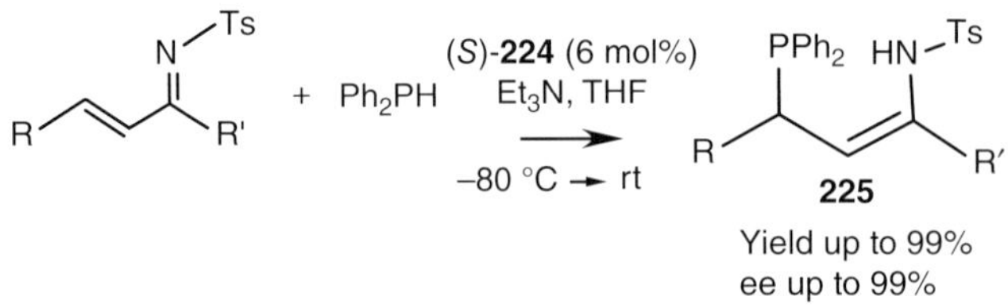

R = Ph	R′ = Ph	99%	ee = 99%
R = 4-ClC$_6$H$_4$	R′ = Ph	96%	ee = 92%
R = 4-FC$_6$H$_4$	R′ = Ph	97%	ee = 97%
R = 4-MeOC$_6$H$_4$	R′ = Ph	98%	ee = 97%
R = Ph	R′ = CH=CHPh	97%	ee = 91%
R = Ph	R′ = 2-thienyl	97%	ee = 91%

Scheme 4.77 Addition of diphenylphosphine to α,β-nonsaturated imines, catalyzed by complex (*S*)-**225**.

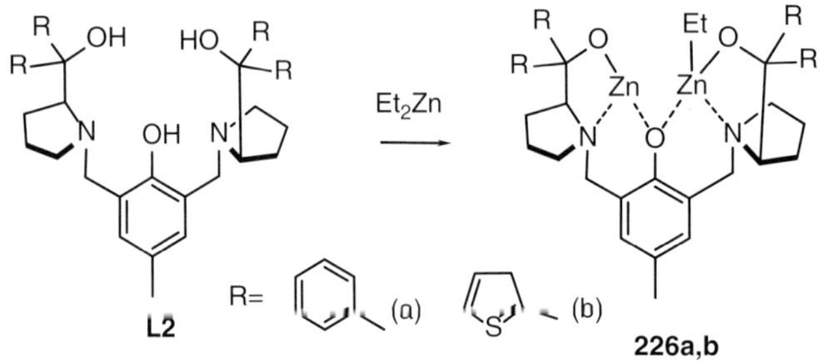

Scheme 4.78 Dinuclear zinc complexes **226**.

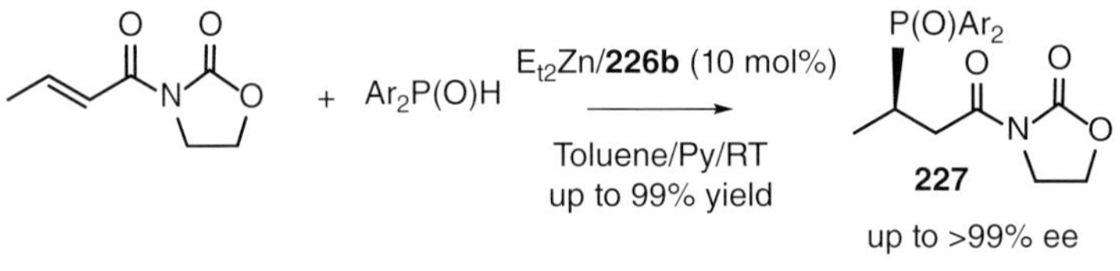

Scheme 4.79

alkyl-substituted enones, which afforded adducts **228** with good enantioselectivities (Schemes 4.79 and 4.80) [135, 144–149].

The Michael addition of dialkyl phosphites to nitroalkenes in the presence of the lithium aluminum bis(binaphthoxide)complex (*S*)-ALB afforded β-nitrophosphonates **229**, precursors of β-aminophosphonic acids with good enantioselectivities. The reaction was performed in toluene at room temperature at 15 mol% loading of the catalyst [150] (Scheme 4.81). The catalytic enantioselective conjugate addition reaction of α-fluoro β-ketophosphonates to nitroalkenes promoted by chiral nickel complexes afforded the corresponding Michael adducts **230** containing fluorinated quaternary stereogenic centers with excellent enantioselectivity (up to >99% ee) (Scheme 4.82) [151].

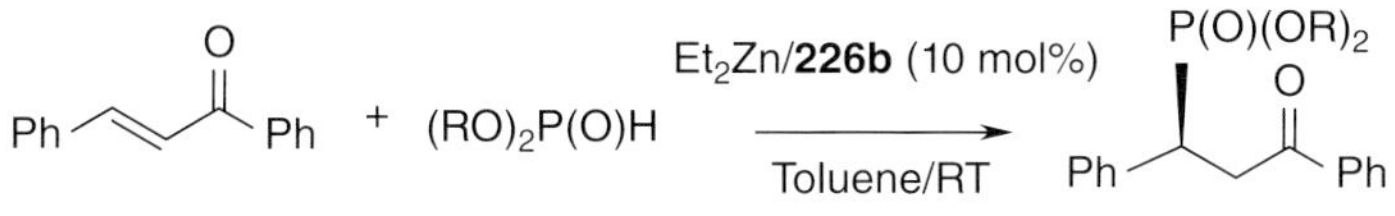

R=Et (a), Me (b), Pr (c)

228a,b

228a	60%	ee = 25%
228a	78%	ee = 40%
34 6b	54%	ee = 75%
228b	55%	ee = 99%
228b	60%	ee = 79%

Scheme 4.80 Asymmetric reactions of 1,4 additions of $R_2P(O)H$ compounds to enones, catalyzed by zinc complexes **226**.

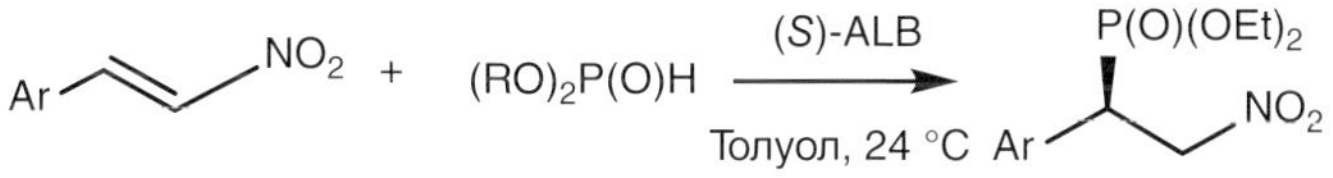

229

Ar = 4-MeOPh	R = Et	74%	ee 97%
Ar = 3,4-(OMe)$_2$Ph	R = Et	54%	ee 84%
Ar = Ph	R = Et	48%	ee 75%
Ar = 4-ClPh	R = Et	52%	ee >99%
Ar = 4-NO$_2$Ph	R = Et	45%	ee >99%
Ar = 2-Furyl	R = Et	20%	ee >99%
Ar = 4-OMePh	R = Me	18%	ee >99%
Ar = Ph	R = Me	24%	ee >99%
Ar = 4-ClPh	R = Me	10%	ee >99%

Scheme 4.81 The Michael addition of dialkyl phosphites to nitroalkenes, catalyzed by (*S*)-ALB.

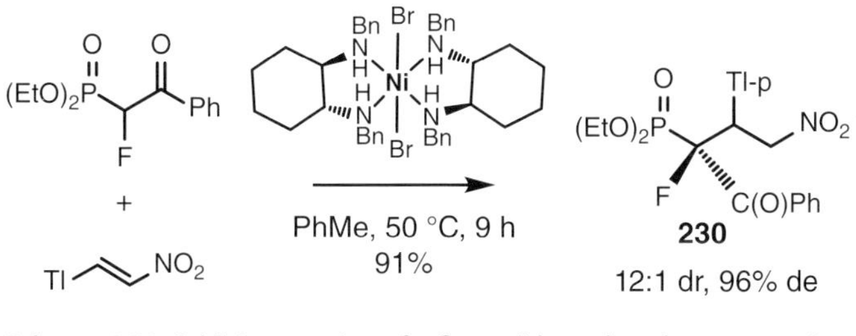

Scheme 4.82 Addition reaction of α-fluoro β-ketophosphonates to nitroalkenes.

4.6 Cycloaddition Reactions

One of the most useful and interesting applications of vinylphosphonates are cycloaddition reactions, which provide easy access to highly functionalized and complex molecules. The Diels–Alder cycloadditions of vinylphosphine oxides are certainly among the most interesting cycloaddition reactions. In this case, the vinylphosphorus group can react as a dienophile, or to be a part of diene. Evans used chiral Lewis

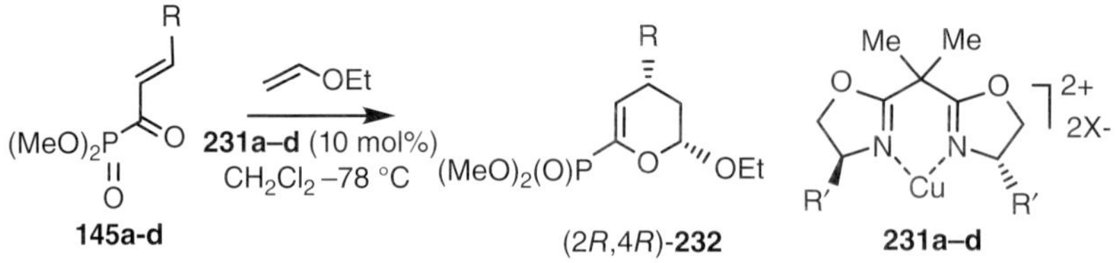

145: R = Me (**a**), *i*-Pr (**b**), OEt (**c**); Ph (**d**); **231**: X = OTf, R′ = *t*-Bu (a),X=SbF_3, R′ = *t*-Bu (b)
X = OTf, R′ = Ph (c); X=SbF_3, R′ =Ph (d)

R	Cat	*endo*/*exo*	ee (%)	Configuration
Me	**231a**	99:1	99	(2*R*,4*R*)
Me	**2131b**	69:1	93	(2*R*,4*R*)
Me	**231c**	32:1	39	(2*S*,4*S*)
Me	**231d**	>99:1	93	(2*S*,4*S*)
Me	**231d**	>99:1	94	(2*S*,4*S*)
i-Pr	**231a**	32:1	93	(2*R*,4*S*)
OEt	**231a**	>99:1	93	(2*R*,4*R*)

Scheme 4.83 The catalytic reaction of crotonyl phosphonates **145** with dienophiles.

acids, in particular C_2-symmetric Cu(II)bis(oxazolin) complexes as Diels–Alder reaction catalysts. The important property of these catalysts is the ability to activate the substrate by chelation with chiral cationic Cu(II) atom. This allows to carry out the enantioselective hetero Diels–Alder reaction leading to the formation of cyclic enol phosphonates **232** with high ee. For example, the reaction of ethylvinyl ester with crotonyl phosphonate **145** in the presence of [Cu((*S*,*S*)-*tert*-Bu-box)](OTf)$_2$ complex **231** afforded cycloadduct **232** in 89% yield, with *endo/exo* isomer ratio 99 : 1 and with stereoselectivity up to 99% ee (Scheme 4.83) [152]. α,β-Unsaturated acyl phosphonates and β,γ-unsaturated α-keto esters and amides are effective heterodienes, while enol ethers and sulfides function as heterodienophiles.

The enantioselective synthesis of dihydropyrans by this method has been shown to be straightforward: cycloadditions may be conducted with as little as 0.2 mol% of the chiral catalyst and are readily run on multigram scale. The reactions exhibited a favorable temperature enantioselectivity profile, with selectivities up to 98% dr at room temperature [153, 154]. The asymmetric hetero Diels–Alder reaction of α,β-unsaturated carbonyls with electron-enriched alkenes catalyzed by the C_2-symmetric bis(oxazoline) Cu(II) complexes **231** provides an expedient entry into enantioenriched dihydropyrans. A number of α,β-unsaturated acyl phosphonates and β,γ-unsaturated α-keto esters and amides have been successfully employed as heterodienes, while enol ethers and sulfides and certain ketone silyl enol ethers have functioned well as heterodienophiles [154, 155] (Scheme 4.84).

The addition reaction of acylphosphonate **145** to cyclopentadiene catalyzed by complex **231** led to the formation of two isomers **236** and **237** in the ratio of 35 : 65. The expected Diels–Alder product **236** was obtained with ratio of *endo/exo* isomers 87 : 13, and with optical purity of *endo*-isomer 84% ee (Scheme 4.85).

Bicyclic adducts **238** were synthesized in good yields and enantioselectivity under catalysis with bis-oxazoline/Cu(II) complexes **231**, especially in the case of more reactive dihydrofuran instead of ethyl vinyl ether. The high diastereoselectivity for

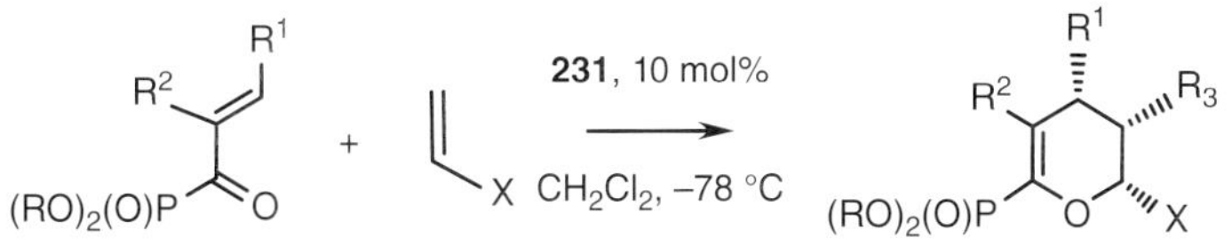

233 234 235

R^1	R^2	X	% yield	*endo*/*exo*	% ee
Me	H	OEt	84	36:1	93
Ph	H	OEt	95	22:1	97
i-Pr	H	OEt	92	22:1	95
OEt	H	OEt	92	44:1	97
Me	Me	OEt	98	25:1	90
Me	H	SEt	75	16:1	96

Scheme 4.84 The hetero Diels–Alder reaction of α,β-unsaturated ketophosphonates.

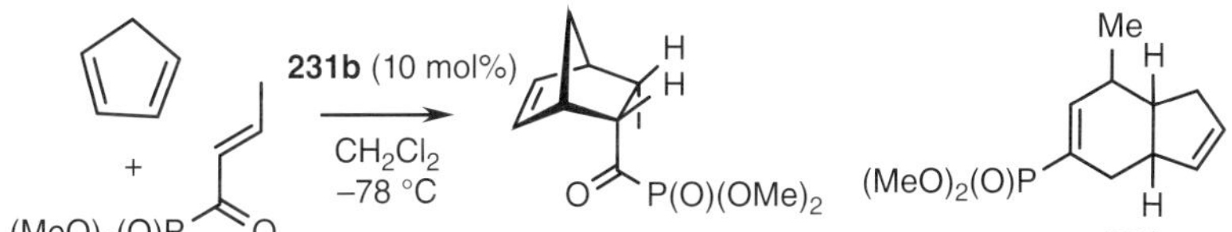

145a

236
endo:*exo* 87:13
84% ee

237
endo:*exo* > 95:5
95% ee

R = Me, Ph, *i*-Pr, OEt

yield 99%, 2:3 = 35:65

Scheme 4.85 Cycloaddition of acylphosphonate to cyclopentadiene, catalyzed by Cu-complex.

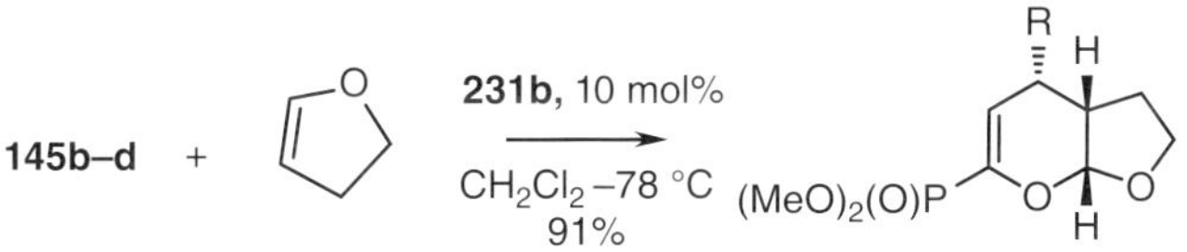

238

endo:*exo* 99:1, 95% ee

R = Ph	99%	(>99:1)	ee = 90%
R = *i*-Pr	79%	(49:1)	ee = 90%
R = OEt	98%	(>99:1)	ee = 97%

Scheme 4.86 The hetero Diels–Alder reaction of acyl phosphonates catalyzed by 2,3-dihydrofurane complexes **231b**.

catalyzed hetero Diels–Alder reactions is a result of frontier orbital control and/or electrostatic effects that preferentially place the OR substituent in proximity with the heterodiene carbonyl carbon (endo orientation). This stereocontrol element has been implicated in related conjugate addition reactions to unsaturated imides and azaimides [153] (Scheme 4.86). The reaction of α-keto-β,γ-unsaturated phosphonates undergo Lewis acid-catalyzed cyclocondensation reactions to give hetero Diels–Alder products

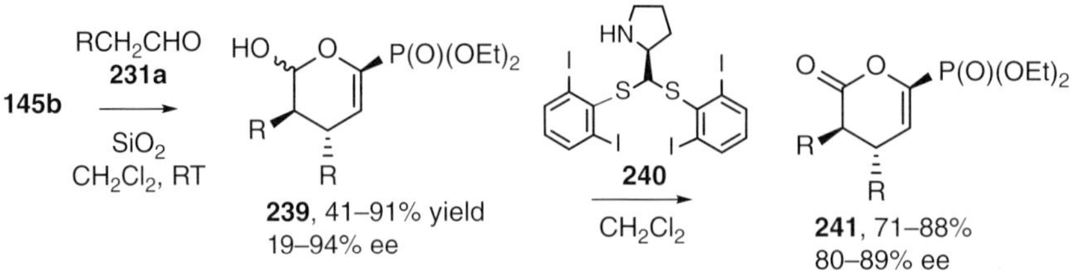

Scheme 4.87 The hetero-Diels–Alder reaction of aldehydes with β,γ-unsaturated α-ketophosphonates.

with cyclopentadiene, cyclohexadiene, dihydrofuran, and dihydropyran with high *endo*-selectivities.

Some prolinal dithioacetal derivatives **240** were studied as catalysts for the hetero-Diels–Alder reaction of enolizable aldehydes with β,γ-unsaturated α-ketophosphonates **145b**. The corresponding 5,6-dihydro-4*H*-pyran-2-ylphosphonates **241** (as mixture of two diastereomers in an 80/20 ratio) were obtained in good ee values (up to 94% ee) as shown in Scheme 4.87. The product **239** was oxidized to the corresponding lactone derivative **241**. The formation of **239** was explained by an inverse-electron-demand hetero-Diels–Alder reaction between the enamine, and proline, and the α,β-unsaturated ketone **145** (Scheme 4.87) [155].

A catalytic asymmetric synthesis of nitrogen-containing gem-bisphoshonates **244** was described. A Lewis acid-Brønsted base bifunctional homodinuclear Ni_2-Schiff base complex **243** promoted catalytic enantioselective conjugate addition of nitroacetates to ethylidenebisphosphonates **242** giving products in up to 93% ee and 94% yield. Transformation of the product into a chiral α-amino ester with a gem-bisphosphonate moiety is also described (Scheme 4.88) [156].

4.7 Summary

The creation of highly effective catalysts based upon chiral organophosphorus compounds and the creation of chiral organophosphorus synthons continue to be important

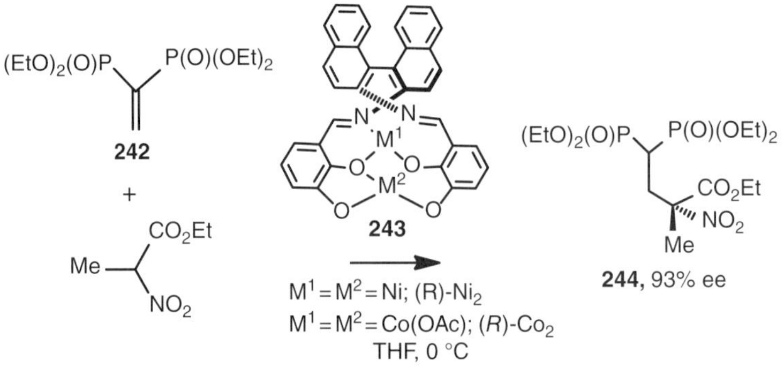

Scheme 4.88 Catalytic enantioselective conjugate addition of nitroacetates to ethylidene bisphosphonates.

problems in modern chemistry. In this chapter, we have considered various versions of asymmetric catalysis and some other techniques that give individual organophosphorus compounds in high enantiomeric excess. It is important to note that despite the impressive progress achieved in the synthesis and studies of properties of chiral organophosphorus compounds, not all problems have been solved. The prospects of chemical modification of chiral phosphorus compounds with the introduction of new and more complex groups including those with a specific configuration at the phosphorus centers are far from exhausted. The problems are the resolution of enantiomers and purification of chiral tertiary phosphines and chiral phosphine oxides. The exact absolute configuration can only be successfully established in limited cases. The problem of the development of enantioselective methods which give easy access to both antipodes of chiral organophosphorus species still remains. The search for enantioselective methods providing rapid access to optically active phosphorus-containing acids, tertiary phosphines, and phosphine oxides remains a topical one. In connection with this, the elaboration of highly efficient methods of organocatalytic, enzymatic, and microbiological synthesis of chiral organophosphorus compounds is of particular importance.

References

1 Börner, A. (ed.) (2008) *Phosphorus Ligands in Asymmetric Catalysis: Synthesis and Applications*, vol. **1 A**, John Wiley & Sons, Inc., 1546 pp.
2 Ojima, I. (ed.) (2000) *Catalytic Asymmetric Synthesis*, 2nd edn, Wiley-VCH Verlag GmbH, New York, Chichester, 998 pp.
3 Kamer, P.C.J. and van Leeuwen, P.W.N.M. (eds) (2012) P-chiral ligands, in *Phosphorus (III) Ligands in Homogeneous Catalysis: Design and Synthesis*, John Wiley & Sons, Ltd, 566 pp.
4 Tang, W. and Zhang, X. (2003) New chiral phosphorus ligands for enantioselective hydrogenation. *Chem. Rev.*, **103** (8), 3029–3069.
5 Kolodiazhnyi, O.I., Kukhar, V.P., and Kolodiazhna, A.O. (2014) Asymmetric catalysis as a method for the synthesis of chiral organophosphorus compounds. *Tetrahedron: Asymmetry*, **25** (12), 865–922.
6 Kolodiazhnyi, O.I. (2015) Recent advances in asymmetric synthesis of P-stereogenic phosphorus compounds, in *Topics in Current Chemistry*, vol. **361** (ed. J.-L. Montchamp), Springer International Publishing, Switzerland, pp. 61–236.
7 Brunel, J.M. and Buono, G. (2002) New chiral organophosphorus catalysts in asymmetric synthesis, in *Topics in Current Chemistry*, (ed. J.-P. Majoral) Springer-Verlag, Berlin-Heidelberg New York, vol. **220**, pp. 80–195.
8 Kolodiazhnyi, O.I. (2014) Advances in asymmetric hydrogenation and hydride reduction of organophosphorus compounds. *Phosphorus, Sulfur Silicon Relat. Elem.*, **189** (7-8), 1102–1131.
9 Demkowicz, S., Rachon, J., Daśko, M., and Kozaka, W. (2016) Selected organophosphorus compounds with biological activity. *Applications in medicine, RSC Adv.*, **6** (9), 7101–7112.
10 Blaser, H.U. and Federsel, H.-J. (2010) *Asymmetric Catalysis on Industrial Scale*, 2nd edn, Wiley-VCH Verlag GmbH, Weinheim, 542 pp.
11 Blaser, H.U. (1999) The chiral switch of metolachlor: the development of a large-scale enantioselective catalytic process. *Chimia*, **53** (6), 275–280.

12 Schollkopf, U., Hoppe, I., and Thiele, A. (1985) Asymmetric synthesis of α–aminophosphonic acids. *Liebigs Ann. Chem.*, **1985** (3), 555–559.

13 Schmidt, U., Oehme, G., and Krause, H. (1996) Catalytic stereoselective synthesis of α-Amino phosphonic acid derivatives by asymmetric hydrogenation. *Synth. Commun.*, **26** (4), 777–781.

14 Selke, R. (1989) Carbohydrate phosphinites as chiral ligands for asymmetric synthesis catalyzed by complexes. *J. Organomet. Chem.*, **370** (1), 241–256.

15 Burk, M.J., Stammers, T.A., and Straub, J.A. (1999) Enantioselective synthesis of α-hydroxy and α-amino phosphonates via catalytic asymmetric hydrogenation. *Org. Lett.*, **1** (3), 387–390.

16 Gridnev, I.D., Yasutake, M., Imamoto, T., and Beletskaya, I.P. (2004) Asymmetric hydrogenation of –unsaturated phosphonates with Rh-BisP* and Rh-MiniPHOS catalysts: scope and mechanism of the reaction. *Proc. Natl. Acad. Sci. U.S.A.*, **101** (15), 5385–5390.

17 Goulioukina, N.S., Dolgina, T.M., Beletskaya, I.P., Henry, J.C., Lavergne, D., Ratovelomanana-Vidal, V., and Genet, J.P. (2001) A practical synthetic approach to chiral α-aryl substituted ethylphosphonates. *Tetrahedron: Asymmetry*, **12** (2), 319–327.

18 Chavez, M.A., Vargas, S., Suarez, A., Alvarez, E., and Pizzano, A. (2011) Highly enantioselective hydrogenation of β-acyloxy and β-acylamino α,β-unsaturated phosphonates catalyzed by rhodium phosphane-phosphite complexes. *Adv. Synth. Catal.*, **353** (14-15), 2775–2794.

19 Wang, D.Y., Hu, X.P., Huang, J.D., Deng, J., Yu, S.B., Duan, Z.C., Xu, X.-F., and Zheng, Z. (2008) Readily available chiral phosphine–aminophosphine ligands for highly efficient Rh catalyzed asymmetric hydrogenation of α-enol ester phosphonates and α-enamido phosphonates Z. *J. Org. Chem.*, **73** (5), 2011–2014.

20 Kondoh, A., Yorimitsu, H., and Oshima, K. (2008) Regio- and stereoselective hydroamidation of 1-alkynylphosphine sulfides catalyzed by cesium base. *Org. Lett.*, **10** (14), 3093–3095.

21 Kadyrov, R., Holz, J., Schaffner, B., Zayas, O., Almena, J., and Boerner, A. (2008) Synthesis of chiral β–aminophosphonates via Rh-catalyzed asymmetric hydrogenation of β-amido-vinylphosphonates. *Tetrahedron: Asymmetry*, **19** (10), 1189–1192.

22 Doherty, S., Knight, J.G., Bell, A.L., El-Menabawey, S., Vogels, C.M., Decken, A., and Westcott, S.A. (2009) Rhodium complexes of (*R*)-Me-CATPHOS and (*R*)-(S)-JOSIPHOS: highly enantioselective catalysts for the asymmetric hydrogenation of (E)- and (Z)-β-aryl-β-(enamido)phosphonates. *Tetrahedron: Asymmetry*, **20** (12), 1437–1444.

23 Wang, D.Y., Hu, X.P., Deng, J., Yu, S.B., Duan, Z.C., and Zheng, Z. (2009) Enantioselective synthesis of chiral alpha-aryl or alpha-alkyl substituted ethylphosphonates via Rh-catalyzed asymmetric hydrogenation with a *P*-stereogenic BoPhoz-type ligand. *J. Org. Chem.*, **74** (11), 4408–4410.

24 Oki, H., Oura, I., Nakamura, T., Ogata, K., and Fukuzawa, S.I. (2009) Modular synthesis of the ClickFerrophos ligand family and their use in Rhodium- and ruthenium-catalyzed asymmetric hydrogenation. *Tetrahedron: Asymmetry*, **20** (18), 2185–2191.

25 Wang, D.-Y., Hu, X.-P., Huang, J.-D., Deng, J., Yu, S.-B., Duan, Z.-C., Xu, X.-F., and Zheng, Z. (2007) Highly enantioselective synthesis of α-hydroxy

phosphonic acid derivatives by Rh-catalyzed asymmetric hydrogenation with phosphine–phosphoramidite ligands. *Angew. Chem. Int. Ed.*, **46** (41), 7810–7813.

26 Duan, Z.C., Hu, X., Wang, D.Y., Huang, J.D., Yu, S.B., Deng, J., and Zheng, Z. (2008) Enantioselective synthesis of optically active alkanephos- phonates via rhodium-catalyzed asymmetric hydrogenation of β-substituted α,β-unsaturated phosphonates with ferrocene-based monophosphoramidite ligands. *Adv. Synth. Catal.*, **350** (13), 1979–1983.

27 Zhang, J., Dong, K., Wang, Z., and Ding, K. (2012) Asymmetric hydrogenation of α- or β-acyloxy α,β-unsaturated phosphonates catalyzed by a Rh(I) complex of monodentate phosphoramidite. *Org. Biomol. Chem.*, **10** (8), 1598–1601.

28 Duan, Z.C., Hu, X.P., Zhang, C., and Zheng, Z. (2010) Enantioselective Rh-catalyzed hydrogenation of 3-Aryl-4-phosphonobutenoates with a *P*-stereogenic BoPhoz-Type ligand. *J. Org. Chem.*, **75** (23), 8319–8321.

29 Fukuzawa, S.I., Oki, H., Hosaka, M., Sugasawa, J., and Kikuchi, S. (2007) ClickFerrophos: new chiral ferrocenyl phosphine ligands synthesized by click chemistry and the use of their metal complexes as catalysts for asymmetric hydrogenation and allylic substitution. *Org. Lett.*, **9** (26), 5557–5560.

30 Duan, Z.C., Hu, X.P., Zhang, C., Wang, D.Y., Yu, S.B., and Zheng, Z. (2009) Highly enantioselective Rh-catalyzed hydrogenation of β,γ-unsaturated phosphonates with chiral ferrocene-based monophosphoramidite ligands. *J. Org. Chem.*, **74** (23), 9191–9194.

31 Cheruku, P., Paptchikhine, A., Church, T.L., and Andersson, P.G. (2009) Iridium-*N,P*-ligand-catalyzed enantioselective hydrogenation of diphenylvinylphosphine oxides and vinylphosphonates. *J. Am. Chem. Soc.*, **131** (23), 8285–8289.

32 Kitamura, M., Tokunaga, M., and Noyori, R. (1995) Asymmetric hydrogenation of beta-KetoPhosphonates: a practical way to fosfomycin. *J. Am. Chem. Soc.*, **117** (10), 2931–2932.

33 Kitamura, M., Tokunaga, M., Pham, T., Lubell, W.D., and Noyori, R. (1995) Asymmetric synthesis of α-amino β-hydroxy phosphonic acids via BINAP-ruthenium catalyzed hydrogenation. *Tetrahedron Lett.*, **36** (32), 5769–5772.

34 Li, Q., Hou, C.-J., Liu, Y.-J., Yang, R.-F., and Hu, X.-P. (2015) Asymmetric hydrogenation of α-keto phosphonates with chiral phosphine–phosphoramidite ligands. *Tetrahedron: Asymmetry*, **26** (12-13), 617–622.

35 Gautier, I., Ratovelomanana-Vidal, V., Savignacs, P., and Genet, J.P. (1996) Asymmetric hydrogenation of β-ketophosphonates and β-ketothiophosphonates with chiral Ru (II) catalysts. *Tetrahedron Lett.*, **38** (43), 7721–7724.

36 Tao, X., Li, W., Ma, X., Li, X., Fan, W., Zhu, L., Xie, X., and Zhang, Z. (2012) Enantioselective hydrogenation of β-ketophosphonates with chiral Ru(II) catalysts. *J. Org. Chem.*, **77** (12), 8401–8409.

37 Tao, X., Li, W., Li, X., Xie, X., and Zhang, Z. (2013) Diastereo- and enantioselective asymmetric hydrogenation of α-amido-β-keto phosphonates via dynamic kinetic resolution. *Org. Lett.*, **15** (1), 72–75.

38 Dayoub, W. and Doutheau, A. (2010) (E)-Enol ethers from the stereoselective reduction of α-alkoxy-β-ketophosphonates and Wittig type reaction. *Sci. China Chem.*, **53** (9), 1937–1945.

39 Hérault, D., Nguyen, D.H., Nuela, D., and Buono, G. (2015) Reduction of secondary and tertiary phosphine oxides to phosphines. *Chem. Soc. Rev.*, **44**, 2508–2528.

40 Meier, C., Laux, W.H.G., and Bats, J.W. (1995) Asymmetric synthesis of chiral, nonracemic dialkyl α–hydroxyarylmethyl and α-, β– and γ-hydroxyalkylphosphonates from keto phosphonates. *Liebigs Ann. Chem.*, **1995** (11), 1963–1979.

41 Meier, C. and Laux, W.H.G. (1995) Asymmetric synthesis of chiral, nonracemic dialkyl-α-, β-, and γ-hydroxyalkylphosphonates via a catalyzed enantioselective catecholborane reduction. *Tetrahedron: Asymmetry*, **6** (5), 1089–1092.

42 Meier, C. and Laux, W.H.G. (1996) Enantioselective synthesis of diisopropyl α-, β-, and γ-hydroxyarylalkylphosphonates from ketophosphonates: a study on the effect of the phosphonyl group. *Tetrahedron*, **52** (2), 589–598.

43 Meier, C., Laux, W.H.G., and Bats, J.W. (1996) Asymmetric synthesis of chiral, Nonracemic dialkyl-α-hydroxyalkylphosphonates via a (−)-Chlorodiisopino-campheylborane (Ipc2B-Cl) reduction. *Tetrahedron: Asymmetry*, **7** (1), 89–94.

44 Guliaiko, I., Nesterov, V., Sheiko, S., Kolodiazhnyi, O.I., Freytag, M., Jones, P.G., and Schmutzler, R. (2008) Synthesis of optically active hydroxyphosphonates. *Heteroat. Chem.*, **2**, 133–139.

45 Nesterov, V.V. and Kolodiazhnyi, O.I. (2006) New method for the asymmetric hydroboration of ketophosphonates and the synthesis of phospho-carnitine. *Tetrahedron: Asymmetry*, **17** (7), 1023–1026.

46 Gryshkun, E.V., Nesterov, V., and Kolodyazhnyi, O.I. (2012) Enantioselective reduction of ketophosphonates using adducts of chiral natural acids with sodium borohydride. *Arkivoc*, **V**, 100–117.

47 Nesterov, V.V. and Kolodiazhnyi, O.I. (2008) Asymmetric syntheses of new phosphonotaxoids. *Phosphorus, Sulfur Silicon Relat. Elem.*, **183** (1-2), 687–689.

48 Kolodiazhnyi, O.I. and Nesterov, V.V. (2008) Dimenthyl (S)-2-hydroxy-3-chloropropylphosphonate- accessible chiron for the asymmetric synthesis of hydroxyphosphonates. *Phosphorus, Sulfur Silicon Relat. Elem.*, **183** (1-2), 681–682.

49 Kolodiazhnyi, O.I., Guliayko, I.V., Gryshkun, E.V., Kolodiazhna, A.O., Nesterov, V.V., and Kachkovskyi, G.O. (2008) New methods, and strategies for asymmetric synthesis of organophosphorus compounds. *Phosphorus, Sulfur Silicon Relat. Elem.*, **183** (1-2), 393–398.

50 Nesterov, V.V. and Kolodiazhnyi, O.I. (2007) Efficient method for the asymmetric reduction of α- and β-ketophosphonates. *Tetrahedron*, **63** (29), 6720–6731.

51 Nesterov, V.V. and Kolodiazhnyi, O.I. (2007) Di(1R,2S,5R)-menthyl 2-hydroxy-3-chloropropyl-phosphonate as useful chiron for the synthesis of α- and β–hydroxyphosphonates. *Synlett*, (15), 2400–2404.

52 Nesterov, V.V. and Kolodyazhnyi, O.I. (2006) Asymmetric synthesis of a phosphorus analog of natural g-amino-b-hydroxybutyric acid. *Russ. J. Gen. Chem.*, **76** (10), 1677–1678.

53 Nesterov, V.V. and Kolodyazhnyi, O.I. (2006) Asymmetic synthesis of a phosphorus analog of natural L-carnitine. *Russ. J. Gen. Chem.*, **76** (9), 1510–1511.

54 Nesterov, V.V. and Kolodyazhnyi, O.I. (2006) Enantioselective reduction of ketophosphonates using chiral acid adducts with sodium borohydride. *Russ. J. Gen. Chem.*, **76** (7), 1022–1030.

55 Corbett, M.T. and Johnson, J.S. (2013) Diametric stereocontrol in dynamic catalytic reduction of racemic acyl phosphonates: divergence from α-keto ester congeners. *J. Am. Chem. Soc.*, **135** (2), 594–597.

56 Son, S.-M. and Lee, H.-K. (2014) Dynamic kinetic resolution based asymmetric transfer hydrogenation of α-alkoxy-β-ketophosphonates. Diastereo- and enantioselective synthesis of monoprotected 1,2-dihydroxyphosphonates. *J. Org. Chem.*, **79** (6), 2666–2681.

57 Barco, A., Benetti, S., Bergamini, P., de Risi, C., Marchetti, P., Pollini, G.P., and Zanirato, V. (1999) Diastereoselective synthesis of β-amino-α-hydroxy phosphonates via oxazaborolidine catalyzed reduction of β-phthalimido-α-ketophosphonates. *Tetrahedron Lett.*, **40** (43), 7705–7708.

58 Vicario, J., Aparicio, D., and Palacios, F. (2011) α-Ketiminophosphonates: synthesis and applications. *Phosphorus, Sulfur Silicon Relat. Elem.*, **186** (4), 638–643.

59 Rassukana, Y.V., Onys'ko, P.P., Kolotylo, M.V., Sinitsa, A.D., Lyzwa, P., and Mikoiajczyk, M. (2009) A new strategy for asymmetric synthesis of aminophosphonic acid derivatives: the first enantioselective catalytic reduction of C-phosphorylated imines. *Tetrahedron Lett.*, **50** (3), 288–290.

60 Guo, W.L., Hou, C.J., Duan, Z.C., and Hu, X.P. (2011) An enantioselective synthesis of 3-aryl-4-phosphonobutyric acid esters via Cu-catalyzed asymmetric conjugate reduction. *Tetrahedron: Asymmetry*, **22** (24), 2161–2164.

61 Thomas, A.A. and Sharpless, K.B. (1999) The catalytic asymmetric aminohydroxylation of unsaturated phosphonates. *J. Org. Chem.*, **64** (22), 8379–8385.

62 Cravotto, G., Giovenzana, G.B., Pagliarin, R., Palmisano, G., and Sisti, M. (1998) A straightforward entry into enantiomerically enriched β-amino-α-hydroxyphosphonic acid derivatives. *Tetrahedron: Asymmetry*, **9** (5), 745–748.

63 Qi, X., Lee, S.H., Kwon, J.Y., Kim, Y., Kim, S.J., Lee, Y.S., and Yoon, J. (2003) Aminobromination of unsaturated phosphonates. *J. Org. Chem.*, **68** (23), 9140–9143.

64 Kobayashi, Y., William, A.D., Tokoro, Y., and Sharpless, K. (2001) Asymmetric dihydroxylation of trans-propenylphosphonate by using a modified AD-mix-α and the synthesis of fosfomycin. *J. Org. Chem.*, **66** (23), 7903–7906.

65 Katsuki, T. and Sharpless, K.B. (1980) The first practical method for asymmetric epoxidation. *J. Am. Chem. Soc.*, **102** (18), 5974–5976.

66 Koprowski, M., Luczak, J., and Krawczyk, E. (2006) Asymmetric oxidation of enol phosphates to α-hydroxy ketones by (salen)manganese(III) complex. Effects of the substitution pattern of enol phosphates on the stereochemistry of oxygen transfer. *Tetrahedron*, **62** (52), 12363–12374.

67 Krawczyk, E., Koprowski, M., Skowronska, A., and Luczak, J. (2004) α-Hydroxy ketones in high enantiomeric purity from asymmetric oxidation of enol phosphates with (salen) manganese(III) complex. *Tetrahedron: Asymmetry*, **15** (17), 2599–2602.

68 Krawczyk, E., Mielniczak, G., Owsianik, K., and Łuczak, J. (2012) Asymmetric oxidation of enol phosphates to α-hydroxy ketones using Sharpless reagents and a fructose derived dioxirane. *Tetrahedron: Asymmetry*, **23** (20–21), 1480–1489.

69 Pawar, V.D., Bettigeri, S., Weng, S.S., Kao, J.Q., and Chen, C.T. (2006) Highly enantioselective aerobic oxidation of α-hydroxyphosphonates catalyzed by chiral vanadyl(V) methoxides bearing *N*-salicylidene-α-aminocarboxylates. *J. Am. Chem. Soc.*, **128** (19), 6308–6309.

70 Anderson, B.J., Glueck, D.S., DiPasquale, A.G., and Rheingold, A.L. (2008) Substrate and catalyst screening in platinum-catalyzed asymmetric alkylation of

Bis(secondary) phosphines. Synthesis of an enantiomerically pure C2-symmetric diphosphine. *Organometallics*, **27** (19), 4992–5001.
71 Glueck, D.S. (2008) Catalytic asymmetric synthesis of chiral phosphanes. *Chem. Eur. J.*, **14** (24), 7108–7117.
72 Glueck, D.S. (2007) Metal-catalyzed asymmetric synthesis of *P*-stereogenic phosphines. *Synlett*, **2007** (17), 2627–2634.
73 Moncarz, J.R., Brunker, T.J., Jewett, J.C., and Orchowski, M. (2003) Palladium-catalyzed asymmetric phosphination. Enantioselective synthesis of PAMP–BH_3, ligand effects on catalysis, and direct observation of the stereochemistry of transmetalation and reductive elimination. *Organometallics*, **22** (16), 3205–3221.
74 Glueck, D.S. (2010) Recent advances in metal-catalyzed C–P bond formation. *Top. Organomet. Chem.*, **31**, 65–100.
75 Scriban, C. and Glueck, D.S. (2006) Platinum-catalyzed asymmetric alkylation of secondary phosphines: enantioselective synthesis of *P*-stereogenic phosphines. *J. Am. Chem. Soc.*, **128** (9), 2788–2789.
76 Guino-o, M.A., Zureick, A.H., Blank, N.F., Anderson, B.J., Chapp, T.W., Kim, Y., Glueck, D.S., and Rheingold, A.L. (2012) Synthesis and structure of platinum Bis(Phospholane) complexes Pt(Diphos*)(R)(X), catalyst precursors for asymmetric phosphine alkylation. *Organometallics*, **31** (19), 6900–6910.
77 Join, B., Mimeau, D., Delacroix, O., and Gaumont, A.C. (2006) Pallado-catalysed hydrophosphination of alkynes: access to enantioenriched *P*-stereogenic vinyl phosphine–boranes. *Chem. Commun.*, **30**, 3249–3251.
78 Moncarz, J.R., Brunker, T.J., Glueck, D.S., Sommer, R.D., and Rheingold, A.L. (2003) Stereochemistry of palladium-mediated synthesis of PAMP–BH_3: retention of configuration at P in formation of Pd–P and P–C bonds. *J. Am. Chem. Soc.*, **125** (5), 1180–1181.
79 Scriban, C., Glueck, D.S., Golen, J.A., and Rheingold, A.L. (2007) Platinum-catalyzed asymmetric alkylation of a secondary phosphine: mechanism and origin of enantioselectivity. *Organometallics*, **26** (7), 1788–1800.
80 Chan, V.S., Chiu, M., Bergman, R.G., and Toste, F.D. (2009) Development of ruthenium catalysts for the enantioselective synthesis of *P*-stereogenic phosphines via nucleophilic phosphido intermediates. *J. Am. Chem. Soc.*, **131** (16), 6021–6032.
81 Chan, V.S., Stewart, I.C., Bergman, R.G., and Toste, F.D. (2006) Asymmetric catalytic synthesis of *P*-stereogenic phosphines via a nucleophilic ruthenium phosphido complex. *J. Am. Chem. Soc.*, **128** (9), 2786–2787.
82 Scriban, C., Glueck, D.S., DiPasquale, A.G., and Rheingold, A.L. (2006) Chiral platinum duphos terminal phosphido complexes: synthesis, structure, phosphido transfer, and ligand behavior. *Organometallics*, **25** (22), 5435–5448.
83 Chapp, T.W., Schoenfeld, A.J., and Glueck, D.S. (2010) Effects of linker length on the rate and selectivity of platinum-catalyzed asymmetric alkylation of the Bis(isitylphosphino)alkanes IsHP(CH_2)nPHIs (Is = 2,4,6-(i-Pr)$_3C_6H_2$, n = 1–5). *Organometallics*, **29** (11), 2465–2473.
84 Blank, N.F., McBroom, K.C., Glueck, D.S., Kassel, W.S., and Rheingold, A.L. (2006) Chirality breeding via asymmetric phosphination. Palladium-catalyzed diastereoselective synthesis of a *P*-stereogenic phosphine. *Organometallics*, **25** (7), 1742–1748.

85 Chan, V.S., Bergman, R.G., and Toste, F.D. (2007) Pd-catalyzed dynamic kinetic enantioselective arylation of silylphosphines. *J. Am. Chem. Soc.*, **129** (49), 15122–15123.
86 Brunker, T.J., Anderson, B.J., Blank, N.F., Glueck, D.S., and Rheingold, A.L. (2007) Enantioselective synthesis of *P*-stereogenic benzophospholanes via palladium-catalyzed intramolecular cyclization. *Org. Lett.*, **9** (6), 1109–1112.
87 Shibata, M., Ikeda, M., Motoyama, K., Miyake, Y., and Nishibayashi, Y. (2012) Enantioselective alkylation of β-keto phosphonates by direct use of diaryl methanols as electrophiles. *J. Chem. Soc., Chem. Commun.*, **48**, 9528–9530.
88 Hou, G., Yu, J., Yu, C., Wu, G., and Miao, Z. (2013) Enantio- and diastereoselective vinylogous mukaiyama aldol reactions of α-keto phosphonates with 2-(Trimethylsilyloxy)-furan catalyzed by Bis(oxazoline)-copper complexes. *Adv. Synth. Catal.*, **355** (2-3), 589–593.
89 (a) Evans, D.A., Fandrick, K.R., Song, H.J., Scheidt, K.A., and Xu, R. (2007) Enantioselective friedel–crafts alkylations catalyzed by Bis(oxazolinyl) pyridine–scandium(III) triflate complexes. *J. Am. Chem. Soc.*, **129** (32), 10029–10041; (b) Takenaka, N., Abell, J.P., and Yamamoto, H. (2007) Asymmetric conjugate addition of silyl enol ethers catalyzed by tethered Bis(8-quinolinolato) aluminum complexes. *J. Am. Chem. Soc.*, **129** (4), 742–743.
90 (a) Jiang, H., Paixao, M.W., Monge, D., and Jørgensen, K.A. (2010) Acyl phosphonates: good hydrogen bond acceptors and ester/amide equivalents in asymmetric organocatalysis. *J. Am. Chem. Soc.*, **132** (8), 2775–2783; (b) Frings, M., Thomé, I., Schiffers, I., Pan, F., and Bolm, C. (2014) Catalytic, asymmetric synthesis of phosphonic γ-(Hydroxyalkyl)butenolides with contiguous quaternary and tertiary stereogenic centers. *Chem. Eur. J.*, **20** (6), 1691–1700.
91 Bernardi, L. and Jorgensen, K.A. (2005) Enantioselective chlorination and fluorination of β-keto phosphonates catalyzed by chiral Lewis acids. *J. Chem. Soc., Chem. Commun.*, (11), 1324–1326.
92 Bachu, P. and Akiyama, T. (2010) Enantioselective Friedel–Crafts alkylation reaction of indoles with α,β-unsaturated acyl phosphonates catalyzed by chiral phosphoric acid. *J. Chem. Soc., Chem. Commun.*, **46** (23), 4112–4114.
93 Sodeoka, M. and Hamashima, Y. (2006) Acid-base catalysis using chiral palladium complexes. *Pure Appl. Chem.*, **78** (2), 477–494.
94 Hamashima, Y., Suzuki, T., Takano, H., Shimura, Y., Tsuchiya, Y., Moriya K-ichi Goto, T., and Sodeoka, M. (2006) Highly enantioselective fluorination reactions of β-ketoesters and β-ketophosphonates catalyzed by chiral palladium complexes. *Tetrahedron*, **62** (30), 7168–7179.
95 Kim, S.M., Kim, H.R., and Kim, D.Y. (2005) Catalytic enantioselective fluorination and amination of β-keto phosphonates catalyzed by chiral palladium complexes. *Org. Lett.*, **7** (17), 2309–2311.
96 Kang, Y.K. and Kim, D.Y. (2010) Recent advances in catalytic enantioselective fluorination of active methines. *Curr. Org. Chem.*, **14** (9), 917–927.
97 Bernardi, L., Zhuang, W., and Jorgensen, K.A. (2005) An easy approach to optically active α-amino phosphonic acid derivatives by chiral Zn(II)-catalyzed enantioselective amination of phosphonates. *J. Am. Chem. Soc.*, **127** (16), 5772–5773.
98 Hornillos, V., Pérez, M., Fañanás-Mastral, M., and Feringa, B.L. (2013) Cu-catalyzed asymmetric allylic alkylation of phosphonates and phosphine oxides with grignard reagents. *Chem. Eur. J.*, **19** (17), 5432–5441.

99 Kolodiazhnyi, O.I. (2006) Chiral hydroxyphosphonates: synthesis, configuration, and biological properties. *Russ. Chem. Rev.*, **75** (3), 227–253.

100 Merino, P., Marques-Lopez, E., and Herrera, R.P. (2008) Catalytic enantioselective hydrophosphonylation of aldehydes and imines. *Adv. Synth. Catal.*, **350** (9), 1195–1208.

101 Nixon, T.D., Dalgarno, S., Ward, C.V., Jiang, M., Halcrow, M.A., Kilner, C., Thornton-Pett, M., and Kee, T.P. (2004) Stereocontrol in asymmetric phospho-aldol catalysis. Chirality relaying in action. *C.R. Chim.*, **7**, 809–821.

102 Sasai, H., Bougauchi, M., Arai, T., and Shibasaki, M. (1997) Enantioselective synthesis of α-hydroxy phosphonates using the $LaLi_3$tris(binaphthoxide) catalyst (LLB), prepared by an improved method. *Tetrahedron Lett.*, **38** (15), 2717–2720.

103 Yamagishi, T., Yokomatsu, T., Suemune, K., and Shibuya, S. (1999) Enantioselective synthesis of α-hydroxyphosphinic acid derivatives through hydrophosphinylation of aldehydes catalyzed by Al-Li-BINOL complex. *Tetrahedron*, **55** (41), 12125–12136.

104 Groaning, M.D., Rowe, B.J., and Spilling, C.D. (1998) New homochiral cyclic diol ligands for titanium alkoxide catalyzed phosphonylation of aldehydes. *Tetrahedron Lett.*, **39** (11), 5485–5488.

105 Nakanishi, K., Kotani, S., Sugiura, M., and Nakajima, M. (2008) First asymmetric Abramov-type phosphonylation of aldehydes with trialkyl phosphites catalyzed by chiral Lewis bases. *Tetrahedron*, **64** (27), 6415–6419.

106 Qian, C., Huang, T., Zhu, C., and Sun, J. (1998) Synthesis of 3,3′-, 6,6′- and 3,3′,6,6′-substituted binaphthols and their application in the asymmetric hydrophosphonylation of aldehydes - An obvious effect of substituents of BINOL on the enantioselectivity. *J. Chem. Soc., Perkin Trans. 1*, (13), 2097–2103.

107 Kolodiazhna, A.O., Kukhar, V.P., Chernega, A.N., and Kolodiazhnyia, O.I. (2004) Double and triple asymmetric induction in phosphaaldol reactions. *Tetrahedron: Asymmetry*, **15** (13), 1961–1963.

108 Gou, S., Zhou, X., Wang, J., Liu, X., and Feng, X. (2008) Asymmetric hydrophosphonylation of aldehydes catalyzed by bifunctional chiral Al(III) complexes. *Tetrahedron*, **64** (12), 2864–2870.

109 Yang, F., Zhao, D., Lan, J., Xi, P., Yang, L., Xiang, S., and You, J. (2008) Self-assembled bifunctional catalysis induced by metal coordination interactions: an exceptionally efficient approach to enantioselective hydrophosphonylation. *Angew. Chem. Int. Ed.*, **47** (30), 5646–5649.

110 Saito, B. and Katsuki, T. (2005) Synthesis of an optically active C1-symmetric Al(salalen) complex and its application to the catalytic hydrophosphonylation of aldehydes. *Angew. Chem. Int. Ed.*, **44** (29), 4600–4602.

111 Suyama, K., Sakai, Y., Matsumoto, K., Saito, B., and Katsuki, T. (2010) Highly enantioselective hydrophosphonylation of aldehydes: base-enhanced aluminum–salalen catalysis. *Angew. Chem. Int. Ed.*, **49** (4), 797–799.

112 Gledhill, A.C., Cosgrove, N.E., Nixon, T.D., Kilner, C.A., Fisher, J., and Kee, T.P. (2010) Asymmetric general base catalysis of the phospho-aldol reaction via dimeric aluminium hydroxides. *J. Chem. Soc., Dalton Trans.*, **39** (40), 9472–9475.

113 Li, W., Qin, S., Su, Z., Yang, H., and Hu, C. (2011) Theoretical study on the mechanism of Al(salalen)-catalyzed hydrophosphonylation of aldehydes. *Organometallics*, **30** (8), 2095–2104.

114 Ito, K., Tsutsumi, H., Setoyama, M., Saito, B., and Katsuki, T. (2007) Enantioselective hydrophosphonylation of aldehydes using an aluminum binaphthyl schiff base complex as a catalyst. *Synlett*, **2007** (12), 1960–1962.
115 Boobalan, R. and Chen, C. (2013) Catalytic enantioselective hydrophosphonylation of aldehydes using the iron complex of a camphor-based tridentate schiff base $[FeCl(SBAIB\text{-}d)]_2$. *Adv. Synth. Catal.*, **355** (17), 3443–3450.
116 Zhou, X., Liu, X., Yang, X., Shang, D., Xin, J., and Feng, X. (2008) Highly enantioselective hydrophosphonylation of aldehydes catalyzed by tridentate schiff base aluminum(III) complexes. *Angew. Chem. Int. Ed.*, **47** (2), 392–394.
117 Zhou, X., Liu, Y., Chang, L., Zhao, J., Shang, D., Liu, X., Lin, L., and Fenga, X. (2009) Highly efficient synthesis of quaternary α-hydroxy phosphonates via lewis acid-catalyzed hydrophosphonylation of ketones. *Adv. Synth. Catal.*, **351** (16), 2567–2572.
118 Zhou, X., Zhang, Q., Hui, Y., Chen, W., Jiang, J., Lin, L., Liu, X., and Feng, X. (2010) Catalytic asymmetric synthesis of quaternary α-hydroxy trifluoromethyl phosphonate via chiral aluminum(III) catalyzed hydrophosphonylation of trifluoromethyl ketones. *Org. Lett.*, **12** (19), 4296–4299.
119 Abell, J.P. and Yamamoto, H. (2008) Catalytic enantioselective pudovik reaction of aldehydes and aldimines with tethered Bis(8-quinolinato) (TBOx) aluminum complex. *J. Am. Chem. Soc.*, **130** (32), 10521–10523.
120 Deng, T. and Cai, C. (2014) Bis(oxazoline)-copper catalyzed enantioselective hydrophosphonylation of aldehydes. *RSC Adv.*, **4** (53), 27853–27856.
121 Maji, B. and Yamamoto, H. (2014) Copper-catalyzed asymmetric synthesis of tertiary α-hydroxy phosphonic acid derivatives with in situ generated nitrosocarbonyl compounds as the oxygen source. *Angew. Chem. Int. Ed.*, **53** (52), 14472–14475.
122 (a) Keglevich, G. and Bálint, E. (2012) The Kabachnik–Fields reaction: mechanism and synthetic use. *Molecules*, **17**, 12821–12835. (b) Cheng, X., Goddard, R., Buth, G., and List, B. (2008) Direct catalytic asymmetric three-component Kabachnik-Fields reaction. *Angew. Chem. Int. Ed.*, **47** (27), 5079–5081.
123 Sasai, H., Arai, S., Tahara, Y., and Shibasaki, M. (1996) Catalytic asymmetric synthesis of alpha-amino phosphonates using lanthanoid-potassium-BINOL complexes. *J. Org. Chem.*, **60** (21), 6656–6657.
124 Schlemminger, I., Saida, Y., Groger, H., Maison, W., Durot, N., Sasai, H., Shibasaki, M., and Martens, J. (2000) Concept of improved rigidity: how to make enantioselective hydrophosphonylation of cyclic imines catalyzed by chiral heterobimetallic lanthanoid complexes almost perfect. *J. Org. Chem.*, **65** (16), 4818–4825.
125 Cheng, X., Goddard, R., Buth, G., and List, B. (2008) Direct catalytic asymmetric three-component Kabachnik–Fields reaction. *Angew. Chem. Int. Ed.*, **47** (27), 5079–5081.
126 Zhou, X., Shang, D., Zhang, Q., Lin, L., Liu, X., and Feng, X. (2009) Enantioselective three-component Kabachnik–Fields reaction catalyzed by chiral scandium(III)–N,N′-dioxide complexes. *Org. Lett.*, **11** (6), 1401–1404.
127 Chen, W., Hui, Y., Zhou, X., Jiang, J., Cai, Y., Liu, X., Lin, L., and Feng, X. (2010) Chiral N,N'-dioxide-Yb(III) complexes catalyzed enantioselective hydrophosphonylation of aldehydes. *Tetrahedron Lett.*, **51** (32), 4175–4178.
128 Saito, B., Egami, H., and Katsuki, T. (2007) Synthesis of an optically active Al(salalen) complex and its application to catalytic hydrophosphonylation of aldehydes and aldimines. *J. Am. Chem. Soc.*, **129** (7), 1978–1986.

129 (a) Kiyohara, H., Nakamura, Y., Matsubara, R., and Kobayashi, S. (2006) Enantiomerically enriched allylglycine derivatives through the catalytic asymmetric allylation of iminoesters and iminophosphonates with allylsilanes. *Angew. Chem. Int. Ed.*, **45** (10), 1615–1617; (b) Kobayashi, S., Kiyohara, H., Nakamura, Y., and Matsubara, R. (2004) Catalytic asymmetric synthesis of α-amino phosphonates using enantioselective carbon–carbon bond-forming reactions. *J. Am. Chem. Soc.*, **126** (21), 6558–6559.

130 Palacios, F., Aparicio, D., Ochoa de Retana, A.M., de los Santos, J.M., Gil, J.I., and Alonso, J.M. (2002) Asymmetric synthesis of 2H-azirines derived from phosphine oxides using solid-supported amines. Ring opening of azirines with carboxylic acids. *J. Org. Chem.*, **67** (21), 7283–7288.

131 Dodda, R. and Zhao, C.G. (2007) Enantioselective synthesis of α-aminopropargylphosphonates. *Tetrahedron Lett.*, **48** (25), 4339–4342.

132 Dodda, R. and Zhao, C.G. (2007) Silver(I) triflate-catalyzed direct synthesis of *N*-PMP protected α-aminopropargylphosphonates from terminal alkynes. *Org. Lett.*, **9** (1), 165–167.

133 Pudovik A. N., Konovalova I. V. Addition reactions of esters of phosphorus(lll) acids with unsaturated systems. *Synthesis* 1979 **1979** (2) 81–96.

134 Scriban, C., Glueck, D.S., Zakharov, L.N., Scott, K.W., DiPasquale, A.G., Golen, J.A., and Rheingold, A.L. (2006) P–C and C–C bond formation by michael addition in platinum-catalyzed hydrophosphination and in the stoichiometric reactions of platinum phosphido complexes with activated alkenes. *Organometallics*, **25** (24), 5757–5767.

135 Zhao, D., Yuan, Y., Chan, A.S., and Wang, R. (2009) Highly enantioselective 1,4-addition of diethyl phosphite to enones using a dinuclear Zn catalyst. *Chem. Eur. J.*, **15** (12), 2738–2741.

136 Sadow, A.D., Haller, I., Fadini, L., and Togni, A. (2004) Nickel(II)-catalyzed highly enantioselective hydrophosphination of methacrylonitrile. *J. Am. Chem. Soc.*, **126** (45), 14704–14705.

137 Sadow, A.D. and Togni, A. (2005) Enantioselective addition of secondary phosphines to methacrylonitrile: catalysis and mechanism. *J. Am. Chem. Soc.*, **127** (48), 17012–17024.

138 Feng, J.J., Chen, X.F., Shi, M., and Duan, W.L. (2010) Palladium-catalyzed asymmetric addition of diarylphosphines to enones toward the synthesis of chiral phosphines. *J. Am. Chem. Soc.*, **132** (16), 5562–5563.

139 Sabater, S., Mata, J.A., and Peris, E. (2013) Chiral palladacycles with *N*-heterocyclic carbene ligands as catalysts for asymmetric hydrophosphination. *Organometallics*, **32** (4), 1112–1120.

140 Huang, Y., Pullarkat, S.A., Teong, S., Chew, R.J., Li, Y., and Leung, P.-H. (2012) Palladacycle-catalyzed asymmetric intermolecular construction of chiral tertiary *P*-heterocycles by stepwise addition of H–P–H bonds to Bis(enones). *Organometallics*, **31**, 4871–4875.

141 Yang, X.-Y., Tay, W.S., Li, Y., Pullarkat, S.A., and Leung, P.-H. (2015) Asymmetric 1,4-conjugate addition of diarylphosphines to α,β,γ,δ-unsaturated ketones catalyzed by transition-metal pincer complexes. *Organometallics*, **34** (20), 5196–5201.

142 Huang, Y., Chew, R.J., Pullarkat, S.A., Li, Y., and Leung, P.H. (2012) Asymmetric synthesis of enaminophosphines via palladacycle-catalyzed addition of Ph2PH to α,β-unsaturated imines. *J. Org. Chem.*, **77** (16), 6849–6854.

143 Hatano, M., Horibe, T., and Ishihara, K. (2013) Chiral magnesium(II) binaphtholates as cooperative brønsted/lewis acid–base catalysts for the highly enantioselective addition of phosphorus nucleophiles to α,β-unsaturated esters and ketones. *Angew. Chem.*, **125** (17), 4647–4651.

144 Zhao, D., Mao, L., Yang, D., and Wang, R. (2010) Zinc-mediated asymmetric additions of dialkylphosphine oxides to α,β-unsaturated ketones and *N*-sulfinylimines. *J. Org. Chem.*, **75** (20), 6756–6763.

145 Zhao, D., Wang, Y., Mao, L., and Wang, R. (2009) Highly enantioselective conjugate additions of phosphites to α,β-unsaturated *N*-acylpyrroles and imines: a practical approach to enantiomerically enriched amino phosphonates. *Chem. Eur. J.*, **15** (41), 10983–10987.

146 Yang, M.J., Liu, Y.J., Gong, J.F., and Song, M.P. (2011) Unsymmetrical chiral PCN pincer palladium(II) and nickel(II) complexes with aryl-based aminophosphine-imidazoline ligands: synthesis via aryl C-H activation and asymmetric addition of diarylphosphines to enones. *Organometallics*, **30** (14), 3793–3803.

147 Zhao, D., Mao, L., Wang, L., Yang, D., and Wang, R. (2012) Catalytic asymmetric construction of tetrasubstituted carbon stereocenters by conjugate addition of dialkyl phosphine oxides to β,β-disubstituted α,β-unsaturated carbonyl compounds. *J. Chem. Soc., Chem. Commun.*, **48**, 889–891.

148 Zhao, D., Mao, L., Wang, Y., Yang, Q., Zhang, D., and Wang, R. (2010) Catalytic asymmetric hydrophosphinylation of α,β-unsaturated *N*-acylpyrroles: application of dialkyl phosphine oxides in enantioselective synthesis of chiral phosphine oxides or phosphines. *Org. Lett.*, **12** (8), 1880–1882.

149 Sun, L., Guo, Q.-P., Li, X., Zhang, L., Li, Y.-Y., and Da, C.-S. (2013) C_2-symmetric homobimetallic zinc complexes as chiral catalysts for the highly enantioselective hydrophosphonylation of aldehydes. *Asian J. Org. Chem.*, **2** (12), 1031–1035.

150 Rai, V. and Namboothiri, I.N.N. (2008) Enantioselective conjugate addition of dialkyl phosphites to nitroalkenes. *Tetrahedron: Asymmetry*, **19** (20), 2335–2338.

151 Sung, H.J., Mang, J.Y., and Kim, D.Y. (2015) Catalytic asymmetric conjugate addition of α–fluoro β-ketophosphonates to nitroalkenes in the presence of nickel complexes. *J. Fluorine Chem.*, **178** (1), 40–46.

152 Evans, D.A. and Johnson, J.S. (1998) Catalytic enantioselective hetero Diels–Alder reactions of α,β-unsaturated acyl phosphonates with enol ethers. *J. Am. Chem. Soc.*, **120** (19), 4895–4896.

153 Evans, D.A., Johnson, J.S., and Olhava, E.J. (2000) Enantioselective synthesis of dihydropyrans. Catalysis of hetero Diels–Alder reactions by bis(oxazoline) copper(II) complexes. *J. Am. Chem. Soc.*, **122** (8), 1635–1649.

154 Evans, D.A., Johnson, J.S., Burgey, C.S., and Campos, K.R. (1999) Reversal in enantioselectivity of tert-butyl versus phenyl-substituted bis(oxazoline) copper(II) catalyzed hetero Diels-Alder and ene reactions. Crystallographic and mechanistic studies. *Tetrahedron Lett.*, **40** (15), 2879–2882.

155 Samanta, S., Krause, J., Mandal, T., and Zhao, C.-G. (2007) Inverse-electron-demand hetero Diels–Alder reaction of β,γ-unsaturated α-ketophosphonates catalyzed by prolinal dithioacetals. *Org. Lett.*, **9** (14), 2745–2748.

156 Kato, Y., Chen, Z., Matsunaga, S., and Shibasaki, M. (2009) Catalytic asymmetric synthesis of nitrogen-containing gem-bisphosphonates using a dinuclear Ni2–schiff base complex. *Synlett*, **2009** (10), 1635–1638.

5
Asymmetric Organocatalysis

5.1 Introduction

Organocatalysis represents a form of catalysis whereby the rate of a chemical reaction is increased by an organic catalyst consisting of carbon, hydrogen, and other nonmetal elements; however, the catalyst does not contain metallic atoms [1, 2]. Organocatalysis is recognized as a third methodology for catalysis, in addition to organometallic and enzymatic catalysis. Organocatalysts have a number of important advantages: they are stable, readily available, and nontoxic. They are inert toward moisture and oxygen in air. Special reaction conditions, for example, inert atmospheres, low temperatures, and absolute solvents are not required in many instances. In this context, simple organic acids can be used as catalyst in water solutions. Because of the absence of transition metals, organocatalytic methods seem to be especially attractive for the preparation of compounds that do not tolerate metal contamination, such as pharmaceutical products. The study of organocatalysis started in the beginning of the twentieth century with Bredig's publications on the use of natural alkaloids as enantioselective catalysts [3]. These studies were continued by Pracejus and Wynberg and the work by Hajos and Wiechert, which used proline as organocatalyst [2–5]. The field of asymmetric organocatalysis has enjoyed phenomenal growth in the past 15 years and during this "golden age" of organocatalysis, many researchers have been involved in this field, with most efforts focused on the search for new effective organocatalysts and development of new asymmetric methodologies. In the last few years, effective methodologies for aldol condensation were created, namely, Michael addition, Mannich-type reactions, studies of aza-Henry and Baylis–Hillman reactions, epoxidation, reduction, acylation, asymmetric synthesis of organophosphorus compounds, and others [6–15]. Organocatalytic strategies were used successfully for the preparation of various biologically active organophosphorus compounds such as α-hydroxy- and α- or β-aminophosphonic acids. Alkaloids, especially quinine and its derivatives and various amino acids, particularly proline, are the most common catalysts used in the organophosphorus chemistry. Sparteine and similar chiral diamines occupy a special place in this field [1, 2].

Asymmetric Synthesis in Organophosphorus Chemistry: Synthetic Methods, Catalysis and Applications, First Edition. Oleg I. Kolodiazhnyi.

5.2 Modes for Catalytic Activation of Substrates in Asymmetric Organocatalysis

The interactions between the substrate and the catalyst in asymmetric organocatalysis are different from mechanisms that act in classical metal-catalyzed reactions. In general, a catalyst interacting with the substrate activates it and creates the chiral environment that is essential in all enantio differentiating reactions. From this point of view, the organocatalysts that are used in organophosphorus chemistry can be divided into several main groups [6–15]:

Cinchona alkaloids and their derivatives [7]. The readily available and inexpensive cinchona alkaloids **1–4** (Table 5.1) having pseudoenantiomeric forms such as quinine, quinidine, cinchonine, and cinchonidine are among the most efficient organocatalysts. Early studies of the asymmetric catalysis of cinchona alkaloids, carried out at a time when a metal-catalyzed asymmetric reaction was still largely an alien concept to most organic chemists, naturally concentrated on Lewis-base/nucleophilic organic catalysis. In 1960, Pracejus reported that methyl phenyl ketene was converted to (*S*)-methyl hydratropate in 74% enantiomeric excess (ee) using *O*-acetylquinine as a catalyst [4]. Wynberg *et al.* [16, 17] carried out interesting studies of the use of cinchona alkaloids as chiral Lewis-base/nucleophilic catalysts. The extensive studies of cinchona alkaloids demonstrated that this class of alkaloids can be used as versatile organocatalysts for various enantioselective transformations. The key structural feature responsible for their synthetic utility is the presence of the tertiary quinuclidine nitrogen, which complements the proximal polar hydroxyl function of the natural compound. The presence of the Lewis acidic and Lewis basic functions makes them bifunctional catalysts. The development of modified cinchona organocatalysts attracted considerable attention from many chemists, one of the most important achievements in this area being the use of dimeric cinchona alkaloid ligands for the asymmetric dihydroxylation of olefins.

TADDOL and its derivatives [8]. TADDOL **5** is one of the oldest, and most extraordinarily versatile, chiral auxiliaries (Table 5.1). The initial design of TADDOL was driven by practical considerations, mainly because it is derived from tartaric acid – the least-expensive chiral starting material with twofold symmetry available from natural sources. The two hydroxyl functions of the genuine molecule can act as a double hydrogen-bond donor, allowing the formation of bidentate complexes. Moreover, these functions can be easily substituted, giving access to a variety of derivatives.

Proline and its derivatives [9]. L-Proline and its derivatives (**8a–c**) are perhaps the most well-known organocatalysts. Although the natural L-form is normally used, proline is available in both enantiomeric forms, this being somewhat of an asset when compared to enzymatic catalysis (Scheme 5.1).

Proline is a natural amino acid possessing secondary amine functionality. The nitrogen atom has a higher pK_a than in other amino acids and so features an enhanced nucleophilicity compared to the other amino acids. Hence, proline is able to act as a nucleophile, in particular with carbonyl compounds or Michael acceptors, to form either an iminium or enamine ion. In these reactions, the carboxylic function of the amino acid reacts as a Brønsted acid that creates the possibility for the proline to act as a bifunctional catalyst. The functional groups of proline can act as acid or base and can also facilitate chemical transformations similar to enzymatic catalysis. The high enantioselectivity of proline-catalyzed reactions can be explained by the capacity of the organocatalyst to form highly organized transition states (TSs) via the formation of strong hydrogen bonds. In proline-catalyzed reactions, the transfer of proton from the amine or the carboxyl group to the forming alkoxide or imide is essential for charge

Table 5.1 Commonly used organocatalysts **1–24**.

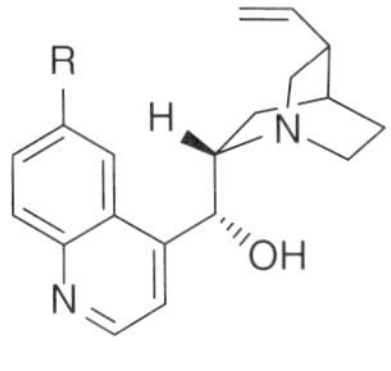

1, R=H, Cinchonidine
2, R=OMe, Quinine

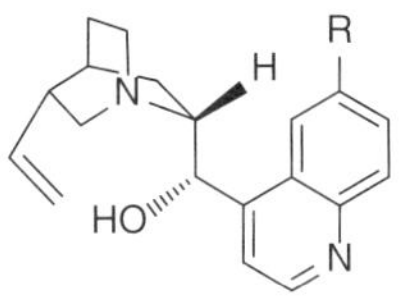

3, R=H, Cinchonine
4, R=OMe, Quinidine

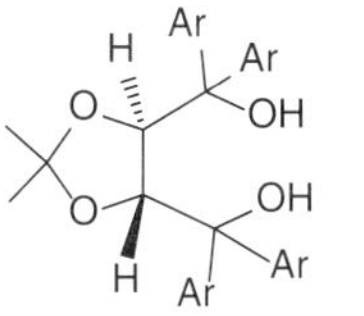

Ar=Ph (a), Nphth (b)
5a,b

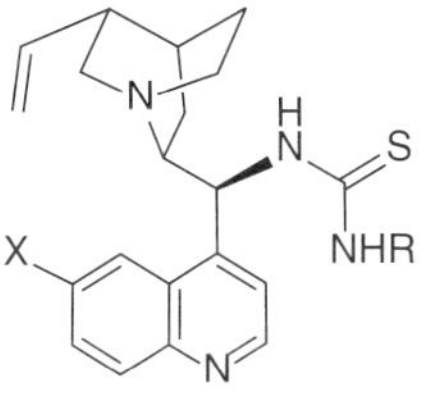

6a–c,7

R = Ph, X=H (**6a**)
R = 2,5-$(CF_3)_2C_6H_3$, X=H (**6b**)
R = 2-$NH_2C_6H_4$, X=OMe (**6c**)

R=
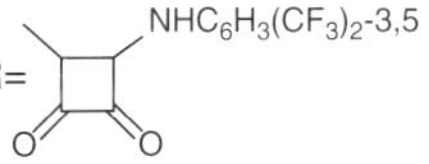

(**7**)

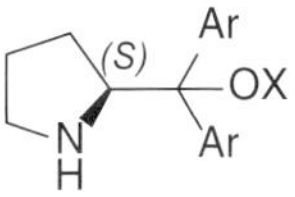

8a–d
Ar = Ph, X=TMS (a)
3,5-$Me_2C_6H_3$, X=TMS (b)
3,5-$(CF_3)_2C_6H_3$, X=TMS (c)
Ar=Ph, X=H (d)

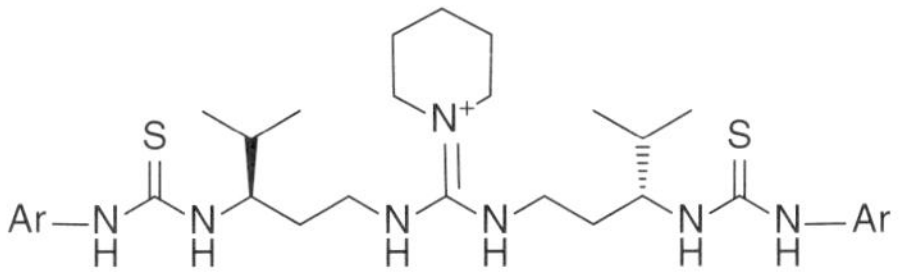

9, Ar=3,5-$(CF_3)_2C_6H_2$

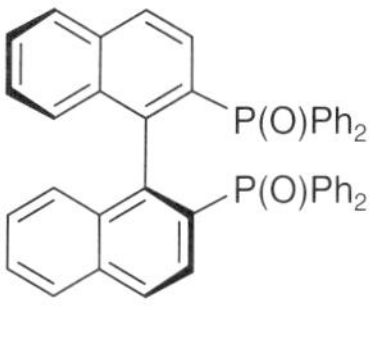

10, Binapo

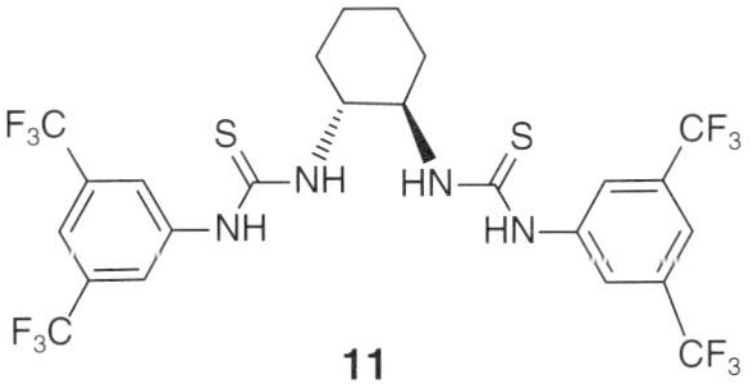

11

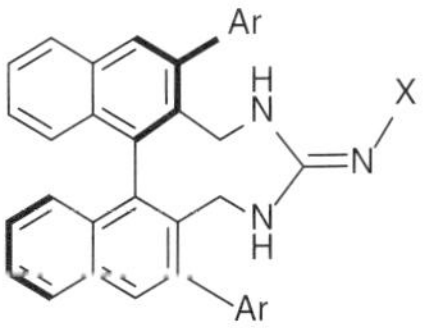

12a–c, X=Me (a), Bn (b), Ph_2CH (c); Ar=3,5-$(CF_3)_2C_6H_3$

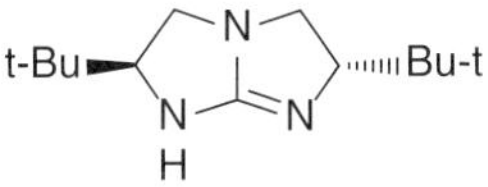

13

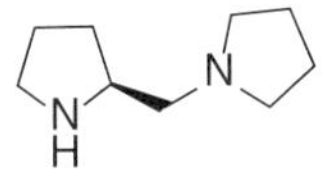

14

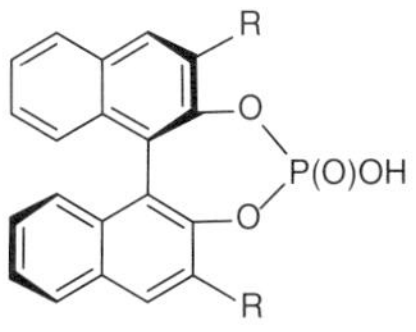

15a,b, R=3,5-$(CF_3)_2C_6H_3$ (a), 2,6-(i-Pr)$_2$-4-(9-Antr)C_6H_2 (b)

Table 5.1 *(continued)*

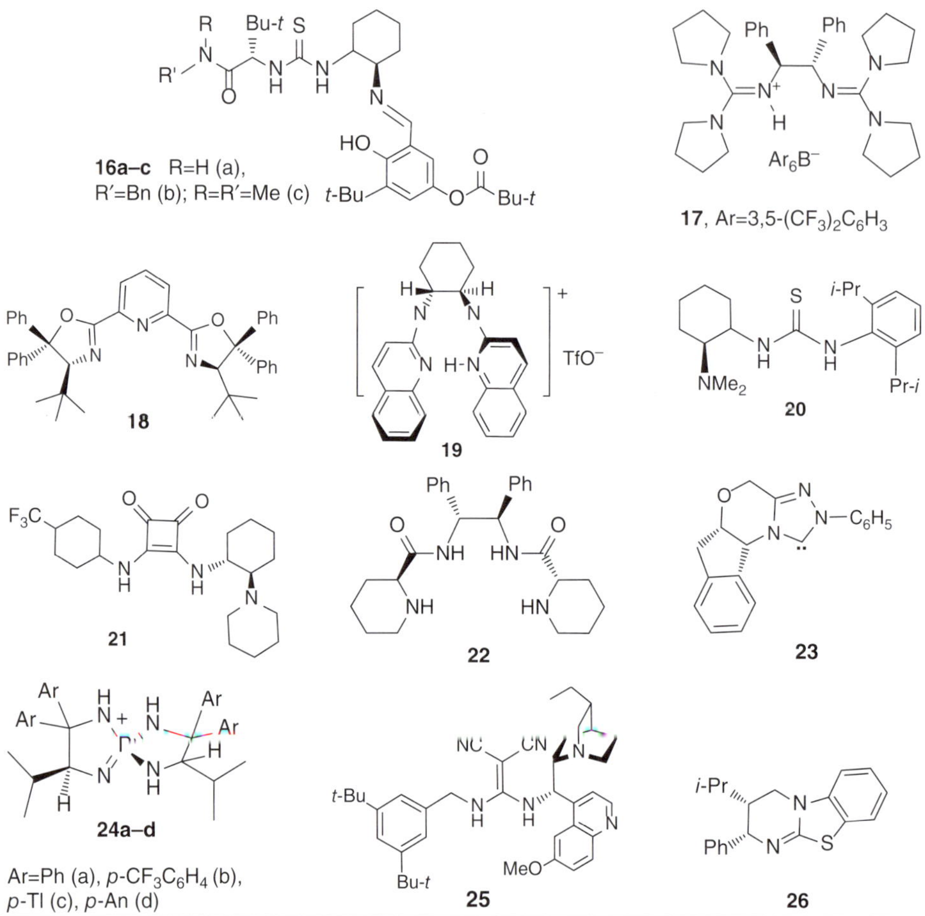

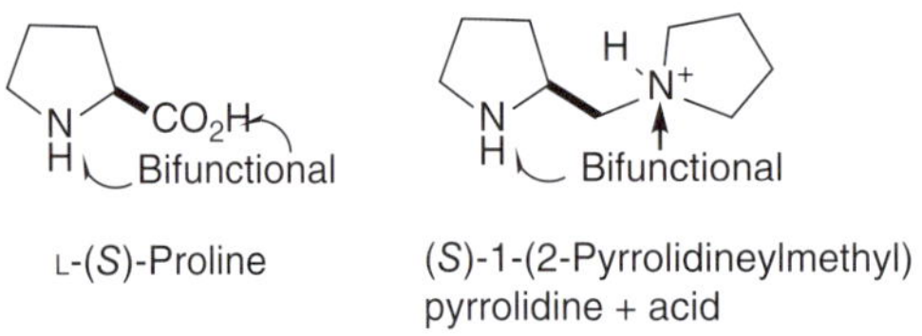

Scheme 5.1

stabilization which facilitates the formation of C–C bonds in the TS. Moreover, there are various chemical reasons that attract interest to proline's role in catalysis.

Thiourea organocatalysis [10]. Urea and thiourea derivatives represent a large and very important group of organocatalysts. These catalytically effective thiourea organocatalysts **6, 9, 12, 15, 20** (Table 5.1) provide explicit double hydrogen-bonding interactions to coordinate and activate H-bond accepting substrates. Thiourea organocatalysis describes the utilization of urea and thiourea derivatives to accelerate organic transformations through predominantly double hydrogen-bonding interactions with the respective substrate(s) (noncovalent organocatalysis). The urea and thiourea act as weak Lewis acids, but operate through explicit double hydrogen bonding instead of covalent bonding known from traditional metal-ion-mediated catalysis and Brønsted acid catalysis. To date, various organic transformations are organocatalyzed through double hydrogen-bonding *N,N′*-bis[3,5-bis(trifluoromethyl)]phenyl thiourea at low catalyst loadings and in good to excellent product yields.

Binaphthol derivatives [11]. The enantiomeric atropoisomers of 1,10-binaphthyl-2,20-diol (BINOL) and their bisdiphenylphosphonate derivatives 2,20-bis(diphenylphosphino)-1,10-binaphthyl (BINAP) are completely synthetic molecules that have been developed to exploit the axial dissymmetry induced by restricted rotation about the biaryl bond. A number of binaphthol derivatives (e.g., **10**, **12**, and **15**) have been synthesized and used as excellent organocatalysts. During the past 15 years, these compounds have become the most widely used ligands for both stoichiometric and catalytic asymmetric reactions, with many analogs and derivatives having been developed recently.

Chiral guanidinium organocatalysts [12]. Salts of organic bases have been shown to be successful in the activation of imines and other anionic intermediates through hydrogen bonding. The guanidinium salts [12] have also demonstrated this potential and have been used elegantly by Uyeda and Jacobsen to catalyze a Claisen rearrangement. Guanidines and guanidiniums **12**, **13**, **17** have been shown to be powerful catalysts for enantioselective reactions. Guanidines are neutral nitrogen compounds that are widely used as strong bases in synthetic organic chemistry. Chiral guanidine derivatives function as asymmetric catalysts by exploiting the high basicity of the guanidine group and the double hydrogen bonding of the guanidinium ion. Tan and co-workers reported the additions of dialkylphosphites and diphenyl phosphonite to various activated alkenes using catalytic amounts of guanidine **1** [13]. Subsequently, it was discovered that chiral bicyclic guanidine **7c** was effective in catalyzing phospha-Michael reactions of nitroalkenes [14]. A series of diaryl phosphine oxides was screened. Excellent enantioselectivities were generally obtained for various nitroalkenes with di-(1-naphthyl) phosphine oxide. Terada and co-workers demonstrated that axially chiral guanidine can catalyze the addition of diphenylphosphite to nitroalkenes [15] with high enantioselectivities.

N-Heterocyclic carbenes. Carbenes belong to the most investigated reactive species in the field of organic chemistry. Since the first reports of stable nucleophilic carbenes by Wanzlick [18] in 1962 and the work by Arduengo *et al.* [19] in the 1990s, the broad application of *N*-heterocyclic carbenes (NHCs) in organic synthesis has been impressively demonstrated. Besides their role as excellent ligands in metal-based catalytic reactions, organocatalytic carbene catalysis has emerged as an exceptionally fruitful research area in synthetic organic chemistry. Over the past several years, scientists have been engaged in developing a family of chiral nucleophilic carbenes and applying them to a variety of transformations. These catalysts, in particular **23**, have proved especially useful in catalytic asymmetric synthesis providing chiral products in typically high yield and enantioselectivity. High enantioselectivity levels were achieved in such reactions as benzoin condensation, the Stetter reaction, and reactions involving enolates and homoenolates[20]. In recent years, NHCs have been used in organophosphorus asymmetric synthesis as organocatalysts.

5.3 Phospha-Aldol Reaction

5.3.1 Catalysis with Cinchona Alkaloids

Organocatalysis in asymmetric synthesis of organophosphorus compounds has received great attention and many effective catalysts have been developed [21]. Over the last few years, several articles and reviews have been devoted to the organocatalysis of the phospha-aldol reaction, initiated by bases of natural origin. Among them, cinchona alkaloids and their derivatives are powerful tools in asymmetric reactions. In the last few years, several articles devoted to organocatalysis of the phospha-aldol reaction, initiated by bases of natural origins were published [16, 17, 22–24]. For example, Wynberg and Smaardijk [16] reported that quinine catalyzes the enantioselective phospha-aldol reaction of dialkylphosphites with *ortho*-nitrobenzaldehyde to give the hydroxyphosphonates **27** with moderate enantioselectivities. The stereoselectivity of reaction increased, when dimentylphosphite reacted with aldehydes, essentially owing to a double asymmetric induction with the formation of crystalline (*R*)-**28**, which were obtained as pure disastereomers after recrystallization (Scheme 5.2) [22, 23].

Di[(1S,2R,5R)-menthyl)] (S)-hydroxy(2-nitrophenyl) methylphosphonate **27d**. To a mixture of 0.01 mol of dimenthylphosphite and 0.01 mol of 2-nitrobenzaldehyde in 5 ml of tetrahydrofuran (THF), 0.002 mol of quinine was added as a catalyst. The mixture was kept for 48 h. In the NMR spectrum of the reaction mixture we observed signals of two diastereomers at δ_P 20.3 and 19.9 ppm in the ratio 87.5 : 12.5. The optically pure (*S*)-diastereomer was isolated by crystallization from acetonitrile. The other diastereomer was isolated by crystallization from hexane (yield 35%, mp 150–159 °C (MeCN), $[\alpha]_D^{20}$ 396 (0.6, $CHCl_3$), ^{31}P NMR spectrum, $CDCl_3$, δ_P, 19.4 ppm).

(S)-[Hydroxy(2-nitrophenyl)methyl]phosphonic acid (*S*)-**28**. A solution of 1 g of hydroxymethylphosphonate II in 50 ml of dioxane was placed in a flask, and 25 ml of 6 N hydrochloric acid was added. The reaction mixture was allowed to stand for 3 days at 80 °C. The hydrolysis progress was followed by ^{31}P NMR spectroscopy. After the reaction was completed, the solvent was removed (yield 85%, $[\alpha]_D^{20}$ −490 (c = 1, MeOH), ^{31}P NMR spectrum (CH_3OH), δ_P, 15 ppm).

The phospha-aldol reaction of diphenylphosphite with N-alkylated isatins **29** catalyzed with quinine **2** and quinidine proceeded with a satisfactory enantioselectivity. By

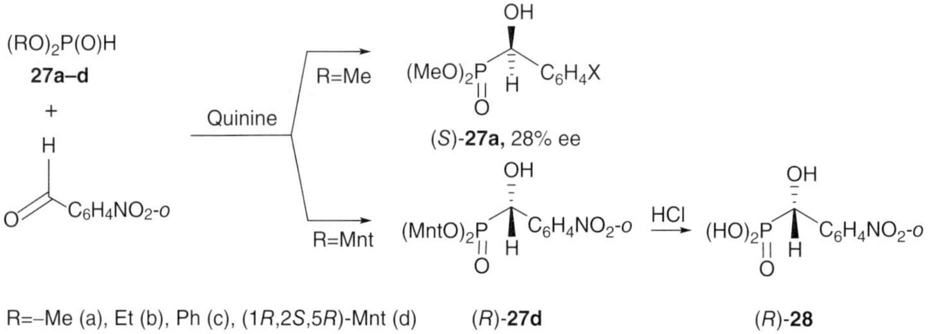

Scheme 5.2 Phospha-aldol reaction of dialkylphosphites with aldehydes catalyzed by quinine.

this method, a number of N-alkylated isatin derivatives **30** were prepared in good yields (up to 99%) and with moderate enantioselectivities (on average, 40–67% ee). However, absolute configurations of compounds **30** were not defined (Scheme 5.3) [24].

Barros and Phillips [25] have described an organocatalytic modified phospha-aldol reaction between dialkyl or diarylphosphites and α-haloketones leading to the formation of β-chloro-α-hydroxyphosphonates **31** in high yields and high stereoselectivity. Aliphatic, aromatic, and cyclic ketones were used in this reaction. It appears that the phosphite nucleophile approaches the carbonyl group from the same side in both enantiomers of the α-haloketone, presumably on the face opposite to the chloro substituent, owing to steric approach control. The presence of a β-chloro functionality allows further synthetic elaboration of the hydroxyphosphonates obtained, which could be useful for target-oriented synthesis, as these compounds have important biological application (Scheme 5.4).

5.3.2 Catalysis with Cinchona-Thiourea

Reactions between aromatic ketoesters and dimethylphosphite, catalyzed by the cinchona-derived thiourea organocatalyst **6a**, led to the formation of hydroxyphosphonates **32** with good yields and enantioselectivities. These organocatalysts acted as bifunctional promoters which can activate an electrophile (reacting as a hydrogen-bond donor) and a nucleophile (reacting as Brönsted base). The best results were obtained with cinchonidine thiourea organocatalysts. Possibly in the TS, **A** ketoester

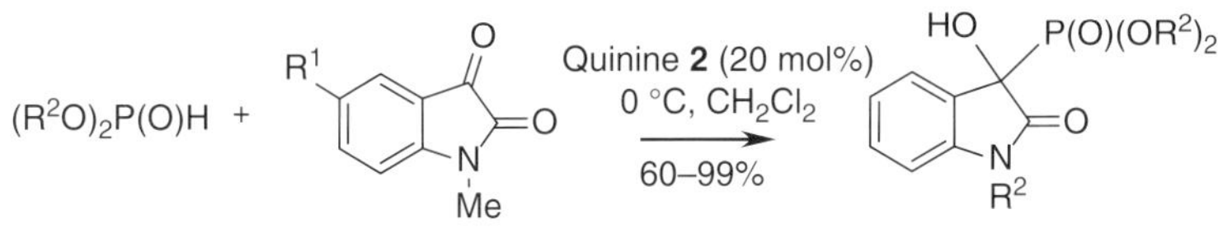

29 **30**, 26–67% ee

R^1=H	R^2=CH_3	99%	ee=67%
R^1=H	R^2=CH_3CH_2	96%	ee=64%
R^1=H	R^2=$PhCH_2$	89%	ee=51%
R^1=CH_3	R^2=CH_3	88%	ee=62%
R^1=F	R^2=CH_3	80%	ee=52%
R^1=Br	R^2=CH_3	60%	ee=44%
R^1=CH_3	R^2=$PhCH_2$	93%	ee=58%
R^1=F	R^2=$PhCH_2$	77%	ee=41%

Scheme 5.3 Enantioselective phospha-aldol reaction of diphenylphosphite with N-alkylated isatins catalyzed by quinine.

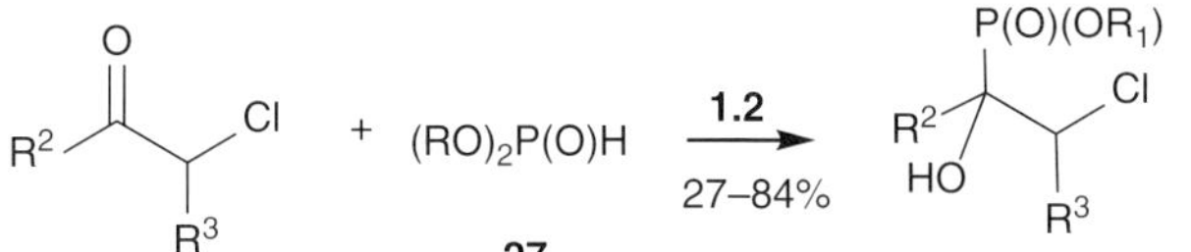

27 **31**, <40% ee

Scheme 5.4 Phospha-aldol reaction between dialkyl- or diarylphosphites and α-haloketones catalyzed by quinine.

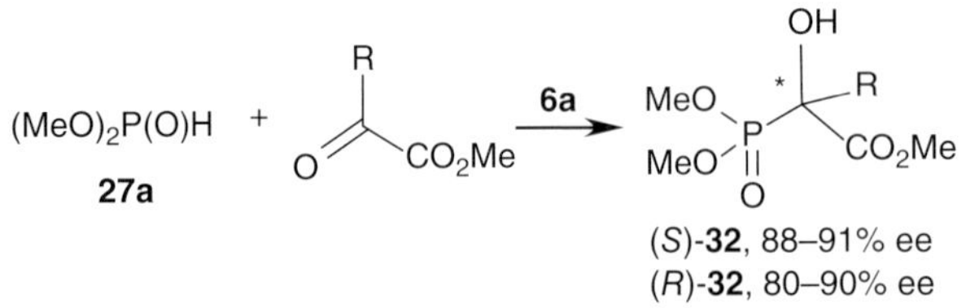

R=Ph	92%	ee= 90%
R=3-$CH_3C_6H_4$	92%	ee= 90%
R=4-$CH_3C_6H_4$	90%	ee= 90%
R=4-$CH_3OC_6H_4$	90%	ee= 91%
R=3-FC_6H_4	94%	ee= 90%
R=4-FC_6H_4	90%	ee= 90%
R=4-ClC_6H_4	85%	ee= 90%
R=4-BrC_6H_4	87%	ee= 90%
R=2-Thiophenyl	91%	ee= 91%
R=2-Naphthyl	86%	ee= 88%

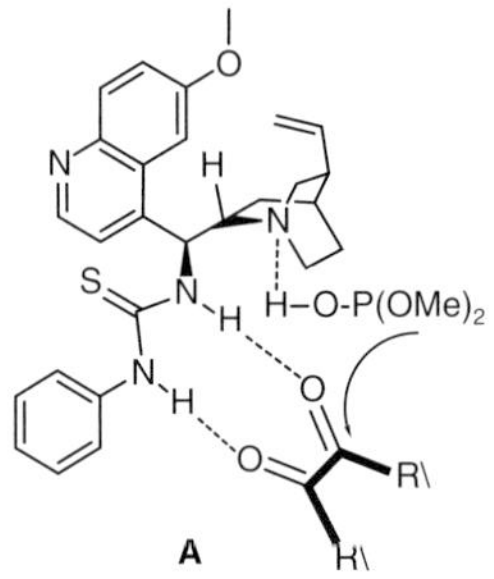

Scheme 5.5 Reaction of ketoesters with dimethylphosphite catalyzed by cinchona-thiourea **6a**.

is activated by a hydrogen bond with a thiourea fragment. Besides, the interaction provides the sufficient discrimination of the two enantiotopic sides of the electrophile that is necessary for enantiodifferentiating reactions. In addition, because of presence of the quinucledine nitrogen atom, the corresponding phosphonate–phosphite equilibrium is shifted toward the phosphite form, which add enantioselectively to the activated electrophile [26, 27] (Scheme 5.5).

5.3.3 Catalysis by Other Organocatalysts

Nakajima reported that chiral phosphine oxides (Lewis bases) catalyze silicon tetrachloride-mediated, enantioselective phosphonylation of aldehydes with trialkylphosphites (Abramov-type reaction), which leads to optically active α-hydroxyphosphonates **33** with moderate enantioselectivities [28]. The BINAPO ligand **10** represents a weak Lewis base, which, together with silicon tetrachloride and a tertiary amine (intermediate **B**), catalyzes the enantioselective phosphonylation of aldehydes with trialkylphosphites in high yields and with moderate enantioselctivities. The catalytic activity of BINAPO **10** has been tested on several aldehydes, mostly containing aromatic substituents. The reaction proceeded easily at low temperature in dichloromethane solution and gave the enantiomerically enriched hydroxyphosphonates **33** (Scheme 5.6).

List reported the enantioselective Abramov reaction of trimethylsilylphosphites with aromatic aldehydes catalyzed by chiral disulfonimides **34**. Several functionalized α-hydroxy phosphonates were synthesized by this method in good yields and with very good enantioselectivity (up to 98% ee). Absolute (*R*)-configurations were established for some compounds by X-ray analysis and by comparing optical rotation values with literature data (Scheme 5.7) [29].

A broad variety of monofuctional and bifunctional chiral imino compounds were developed as organocatalysts to accelerate various synthetically useful organic transformations employing H-bond accepting substrates, for example, carbonyl compounds, imines, and nitroalkenes as the starting materials included in the phospha-aldol

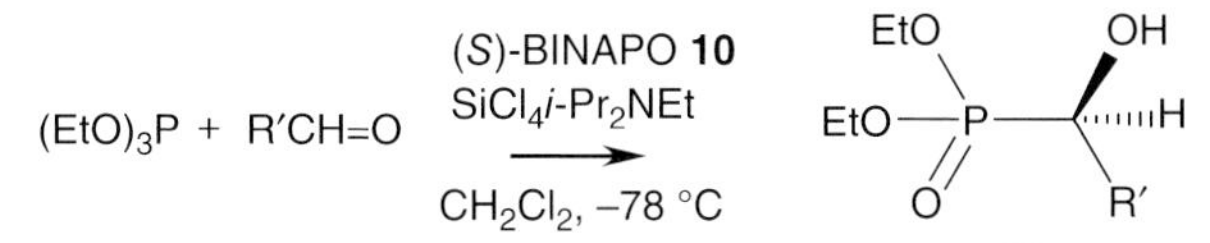

33

R=p-MeOC$_6$H$_4$	90%	ee=40% (*R*)
R=p-BrC$_6$H$_4$	87%	ee=22% -
R=2-Naphthyl	98%	ee=33% -
R=(*E*)-PhCH=CH	89%	ee=49% (R)
R=(*E*)-PhCH=CH	64%	ee=52% (S)*
R=PhCH$_2$CH$_2$	65%	ee=17% -
R=Cyclohexyl	49%	ee=23% -

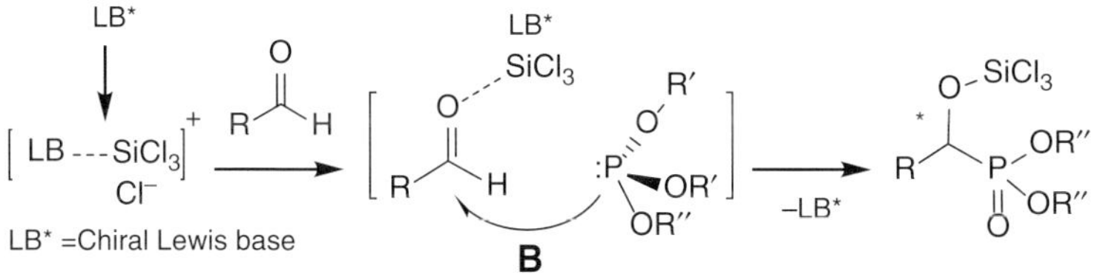

Scheme 5.6 BINAPO catalysis of phospha-aldol reaction.

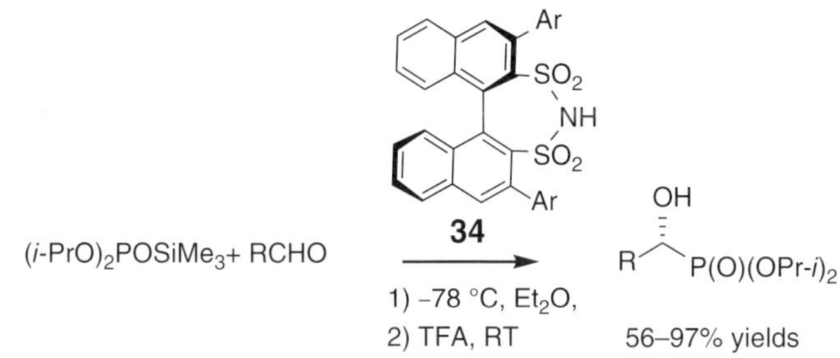

Ar=3,5-(CF$_3$)$_2$C$_6$H$_4$ (a), 3,5-(*i*-C$_3$F$_7$)$_2$C$_6$H$_4$ (b),
R=XC$_6$H$_4$: X=H, 3-Me, 4-Me, 4-*t*-Bu, 3-MeO; X=3,5-Me$_2$,
3,5-Et$_2$, 3,5-MeO$_2$, X-Nphth: X=5-Br, 5-MeO, 5-OH, furyl, thiophen

Scheme 5.7

reaction. Thus, the tris-aminoiminophosphorane **24**, generated *in situ* from chiral P-spiro-tetraaminophosphonium salts and potassium *tert*-butoxide is a very effective catalyst for phospha-aldol reactions The iminophosphoranes **24**, which contain electron-donating substituents on the benzene ring, showed the highest catalytic activity even when reduced to 1 mol% and at 98 °C. They catalyzed the reaction of dimethylphosphite with aliphatic, heteroaromatic, and aromatic aldehydes to give the corresponding α-hydroxyphosphonates **35** in high yields (92–94%) and with enantioselectivities up to 99% ee. The authors came to a conclusion that the reaction proceeds through the formation of highly active dimethylphosphite salt with a chiral tetraaminophosphonium cation which is responsible for the stereochemistry of the addition reaction [30, 31]. The generation of chiral tetraaminophosphonium phosphite has been detected by low-temperature NMR analysis, and its synthetic relevance has been successfully demonstrated by its application to the establishment of highly efficient and enantioselective hydrophosphonylation of aldehydes. Probably, the most

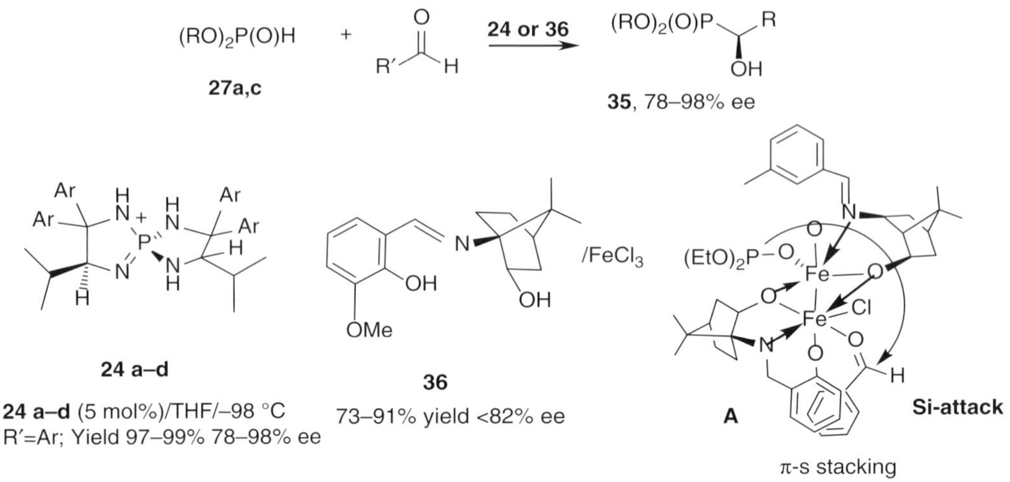

Scheme 5.8 Catalysis of phospha-aldol reaction with catalyst **36** and a possible intermediate **C** for the generation of phosphonate.

stable TS results from the formation of two hydrogen bonds between the catalyst and reactant with minimal distortion of the catalyst structure. In a less stable TS, the catalyst must undergo geometric distortion (Scheme 5.8) [31a]. Chiral camphor Schiff bases together with $FeCl_3$ were also used as catalysts for the asymmetric hydrophosphonylation of aldehydes. The electron-rich and electron-deficient aromatic aldehydes reacted readily with dialkylphosphites to give (*S*)-hydroxyphosphonates with good enantiomeric excesses (up to 82% ee). The intermediate **C** was proposed to explain the absolute configurations of the obtained phosphonates (predominantly *S*). The trivalent phosphorus attacks the *Si*-face of the benzaldehyde, as there is a steric hindrance from the bridge dimethyl group of the camphor moiety if the addition occurred on the *Re*-face of the benzaldehyde [31b].

The diaminomethylenemalononitrile (DMM) catalyst **26** (Table 5.1) was effective in the asymmetric hydrophosphonylation of aldehydes to provide (*S*)-hydroxyphosphonates with high yields and very good enantioselectivities (up to 98% yield and 96% ee) [32]. The addition of diphenyl phosphonate **27c** to aldehydes using DMM organocatalyst **26** proceeded via the TS **A**. The tertiary amine group traps a proton from the hydroxy group in the phosphite form, which is generated by an equilibrium donor. Protons from the DMM motif then successfully interact with the oxygen atoms in the aldehydes to direct the approach of the phosphite (an attack on the *Si*-face of the aldehydes) because of steric repulsion by the 3,5-di-*tert*-butyl benzyl group. The TS **D** ultimately affords the corresponding α-hydroxy phosphonates with excellent enantioselectivity. Herrera *et al.* [33] came to the analogous conclusion concerning the squaramide-catalyzed hydrophosphonylation of aldehydes (Scheme 5.9).

5.4 Phospha-Mannich Reactions

5.4.1 Organocatalysis by Cinchona Alkaloids

Nakamura *et al.* [34] used quinine and quinidine as catalysts in the enantioselective hydrophosphonylation of aldimines. *N*-(6-methyl-2-pyridilsulfonyl)imines prepared

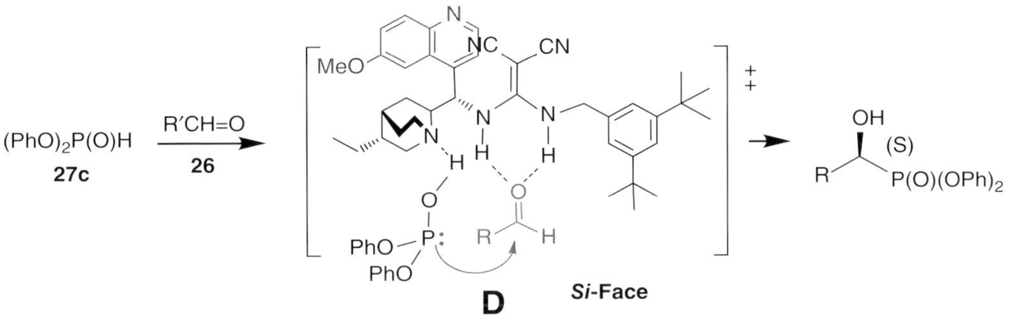

R	Yield (%)	ee (%)
4-NCC_6H_4	95	96
4-ClC_6H_4	98	93
2-ClC_6H_4	61	94
R=C_6H_5	97	90
1-Naphthyl	70	93
4-MeC_6H_4	80	91
4-$MeOC_6H_4$	53	92
Cyclohexyl	99	67

Scheme 5.9 Phospha-aldol reaction catalyzed by **26** and transition state **D**.

from aromatic aldehydes reacted with diphenylphosphite to give products **37** with quantitative yields and high enantioselectivities. Methyl hydrocupreine and hydroquinidine used as organocatalysts led to the formation of both enantiomers of compounds **37** with comparable enantioselectivity. It was supposed that in the TS, the hydroxyl group of the catalyst activates the imine because of hydrogen bond formation. Besides, the quinuclidine nitrogen atom of the the Brönsted base as catalyst activates phosphite (structure **E**). Desulfuration of hydrophosphonylation products and the subsequent deprotection of the phosphonate group led to the formation of optically active α-aminophosphonic acids (Scheme 5.10).

Pettersen *et al.* [35] studied the effect of chiral bases on the phospha-Mannich reaction of N-protected arylimines with diethylphosphite. The reaction was carried out in xylene at ambient temperature or at 20 °C. At low temperatures, the yields of reaction products **38** were reduced and the reaction rate decelerated, while the enantiomeric purity of the aminophosphonates decreased. Among the catalysts studied, the biggest efficacy was displayed by quinine. Electron-donating and electron-accepting substituents on a benzene ring had no significant effect on the reactivity and enantioselectivity. The authors have assumed that the free hydroxyl group at the C-9 atom of the catalyst enforced imine activation owing to hydrogen bond formation (Scheme 5.11).

5.4.2 Organocatalysis by Imines

Electron-rich aldimines showed higher activity, while the reaction of electron-deficient aldimines proceeded with low enantioselectivity, although with high yields of aminophosphonates [36]. Cyclic (*R*)-BINOL phosphoric acids used as chiral Brønsted

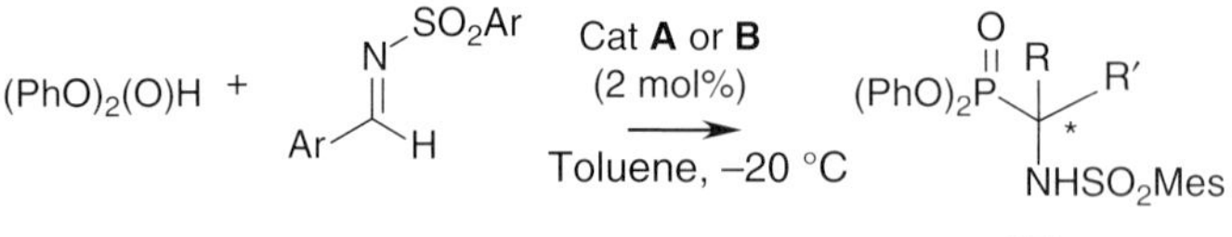

37

Ar = 6-Methyl-2-pyridyl
Cat A = Hydroquinine, Cat B = Hydroquinidine

R^1	R^2	cat	Yield (%)	ee (%)	Configuration
Ph	Me	A	99	97%	(*S*)-
p-Tolyl	Me	A	97	96%	(*S*)-
p-$MeOC_6H_4$	Me	A	99	97%	(*S*)-
p-ClC_6H_4	Me	A	99	94%	(*S*)-
p-FC_6H_4	Me	A	99	97%	(*S*)-
2-Naphthyl	Me	A	99	96%	(*S*)-
Cyclohexyl	Me	A	97	75%	(*S*)-
p-$MeOC_6H_4$	Me	B	91	94%	(*R*)-
2-Naphthyl	Me	B	91	93%	(*R*)-
Ph	Et	B	92	92%	(*R*)-
$R^1 + R^2$ = 1-Indanone		B	86	82%	(*R*)-

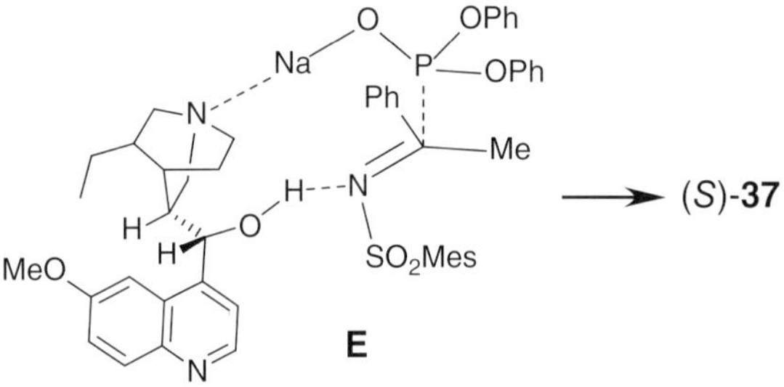

E

Scheme 5.10 Enantioselective hydrophosphonylation of aldimines catalyzed by hydroquinine or hydroquinidine.

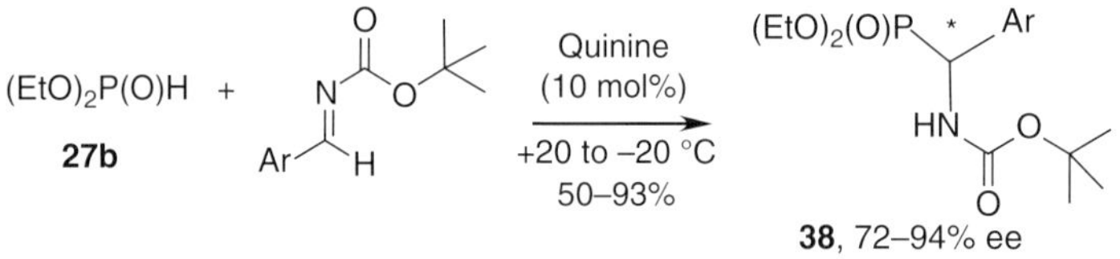

38, 72–94% ee

Ar=Ph, *p*-Tl, *m*-Tl, *p*-An, 2,5-$Me_2C_6H_3$, 2-naphthyl, *p*-ClC_6H_4, 1-naphthyl, 2-naphthyl

Scheme 5.11 Phospha-Mannich reaction of *N,N*-protected arylimines with diethylphosphite.

acids (10 mol%) catalyzed the hydrophosphonylation of aldimines with diisopropylphosphite at room temperature to give aminophosphonates with enantioselectivities ranging from good to high. Jacobsen and Joly [37] studied nucleophilic addition of di(*o*-nitrobenzyl)phosphite to *N*-benzylimines **39** in the presence of chiral thioureas **16** as catalyst. This reaction provides general and convenient access to a wide range of highly enantiomerically enriched α-aminophosphonates **40**. High enantioselectivities were obtained across a wide range of both aliphatic and aromatic substrates. In general, the best reaction rates were realized with aliphatic imines, while electron-poor aromatic substrates required longer reaction times and, in certain cases, elevated temperatures. Products of hydrophosphonylation **40** were transformed into α-aminophosphonic acids. The treatment of adducts with 20 mol% Pd/C under an atmosphere of hydrogen afforded enantiomerically enriched α-amino phosphonic acids with high yields and with retention of optical purity (Scheme 5.12).

5.4.3 Organocatalysis by Iminium salts

Tan *et al.* [38] have reported the phospha-Mannich reaction of H-phosphines with *N*-tosylimines **42** for the synthesis of α-aminophosphine oxides and phosphinates **43**. Guanidinium salt **16** showed the high catalytic activity in the reaction of P–C bond formation. The reaction was performed with threefold excess of *N*-phosphinates with K_2CO_3 as an additive achievement of high level of stereoinduction and comprehensible reaction rates. Aminophosphinates **43** with asymmetric phosphorus atom were obtained as mixture of two diastereoisomers with predominance of the *syn*-isomer (Scheme 5.13 and Table 5.2).

5.4.4 Organocatalysis by Chiral Brønsted acids

Reference [39] described an interesting example of the asymmetric Kabachnik–Fields reaction. The reaction of three-component mixture consisting of an aldehyde, *P*-anisidine, and di-3-pentylphosphite, catalyzed by chiral atropoisomeric acid **15b** (*p*-anthracenyl replaced the analog TRIP), led to the formation of aminophosphonates **44**, **45** in high yield, good stereoselectivity. Some derivatives of L-proline also effectively catalyze the asymmetric Kabachnik–Fields reaction. It was found that bulky alkyl substituents affect the stereoselectivity of reaction. The highest levels of stereoselectivity were attained in the case of aldehydes containing branched alkyl substituents (*i*-Pr, *c*-C_5H_9, *c*-C_6H_{11}). On the contrary, aldehydes containing R = Me or Et, reacted with low stereoselectivity (Scheme 5.14) [40]. Akiyama *et al.* [41] also studied the

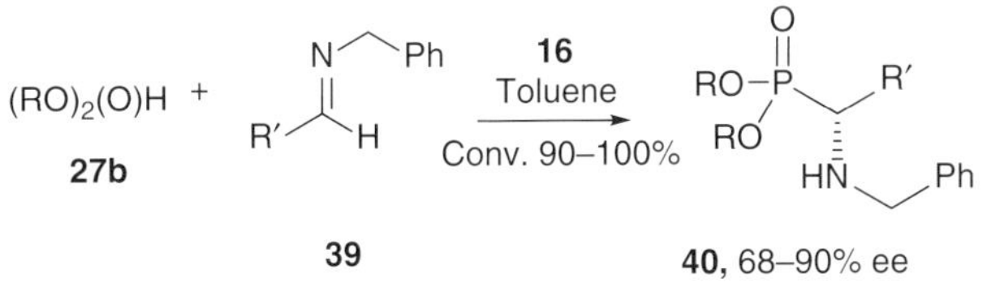

R=Ph, 2,2,2-$CF_3C_2H_4$, 2-NCC_2H_4, 2-ClC_2H_4, *p*-$NO_2C_6H_4$, *o*-$NO_2C_6H_4$; R′=Ph, R′=3-C_5H_{11}

Scheme 5.12 The addition of di(*o*-nitrobenzyl)phosphite to *N*-benzylimines catalyzed by chiral thioureas **15**.

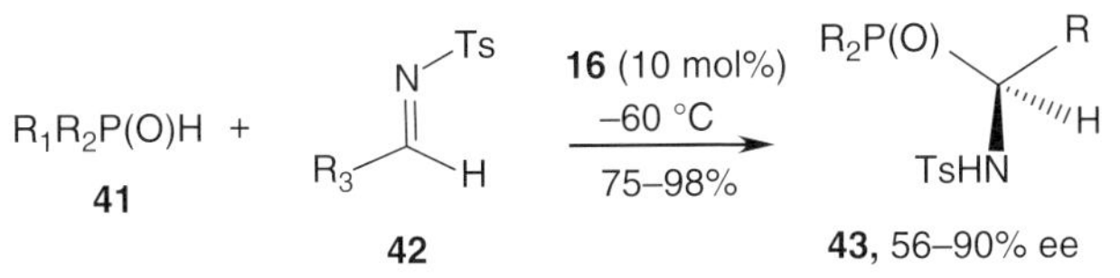

Scheme 5.13 Reaction of *H*-phosphines with *N*-tosylimines catalyzed by guanidinium salts **16**.

Table 5.2 Preparation of aminophosphinates **43**.

R[1]	R[2]	R[3]	Yield (%)	ee (%)
1-Nphth	1-Nphth	4-MeC_6H_4	98	92
1-Nphth	1-Nphth	4-FC_6H_4	97	90
1-Nphth	1-Nphth	2-Naphthyl	98	92
1-Nphth	1-Nphth	2-Furyl	92	87
1-Nphth	1-Nphth	Cy	95	70
1-Nphth	1-Nphth	*t*-Bu	89	91
1-Nphth	1-Nphth	*trans*-PhCH=CH	89	90
Ph	Ph	Ph	75	56
2-$CF_3C_6H_6$	2-$CF_3C_6H_6$	Ph	93	82
Ph	1-Nphth	Ph	90	75

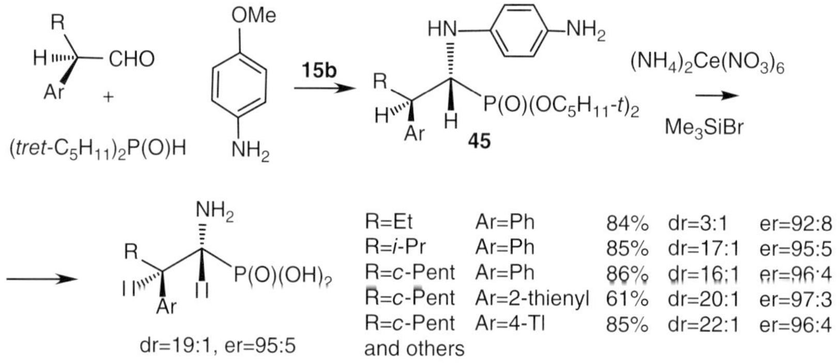

Scheme 5.14 Catalytic asymmetric three-component Kabachnik–Fields reaction.

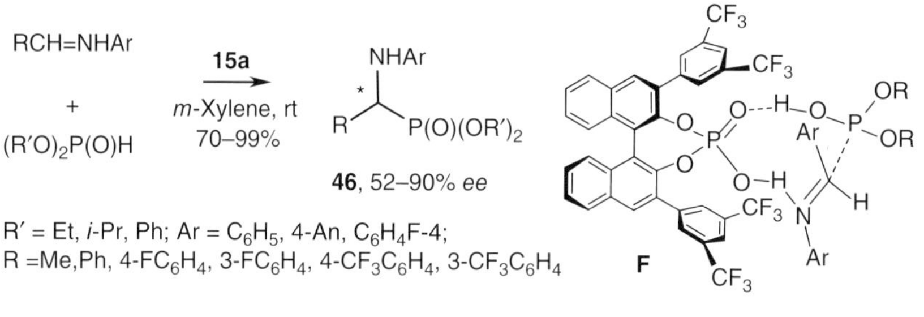

Scheme 5.15 Asymmetric reaction of achiral imines with dialkylphosphites.

reaction of imines with dialkylphosphites, catalyzed by the chiral Brønsted acid **15a**. The reaction led to the formation of α-aminophosphonates **44** with moderate yields and good enantioselectivities up to 90% ee (Scheme 5.15). The authors proposed a nine-membered TS **F** to explain the stereoselectivity of reaction. They came to the conclusion that phosphoric acid works as a bifunctional organocatalyst bearing both Brønsted acidic and Brønsted basic sites [42]. The substrate structure is the important

Table 5.3 Asymmetric reaction of achiral imines with dialkylphosphites (Scheme 5.12).

Entry	R	R	R′	Yield (%)	ee (%)	References
1	C_6H_5	MeO	Et	99	43	[41]
2	$C_6H_5CH{=}CH$	MeO	Et	70	73	[41]
3	*o*-$CH_3C_6H_4$	MeO	*i*-Pr	76	69	[41]
4	*o*-$NO_2C_6H_4$	MeO	*i*-Pr	72	77	[41]
5	$C_6H_5CH{=}CH$	MeO	*i*-Pr	92	84	[41]
6	*p*-$CH_3C_6H_4CH{=}CH$	MeO	*i*-Pr	88	86	[41]
7	*p*-$ClC_6H_4CH{=}CH$	MeO	*i*-Pr	97	83	[41]
8	*o*-$CH_3C_6H_4CH{=}CH$	MeO	*i*-Pr	80	82	[41]
9	*o*-$ClC_6H_4CH{=}CH$	MeO	*i*-Pr	82	87	[41]
10	*o*-$NO_2C_6H_4CH{=}CH$	MeO	*i*-Pr	92	88	[41]
11	*o*-$CF_3C_6H_4CH{=}CH$	MeO	*i*-Pr	86	90	[41]
12	1-Naphthyl-CH=CH	MeO	*i*-Pr	76	81	[41]
13	*m*-$CF_3C_6H_4$	PhCH=CH	Et	64	83.6	[43]
14	*p*-$CF_3C_6H_4$	PhCH=CH	*n*-Pr	65	82.8	[43]
15	*m*-$CF_3C_6H_4$	PhCH=CH	*n*-Pr	68	88.	[43]
16	*p*-$CF_3C_6H_4$	PhCH=CH	*n*-Bu	71	83.7	[43]
17	*m*-$CF_3C_6H_4$	*p*-$FC_6H_4CH{=}CH$	*n*-Bu	73	90.6	[43]

factor influencing the result of reaction, because the reaction of diisopropylphosphite with aldimines, containing electron-accepting groups in the *ortho*-position of the phenyl ring (CF_3, NO_2, Cl) proceeded with the highest enantioselectivities (Table 5.3). Yamanaka and Hirata [44] has performed density functional theory (DFT) theoretical studies (BH and HLYP/6-31G*, Gaussian 98 package) on the mechanism of this reaction. The calculations confirmed that the reaction proceeded via the nine-membered zwitterionic TS with the chiral phosphoric acid, where the aldimine and phosphite could be activated by the Brønsted acidic site and Lewis basic site, respectively. The *Si*-facial attacking TS was less favored by the steric repulsion of the 3,3′-aryl groups on the chiral phosphoric acid with the bulky phosphite. When using the aldimine derived from benzaldehyde, the *Re*-facial attacking TS was destabilized to decrease the enantioselectivity in agreement with experiment. Bhadury and Li [43] synthesized fluorinated aminophosphonates **46** possessing antibacterial properties (Scheme 5.15). They confirmed that the reaction of imines with dialkylphosphites in the presence of chiral acids **15a** in xylene proceeds with high enantioselectivity to give the α-aminophosphonates **46** with greater than 90% ee and yields of 30–65% (Table 5.3). The phospha-Mannich reaction catalyzed by a chiral magnesium BINOL phosphate salt was also reported [45].

Bhusare *et al.* [46] reported an enantioseletive one-pot synthetic method for the syntheses of optically active α-aminophosphonates by employing an organocatalyst. Some new organocatalysts were synthesized and studied for their reactivity and enantioselectivity (Scheme 5.16). The organocatalyst **46** effectively catalyzed the reaction and provided the corresponding α-aminophosphonates in high yields (71–90%) with excellent enantiomeric excess values (73–92%). Although several similar organocatalysts were able to catalyze the synthesis of the α-aminophosphonate efficiently, the organocatalyst (*S*)-1-acetyl-*N*-tosylpyrrolidine-2-carboxamide **46** was the best and provided very high yields for a wide range of α-aminophosphonates with good enantioselectivies (Scheme 5.16).

A highly enantioselective addition of diphenylphosphite to ketimines derived from isatins has been achieved using a bifunctional organocatalyst, quinine-derived squaramide catalyst 7. This method works efficiently with several ketimines to produce the corresponding 3-amino-2-oxoindolin-3-yl-phosphonates **47** in excellent yields and with high enantioselectivity (up to 98% ee) [47, 48]. A variety of chiral 3-amino-2-oxoindolin-3-ylphosphonates were prepared in good yields and with high optical purity (up to 98% ee) using this method. Among various N-protected isatin imines, the best enantiomeric excesses were obtained with ketamine 7 derived from *N*-benzylisatin. Chimni has developed a highly enantioselective addition of diphenylphosphite to ketimines derived from isatins employing bifunctional thiourea–tertiary amine organocatalysts. Isatin-derived ketimines react readily with diphenylphosphite in the presence of Cinchona-derived thiourea (epiCDT) to provide biologically important chiral 3-substituted 3-amino-2-oxindoles in good yields (up to 88%) and very high enantioselectivity (up to 97% ee) [48] (Scheme 5.17).

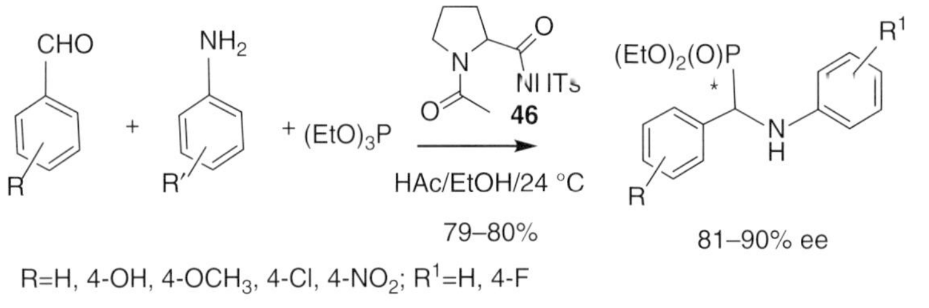

Scheme 5.16 Organocatalytic asymmetric three-component Kabachnik–Fields reaction.

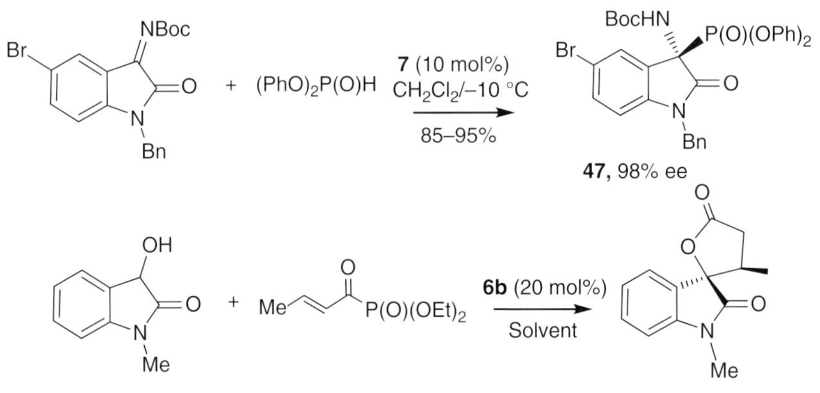

Scheme 5.17 Phospha-Mannich reaction of diphenylphosphite with *N*-Boc ketimines.

5.5 Phospha-Michael Reaction

Asymmetric Michael addition of trivalent phosphorus acids to enones can be catalyzed by various organocatalyst [49–55]. For example, chiral alkaloids and amino acids are effective catalysts of the phospha-Michael reaction. Quinine, dihidroquinine, and their derivatives, especially thioureas of cinchona alkaloids, were used successfully as organocatalysts of phospha-Michael addition.

5.5.1 Organocatalysis by Cinchona Alkaloids

Lattanzi and Russo [50] reported enantioselective asymmetric addition of diphenylphosphine oxides to chalcones, catalyzed by dihydroquinine. The reaction proceeded in good yields and optical purity, with the products **48** attaining enantioselectivity of 89% ee. The crystallization of enantiomerically enriched adducts **48** allowed to obtain enantiomerically pure compounds. Other alkaloids: cupreine, cupreidine, and thiourea derivatives of cinchona alkaloids were inefficient catalysts. On the contrary, dihydroquinine efficiently catalyzed the Michael addition reaction. On the basis of the absolute configuration of adducts **48**, a transition-state **G** was proposed for the dihydroquinine-catalyzed phospha-Michael addition. The diarylphosphane oxide was activated by the quinuclidine nitrogen atom by shifting the equilibrium toward the reactive phosphinous acid form and the preferential attack of the activated nucleophile toward the *Re* face of chalcone led to the (*R*)-absolute configuration of the reaction products (Scheme 5.18). The Michael addition of secondary phosphines and phosphites to nitroalkenes represents a convenient synthetic route to optically active β-nitrophosphonates which, because of the high synthetic variability of the nitro group, can be easily transformed to chiral functionalized phosphonates. For example, the Michael addition of diphenylphosphite to nitroalkenes, catalyzed by quinine, represents a convenient method for the synthesis of enantiomerically enriched β-nitroalkylphosphonates **49** which were converted to chiral β-aminophosphonates.

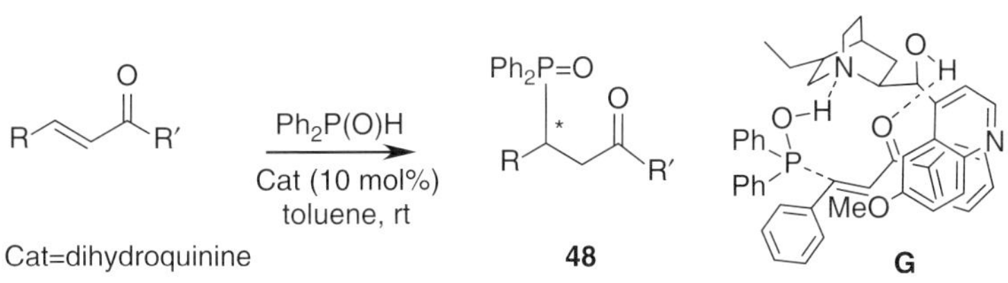

Ph	Ph	91%	ee= 80%
4-MeC_6H_4	Ph	75%	ee= 78%
3-BrC_6H_4	Ph	98%	ee= 76%
4-$NO_2C_6H_4$	Ph	99%	ee= 65%
2-$MeOC_6H_4$	Ph	87%	ee= 89%
2-ClC_6H_4	Ph	80%	ee= 79%
Ph	4-$MeOC_6H_4$	99%	ee= 77%
Ph	4-BrC_6H_4	95%	ee= 82%
Ph	4-ClC_6H_4	85%	ee= 79%
Cyclohexyl	Ph (p)	30%	ee= 60%

Scheme 5.18 1,4-Addition of diarylphosphines to enones catalyzed by dihydroquinine.

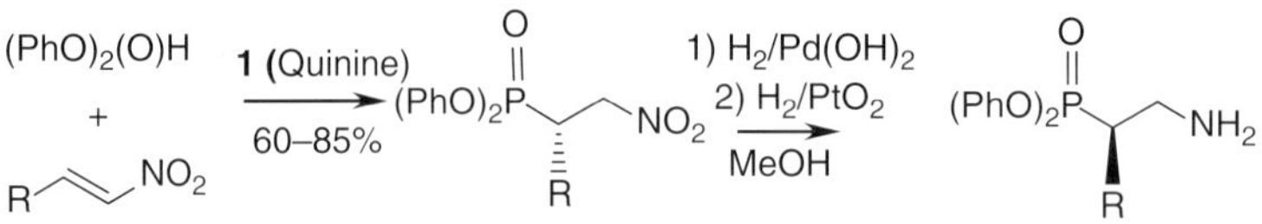

49, 40–88% ee   48%

R=Ph	82%	ee = 70%
R=4-F-C_6H_4	85%	ee = 77%
R=4-Cl-C_6H_4	82%	ee = 72%
R=4-Me-C_6H_4	83%	ee = 80%
R=4-MeO-C_6H_4	78%	ee = 75%
R=3-BnO-4-MeO-C_6H_3	82%	ee = 82%
R=2-Thiophene	79%	ee = 88%
R=2-Furan	67%	ee = 72%
R=$PhCH_2CH_2$	68%	ee = 63%
R=*n*-C_5H_{11}	62%	ee = 60%

Scheme 5.19 Reaction of diphenylphosphite with nitroalkenes catalyzed by quinine.

A number of enantiomerically enriched β-nitrophosphates and aminophosphonates bearing different aromatic and heterocyclic groups at the α-carbon atom were obtained by this method (Scheme 5.19) [51]. The addition of molecular sieves (MSs, 4 or 3 Å) to the reaction mixture improves the yields and enantioselectivities. The MS act as a scavengers of water and acid to remove the impurities in phosphite samples. After such a modification of the reaction procedure, high and reproducible yields of the β-nitrophosphonates **49** were obtained with good enantioselectivities (up to 88% ee) [52].

5.5.2 Organocatalysis by Thiourea

The development of *N,N*-dialkylthyourea derivatives as effective bifunctional organocatalysts for the Michael addition of chalcones and malonates to nitroolefins [53–55] led Melchiorre to apply these catalysts to the asymmetric addition of phosphines to nitroolefins [53]. This organocatalytic approach, which provides a direct route to potentially useful enantiopure P,N-ligands, constitutes a bridge between the two complementary areas of asymmetric catalysis: organo- and metal-catalyzed transformations. Various chiral *N,N*-dialkylthiourea catalysts were tested, including $(DHQ)_2$PHAL, thiourea-based derivatives of aminonaphthalene, and others. However, only the thiourea-based derivatives of cinchona alkaloid derivatives **6b** allowed to obtain satisfactory results: 86% yield and 67% ee of the addition product **50**. The reduction of nitro compounds **50** to amines followed by crystallization of the products increased the enantiomeric excess of the aminophosphines **51** up to 99% ee (Scheme 5.20).

High efficiency in phospha-Michael reactions have been found with urea derivatives of cinchona alkaloids [56, 57]. For example, Wen *et al.* [56] used thiourea–quinones **6** for the enantioselective organocatalytic phospha-Michael reaction of cyclic β-unsaturated ketones with diarylphosphine oxides. Optically active products **52** and **53**, bearing quaternary chiral carbon stereocenters were prepared in high yields and with enantioselectivities up to 98% ee (Scheme 5.21).

The Michael addition of α-nitrophosphonates to enons and to α-substituted nitroolefins, catalyzed by a cinchona-based organocatalyst, proceeded with high

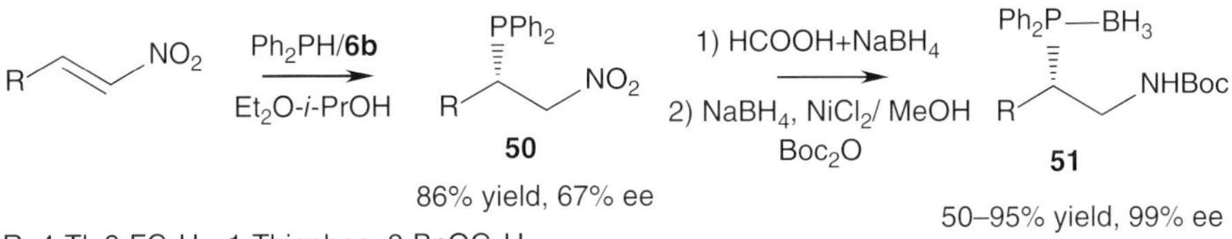

R=4-Tl, 2-FC$_6$H$_4$, 1-Thiophen, 2-BnOC$_6$H$_4$

Scheme 5.20 Asymmetric hydrophosphination of nitroalkenes.

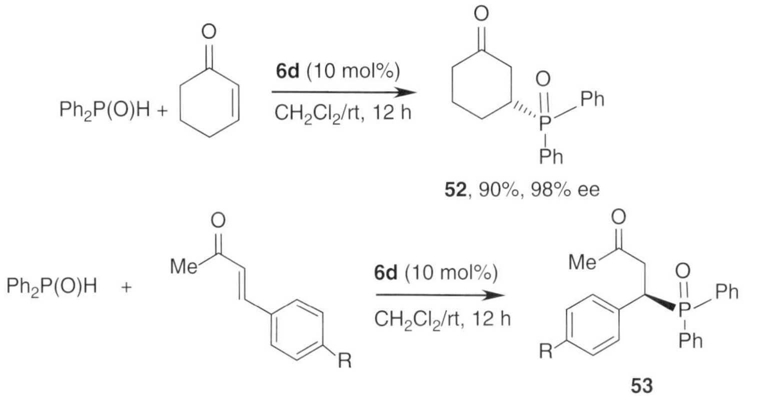

R=NO$_2$, Yield 90%, ee 92%; R=MeO, Yield 97%, ee 85%; R=Br, Yield 94%, ee 94%

Scheme 5.21 Phospha-Michael reaction catalyzed by urea derivatives of cinchona.

diastereo- and enantioselectivity leading to the formation of nitroalkylphosphonates **54**, **55**, bearing quaternary and tertiary stereocenters [58, 59]. The addition of α-nitroethylphosphonates to acrylic acid aryl esters, catalyzed by cinchona-based squaramides, resulted in nitroalkylphosphonates **55** bearing quaternary stereocenters in good yields and high enantioselectivities. Nitrophosphonates **55** were converted into cyclic quaternary α-aminophosphonates **56** by intramolecular reduction–cyclization or Baeyer–Villiger oxidation, as shown in Scheme 5.22 [58]. The additions of α-substituted nitrophosphonates to various nitroolefins proceeded also with high diastereo- and enantioselectivity when catalyzed by a quinine-derived thiourea–tertiary amine bifunctional catalyst **5b** and generated α,γ-diaminophosphonic acid precursors **57** with contiguous quaternary and tertiary stereocenters [60]. The Michael addition of α-nitrophosphonates to vinyl ketones initiated by quinine-squaramide catalyst **5c** afforded the addition products **58** with the highest yields and enantioselectivity [61].

α-Nitro-γ-sulfonyl phosphonates **36** with a key tetrasubstituted chiral α-carbon center have been synthesized in high yield and enantioselectivity through a quinine-squaramide-catalyzed conjugate addition of α-nitro phosphonates to aryl vinyl sulfones with good enantioselectivity. The enantioselectivities were 90–98% ee, when the aryl was benzenoid and 74–79% ee with a tetrazolyl group. The reduction of nitro and sulfonyl groups led to the formation of aminophosphonates with quaternary α-carbon atom (Scheme 5.23) [61a].

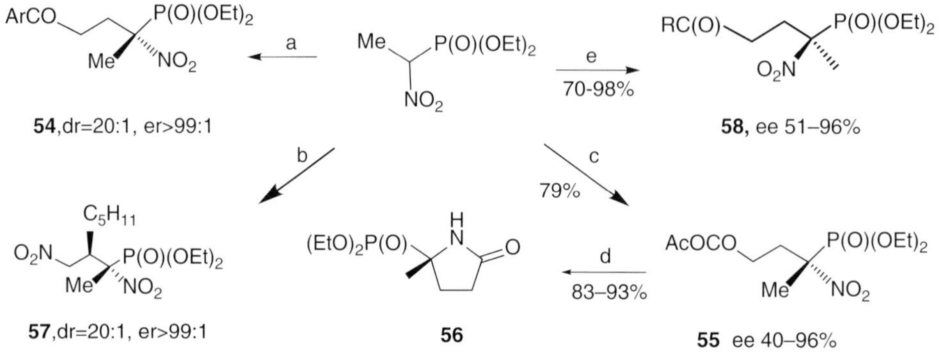

Reagents and conditions: (a) 3,4-$(MeO)_2C_6H_3COCH{=}CH_2$/**5a**; (b) $O_2NCH{=}CHC_5H_{11}$/**5b**; (c) $ArOCOCH{=}CH_2$/**5b**/$PhCF_3$/-10 °C; (d) H_2/Pd-C; (e) RC(O)$CH{=}CH_2$ (R=Ph, Tl, An, 2-furyl, 2-anisyl, 1-Nphth, and others), **5c**, mesitylene, xylene, –65 °C

Scheme 5.22 Michael addition of α-nitrophosphonates to enons catalyzed by **5a–d**.

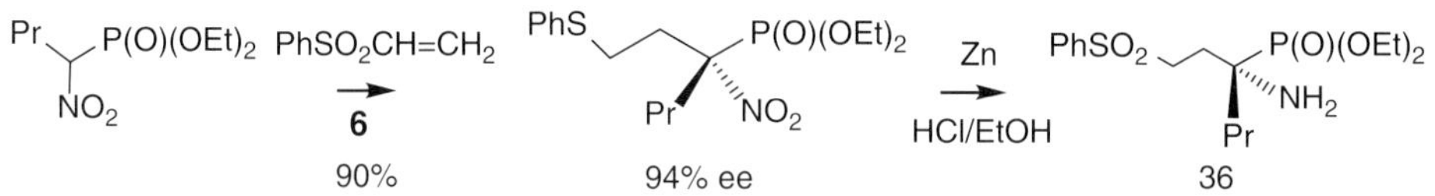

Scheme 5.23 Synthesis of chiral aminophosphonates with tetrasubstituted α-carbon atom.

5.5.3 Organocatalysis by Iminium salts

Enantioselective phospha-Michael addition of diphenyl phosphonate to nitroolefins was also achieved by means of conformationally flexible 1,3-diamine-tethered guanidinium/bisthiourea organocatalyst **9** (Table 5.1). Nitroolefins bearing various aromatic and aliphatic substituents reacted with nitroalkenes to afford the phospha-Michael adducts **59** with 90–98% ee. Monomeric or oligomeric catalysts can be used, depending on the presence or absence of water. The addition of water improved the enantioselectivity up to 98% ee. Among the solvents tested, the highest enantioselectivities (89% ee) were observed in toluene. The reaction proceeded at 1 mol% of loading catalyst to give the addition products **59** with 99% yield and 95% ee. The absolute configurations of the products were determined to be (*R*) on the basis of comparison of optical rotations of the synthesized products with known compounds (Scheme 5.24) [62, 63].

Terada *et al.* [15] have applied axial chiral derivatives of guanidine **12** for asymmetric initiation of the phospha-Michael reaction. This type of organocatalysts is very convenient for asymmetric addition of malonates to nitroolefins and enantioselective aminating of cyclic ketones. The catalyst **12** as Brönsted base catalyzes enantioselective reactions via deprotonation of 1,3-dicarbonyl compounds. Axially chiral guanidine **12** catalyzed the Michael addition of diphenylphosphite to nitroalkenes with the formation of β-nitrophosphonates **60** with enantioselectivities of 87–97% ee. Increasing the volume of the *N*-alkyl or aryl substituents in guanidines **12a–c** increased the enantioselectivity of the catalyst at low loading (1 mol%). The phospha-Michael reaction allowed to phosphonylate aromatic, heteroaromatic, and aliphatic nitroalkenes. Authors

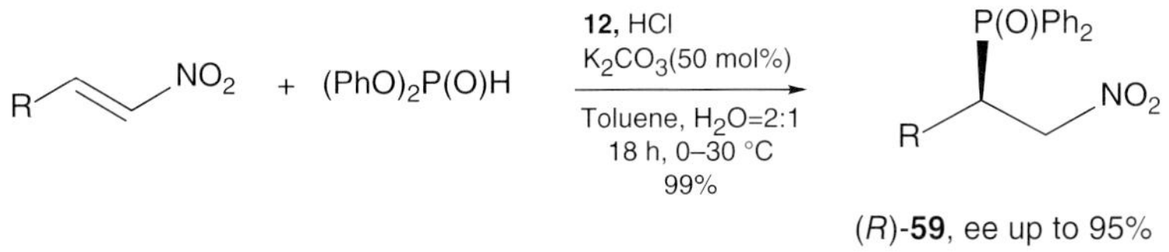

R=4-MeC$_6$H$_4$	88%	ee=91%
R=4-ClC$_6$H$_4$	88%	ee=90%
R=4-MeOC$_6$H$_4$	90%	ee=96%
R=2-Naphthyl	81%	ee=98%
R=2-Thienyl	77%	ee=95%
R=2-Furyl	78%	ee=95%
R=*c*-C$_6$H$_{11}$	98%	ee=92%
R=*t*-Bu	95%	ee=92%
R=PhCH$_2$CH$_2$	95%	ee=92%

Scheme 5.24 Phospha-Michael reaction with nitroolefins catalyzed by guanidinium/bisthiourea.

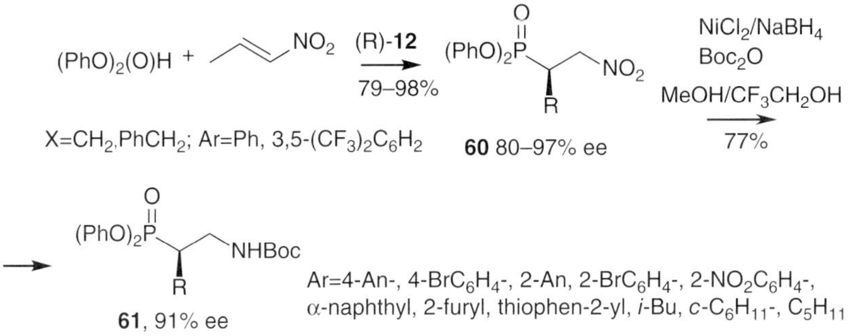

Scheme 5.25 Phospha-Michael phosphonylation of nitroalkenes catalyzed by guanidine (*R*)-**12**.

showed the applicability of Michael adducts for the synthesis of biologically important β-aminophosphonates **61** (Scheme 5.25).

Chiral guanidines and guanidinium salts are excellent enantioselective catalysts for a variety of reactions including the Strecker, Diels–Alder, and Michael reactions. Tan *et al.* [14] reported that chiral guanidines **13** catalyze the reaction of diarylphosphine oxides **62** with nitroalkenes **61** resulting in the formation of chiral β-aminophosphine oxides **63** with high enantioselectivities. They found that with 10 mol% of chiral bicyclic guanidine, the addition of diphenyl phosphine oxide to β-nitrostyrene proceeded smoothly in various solvents. The reaction between di(1-naphthyl) phosphine oxide and β-nitrostyrene was optimized to 91% ee by lowering the reaction temperature to 40 °C. The enantiomeric purity of compounds **63** was improved to 99% ee after a recrystallization from MeOH or *t*-BuOMe–CH_2Cl_2. The reaction proceeded easily even at a catalyst loading of 1 mol%, and despite such a low loading of the catalyst, aromatic, heteroaromatic, and aliphatic nitroalkenes reacted with dialkylphosphites, furnishing the adducts **63** in high yields and enantioselectivities. The method was used for the synthesis of biologically important β-aminophosphonates **64**. The reduction of β-nitroalkylphosphine oxide **63** with zinc in muriatic acid and then with trichlorosilane led to the formation of enantiomerically pure β-aminophosphines **64** with 99% ee (Scheme 5.26).

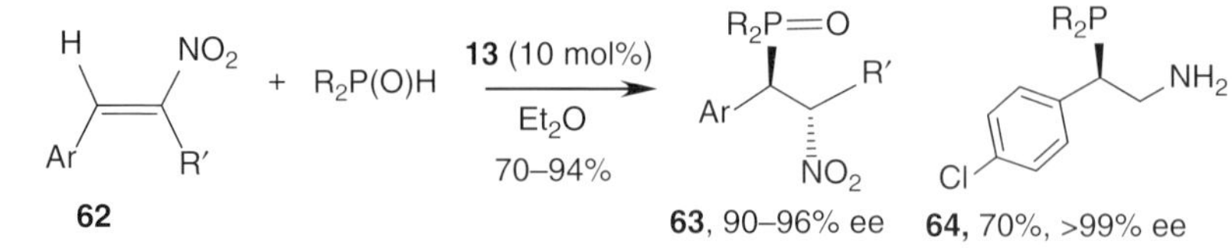

R=1-Naphthyl

Ar=Ph, 4-ClC_6H_4, 3-ClC_6H_4, 2-ClC_6H_4, 4-BrC_6H_4, 3-$NO_2C_6H_4$, 2-$NO_2C_6H_4$, 4Tl, 2-Nphth

Scheme 5.26 Reaction of diarylphosphine oxides with nitroalkenes catalyzed by chiral guanidines **13**.

5.5.4 Organocatalysis by N-Heterocyclic Carbenes

Since the first reports of stable nucleophilic carbenes by Wanzlick [18] and Arduengo *et al.* [19], the broad application of NHCs in organic synthesis has been impressively demonstrated. Besides their role as excellent ligands in metal-based catalytic reactions, organocatalytic carbene catalysis has emerged as an exceptionally fruitful research area in synthetic organic chemistry, including organophosphorus chemistry. Thus, Cullen and Rovis [64] have developed an intramolecular asymmetric Stetter reaction employing vinylphosphine oxides and vinylphosphonates as electrophilic acceptors (Scheme 5.27). Both aromatic and aliphatic substrates were tolerated providing cyclic ketophosphonates and phosphine oxides **66** and **68**. The treatment of aldehydes **65** and **67** with NHC catalyst **23** led to the addition of an acyl anion equivalent to a vinylphosphine oxide (Scheme 5.27) or a vinylphosphonate Michael acceptor and the formation of compounds **66**, **67** in good to excellent yields and enantioselectivities. The addition of α,β-unsaturated aldehydes to α-ketophosphonates under carbene catalysis afforded enantioenriched lactones **68** [65] (Scheme 5.28).

5.5.5 Organocatalysis Using Proline Derivatives

Various proline derivatives, in particular the compounds **8a–c**, displayed high organocatalytic activity in phospha-aldol reactions. For example, the asymmetric addition of diphenylphosphine to enals, catalyzed by pyrrolidine derivatives **8a–d**, led to the formation of chiral tertiary phosphines **69** containing aryl, heteroaryl, alkyl, and alken (Scheme 5.29) [66, 67]. The reaction proceeded as a 1,4-addition of diphenylphosphine to conjugated bonds of unsaturated substrates. The highest enantioselectivities were obtained in the reaction of enals with diphenylphosphine, catalyzed by diarylprolinol **8c** bearing 3,5-bis(trifluoromethyl)phenyl groups, in toluene

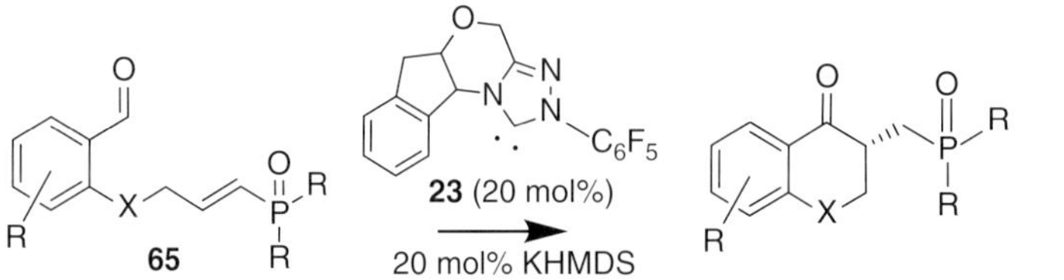

X=O, S, CH_2
R=EtO, PhO, Ph

66, 65–99%; up to 95% ee

Scheme 5.27 Intramolecular asymmetric Stetter reaction of vinylphosphineoxides **65**.

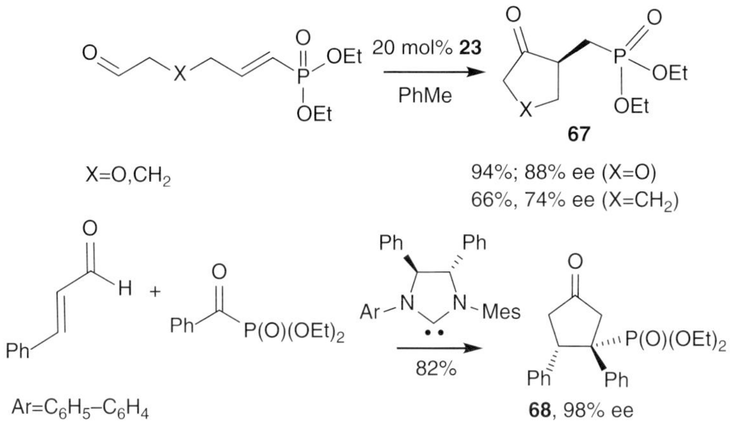

Scheme 5.28 Intramolecular asymmetric Stetter reaction of vinylphosphonates **68**.

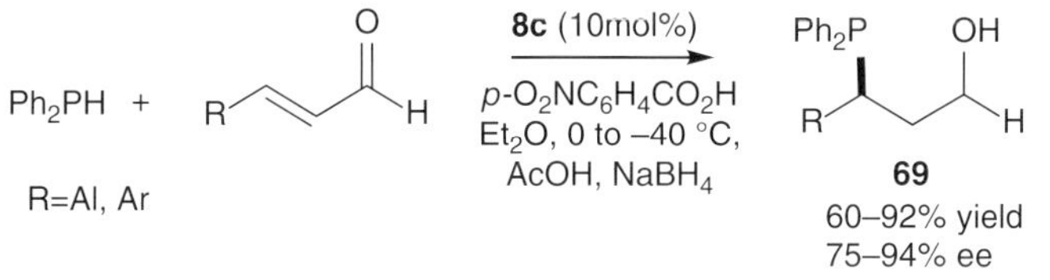

Scheme 5.29 Phospha-Michael reaction catalyzed by urea derivatives of cinchona.

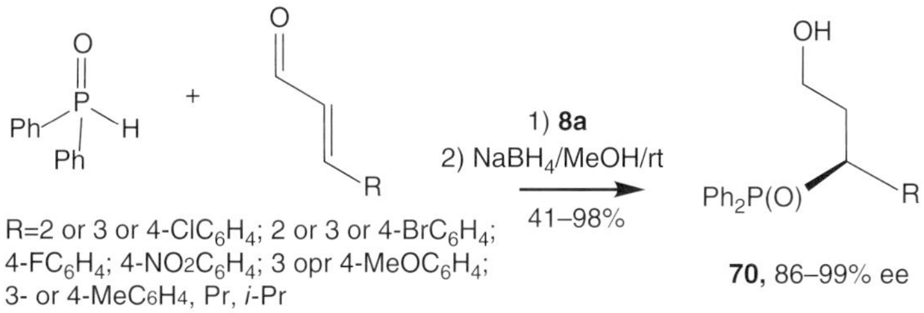

Scheme 5.30 Reaction of enals with diphenylphosphine.

or chloroform in the presence of additives of 2-fluorobenzoic acid or 4-nitrobenzoic acid [66]. The stereoselectivity of the reaction was explained by the formation of iminium complexes with a *trans*-configuration. The stereoselectivity has its origins in a combination of the orientations of the iminium complexes with *E-trans*-configuration, and efficient steric shielding of the *Re*-face of this iminium complex by the bulky chiral group of the protected diarylprolinol catalyst (Scheme 5.30) [68].

The enantioselective phospha-Michael reaction of α,β-unsaturated aldehydes with diaryl phosphine oxides catalyzed by (*S*)-2-(diphenyl(trimethylsilyloxy)methyl)pyrrolidine **8a** was reported by Ye and co-workers [56]. The catalytic reaction afforded 1,4-addition adducts **71, 72** in good to excellent enantioselectivities and yields for a broad range of enals, including aromatic and aliphatic α,β-unsaturated aldehydes (Scheme 5.31).

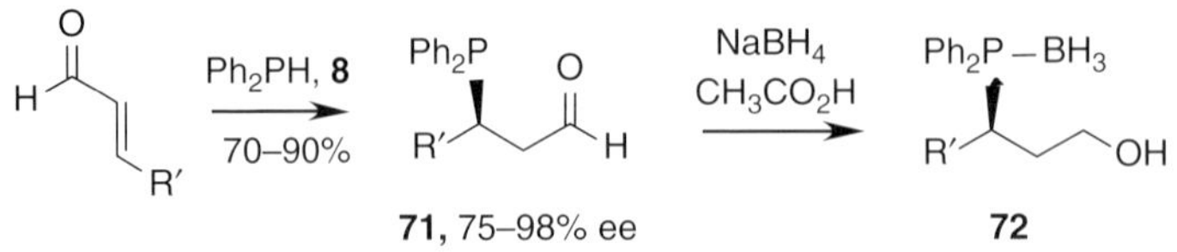

R = Me , Et, *i*-Pr, CHCH=CH, Ph, 4-MeOC$_6$H$_4$, 4-NO$_2$C$_6$H$_4$, 3-NO$_2$C$_6$H$_4$, 4-ClC$_6$H$_4$, *o*-Cl-C$_6$H$_4$, 4-BrC$_6$H$_4$, 3-BrC$_6$H$_4$, 2-Naph, 2-furyl, BnOCH$_2$CH$_2$CH$_2$, PhC(=O)CH$_2$CH$_2$CH$_2$

Scheme 5.31 Reaction of enals with diphenylphosphine oxide.

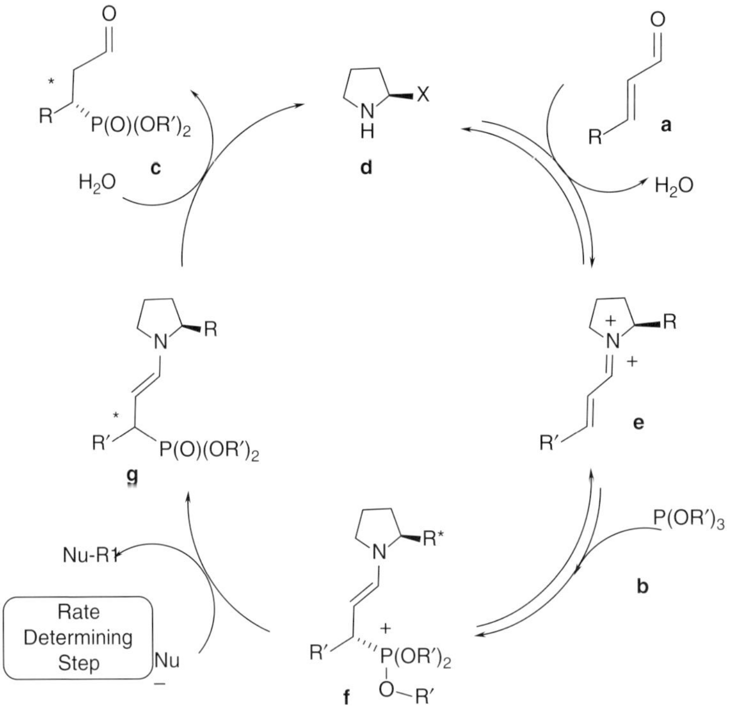

Scheme 5.32 Mechanism of asymmetric hydrophosphination of α,β-unsaturated adehydes.

Cordova *et al.* [69] studied the mechanism of asymmetric hydrophosphination of α,β-unsaturated aldehydes. They performed DFT calculations on the P–C bond-forming step, using the Gaussian 03 software package (Scheme 5.32) [68]. The TS leading to the (*S*)-product was calculated to have the lowest energy. The attack takes place at the face of the *E*-iminium ion that is not shielded by the bulky group of the catalyst. The attack on the shielded face of the *E*-iminium ion which yields the (*R*) product, is 1.5 kcal mol^{-1} higher in energy. They came to a conclusion that the bulky substituent shields *Re*-face (R = Ar) of the *E*-iminium ion leading to *Si*-facial attack. The rate-determining step for the reaction is the conversion of P(III) to P(V), which occurs via a nucleophilic S$_N$2-type dealkylation. The mechanistic studies of this reaction [70, 71] showed that the first step of the catalytic process, after the iminium ion **e** formation, is the addition of phosphite **d** to the β-carbon atom of **a**, leading to the phosphonium ion-enamine intermediate **f**.

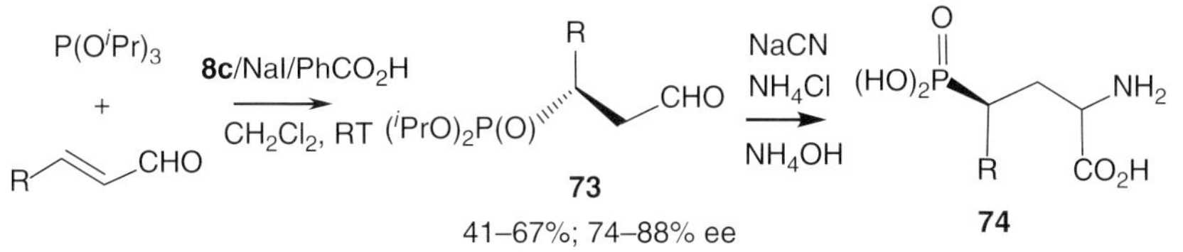

R = Pr, Et, Ph, *p*-An, *p*-ClC$_6$H$_4$, *p*-NMe$_2$C$_6$H$_4$, (*Z*)-*n*-hex-3-en, CH$_2$CH$_2$OTBDMS, 2-furyl, 2-thiophene

Scheme 5.33 Enantioselective phosphonylation of α,β-unsaturated aldehydes with trialkylphosphites catalyzed by diarylprolinol **8c**.

The next step is a transformation of P(III) to P(V), which was performed via a nucleophilic substitution on the alkyl chain in the α-position to the oxygen atom of **f**, leading to the phosphonate-enamine intermediate **g**. The overall result of these two steps is the conversion of trivalent phosphorus into pentavalent phosphorus. Hydrolysis of **g** regenerates the catalyst and liberates the optically active phosphonate **c**.

Jørgensen reported an enantioselective phosphonylation of α,β-unsaturated aldehydes with trialkylphosphites, catalyzed by diarylprolinol **8c** in combination with a Brønsted acid and a nucleophile. Organocatalytic enantioselective reaction of β-phosphonylated aliphatic α,β-nonsaturated aldehydes proceeded with good yields and enantioselectivities in average on the level of 84–85% ee. Optically active β-aldehydophosphonates **73**, formed as a result of β-phosphonylation of β-nonsaturated aldehydes, are chiral synthons and can be used for the preparation of biologically active molecules **74** [70]. In general, the reaction furnished the phosphonates in good yields and, with high enantioselectivities in case of various enals, which are derivatives of aliphatic, aromatic, heteroaromatic α,β-nonsaturated aldehydes. The reaction provides also good results with aromatic α,β-unsaturated aldehydes, such as cinnamaldehyde and its *para*-substitution derivates. For these aromatic α,β-unsaturated aldehydes, the corresponding phosphonates were obtained with satisfactory yields and enantioselectivities of 41–88% ee (Scheme 5.33).

The application of organocatalysis in the Michael asymmetric addition of nucleophiles to vinyl-bis-phosphonates, activated by electronegative phosphonous groups attracted considerable attention. Alexakis *et al.* [72, 73] have reported that, in the presence of chiral diphenylprolinol, **8d** aldehydes add to tetraethyl methylene-bis-phosphonates **76** with the formation of geminal bis-phosphonates **77** (Scheme 5.34). The reaction was completed in 12 h at ambient temperature in the presence of 20% of catalyst in $CHCl_3$ and afforded the Michael adducts **77** in yield of 80% and with enantioselectivities up to

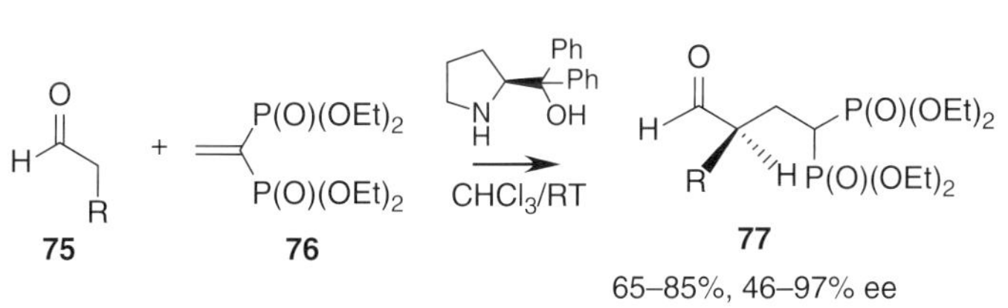

R=*i*-Pr, Yield 80%, ee 90%; R=Me, Yield 75%, ee 75%; R=*t*-Bu, Yield 85%, ee 97%; R=Pr, Yield 75%, ee 86%; R=*i*-Pr, Yield 71%, ee 91%; R=Bn, Yield 81%, ee 85%

Scheme 5.34 Asymmetric conjugate addition (ACA) of aldehydes to vinyl bis-phosphonates.

90% ee. The enantioselectivity decreased from 90% to 80% ee on lowering the temperature. Reduction of loading of catalyst to 10% practically did not influence the enantioselectivity level. The determination of the absolute configuration of **77** allowed to postulate a Michael acceptor attack from the *Si*-face of the *E*-enamine (Scheme 5.35). An enantioselective method for the synthesis of α-methylene-δ-lactones and δ-lactams catalyzed by diarylprolinol **8c** was proposed. This methodology used the Michael addition of unmodified aldehydes to ethyl 2-(diethoxyphosphoryl)acrylates as a key step with formation of enantiomerically enriched adducts **81** that were transformed into compounds **83**, with retention of enantioselectivity. The methodology allowed the preparation of optically active γ-substituted α-methylene-δ-lactones **82**, and δ-lactams **84** (Scheme 5.36) [74].

Jørgensen reported the addition of cyclic β-ketoesters **85**, **87** to ethylene-bis-phosphonate **76**, catalyzed by dihydroquinine leading to optically active geminal bis-phosphonates **86**, **88**, bearing the stereocenter with quaternary carbon atom, in good yields and with high enantioselectivities (Schemes 5.37 and 5.38) [70].

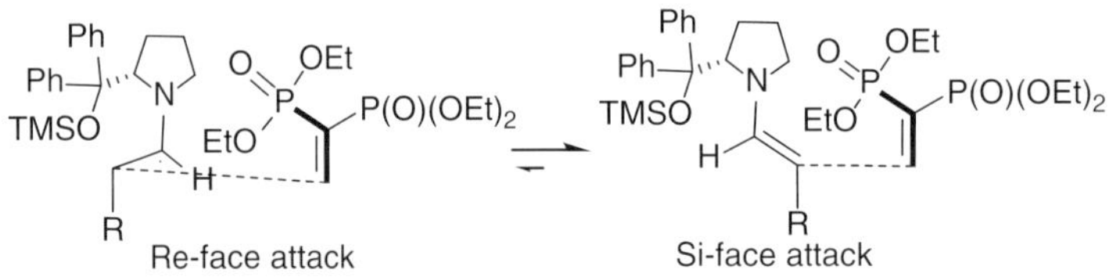

78 **79**

Scheme 5.35 Transition state for a Michael acceptor attack from the *Si*-face of the *E*-enamine.

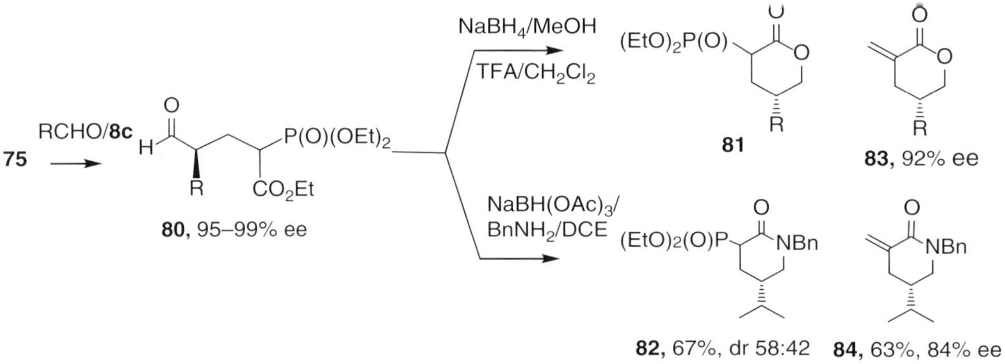

Scheme 5.36 Michael addition of aldehydes to ethyl 2 (diethoxyphosphoryl)acrylates.

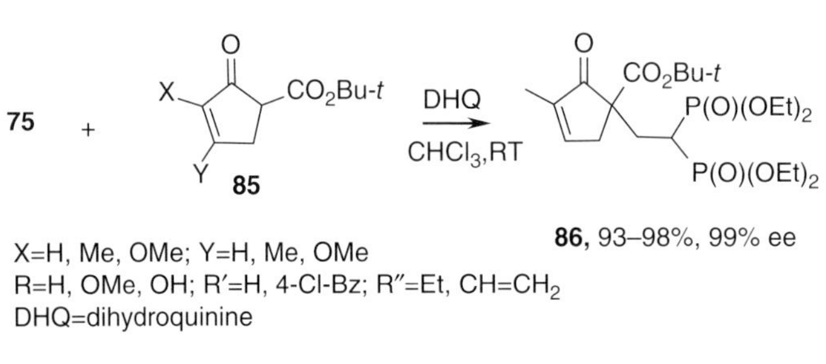

Scheme 5.37

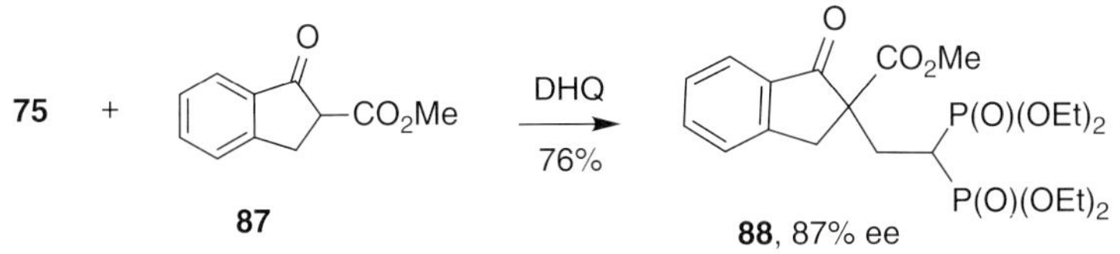

Scheme 5.38 Addition of cyclic β-ketoesters **86, 88** to ethylene-bis-phosphonate **85**.

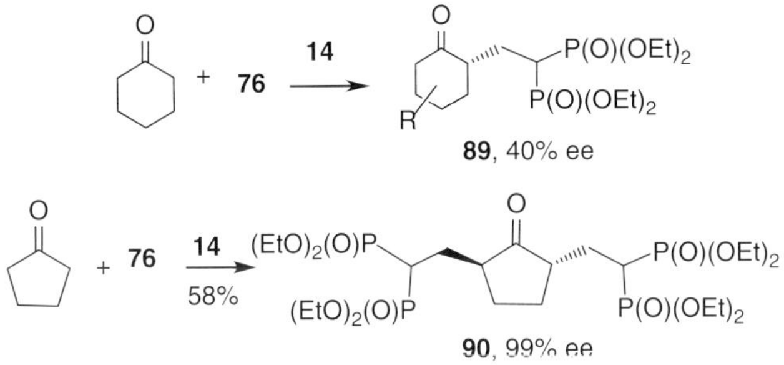

Scheme 5.39 Synthesis of chiral α-keto-bis-phosphonates.

Barros and Phillips synthesized chiral γ-keto-bis-phosphonates by the Michael addition of cyclic ketones to vinyl-bis-phosphonates catalyzed with 0.1 mol equivalent of (*S*)-(+)-1-(2-pyrrolidinyl)-pyrrolidine and benzoic acid as a cocatalyst [75]. All reactions proceeded with good enantioselectivities, with high diastereoselectivity (*cis*:*trans*) in the case of geminal bis-phosphonate (dr = 1 : 99), and yields of up to 86%. Cyclohexanone and its derivatives provided monoalkylated products **89**; however, the reaction with cyclopentanone led to the formation of 2,5-dialkylated products **90** with 99% ee (Scheme 5.39).

Albrecht, Jørgensen *et al.* [49] have described enantio- and diastereoselective Michael addition initiated by the organocatalyst **8c** and proceeding with the formation of optically active 6-substituted-3-diethoxyphosphoryl-2-oxocyclohex-3-en-carboxylates **91**, as shown in Scheme 5.40.

The methodology used the reaction of ethyl 4-diethoxyphosphoryl-oxobutanoate with β-unsaturated aldehydes that was catalyzed by chiral diarylprolinol ethers. Cyclohexencarboxylates **92–97** are especially convenient for the preparation of functionalized tetrahydrobenzenes and cyclohexane derivatives, bearing four chiral centers and high levels of stereocontrol (Table 5.4).

Squaramides effectively catalyzed enantioselective Michael addition of diphenylphosphite to nitroalkenes (90–98% ee) [13]. This reaction provides a simple, highly enantioselective synthesis of chiral β-nitro phosphonates, which are precursors to biologically active β-amino phosphonic acids. The high yields and uniformly excellent enantioselectivities obtained for both aryl- and alkyl-substituted nitroalkenes, including those bearing acidic protons or sterically-demanding substituents, point to the unique capability of the squaramide scaffold. The introduction of 3,5-substituents onto the phenyl ring of the Ar substituents in guanidine **12a–c** was the most effective in enhancing both the enantioselectivity and catalytic efficiency (43–92% ee).The phospha-Michael reaction allowed phosphonylation of aromatic, heteroaromatic, and

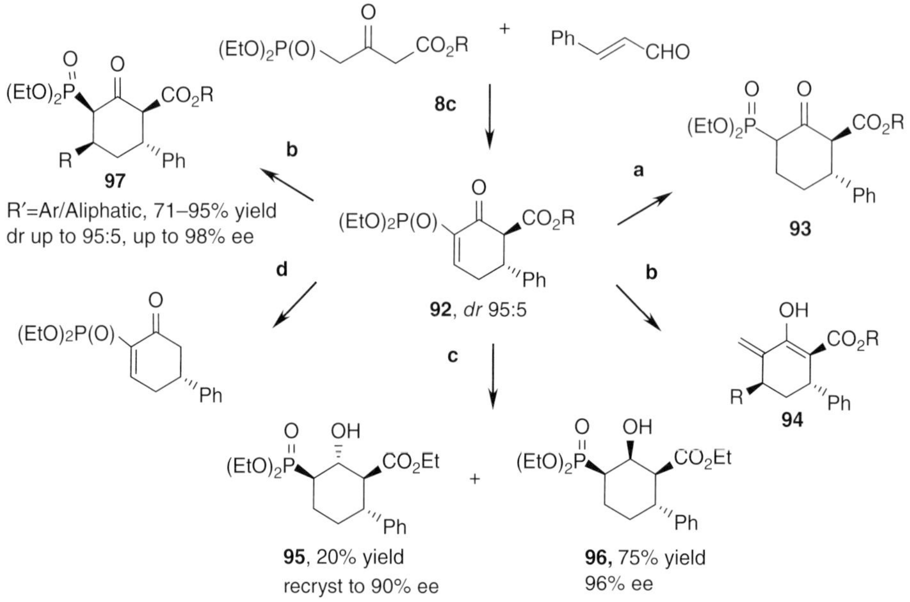

Scheme 5.40 Synthesis of diethoxyphosphoryl-2-oxocyclohex-3-en-carboxylates **92–97**.

Table 5.4 Organocatalytic reaction of cinnamaldehyde with ethyl 4-diethoxyphosphoryl-3-oxobutanoate.

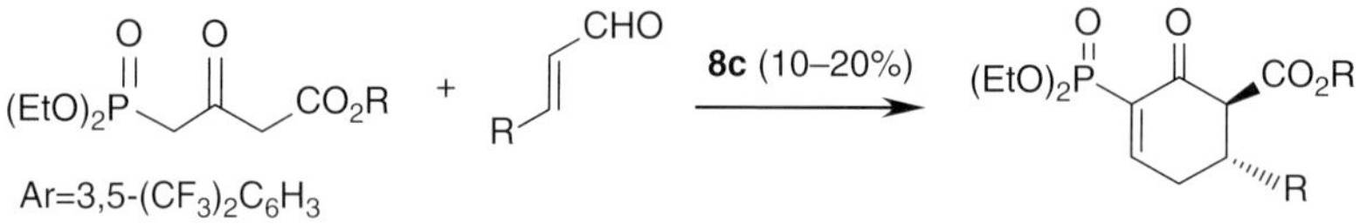

Entry	R	Additives	Solvent	*T* (°C)	Yield (%)	ee (%)	dr
1	Ph	—	CH_2Cl_2	RT	66	94	>95 : 5
2	Ph	—	$CHCl_3$	RT	67	94	>95 : 5
3	Ph	—	EtOH	RT	55	96	>95 : 5
4	Ph	—	Toluene	RT	42	94	>95 : 5
5	Ph	—	CH_2Cl_2	−30	76	98	>95 : 5
6	C_6H_5	$PhCO_2H$	CH_2Cl_2	−30	79	98	>95 : 5
7	4-NO_2-C_6H_4-	$PhCO_2H$	CH_2Cl_2	−30	95	98	87 : 13
8	4-CF_3-C_6H_4-	$PhCO_2H$	CH_2Cl_2	−30	81	98	>95 : 5
9	2-CH_3O-C_6H_4-	$PhCO_2H$	CH_2Cl_2	−30	94	97	92 : 8
10	3-CH_3O-C_6H_4-	$PhCO_2H$	CH_2Cl_2	−30	76	97	>95 : 5
11	Biphenyl	$PhCO_2H$	CH_2Cl_2	−30	78	98	>95 : 5
12	2-Furyl	$PhCO_2H$	CH_2Cl_2	−30	71	97	90 : 10
13	C_6H_5-	$PhCO_2H$	CH_2Cl_2	−30	76	98	>95 : 5

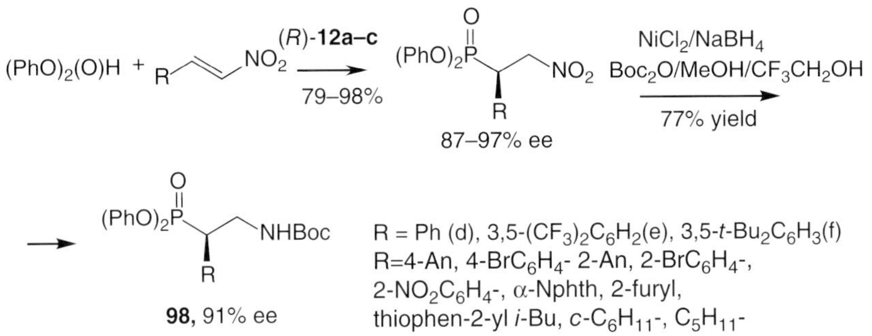

Scheme 5.41 Michael addition catalyzed by axially chiral guanidines **11**.

aliphatic nitroalkenes. The authors showed also the applicability of the Michael adducts for the synthesis of biologically important β-aminophosphonates **98** (Scheme 5.41).

An interesting organocatalytic methodology was developed for the asymmetric synthesis of bis-phosphonate derivatives, a class of pharmaceutically important molecules. Cheap and commercially available dihydroquinine effectively catalyzed conjugate additions of cyclic β-ketoesters to ethylidene-bis-phosphonate esters, leading to optically active geminal bis-phosphonates, bearing an all-carbon-substituted quaternary stereocenter, in high yields and enantioselectivities of up to 99% ee.

Jørgensen reported on the addition of cyclic β-ketoesters **99** to ethylene-bis-phosphonate catalyzed by dihydroquinine leading to optically active geminal bis- phosphonates **100** bearing a stereocenter with a quaternary carbon atom in good yields and with high enantioselectivity [70] (Scheme 5.42).

High yields and enantioselectivities were achieved under simple conditions for a wide range of indanone-based β-ketoesters, as well as various unprecedented 5-*tert*-butyloxycarbonyl cyclopentenones.

Zhao and co-workers [76] described the organocatalyzed asymmetric Michael reaction of β-aryl-α-ketophosphonates and nitroalkenes. The enantioselective Michael reaction of β-aryl-α-ketophosphonates and nitroalkenes has been realized by using a new bifunctional Takemoto-type thiourea catalyst **20**. The primary Michael adducts obtained were converted *in situ* to the corresponding amides through aminolysis. High yields, excellent diastereoselectivities (>95 : 5 dr), and good enantioselectivities (up to 81% ee) have been achieved for the corresponding α,β-disubstituted-nitroamides **101**. This reaction demonstrated that α-ketophosphonates are interesting pronucleophiles that can be used as amide surrogates in organocatalyzed reactions (Scheme 5.43).

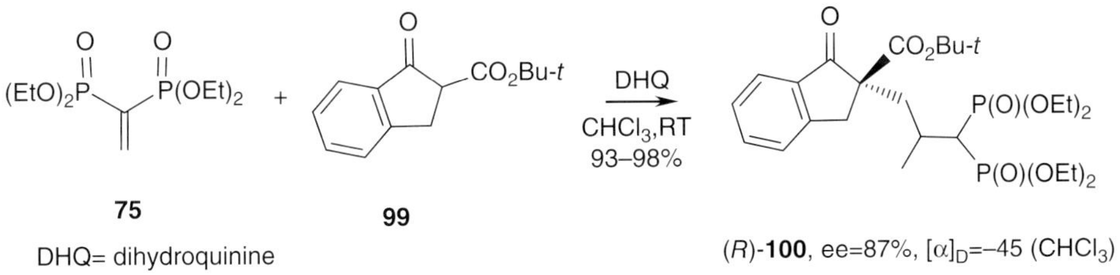

Scheme 5.42 Addition of cyclic β-ketoesters to ethylene-bis-phosphonate.

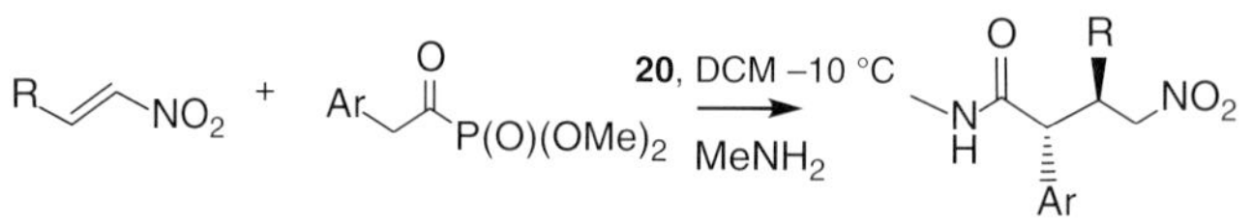

101, >95:5 dr, 81% ee

Scheme 5.43 Enantioselective Michael reaction of β-aryl-α-ketophosphonates with nitroalkenes.

5.6 Organocatalytic Addition to Ketophosphonates

5.6.1 Proline, Amino Acids, and Their Derivatives

Proline and its derivatives catalyze the reaction of α-ketophosphonates with enolizable ketones to result in the formation of chiral α-hydroxy-γ-ketophosphonates **105** [77–79]. The reaction of acetone with methyl or isopropyl ketophosphonates **102** provides hydroxyketophosphonates **105** with the highest optical yields; however, the reaction of butanone or methoxyacetone with ketophosphonates affords phosphonates **105** with moderate enantioselectivities (Scheme 5.44 and Table 5.5).

Diethyl (1-hydroxy-3-oxo-1-phenylbutyl)phosphonate **105**. L-Proline (0.25 mmol) at −30 °C was added to a stirred solution of the diethyl α-ketophosphonate **102** (0.5 mmol) and dry acetone (2.0 ml). The reaction mixture was stirred at this temperature, monitored by TLC, and then quenched by few drops of water. The mixture was extracted with ethyl acetate (3 × 10 ml), and the combined extracts were washed with brine solution (2 ml), dried over $MgSO_4$, and evaporated to give the crude product. The crude product was purified by column chromatography over silica gel (1 : 1 ethyl acetate/hexane) to furnish the α-hydroxy phosphonate as a pure compound (yield 60%, 85% ee).

The organocatalytic reaction of ketones with α-ketophosphonates having an asymmetric phosphorus atom was also studied. However, because of the stereogenic center on the phosphorus atom, the reaction led to the formation of two diastereomers (*R*,*R*)-**106** and (*R*,*S*)-**107** in a ratio of 1 : 1, which were separated by crystallization. The enatiomeric purity of these two diastereomers was high (81–99% ee for (*R*,*R*)-**106** and 61–91% ee for (*R*,*S*)-**107**). The absolute configurations were defined by X-ray crystal analysis (Scheme 5.45) [77, 78].

Zhang and Zhao [79] reported that the 9-amino-9-deoxyepiquinine derivatives **108** catalyze the cross-aldol reaction between aldehydes **109** and arylketophosphonates **110** (Scheme 5.46). The reaction catalyzed by this organocatalyst led to the

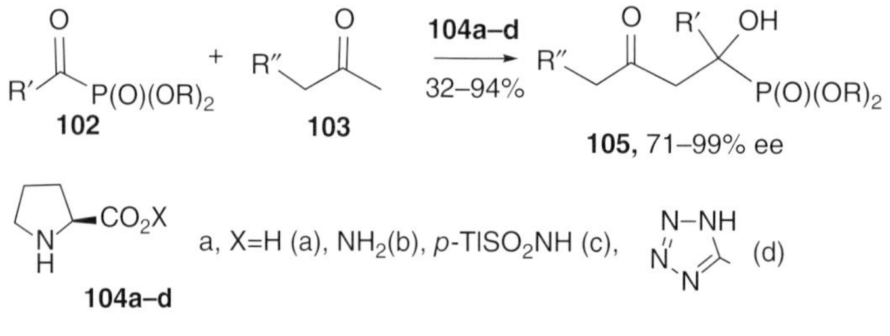

Scheme 5.44 Reaction of ketophosphonates **102** with ketones catalyzed by proline.

Table 5.5 Reaction of ketophosphonates with ketones catalyzed by proline.

Entry	R	R′	R″	104	Yield (%)	ee (%)
1	Et	Ph	H	**a**	65	87
2	Me	Ph	H	**a**	66	95
3	*i*-Pr	Ph	H	**a**	60	96
4	Et	*p*-ClC_6H_4	H	**a**	68	91
5	*i*-Pr	*p*-ClC_6H_4	H	**a**	63	95
6	Et	*p*-FC_6H_4	H	**a**	47	80
7	*i*-Pr	*p*-FC_6H_4	H	**a**	68	96
8	Et	*p*-BrC_6H_4	H	**a**	66	99
9	Et	*p*-IC_6H_4	H	**a**	67	94
10	Et	*p*-MeC_6H_4	H	**a**	63	85
11	Et	*p*-$MeOC_6H_4$	H	**a**	32	86
12	Et	Me	H	**a**	91	97
13	Et	$PhCH_2$	H	**a**	86	92
14	Me	Ph	Me	**b**	87	69
15	Et	Ph	Me	**b**	83	69
16	*i*-Pr	Ph	Me	**b**	82	74
17	*i*-Pr	Ph	MeO	**b**	93	85

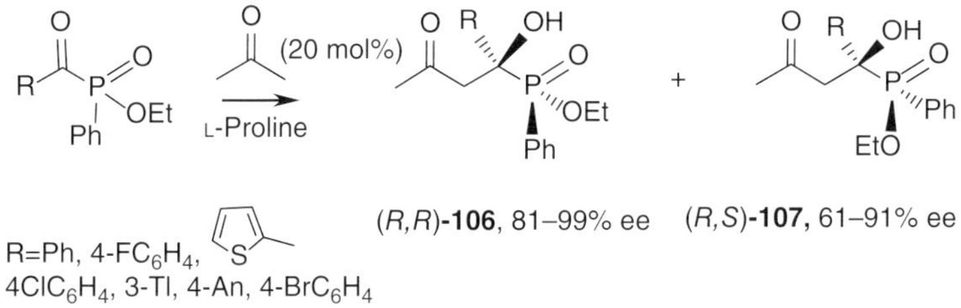

Scheme 5.45 Reaction of asymmetric ketophosphonates with acetone catalyzed by proline.

formation of the β-formyl-α-hydroxyphosphonates **111** in yields of 35–75% and with an enantioselectivity of 68–99% ee. The (*R*)-absolute configuration of several hydroxyphosphonates **111** was proven by X-ray monocrystal analysis. The reaction works especially well with acetaldehyde, which is a tough substrate for organocatalyzed cross-aldol reactions. The products were demonstrated to have anticancer activities. Some of the β-formyl-α-hydroxyphosphonate products suppressed the proliferation of human and murine tumor cells (Scheme 5.46).

The reaction of diethyl formylphosphonates with ketones catalyzed by L-prolinamide **104b** led to the formation of secondary hydroxyphosphonates **112** in high yields. Various ketones reacted smoothly under optimal reaction conditions to afford enantioenriched secondary hydroxyphosphonates **112** (Scheme 5.47). The authors have supposed that the addition proceeds through the transient states **A**, **B** having a nine-membered chair conformation in which the phosphonate occupies the pseudo-equatorial position. The enamine undergoes the addition from the *Si*-stereo face side of diethyl formylphosphonate to afford observable configuration of products (Scheme 5.48) [78].

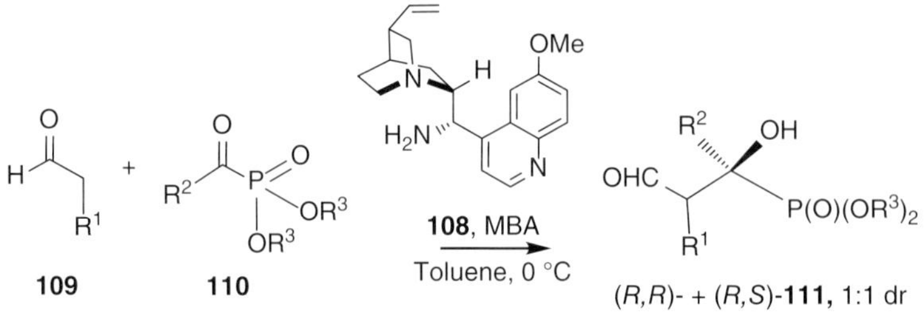

Scheme 5.46 Reaction of aldehydes with arylketophosphonates catalyzed by **110**.

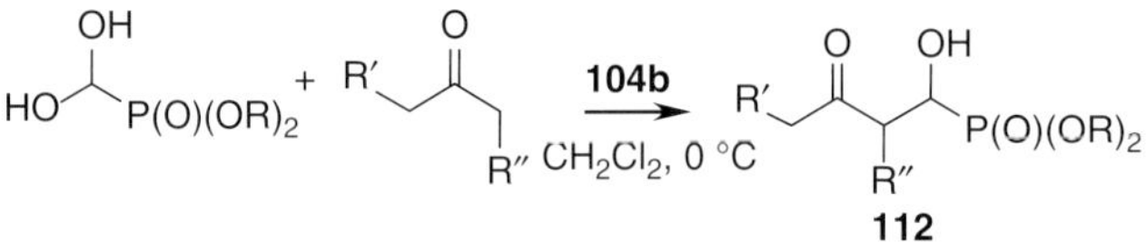

R′=H	R″=H	95%	—	94%
R′=H	R″=Me	20%	dr=95:5	99%
R′=Et	R″=H	58%	—	97%
R′=Cl	R″=H	65%	—	90%
R′=OMe	R″=H	77%	—	85%
R′=AcCH$_2$	R″=H	60%	—	86%
R′+R=–(CH$_2$)$_2$–		94%	dr=95:5	>99%
R′+R=–(CH$_2$)$_3$–		82%	dr=65:35	93%
R′+R=–(CH$_2$)$_3$		[illegible]	dr=[illegible]	90%
R′+R=–CH$_2$OCH$_2$–		85%	dr=70:30	97%
R′+R=–CH$_2$SCH$_2$–		88%	dr=70:30	93%

Scheme 5.47 Reaction of diethyl formylphosphonates with ketones catalyzed by L-**115b**.

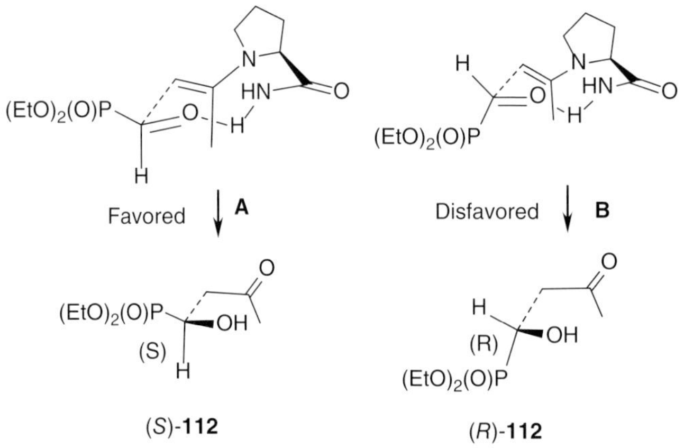

Scheme 5.48 The reaction of diethyl formylphosphonates with ketones catalyzed by L-prolinamide **104b** .

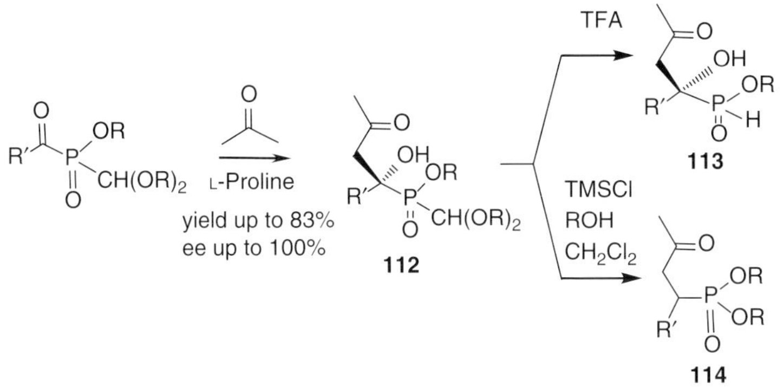

Scheme 5.49 Phosphaaldol reaction of L-α-acylphosphinates with acetone, catalyzed by proline.

A highly enantioselective synthesis of α-hydroxyphosphinates **112** was achieved based on the L-proline-catalyzed aldol reaction of α-acylphosphinates and acetone [80]. Owing to the preexisting chirality at the phosphorus center, mixtures of two diastereomers of the α-hydroxyphosphinates were obtained with high enantioselectivity for both diastereomers. α-Hydroxyphosphinates **112** were converted into α-hydroxy-*H*-phosphinates **113** by treatment with trifluoroacetic acid (TFA). An oxidation–reduction reaction of α-hydroxyphosphinates or α-hydroxy-*H*-phosphinates, with the formation of phosphonates **114**, was observed in the presence of trimethylchlorosilane and alcohol as shown in Scheme 5.49 [80].

The cinchonine-based thiourea **6c** effectively catalyzed the highly enantioselective Michael addition of 2-hydroxy-1,4-naphthoquinone to β,γ-unsaturated α-ketophosphonates **115** with the formation of the corresponding ketophosphonates, which, after treatment with 1,8-diazobicyclo[5.4.0]undec-7-ene (DBU) and MeOH, were converted into β-substituted carboxylates **116** in good yields and with high enantioselectivities (94–99% ee) (Scheme 5.50). Cinchonine-based thiourea **6c** was proved to be an efficient catalyst for the asymmetric Michael addition of 2-hydroxy-1,4-naphthoquinones to α,β-unsaturated acyl phosphonates [81] (Scheme 5.50). Isothiourea **117** catalyzed the asymmetric Michael addition/lactonization of aryl- and alkenylacetic acids using α-keto-β,γ-unsaturated

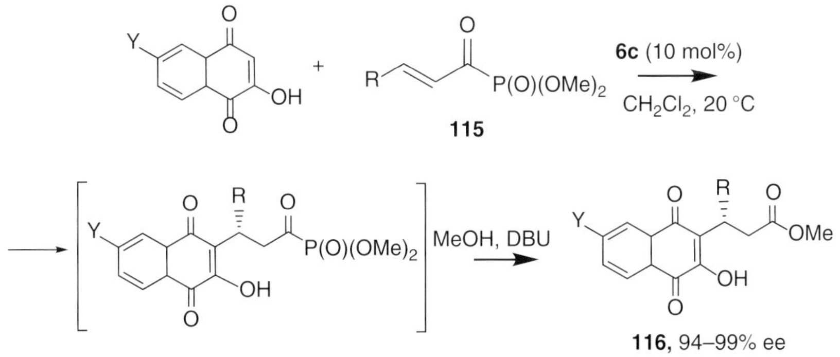

Scheme 5.50 Enantioselective Michael additions of 2-hydroxy-1,4-naphthoquinone to α-ketophosphonates **115** catalyzed by cinchonine-based thiourea **6c**.

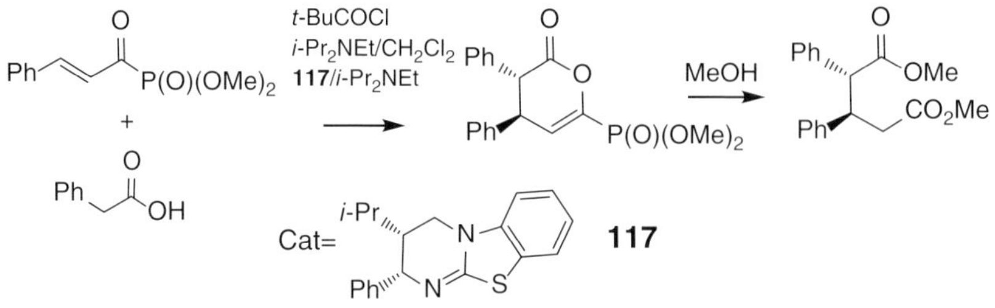

Scheme 5.51 Asymmetric Michael addition/lactonization reaction catalyzed by Isothiourea.

phosphonates as α,β-unsaturated ester surrogates, giving access to a diverse range of stereodefined lactones or enantioenriched functionalized diesters upon ring opening [82] (Scheme 5.51).

Rawal reported the enantioselective Mukaiyama aldol reactions of *N,O*-ketene acetals with acyl phosphonates catalyzed by commercially available TADDOL (Scheme 5.52) [83].

On the basis of enantioselective proline-catalyzed reactions between NH-iminotrifluoroethylphosphonate **119** and acetone, an effective synthetic approach to both (*S*)- and (*R*)-enantiomers of α-amino-α-trifluoromethyl-γ-oxobutylphosphonate was developed. The synthetic potential of these chiral synthons was illustrated by their cyclocondensation reactions with 4-chlorophenylisocyanate and 2,5-dimethoxy-tetrahydrofuran to afford the phosphorylated 3,4-dihydropyrimidin-2-ones or 3*H*-pyrrolizines, incorporating pharmacophoric optically active fragments of phosphonotrifluoroalanine. It was found that *O,O*-diethyl α-iminotrifluoroethyl-phosphonate reacts with acetone in the presence of L-proline at room temperature in DMSO solution to furnish (*R*)-diethyl α-amino-α-trifluoromethyl-γ-oxobutyl-phosphonate **120** in 81% isolated yield and ee greater than 90%. Under the same conditions, but using D-proline as catalyst, it was obtained (*S*)-**120** in yield 80% and with ee greater than 90% [84] (Scheme 5.53).

Zhao and Dodda have developed an enantioselective method for the synthesis of enantioenriched α-aminopropargylphosphonates **121**. High yields and good enantioselectivities (60–81% ee) were achieved using a complex of monovalent copper and pybox

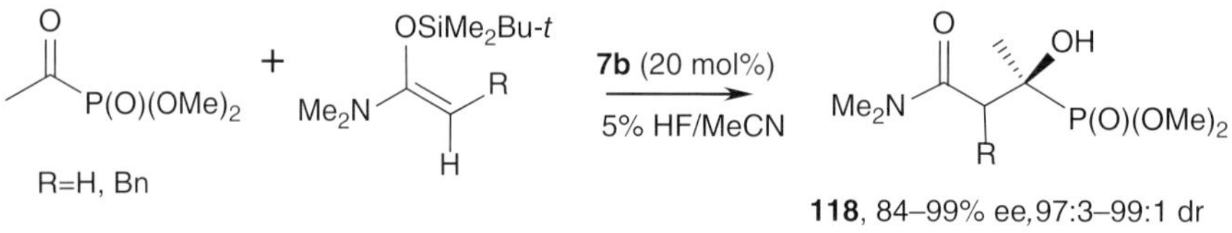

Scheme 5.52 Enantioselective Mukaiyama aldol reactions of *N,O*-ketene acetals with acyl phosphonates.

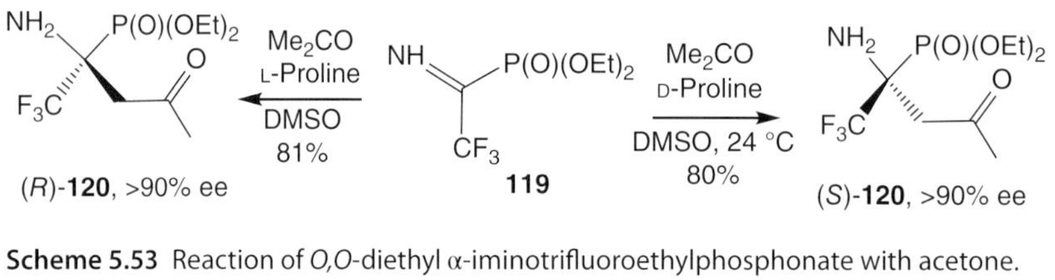

Scheme 5.53 Reaction of *O,O*-diethyl α-iminotrifluoroethylphosphonate with acetone.

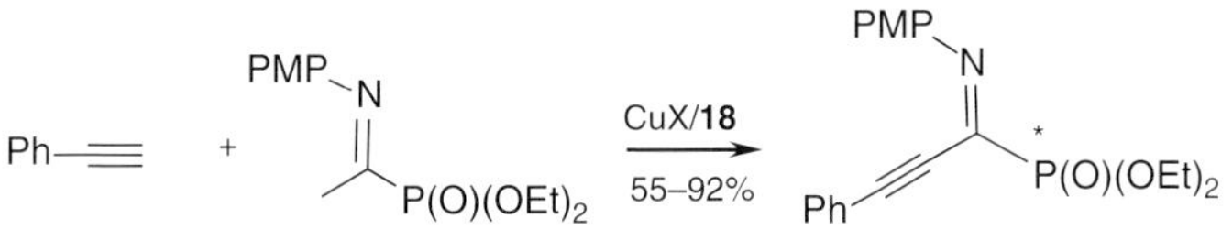

121, 74% ee

Scheme 5.54 Enantioselective synthesis of enantioenriched α-aminopropargylphosphonates **130**.

ligand **18** as the catalyst. A loading of 2 mol% of catalyst was sufficient enough to give a reasonable conversion. Under the optimized reaction conditions (2 mol% of loading catalyst, $CHCl_3$ as solvent, room temperature), the reaction was successful with various terminal alkynes (Scheme 5.54) [85, 86].

Palacios *et al.* [87] has described a simple asymmetric synthesis of 2*H*-azirin-2-phosphine oxides **122** from easily accessible oximes, using chiral amines immobilized on a polymer. These heterocycles are useful intermediates for the synthesis of α-ketamides and phosphorylated oxazoles. The key step is a solid-phase bound achiral or chiral amine-mediated Neber reaction of ketoxime tosylates derived from phosphine oxides. The reaction of 2*H*-azirines with carboxylic acids yielded phosphorylated ketamides. The ring closure of the ketamides with triphenylphosphine and hexachloroethane in the presence of triethylamine led to the formation of phosphorylated oxazoles **123** (Scheme 5.55) [86].

Palacios developed an accessible method for the preparation of α-iminophosphonates **124** derived from ketones, which were used for the enantioselective synthesis of tetrasubstituted α-aminophosphonic acid derivatives **125**. The key step in this method was an asymmetric cyanation of α-ketiminophosphonates **125** catalyzed by cinchona alkaloids. The use of tosyl cyanide or diethyl cyanophosphonate as a source of cyanide with a strong electron-withdrawing group was needed in this reaction. The absolute configuration of α-cyano-α-aminophosphonate was determined by X-ray analysis [87] (Scheme 5.56).

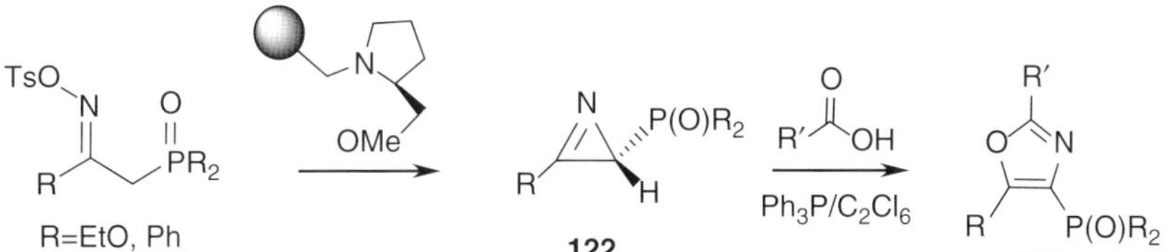

122

123

Scheme 5.55 Example of diastereoselective phospha-Mannich additions.

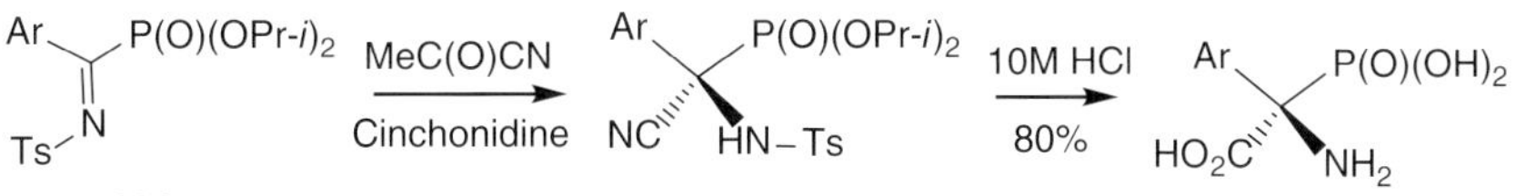

124

(*S*)-**125**

Ar-Ph, 80%, 99% ee; Ar=4-$F_3C_6H_4$78%, 95% ee; 4-$O_2NC_6H_4$, 75%, 82% ee; Ar=4-An, 78%, 95% ee; Ar-4-ClC_6H_4, 77%, 98% ee

Scheme 5.56 Organocatalytic method of asymmetric cyanation of ketimines.

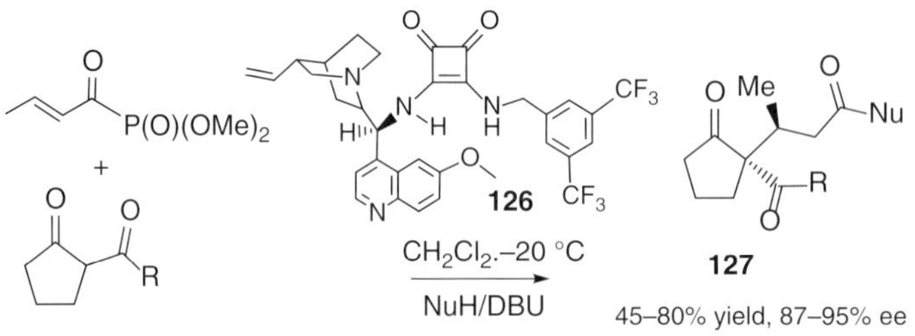

Scheme 5.57 Addition of 1,3-dicarbonyl compounds to (*E*)-but-2-enoylphosphonates.

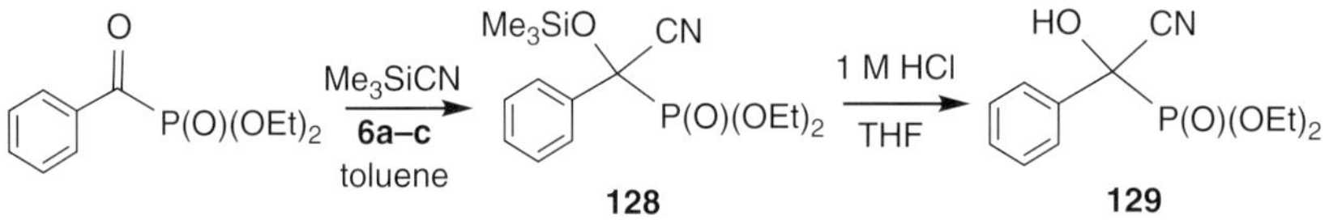

Scheme 5.58 Catalytic asymmetric addition reaction between ketophosphonates and Me_3SiCN.

5.6.2 Organocatalysis by Thiourea

Stereoselective conjugate additions of carbon-based nucleophiles (oxazolones, indoles, and 1,3-dicarbonyl compounds) to α,β-unsaturated acyl phosphonates catalyzed by chiral thioureas or squaramides **126** afforded products **127** in satisfactory yields and with enantioselectivities of 72–90% ee (Scheme 5.57) [88]. The stereoselectivity of the 1,3-dicarbonyl addition to acyl phosphonates is believed to originate from bifunctional coordination of the nucleophilic and electrophilic reaction partners to the quinine-derived catalyst. It was proposed that the acyl phosphonate is hydrogen bonded to the squaramide motif, placing the alkene side chain in the sterically less-demanding area away from the C-9 center of the catalyst, while the 1,3-dicarbonyl compound is deprotonated and directed for nucleophilic attack by the tertiary nitrogen atom of the catalyst. The R-group of the nucleophile is oriented away from the reaction site and the subsequent conjugate (Scheme 5.57). The carbohydrate/cinchonine-based thiourea **6a–c** catalyzed asymmetric addition of Me_3SiCN to α-ketophosphonates [89]. The initial product, the Me_3Si-ether cyanophosphonate **128**, was transformed into cyanohydrin phosphonate **129** by mild acidic hydrolysis. The reaction of the acyl phosphonate with silylcyan in the presence of the catalyst **6a–c** in toluene solution at −78 °C afforded the addition product **128** with 29% ee and in 89% yield (Scheme 5.58). Enantioselective Michael/cyclization reactions of 3-hydroxyoxindoles or 3-aminooxindoles with α,β-unsaturated acyl phosphonates **130** catalyzed by cinchonine or lactams afforded **131** in good yields (up to 98%) and good diastereo- and enantioselectivities (up to >99 : 1 dr and 97% ee) [90]. The intermediate aldol product was converted into the corresponding ester **132** through methanolysis or aminolysis [91] (Scheme 5.59).

5.7 Phospha-Henry Reaction

The phospha-Henry (phospha-nitroaldol) reaction between nitroalkanes and carbonyl compounds is an important carbon–carbon bond-forming method in organic

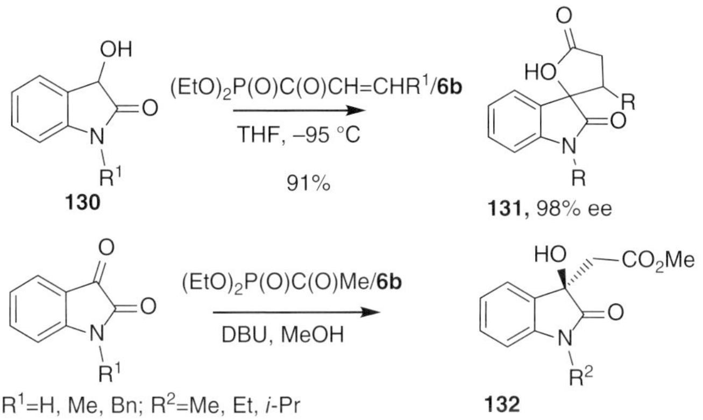

Scheme 5.59 Enantioselective reactions of acetylphosphonates with *N*-alkylisatin.

synthesis [92]. This process represents a useful tool for the synthesis of valuable β-nitro-α-hydroxyphosphonates, which can be further transformed into a number of important nitrogen and oxygen-containing phosphonates, for example β-amino-α-hydroxyphosphonates. The *P*-nitroaldol reaction may be promoted under many different conditions and using diverse catalytic systems providing moderate to good enantioselectivities [93–97].

Zhao *et al.* [93] described the first organocatalytic highly enantioselective *P*-nitroaldol reaction of α-ketophosphonates and nitromethane using cupreine **134a** or 9-*O*-benzylcupreine **134**b as catalyst at a low catalyst loading (5 mol%) (Scheme 5.60). The resulting α-hydroxy-β-nitrophosphonates **133**, obtained with good yields and excellent enantioselectivities, were transformed in β-amino-α-hydroxyphosphonates **135** without loss of the enantioselectivity. The ketophosphonates bearing aromatic, heteroaromatic, and aliphatic substituents entered readily into this reaction to afford hydroxyphosphonates in good yields and high enantioselectivities. The reaction was carried out at 0 °C, and additives of 2,4-dinitrophenol and excess $MeNO_2$ in the mixture of *t*-BuOMe/PhOMe at +20 °C increased yields as well as ee. Theoretical calculations supported hydrogen-bond interactions between catalyst **134** and the substrates, which could be crucial for the reactivity and enantioselectivity of this process. The authors came to a conclusion that the acidic additives protonates the piperidine moiety

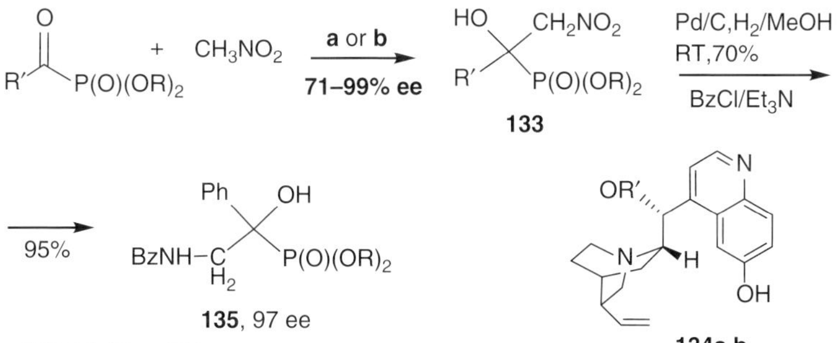

a= **134 a,b** (5 mol%);

b= **22** (5 mol%),2,4-dinitrophenol (6 mol%)–20 °C, *t*-BuOMe+PhOMe (2:1)

Scheme 5.60 Organocatalytic nitroaldol reaction of acyl phosphonates with nitromethane.

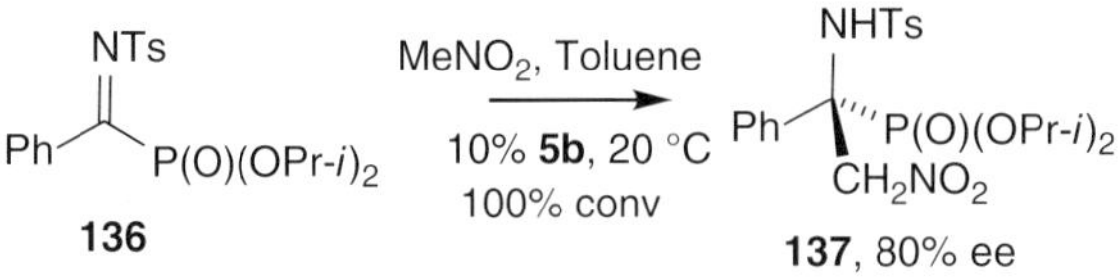

Scheme 5.61 Organocatalytic enantioselective aza-Henry reaction of iminophosphonates.

of the catalyst that activates the acyl phosphonate via hydrogen bonding [84]. The secondary amine–amide **22** (Table 5.1) is the most efficient catalyst for asymmetric Henry reactions of α-ketophosphonates with nitroalkanes under mild conditions [91, 94]. In the presence of 5 mol% organocatalyst **22**, very good enantioselectivities were achieved for hydroxyphosphonates **133** (up to 99% ee) as well as for most substrates. A theoretical study on the TSs revealed that this secondary amine-amide catalyst could be involved in hydrogen-bond interactions, which is important for the reactivity and enantioselectivity of this reaction. A hydrogen bond between the phosphorus oxygen and the amide moiety contributes greatly to the stability of the TS leading to the formation of the major (*R*)-product in accordance with the experimental results (Scheme 5.60) [91, 97].

Bifunctional cinchona alkaloid thioureas **6b** efficiently catalyze asymmetric nucleophilic addition of nitromethane to ketimines to afford tetrasubstituted α-amino-β-nitro-phosphonates **137**. Catalytic hydrogenation of (*S*)-α-amino-β-nitro-phosphonate **137** gave enantiopure (*S*)-α,β-diaminophosphonate which are susceptible for further transformations into enantiopure α,β-diaminophosphonates (Scheme 5.61) [96].

5.8 Organocatalytic Modification of *P*-ylids

Phosphorus ylides are capable of carrying out nucleophilic attack on activated imines with the formation of functionalized *P*-ylids containing a chiral center. In recent years, several interesting conversions of *P*-ylides initiated by organocatalysts were reported. For example, Chen developed an interesting organocatalytic approach to enantiomeric enriched *N*-Boc-β-amino-α-methylenecarboxilates **140**, which were earlier obtained by the aza-Morita–Baylis–Hillman reaction [98]. According to this methodology, the bis-thiourea **11** catalyzed the Mannich-type reaction between ylids and *N*-Boc-imines **139** with the subsequent olefination of **140** with formaldehyde. The catalyst **11** was regenerated by flash-chromatography and reused without reduction in activity. A tandem Mannich/Wittig-type reaction was applied for the preparation of various chiral esters *N*-Boc-α-amino-α-methylenecarbonic acids **140** (Scheme 5.62).

The stereoselective synthesis of *cis*-5-nitro-4,6-diphenylcyclohexen-1-carbonates **142–144** was attained by organocatalytic enantioselective cascade nitro-Michael/Michael-Wittig reaction with certain evidence of dynamic kinetic asymmetric transformation (DYCAT) [99]. The reaction of triphenylphosphonium ylide with nitrostyrene catalyzed by chiral pyrrolidine derivatives **8a** provided addition products **141** with enantioselectivities 92–99% ee. The addition of cinnamaldehyde to **141** afforded the cyclohexenecarboxylates **142–144** in a ratio of dr 4 : 1 : 3-6 : 1 : 2. An

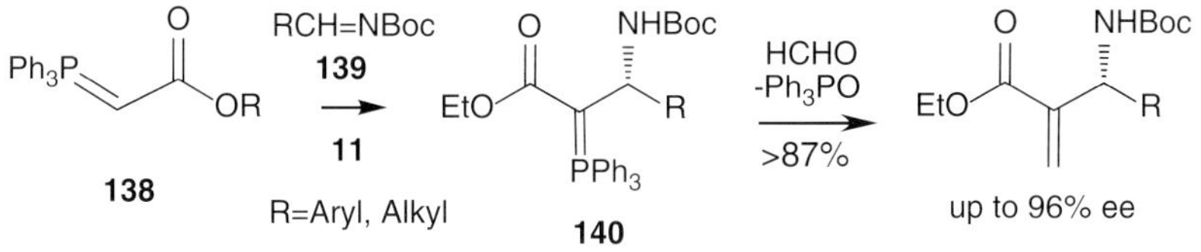

R′=Ph	87%	ee 89%
R′=*p*-F-Ph	67%	ee 87%
R′=*p*-Cl-Ph	65%	ee 90%
R′=*p*-Br-Ph	80%	ee 94%
R′=*p*-Me-Ph	78%	ee 95%
R′=*m*-Me-Ph	84%	ee 93%
R=*p*-MeO-Ph	84%	ee 91%
R′=2-Thienyl	84%	ee 68%
R′=*n*-Propyl	70%	ee 57%
R′=Cyclohexyl	52%	ee 91%
R′=*i*-Propyl	35%	ee 96%

Scheme 5.62 Asymmetric synthesis of *N*-Boc-α-amino-α-methylencarbonates via Mannich/Wittig tandem-type reaction.

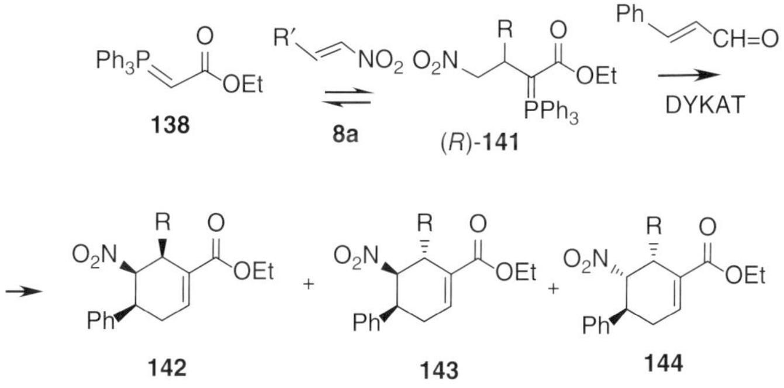

R′=Ph	55%	dr 6:1:2	ee 95%
R′=4-BrC_6H_4	41%	dr 4:1:3	ee 99%
R′=4-FC_6H_4	52%	dr 4:1:2	ee 96%
R =4-MeOC_6H_4	50%	dr 5:1:3	ee 99%
R′=3,4-(OCH_2O)C_6H_3	39%	dr 5:1:3	ee 99%

Scheme 5.63 Organocatalytic enantioselective cascade nitro-Michael/Michael-Wittig reaction.

increase in diastereoselectivity was achieved by organocatalytic asymmetric conjugate addition of phosphonium ylide to nitrostyrenes with the noncovalent thiourea catalyst **11**. This annulation reaction provides a simple protocol for the stereoselective construction of trisubstituted cyclohexenecarboxylates **142–144** containing three contiguous chiral centers with *all-cis* stereochemistry and with high enantioselectivity (up to 99% ee) (Scheme 5.63). The asymmetric organocatalytic Michael reaction of phosphorus ylides **145** with nitroolefins catalyzed by chiral Brönsted acids proceeded with the formation of γ-α-nitro-β-aryl-α-methylenecarboxylated **146**. In particular, chiral thioureas **11** effectively catalyzed this reaction with the formation of optically pure γ-α-nitro-β-aryl-α-methylene carboxylates **146** through an

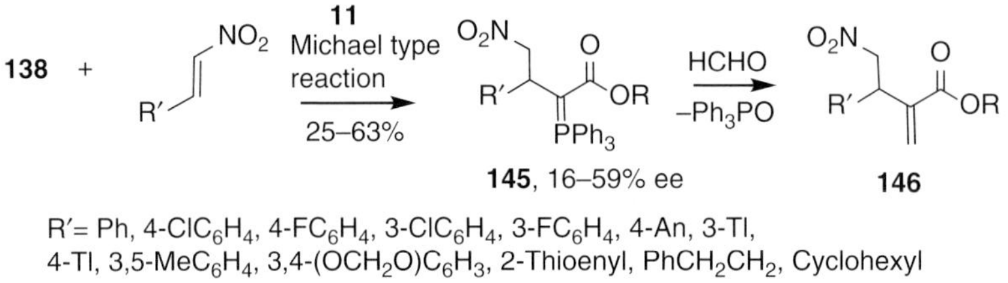

Scheme 5.64 Asymmetric organocatalytic Michael reaction of phosphorus ylides with nitroolefins, catalyzed by chiral Brönsted acids.

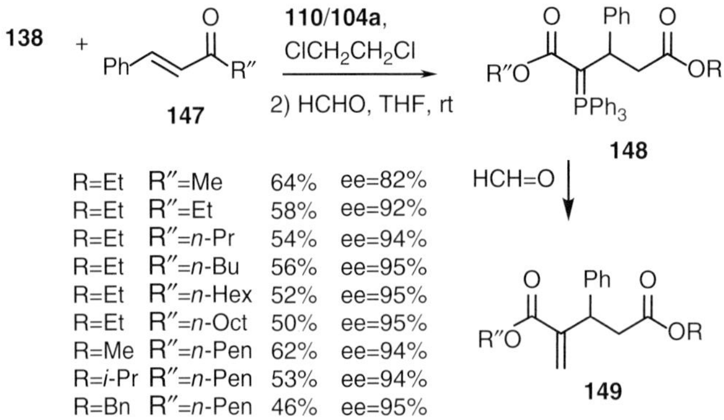

Scheme 5.65 Asymmetric catalytic reaction of phosphorus ylids with α,β-unsaturated ketones.

enantioenriched phosphorane intermediate. Besides, the reaction allowed to obtain highly functionalized γ-α-nitro carbonyl compounds, which cannot be obtained by the classic Morita–Baylis–Hillman (MBH) reaction of nitroolefins with acrylates (Scheme 5.64) [100].

The asymmetric catalytic reaction of phosphorus ylides **138** with α,β-unsaturated ketones **147** via the formation of addition products **148** resulted in α,β-unsaturated ketones **149** (Scheme 5.65). The addition of phosphorus ylids to α,β-unsaturated ketones, was carried out using a chiral ion-pair catalyst. The ion-pair catalyst containing a chiral counterion, was synthesized by simple mixing of 9-amino-(9-deoxy)-*epi*-quinine **108** with L-proline **104a** [100]. The reaction allowed to obtain a number of α, β-unsaturated ketones with high enantioselectivity (up to 95% ee). The Wittig reaction product **148** was introduced into the reaction with formaldehyde, to obtain a number of chiral α-methylene-δ-ketoesters **149** [101].

5.9 Asymmetric Catalysis with Chiral Diamines

Since Evans *et al.* [102] discovered that prochiral alkyl(dimethyl)phosphine boranes can undergo the enantioselective deprotonation of one methyl group using butyllithium and (−)-sparteine, these compounds have been widely used for the synthesis of P-chirogenic borane phosphines [103]. Lithium alkyls form chiral complexes **150** with sparteine and related chiral diamines, which were investigated by single-crystal X-ray analysis [103, 104].

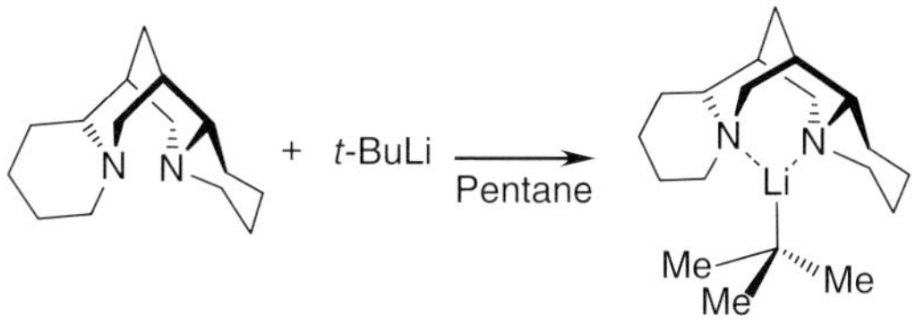

(–)-Sparteine

150

Prochiral dimethylarylphosphine boranes **151** react with a chiral sparteine-alkyllithium complex to undergo enantioselective deprotonation of one of methyl groups with the formation of lithium derivatives **152**, which react with electrophiles to give P-chiral compounds **153** (Scheme 5.66). Sparteine effectively complexes the lithium atom while deprotonation takes place and, in this chiral environment, *sec*-BuLi differentiates between the two enantiotopic methyl groups.

P-Chirogenic phosphine ligands **154** were prepared via desymmetrization of prochiral phosphine boranes **152** using a *s*-BuLi/(–)-sparteine complex [105–107]. The reaction of lithium derivative **152** with benzophenone led to the formation of alcohol (*S*)-**155** in satisfactory yields and enantioselectivity (Scheme 5.67) [103]. Enantioselective deprotonation of tertiary dimethylphosphines **151** can be achieved with *s*-BuLi in the presence of accessible and cheap derivatives of alkaloid (–)-cytisine **156**. The derivatives

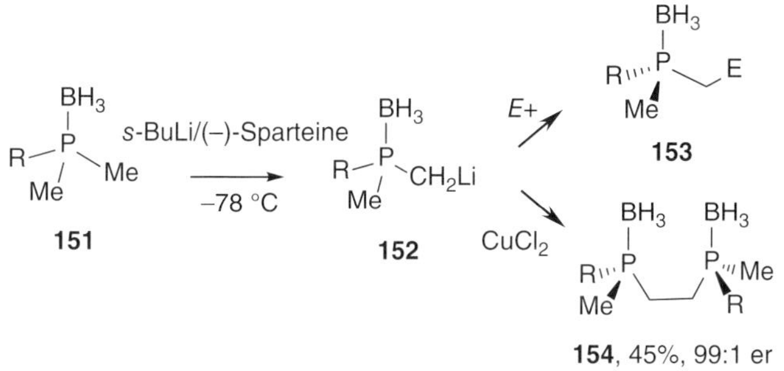

Scheme 5.66 Enantioselective deprotonation of dimethylarylphosphine boranes **3** with a chiral sparteine-alkyllithium complex.

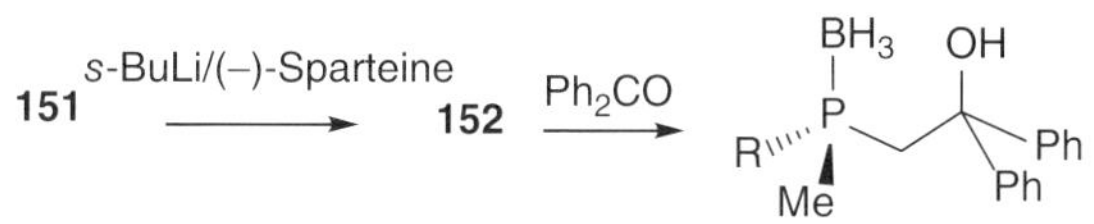

R	yield (%)	ee (%)
Ph	88	79
o-An	81	83
o-Tol	84	87
1-Nphth	86	82

Scheme 5.67 The reaction of lithium derivative **144** with benzophenone.

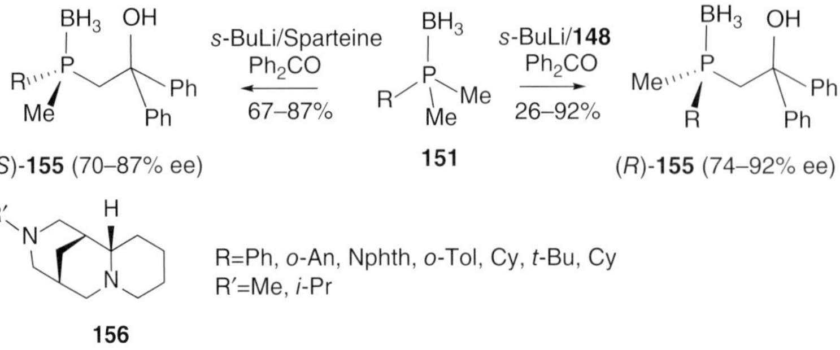

Scheme 5.68 Enantioselective deprotonation of tertiary dimethylphosphines with (−)-sparteine or cytisine derivatives **156**.

of cytisine **156** are useful sparteine surrogates for the desymmetrization of prochiral phenyl-, cyclohexyl-, and *tert*-butyl dimethyl phosphine boranes, yielding chiral phosphine boranes in up to 92% ee [108, 109]. Genet reported that increase of stoichiometric amount of chiral diamine augments the enantioselectivity of reaction (Scheme 5.68) [105].

*(R)-P-(2,2-Diphenyl-2-hydroxyethyl)-P-methyl-tert-butylphosphine (R)-***155** [103]

a) *s*-BuLi (0.4 mmol) was added dropwise to a stirred solution of (−)-sparteine (0.4 mmol, 0.3 equiv.) in Et_2O or toluene (5 ml) at −78 °C under argon. After stirring for 15 min, a solution of dimethylphenylphosphine borane **151** (1.36 mmol) in Et_2O or toluene (2 ml) was added dropwise over 30 min using a syringe pump. The resulting solution was stirred for 30 min. Following this, *s*-BuLi (0.36 ml of a 1.3 M solution in cyclohexane, 0.45 mmol) was added dropwise and the resulting solution was stirred for 72 min to give a solution of the lithiated intermediate in Et_2O. Next a solution of benzophenone (383 mg, 2.0 mmol) in Et_2O (3 ml) was added dropwise to the reaction solution and the mixture was allowed to warm to r.t. over 16 h. Then 10 ml of 0.1 N HCl and EtOAc were added and the two layers were separated. The aqueous layer was extracted with EtOAc (310 ml) and the combined organic layers were washed with brine (10 ml), dried (Na_2SO_4) and evaporated under reduced pressure to give the crude product. Purification was carried out by flash column chromatography using hexane-EtOAc as eluent. The adduct (*R*)-**155** was obtained as a white solid (yield 83%, 91% diastereomeric excess (de), $^{31}P\{^1H\}$ NMR ($CDCl_3$) δ 20.6 ppm).

Enantioselective deprotonation of dimethylphosphine boranes **151** by a *s*-BuLi/(−)-sparteine complex, and subsequent oxidation with molecular oxygen in the presence of triethylphosphite led to the formation of alkyl(hydroxymethyl)methylphosphine boranes **158** in 91–93% ee in case of bulky alkyl groups and 75–81% ee in case of cyclohexyl or phenyl groups [106–110]. The treatment of the carbanion with CO_2, led to the formation of phosphorylated carboxylic acids. The reduction of the carboxyl group with borane and reaction with tosyl chloride provided the (*R*)-tosylates **157** in 90% ee and good yields. P-Chirogenic phosphine-sulfide borane ligands **160** were prepared by reaction of **151** with *s*-BuLi/(−)-sparteine and phenyl disulfide (Scheme 5.69) [111, 112].

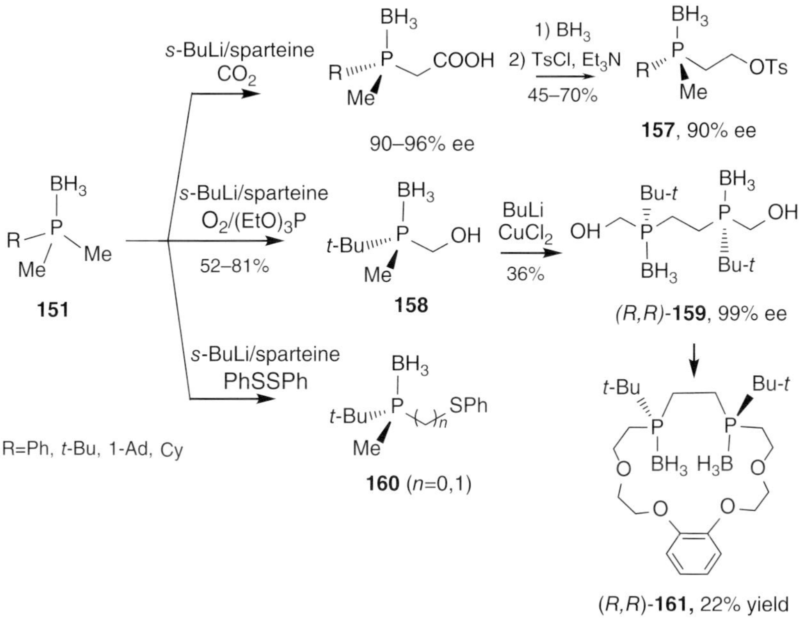

Scheme 5.69 Asymmetric synthesis of alkylidenedimethylphosphine borane derivatives.

The syntheses of α-alkoxyphosphine boranes as potential ligands for asymmetric organometallic reactions were developed via deprotonation of chiral hydroxymethylphosphine precursors **158**, followed by alkylation with various electrophiles and quenching with a polymer-bound scavenger. Enantiopure P-stereogenic secondary bisphosphines **159** were prepared starting from chiral hydroxymethylphosphine **158**. They were used as key building blocks for the preparation of P-stereogenic benzodiphosphacrowns **161** [106, 107].

Lithiation-oxygenation of tert-butyldimethyl phosphine borane **151**

a) *s*-BuLi (1.92 ml of a 1.20 M solution in cyclohexane, 2.21 mmol, 1.1 equiv.) was added dropwise to a stirred solution of (−)-sparteine (0.2–1.2 equiv.) in Et_2O (10 ml) at −78 °C under Ar. After stirring for 15 min at −78 °C, a solution of phosphine borane **151** (265 mg, 2.01 mmol) in Et_2O (5 ml) was added dropwise over 10 min via a cannula. The resulting solution was stirred at −78 °C for 3 h. Then dry air was bubbled through reaction mixture with stirring for 1 h, and the reaction mixture was placed to freeze for 16 h to complete the reaction and then the mixture was warmed to r.t. 1 M HCl(aq) (10 ml) and then EtOAc (15 ml) were the added and the layers separated; the aqueous layer was extracted with EtOAc (3 × 10 ml). The combined organic layers were washed with 1 M HCl(aq) (10 ml), water (10 ml), and brine (10 ml), and then dried ($MgSO_4$) and evaporated under reduced pressure to give the crude product as a white solid [106].
b) *1,2-Bis[(boranato)(tert-butyl)methylphosphino)]ethane* **154**. *tert*-Butyldimethyl phosphine borane **151** and *s*-BuLi (3.48 mmol, 1.1 equiv.) in cyclohexane were added dropwise to a stirred solution of (−)-sparteine (0.2–0.5 equiv.) in Et_2O (10 ml) at −78 °C. After stirring for 15 min at −78 °C, a solution of phosphine borane **151** (416 mg, 3.16 mmol) in Et_2O (5 ml) was added dropwise over 10 min.

The resulting solution was stirred at −78 °C for 3 h. Then, anhydrous $CuCl_2$ (5 mmol) was added and the resulting solution was stirred at −78 °C for 1 h and then allowed to warm to r.t. over 16 h. 25% NH_3 (aq) (3.0 ml) was added and the layers were separated. The aqueous layer was extracted with EtOAc (3 × 10 ml) and the combined organic extracts washed with 5% NH_3(aq) (10 ml), 1 M HCl(aq) (10 ml), dried (Na_2SO_4), and evaporated under reduced pressure to give the crude product as a white solid. Purification by flash chromatography with 9 : 1 petrol-EtOAc as eluent gave the adduct *meso*-**7** (31 mg, 15%, mp 178–180 °C) as a white solid.

Johansson *et al.* [110] reported the preparation of P-chirogenic aldehydes **162** via the desymmetrization of prochiral phosphine boranes **151**. Enantioenriched formyl phosphine boranes were obtained in good yields by reaction of asymmetric lithium derivatives **152** with DMF at −78 °C or by deprotonation of phenyl-*m*-anisyl-methylphosphine borane (borane-protected PAMP) with *s*-BuLi and subsequent quenching with DMF. The formyl phosphine boranes **162** were transformed to the β-aminophosphine boranes **163** employing reductive amination under microwave irradiation. This methodology gives access to P-chirogenic compounds **155–159** that are versatile building blocks for the design and construction of new chiral phosphine ligands. For example, the ligands **155–159** were evaluated in the asymmetric conjugate addition of diethylzinc to *trans*-nitrostyrene (Scheme 5.70).

Enantiomerically pure bisphosphine (BisP*) **166** [111] and tris-phosphine (MT-Siliphos) **167** ligands were obtained in high yields and used for preparation of various complexes of transition metals (Pd, Pt, Cu, Rh, Ru) (Scheme 5.71).

Cationic rhodium complexes of bis-phosphines **166** were used as catalysts in asymmetric hydrogenation of (acylamino)acrylates with enantioselectivity up to 99.9% ee. *cis*- and *trans*-1,4-Diphosphacyclohexanes **160** were synthesized by stereospecific intramolecular coupling reaction of bis-phosphines **154**. The coupling reaction of optically active **154** resulted in the *trans*-isomer **168**; in the meantime, a *cis*-isomer **168** was prepared along with the *trans*-isomer from a mixture of *rac*- and *meso*-bisphosphines **154** (Scheme 5.72) [110, 113].

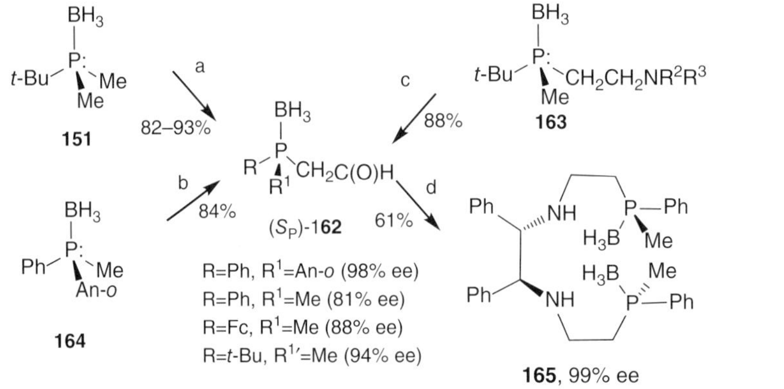

(a) *s*-BuLi/(−)-Sparteine/DMF, −78 °C; (b) *s*-BuLi/DMF, −78 °C
(c) NR^2R^3=PhCH(Me)NH_2-(*S*); NaBH$(OAc)_3$, SCX-2, DCE, MW (Reductive amination)
(d) C_2-symmetric diamine, NaBH$(OAc)_3$, SCX-2, DCE

Scheme 5.70 Versatile α-formyl phosphine intermediate.

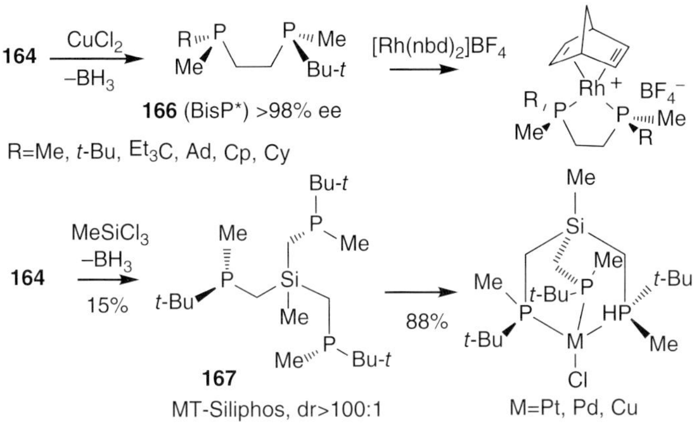

Scheme 5.71 Enantiomerically pure (BisP*) ligands **161** and MT-Siliphos with bulky substituents.

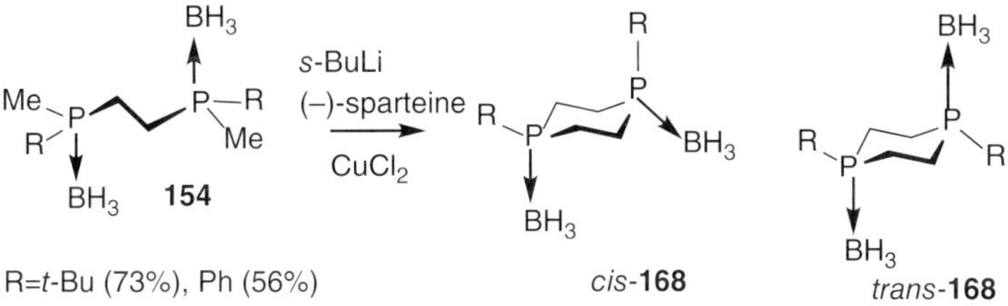

Scheme 5.72 *cis-* and *trans*-1,4-Diphosphacyclohexanes.

P-Chirogenic BisP* ligands **169** were prepared from alkyl(hydroxymethyl) methylphosphine boranes **158** in good isolated yield and with high optical purity as shown in the Scheme 5.73. Rhodium catalysts with BisP* ligands **169** have demonstrated high enantioselectivity (up to 98% ee) in the hydrogenation of α-dehydroamino acid derivatives (Scheme 5.73) [107]. Photoresponsive polymers **170** having chiral phosphine in the main chain were prepared by lithiation of (*S*,*S*)-**169** with BuLi and subsequent reaction of the prepared dilithium derivative with azobenzene derivative. According to GPC analysis, the number average-molecular weight (Mn) and the average molecular weight (Mw) of **170** were found to be 3000 and 5000, respectively The polymer isomerized from the *trans-* to *cis*-form upon UV irradiation and reverted to the *trans*-form reversibly. The polymer was able to coordinate to platinum and the resulting polymer complex exhibited the Cotton effect owing to the chirality of the phosphorus atoms. The polymer chain was induced to rotate helically when complexed with transition metals through the chiral phosphorus atoms (Scheme 5.74) [114, 115, 116].

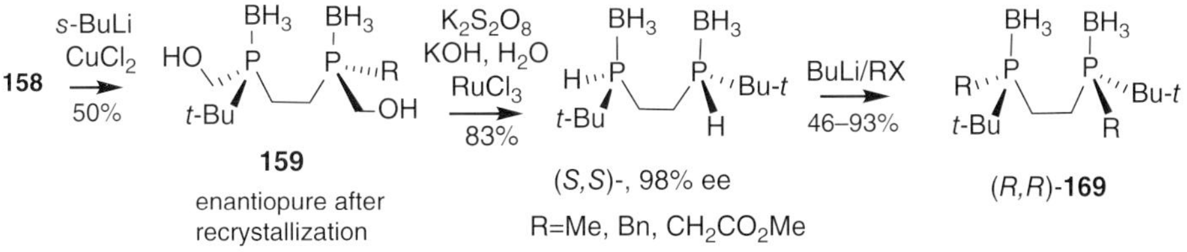

Scheme 5.73 The synthesis of BisP* borane ligands **169**.

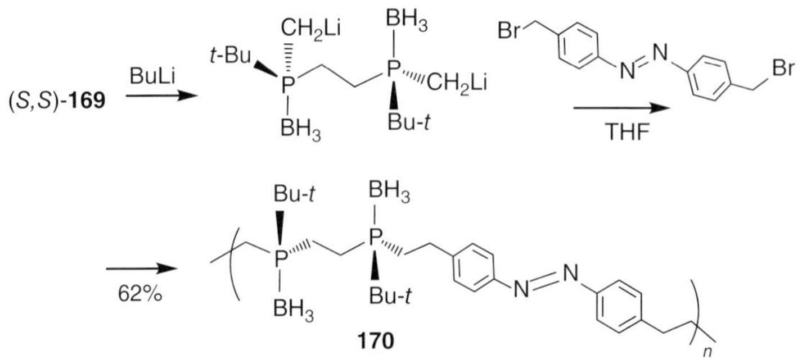

Scheme 5.74 Synthesis of photoresponsive polymers.

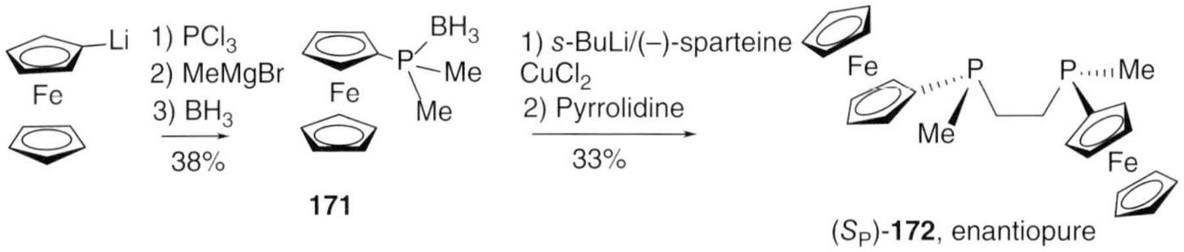

Scheme 5.75 Preparation of ethylene bridged *P*-chirogenic diferrocene diphosphines (S_P)-**168**.

Imamoto *et al.* [117] used diastereoselective deprotonation of dimethylferrocenyl borane for the preparation of ethylene-bridged P-chirogenic diphosphines (S_P)-**172** containing a ferrocenyl moiety. One of the enantiotopic methyl groups of **171** was deprotonated with a (−)-sparteine/*sec*-BuLi complex and the resulting carbanion subjected to oxidative dimerization by treatment with copper(II) chloride to give the chiral diphosphine borane with small impurity of the *meso*-product. Recrystallization from toluene allowed the removal of the *meso*-isomer and to obtain enantiomerically pure product (S_P)-**172** in 33% yield. The ligand was used in the rhodium-catalyzed asymmetric hydrogenation of dehydroamino acid derivatives (up to 77% ee) and in the palladium-catalyzed asymmetric allylic alkylation of 1,3-diphenyl-2-propenyl acetate (up to 95% ee) (Scheme 5.75). Hoge reported P-chiral ligands bearing phospholane rings [118]. The P-chiral phospholane ligand **174** was prepared starting from (1*R*,2*S*,5*R*)-menthyl dichlorophosphite. The treatment of (1*R*,2*S*,5*R*)-menthyl dichlorophosphite with a bis-Grignard reagent generated from 1,4-dibromobutane, followed by complexation of the free phosphine with borane-dimethyl sulfide complex led to the formation of phosphinite borane **173**, which was then converted to enantiomerically enriched phospholane borane **174**. The subsequent reaction of methyl phospholane **173** with a *sec*-butyllythium/sparteine complex and copper chloride led to the formation of P-chiral diphosphine-borane **174**, which after recrystallization and deprotection with fluoroboronic acid was obtained with 99% ee (Scheme 5.76). Asymmetric hydrogenation of acetamidoacrylic acid derivatives using Rh catalyst with diphosphine ligand **174**, provided enantioselectivity of 77–95% ee under low H_2 pressure [118].

Sterically hindered tertiary diphosphines **176**, **177** were synthesized with moderate diastereoselectivity (Scheme 5.77) [119]. Dihydroboronium derivatives of *t*-Bu-BisP* with different counteranions were prepared, as shown in Scheme 5.78. The reaction

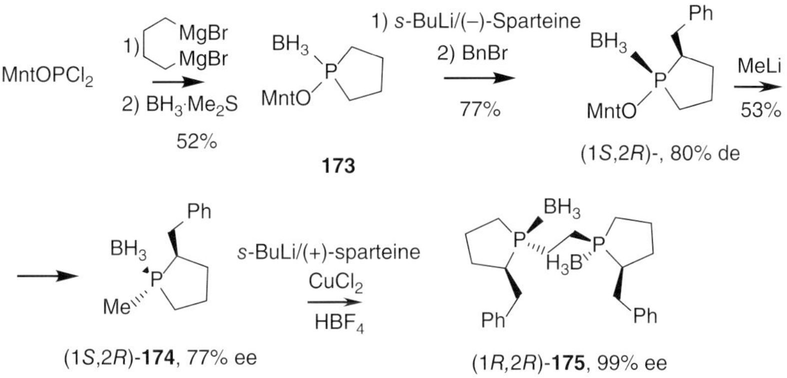

Scheme 5.76 P-Chiral ligands **169** bearing phospholane cycles.

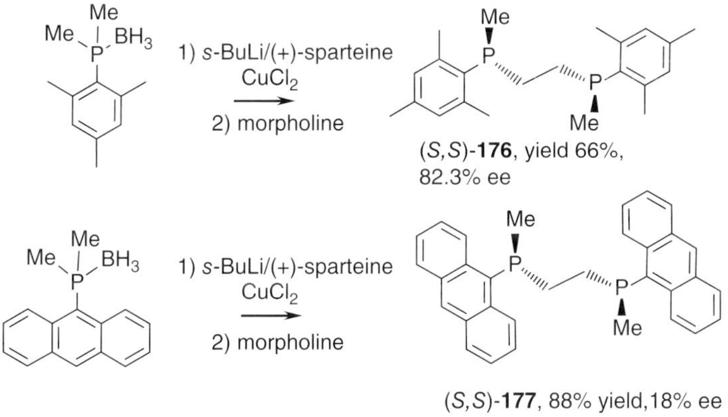

Scheme 5.77 Sterically hindered diphosphines **176**, **177**.

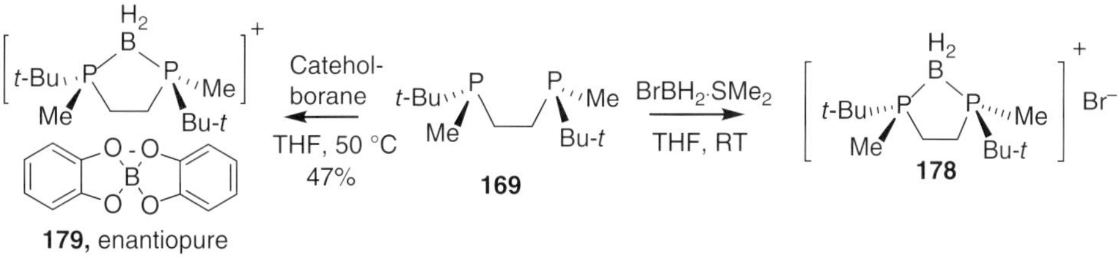

Scheme 5.78 (*S,S*)-1,2-Bis(*tert*-butylmethylphosphino)ethane boranes **178**, **179** (*t*-Bu-BisP*).

of BisP* with BH_2Br afforded the boronium salt **178** possessing a bromide ion. The dihydroboronium derivative of (*S,S*)-1,2-bis(*tert*-butylmethylphosphino)ethane **179** (*t*-Bu-BisP*) was used as chiral diphosphine ligand precursor in Rh-catalyzed asymmetric hydrogenated of methyl (*Z*)-acetamidocinnamate to afford the hydrogenation product in up to 94% ee [120] (Scheme 5.78).

Imamoto has reported the synthesis of P-chiral diphosphines with a methylene bridge and bulky alkyl groups on each phosphorus atom. These ligands named MiniPHOS **185** were synthesized using a phosphine borane **147** as an intermediate. The subsequent reaction of **152** with $RPCl_2$, methylmagnesium bromide, and borane afforded the

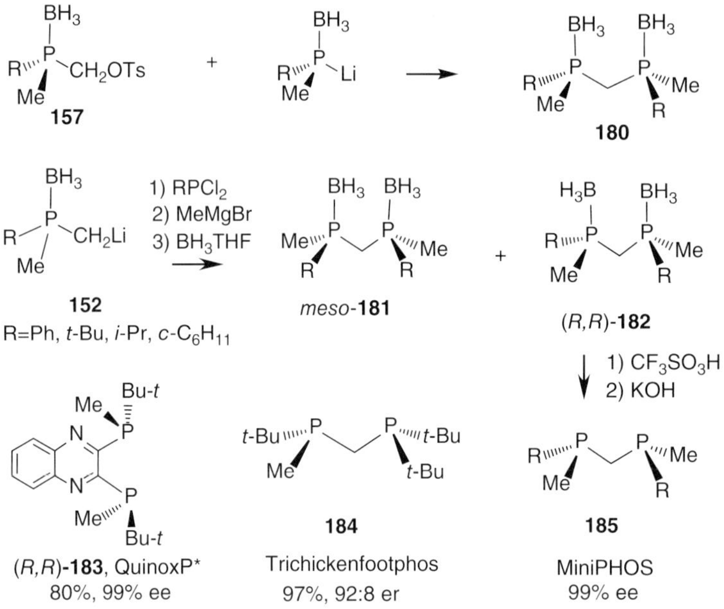

Scheme 5.79 P-Chiral diphosphines with methylene bridge.

diphosphine boranes (*R,R*)-**181** and *meso*-**181**. The purification of the reaction mixture by crystallization and deboration resulted in the pure MiniPHOS **185** in yield of 13–28% and with 99% ee (Scheme 5.79) [111, 112, 121].

Improved synthetic routes to methylene-bridged P-chiral diphosphine ligands **185** via tertiary phosphine–boranes **151** without the formation of *meso*-isomers was also reported [122]. The use of (−)-sparteine or (+)-sparteine surrogate as chiral catalysts facilitates access to P-stereogenic phosphines with opposite configuration. The method was exemplified by the catalytic asymmetric synthesis of each enantiomer of precursors to *t*-Bu-QuinoxP* **183**, Trichickenfootphos **184**, and Mini-PHOS **185** (R = *t*-Bu). The ligand **182** exhibited very good asymmetric induction in Pd-catalyzed asymmetric allylic substitution of 1,3-diphenyl-2-propenyl acetate (up to 98.7% ee) and in Ru-catalyzed asymmetric hydrogenation of ketones (up to 99.9% ee) (Scheme 5.79) [111, 112]. Interesting examples of the asymmetric synthesis of P-stereogenic vinylic phospholene boranes **186** using sparteine catalysis and Grubbs catalysts ring-closing metathesis have been described (Scheme 5.80) [123].

BisP* ligands contain two methylene groups in the backbone and therefore their metal ligand complexes are conformationally flexible. Zhang and Tang prepared the

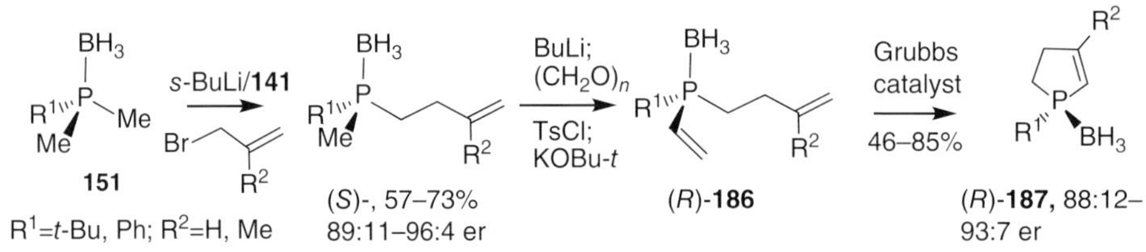

Scheme 5.80 P-Stereogenic vinylic phospholene boranes **186**.

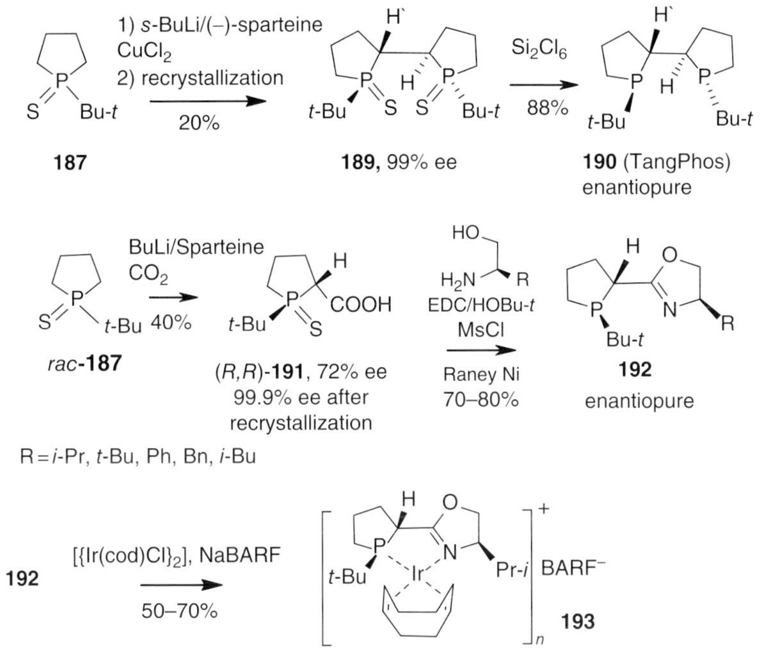

Scheme 5.81 Preparation of TangPhos ligand **186** and phospholane–oxazoline ligands **188**.

TangPhos ligand **190** having a well-defined rigid conformation through two additional five-membered rings on the backbone, which exhibited a more rigid chiral environment than the BisP* ligands. The TangPhos ligand was prepared in three steps using phosphine sulfides **188** and **189** as intermediates (Scheme 5.81). This ligand was used for the Rh-catalyzed asymmetric hydrosilylation of α-(acylamino)acrylic acids and α-arylenamides with the formation of optically active amides with 98–99% ee [121, 124, 125]. Zhang *et al.* [124] have developed a convenient method for the synthesis of P-chiral phospholane–oxazolines ligands **192** based on phosphine sulfides as intermediate compounds. Selective deprotonation of **187** by *n*-butyllithium in the presence of (−)-sparteine followed by reaction with CO_2 provided acid **191** with 72% ee. Recrystallization of the acid from ethanol yielded the enantiomerically pure (*R*,*R*)-**191** in moderate yield. The condensation of **191** with chiral amino alcohols by using EDC/HOBu-*t* proceeded smoothly to yield the coupling products, which were subsequently treated with MsCl to form the oxazoline compounds. Desulfurization of oxazoline compounds using Raney Ni provided phospholane–oxazoline ligands **192** in excellent yields. (*S*,*S*)-**192** was used for Ir-complex **193**-catalyzed asymmetric hydrogenation of β-methylcinnamic esters and methylstilbene derivatives. A variety of chiral 3-arylbutyric esters and diaryl(methyl)ethanes were obtained from moderate to very high enantioselectivity (up to 99% ee) (Scheme 5.81).

Imamoto and Crepy [122] obtained access to enantiomerically pure diphosphine dioxides **195** by oxidative dimerization of *rac*-1-*tert*-butylbenzophosphine oxide **194** by treatment with *s*-BuLi/$CuCl_2$, and subsequent resolution with (+)- or (−)-DBTA. The reduction of **195** with hexachlorosilane led to the formation of diphosphine **196** with retention of absolute configuration. The ligand was used as a rhodium complex

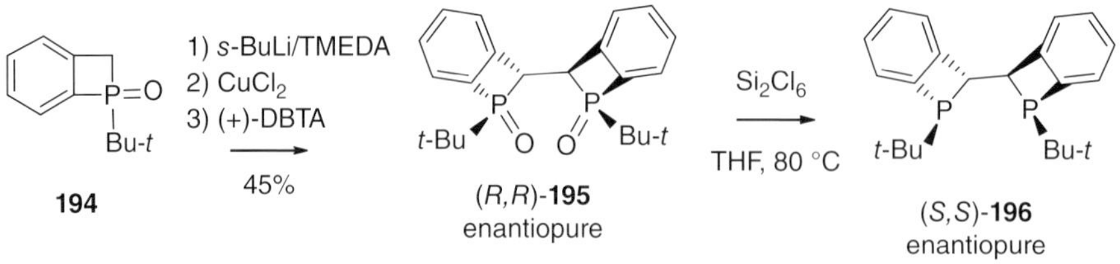

Scheme 5.82 Synthesis of enantiomerically pure diphosphine dioxides **192**.

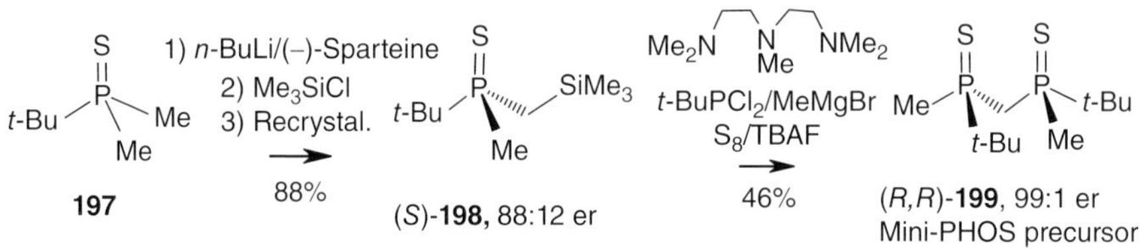

Scheme 5.83

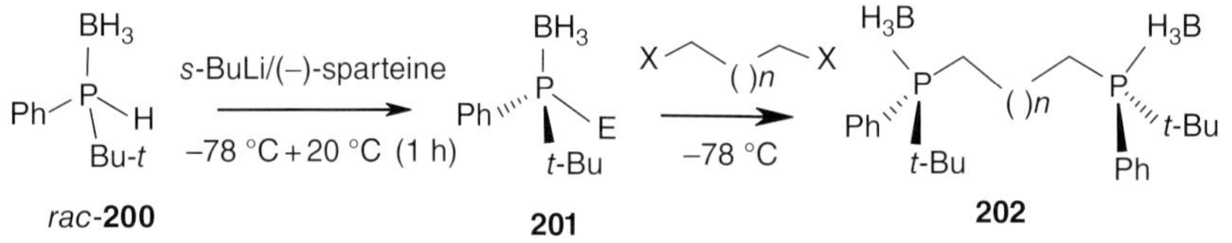

Scheme 5.84 Dynamic resolution with *sec*-butyl/spartein-complex.

directly after reduction for the hydrogenation of α-acetamidocinnamate with 96% ee (Scheme 5.82) [126].

Starting from a trimethylsilyl-substituted phosphine sulfide **198** (generated by *n*-BuLi/(−)-sparteine-mediated asymmetric lithiation of a dimethylphosphine sulfide **197**), a two-step process of regioselective lithiation-trapping and silyl group removal has been used to prepare a range of P-stereogenic compounds, including precursors to diphosphine ligands (e.g., Mini-PHOS). This two-step protocol delivers products **199** with the opposite configuration to that obtained by direct asymmetric lithiation-trapping of a dimethylphosphine sulfide **197** using *n*-BuLi/(−)-sparteine (Scheme 5.83) [127].

Livinghouse and Wolfe [128] reported that the phosphide obtained by treatment of *tert*-butylphenylphosphine borane **200** with *s*-BuLi/(−)-Spartein complex, can be dynamically resolved; the observed enantioselectivity depends on time and temperature. It was found that stirring the suspended (−)-sparteine-lithium complex **201** for 1 h at room temperature prior to alkylation resulted in an increase in ee up to 95% in case of monodentate phosphines, and to diastereoisomeric ratios of 22 : 1 in case of diastereoisomeric bidentate phosphines The subsequent reaction with $[(CH_2)_nX)]_2$ led to the formation of enantiomerically pure diphosphines **202** (Scheme 5.84).

P-OP-directed asymmetric deprotonation of benzylic amines **203** using *n*-BuLi/(−)-spartein complex provides an efficient method for the synthesis of chiral NC-α- and NC-α,α-derivatives **204**, **206** with total selectivity with respect to competing allylic and

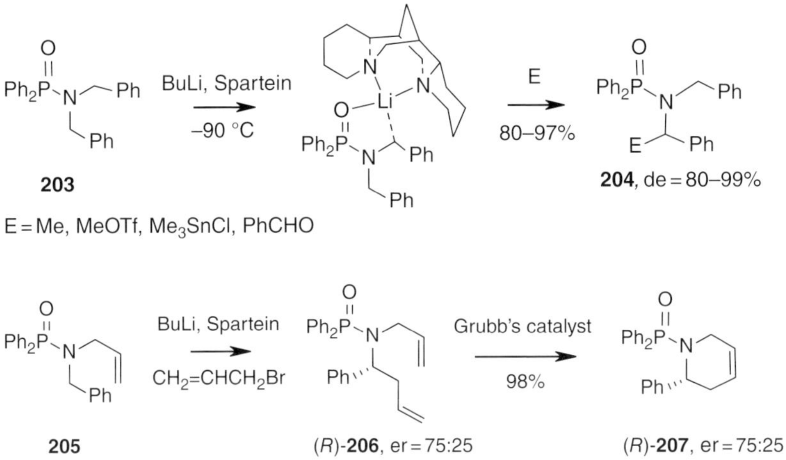

Scheme 5.85 Synthesis of chiral N-POP-protected nitrogen heterocycles **207**.

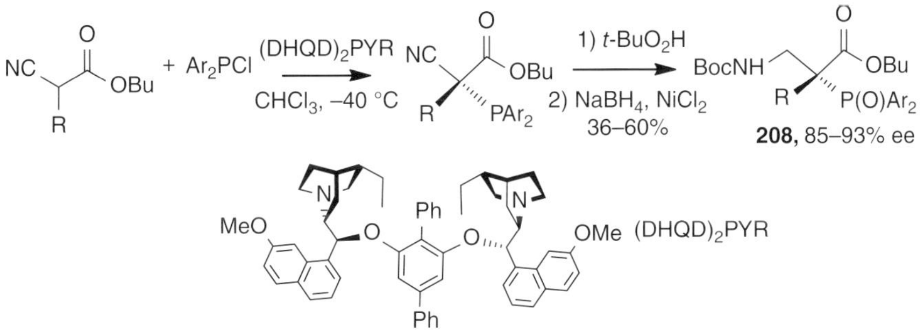

Scheme 5.86 Organocatalysis with $(DHQD)_2PYR$).

ortho-phenyl lithiation. The reaction represents a convenient method for the synthesis of chiral N-POP-protected nitrogen heterocycles (*R*)-**207** (Scheme 5.85) [129].

5.10 Miscellaneous

The asymmetric catalytic C*–P bond formation employing electrophilic phosphorus compounds and catalytic amounts of $(DHQD)_2PYR$ in combination with the proton sponge led to the formation of α-quaternary α-phosphino β-amino acids with high stereoselectivities and good yields. ^{31}P NMR experiments allowed to propose a reaction mechanism with a cinchona alkaloid-catalyzed nucleophilic activation of the phosphorus electrophile (Scheme 5.86) [130].

A stereoselective methodology for the synthesis of optically active dihydropyran phosphonates bearing three contiguous stereogenic centers was developed using a bifunctional squaramide-containing aminocatalyst. In general, aliphatic and aromatic α,β-unsaturated acyl phosphonates having electron-withdrawing or electron-donating substituents were successfully applied in the asymmetric inverse-electron-demand hetero-Diels–Alder reaction. A range of α,β-unsaturated aldehydes were shown to be compatible with the presented protocol, providing good yields ranging from 64%

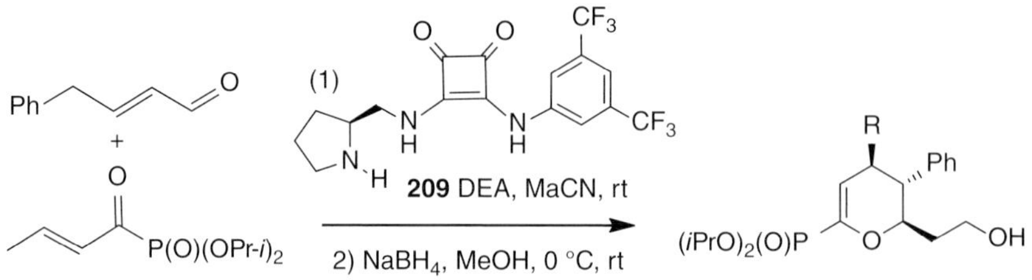

Scheme 5.87

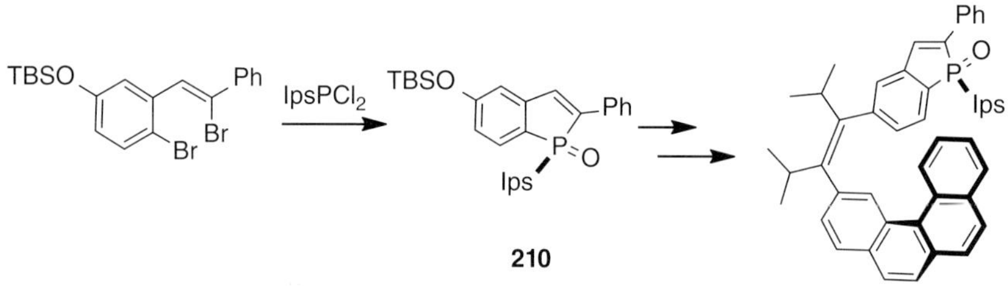

Scheme 5.88 Chiral phospha helicenes.

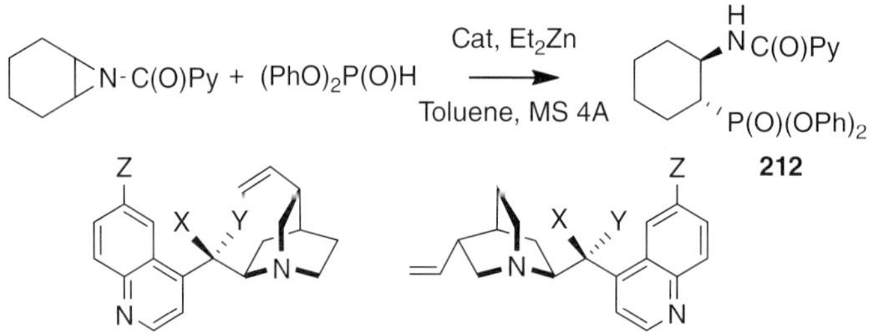

Scheme 5.89 Enantioselective desymmetrization of aziridines with phosphites.

to 84%. The derived cycloadducts were transformed into useful chiral and complex building blocks (Scheme 5.87) [131].

Chiral phosphahelicenes **211** bearing an isopinocampheyl group on phosphorus were prepared. These phosphahelicenes demonstrated the high potential in enantioselective nucleophilic organocatalysis by the development of [3 + 2] cyclization reactions between activated olefins and γ-substituted allenes with 97% ee (Scheme 5.88) [132].

The enantioselective desymmetrization of aziridines with phosphites was developed using catalysts derived from 9-amino-9-deoxy-*epi*-cinchona alkaloids **214**. Excellent yields and enantioselectivities were observed in this reaction (Scheme 5.89) [133].

Phillips and Barros reported recently an interesting synthesis of α-cyclopropylphosphonates through a domino Michael addition/intramolecular alkylation reaction of α,β-unsaturated aldehydes with bromophosphonoacetates. Highly functionalized cyclopropylphosphonates **215** containing three chiral centers, one of which being a quaternary carbon atom, were obtained with good diastereoselectivities of up to 83 : 17 and very high enantioselectivities of up to 99% (Scheme 5.90) [134, 135].

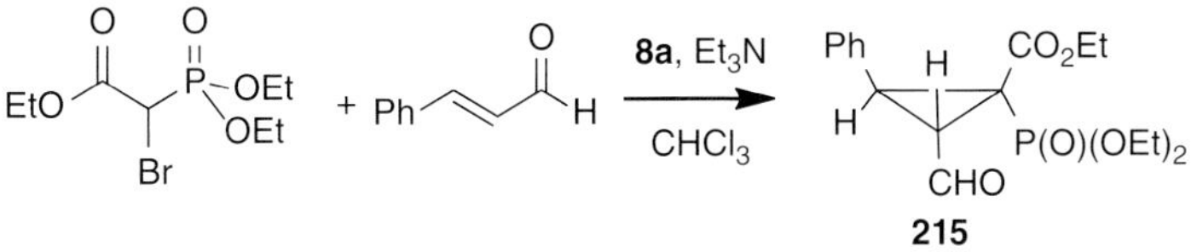

Scheme 5.90 Organocatalytic cyclophosphonation reaction.

References

1 List, B. (ed.) (2010) *Asymmetric Organocatalysis*, Topics in Current Chemistry, Springer, Heidelberg. vol. **291**, 460 pp.
2 Berkessel, A. and Gröger, H. (2005) *Asymmetric Organocatalysis – From Biomimetic Concepts to Applications in Asymmetric Synthesis*, Wiley-VCH Verlag GmbH, Weinheim, 440 pp.
3 Bredig, G. and Fiske, P.S. (1912) Durch Katalysatoren bewirkte asymmetrische Synthese. *Biochem. Z.*, **46**, 7–23.
4 Organische, P.H. and Katalysatoren, L.X.I. (1960) Asymmetrische synthesen mit ketenen, 1 Alkaloid-katalysierte asymmetrische Synthesen von α–Phenyl-propionsaureestern. *Justus Liebigs Ann. Chem.*, **634**, 9–22.
5 List, B., Lerner, R.A., and Barbas, C.F. (2000) Proline-catalyzed direct asymmetric aldol reactions. *J. Am. Chem. Soc.*, **122** (10), 2395–2396.
6 Gaunt, M.J., Johansson, C.C.C., McNally, A., and Vo, N.T. (2007) Enantioselective organocatalysis. *Drug Discovery Today*, **12** (1-2), 8–27.
7 Tommaso, M. (2011) Organocatalysis: cinchona catalysts, in *Wiley Interdisciplinary Reviews: Computational Molecular Science*, vol. **1**, John Wiley & Sons, Ltd, pp. 142–152.
8 Gratzer, K., Gururaja, G.N., and Waser, M. (2013) Towards tartaric-acid-derived asymmetric organocatalysts. *Eur. J. Org. Chem.*, **21**, 4471–4482.
9 List, B. (2002) Proline-catalyzed asymmetric reactions. *Tetrahedron*, **58** (28), 5573–5590.
10 Wikipedia Thiourea Organocatalysis, https:en.wikipedia.org/wiki/Thiourea_organocatalysis (accessed 18 July 2015).
11 Schenker, S., Zamfir, A., Freund, M., and Tsogoeva, S.B. (2011) Developments in chiral binaphthyl-derived brønsted/lewis acids and hydrogen-bond-donor organocatalysis. *Eur. J. Org. Chem.*, **12**, 2209–2222.
12 Leow, D. and Tan, C.H. (2009) Chiral guanidine catalyzed enantioselective reactions. *Chem. Asian J.*, **4** (4), 488–507.
13 Zhu, Y., Malerich, J.P., and Rawal, V.H. (2010) Squaramide-catalyzed enantioselective Michael addition of diphenyl phosphite to nitroalkenes. *Angew. Chem.*, **122** (1), 157–160.
14 Fu, X., Jiang, Z., and Tan, C.H. (2007) Bicyclic guanidine-catalyzed enantioselective phospha-Michael reaction: synthesis of chiral β-aminophosphine oxides and β-aminophosphines. *J. Chem. Soc., Chem. Commun.*, (47), 5058–5060.
15 Terada, M., Ikehara, T., and Ube, H. (2007) Enantioselective 1,4-addition reactions of diphenyl phosphite to nitroalkenes catalyzed by an axially chiral guanidine. *J. Am. Chem. Soc.*, **129** (46), 14112–14113.

16 Wynberg, H. and Smaardijk, A.A. (1983) Asymmetric catalysis in carbon-phosphorus bond formation. *Tetrahedron Lett.*, **24** (52), 5899–5900.
17 Smaardijk, A.A., Noorda, S., van Bolhuis, F., and Wynberg, H. (1985) The absolute configuration of α-hydroxyphosphonates. *Tetrahedron Lett.*, **26** (4), 493–496.
18 Wanzlick, H.W. (1962) Aspects of nucleophilic carbene chemistry. *Angew. Chem. Int. Ed. Engl.*, **1** (20), 75–80.
19 Arduengo, A.J., Harlow, R.L., and Kline, M. (1991) A stable crystalline carbene. *J. Am. Chem. Soc.*, **113** (1), 361–363.
20 Flanigan, D.M., Romanov-Michailidis, F., White, N.A., and Rovis, T. (2015) Organocatalytic reactions enabled by N-heterocyclic carbenes. *Chem. Rev.*, **115** (17), 9307–9387.
21 Dzięgielewski, M., Pięta, J., Kamińska, E., and Albrecht, Ł. (2015) Organocatalytic synthesis of optically active organophosphorus compounds. *Eur. J. Org. Chem.*, **2015** (4), 677–702.
22 Kolodiazhna, A.O., Kukhar, V.P., and Kolodiazhnyi, O.I. (2008) Modified alkaloids as organocatalysts for the asymmetric synthesis of organophosphorus compounds. *Phosphorus, Sulfur Silicon Relat. Elem.*, **183** (1-2), 728–729.
23 Kolodiazhna, A.O., Kukhar, V.P., and Kolodiazhnyi, O.I. (2008) Organocatalysis of phosphaaldol reaction. *Russ. J. Gen. Chem.*, **78** (11), 2043–2051.
24 Peng, L., Wanga, L.L., Bai, J.F., Jia, L.N., Yang, Q.C., Huang, Q.C., Xu, X.Y., and Wang, L.X. (2011) Highly effective and enantioselective Phospho-Aldol reaction of diphenyl phosphite with N-alkylated isatins catalyzed by quinine. *Tetrahedron Lett.*, **52** (11), 1157–1160.
25 Barros, M.T. and Phillips, A.M.F. (2011) Organocatalyzed synthesis of tertiary α-hydroxyphosphonates by a highly regioselective modified Pudovik reaction. *Eur. J. Org. Chem.*, **2011** (20–21), 4028–4036.
26 Connon, S.J. (2008) Asymmetric catalysis with bifunctional cinchona alkaloid-based urea and thiourea organocatalysts. *Chem. Commun.*, 2499–2510.
27 Wang, F., Liu, X., Cui, X., Xiong, Y., Zhou, X., and Feng, X. (2009) Asymmetric hydrophosphonylation of α–ketoesters catalyzed by cinchona-derived thiourea organocatalysts. *Chem. Eur. J.*, **15** (3), 589–592.
28 Nakanishi, K., Kotani, S., Sugiura, M., and Nakajima, M. (2008) First asymmetric Abramov-type phosphonylation of aldehydes with trialkyl phosphites catalyzed by chiral Lewis bases. *Tetrahedron*, **64** (27), 6415–6419.
29 Guin, J., Wang, Q., van Gemmeren, M., and List, B. (2015) The catalytic asymmetric abramov reaction. *Angew. Chem. Int. Ed.*, **54** (1), 355–358.
30 Uraguchi, D., Ito, T., and Ooi, T. (2009) Generation of chiral phosphonium dialkyl phosphite as a highly reactive P-nucleophile: application to asymmetric hydro-phosphonylation of aldehydes. *J. Am. Chem. Soc.*, **131** (11), 3836–3837.
31 (a) Simón, L. and Paton, R.S. (2015) Origins of asymmetric phosphazene organocatalysis: computations reveal a common mechanism for nitro- and phospho-aldol additions. *J. Org. Chem.*, **80** (5), 2756–2766; (b) Xu, F., Liu, Y., Tu, J., Lei, C., and Li, G. (2015) Fe(III) catalyzed enantioselective hydrophosphonylation of aldehydes promoted by chiral camphor Schiff bases. *Tetrahedron: Asymmetry*, **26** (17), 891–896.
32 Hirashima, S.-i., Arai, R., Nakashima, K., Kawai, N., Kondo, J., Koseki, Y., and Miuraa, T. (2015) Asymmetric hydrophosphonylation of aldehydes using a

cinchona–diaminomethylenemalononitrile organocatalyst. *Adv. Synth. Catal.*, **357** (18), 3863–3867.

33 Alegre-Requena, J.V., Marqués-López, E., Sanz Miguel, P.J., and Herrera, R.P. (2014) Organocatalytic enantioselective hydrophosphonylation of aldehydes. *Org. Biomol. Chem.*, **12** (8), 1258–1264.

34 Nakamura, S., Hayashi, M., Hiramatsu, Y., Shibata, N., Funahashi, Y., and Toru, T. (2009) Catalytic enantioselective hydrophosphonylation of ketimines using cinchona alkaloids. *J. Am. Chem. Soc.*, **131** (51), 18240–18241.

35 Pettersen, D., Marcolini, M., Bernardi, L., Fini, F., Herrera, R.P., Sgarzani, V., and Ricci, A. (2006) Direct access to enantiomerically enriched α-amino phosphonic acid derivatives by organocatalytic asymmetric hydrophosphonylation of imines. *J. Org. Chem.*, **71** (16), 6269–6272.

36 Duxbury, J.P., Cawley, A., Thornton-Pete, M., Wantz, L., Warne, J.N.D., Greatrex, R., Brown, D., and Kee, T.P. (1999) Chiral Aluminium Complexes as Phospho-Transfer Catalysts. *Tetrahedron Lett.*, **40** (23), 4403–4406.

37 Joly, G.D. and Jacobsen, E.N. (2004) Synthesis of an optically active Al(salalen) complex and its application to catalytic hydrophosphonylation of aldehydes and aldimines. *J. Am. Chem. Soc.*, **126** (7), 4102–4103.

38 Fu, X., Loh, W.T., Zhang, Y., Chen, T., Ma, T., Liu, H., Wang, J., and Tan, C.H. (2009) Chiral guanidinium salt catalyzed enantioselective phospha-mannich reactions. *Angew. Chem. Int. Ed.*, **48** (40), 7387–7390.

39 Cheng, X., Goddard, R., Buth, G., and List, B. (2008) Direct catalytic asymmetric three-component Kabachnik–Fields reaction. *Angew. Chem. Int. Ed.*, **47** (27), 5079–5081.

40 Thorat, P.B., Goswami, S.V., Magar, R.L., Patil, B.R., and Bhusare, S.R. (2013) An efficient organocatalysis: a one-pot highly enantioselective synthesis of α-aminophosphonates. *Eur. J. Org. Chem.*, **24** (24), 5509–5516.

41 Akiyama, T., Morita, H., Itoh, J., and Fuchibe, K. (2005) Chiral brønsted acid catalyzed enantioselective hydrophosphonylation of imines: asymmetric synthesis of α-amino phosphonates. *Org. Lett.*, **7** (13), 2583–2585.

42 Shi, F.Q. and Song, B.A. (2009) Origins of enantioselectivity in the chiral Brønsted acid catalyzed hydrophosphonylation of imines. *Org. Biomol. Chem.*, **7**, 1292–1298.

43 Bhadury, P.S. and Li, H. (2012) Organocatalytic asymmetric hydrophosphonylation/Mannich reactions using thiourea, cinchona and brønsted acid catalysts. *Synlett*, (8), 1108–1131.

44 Yamanaka, M. and Hirata, T.J. (2009) DFT study on bifunctional chiral brønsted acid-catalyzed asymmetric hydrophosphonylation of imines. *J. Org. Chem.*, **74** (9), 3266–3271.

45 Ingle, G.K., Liang, Y., Mormino, M.G., Li, G., Fronczek, F.R., and Antilla, J.C. (2011) Chiral magnesium BINOL phosphate-catalyzed phosphination of imines: access to enantioenriched r-amino phosphine oxides. *Org. Lett.*, **13** (8), 2054–2057.

46 Thorat, P.B., Goswami, S.V., Magar, R.L., Patil, B.R., and Bhusare, S.R. (2013) An efficient organocatalysis: a one-pot highly enantioselective synthesis of α-aminophosphonates. *Eur. J. Org. Chem.*, **2013** (24), 5509–5516.

47 George, J., Sridhar, B., and Reddy, B.V.S. (2014) First example of quinine-squaramide catalyzed enantioselective addition of diphenyl phosphite to ketimines derived from isatins. *Org. Biomol. Chem.*, **12** (10), 1595–1602.

48 Kumar, A., Sharma, V., Kaur, J., Kumar, V., Mahajan, S., Kumar, N., and Chimni, S.S. (2014) Cinchona-derived thiourea catalyzed hydrophosphonylation of ketimines – an enantioselective synthesis of α-amino phosphonates. *Tetrahedron*, **70** (39), 7044–7049.

49 Albrecht, Ł., Richter, B., Vila, C., Krawczyk, H., and Jørgensen, K.A. (2009) Organocatalytic domino Michael–Knoevenagel condensation reaction for the synthesis of optically active 3-diethoxyphosphoryl-2-oxocyclohex-3-enecarboxylates. *Chem. Eur. J.*, **15** (13), 3093–3102.

50 Russo, A. and Lattanzi, A. (2010) Asymmetric organocatalytic conjugate addition of diarylphosphane oxides to chalcones. *Eur. J. Org. Chem.*, **2010** (35), 6736–6739.

51 Wang, J., Heikkinen, L.D., Li, H., Zu, L., Jiang, W., Xie, H., and Wang, W. (2007) Quinine-catalyzed enantioselective Michael addition of diphenyl phosphite to nitroolefins: synthesis of chiral precursors of α–substituted β-aminophosphonates. *Adv. Synth. Catal.*, **349** (7), 1052–1056.

52 Abbaraju, S., Bhanushali, M., and Zhao, C.G. (2011) Quinidine thiourea-catalyzed enantioselective synthesis of β-nitrophosphonates: beneficial effects of molecular sieves. *Tetrahedron*, **67** (39), 7479–7484.

53 Bartoli, G., Bosco, M., Carlone, A., Locatelli, M., Mazzanti, A., Sambri, L., and Melchiorre, P. (2007) Organocatalytic asymmetric hydrophosphination of nitroalkenes. *J. Chem. Soc., Chem. Commun.*, (7), 722–724.

54 Connon, S. (2006) Organocatalysis mediated by (Thio)urea derivatives. *Chem. Eur. J.*, **12** (21), 5418–5427.

55 Taylor, M.S. and Jacobsen, E.N. (2006) Asymmetric catalysis by chiral hydrogen-bond donors. *Angew. Chem. Int. Ed.*, **45** (10), 1520–1543.

56 Wen, S., Li, P., Wu, H., Yu, F., Liang, X., and Ye, J. (2010) Enantioselective organocatalytic phospha-Michael reaction of α,β-unsaturated ketones. *J. Chem. Soc., Chem. Commun.*, **46** (26), 4806–4808.

57 Li, P., Wen, S., Yu, F., Liu, Q., Li, W., Wang, Y., Liang, X., and Ye, J. (2009) Enantioselective organocatalytic Michael addition of malonates to α,β-unsaturated ketones. *Org. Lett.*, **11** (3), 753–756.

58 Pham, T.S., Gönczi, K., Kardos, G., Süle, K., Hegedus, L., Kállay, M., Kubinyi, M., Szabó, P., Petneházy, I., Toke, I., and Jászay, Z. (2013) Cinchona based squaramide catalysed enantioselective Michael addition of α–nitrophosphonates to aryl acrylates: enantioselective synthesis of quaternary α–aminophosphonates. *Tetrahedron: Asymmetry*, **24** (24), 1605–1614.

59 Bera, K. and Namboothiri, I.N.N. (2012) Enantioselective synthesis of quaternary α-aminophosphonates via conjugate addition of α-nitrophosphonates to enones. *Org. Lett.*, **14** (4), 980–983.

60 Tripathi, C.B., Kayal, S., and Mukherjee, S. (2012) Catalytic asymmetric synthesis of α,β-disubstituted α,γ-diaminophosphonic acid precursors by Michael addition of α-substituted nitrophosphonates to nitroolefins. *Org. Lett.*, **14** (13), 3296–3299.

61 (a) Bera, K. and Namboothiri, I.N.N. (2015) Quinine-derived thiourea and squaramide catalyzed conjugate addition of α-nitrophosphonates to enones: asymmetric synthesis of quaternary α-aminophosphonates. *J. Org. Chem.*, **80** (3), 1402–1413; (b) Bera, K. and Namboothiri, I.N.N. (2013) Enantioselective synthesis of α-amino-γ-sulfonyl phosphonates with a tetrasubstituted chiral α-carbon via quinine–β–catalyzed Michael addition of nitrophosphonates to vinyl sulfones. *Adv. Synth. Catal.*, **355** (7), 1265–1270.

62 Sohtome, Y., Horitsugi, N., Takagi, R., and Nagasawa, K. (2011) Enantioselective Phospha-Michael reaction of diphenyl phosphonate with nitroolefins utilizing conformationally flexible guanidinium/bisthiourea organocatalyst: assembly-state tunability in asymmetric organocatalysis. *Adv. Synth. Catal.*, **353** (14-15), 2631–2636.
63 Sohtome, Y., Hashimoto, Y., and Nagasawa, K. (2005) Guanidine-thiourea bifunctional organocatalyst for the asymmetric Henry (nitroaldol) reaction. *Adv. Synth. Catal.*, **347** (11-13), 1643–1648.
64 Cullen, S.C. and Rovis, T. (2008) Catalytic asymmetric Stetter reaction onto vinyl-phosphine oxides and vinylphosphonates. *Org. Lett.*, **10** (14), 3141–3144.
65 Jang, K.-P., Hutson, G.E., Johnston, R.C., McCusker, E.O., Cheong, P.H.-Y., and Scheidt, K.A. (2014) Asymmetric homoenolate additions to acyl phosphonates through rational design of a tailored N-heterocyclic carbene catalyst. *J. Am. Chem. Soc.*, **136** (1), 76–79.
66 Carlone, A., Bartoli, G., Bosco, M., Sambri, L., and Melchiorre, P. (2007) Organocatalytic Asymmetric Hydrophosphination of a,b-Unsaturated Aldehydes. *Angew. Chem. Int. Ed.*, **46** (24), 4504–4506.
67 Lu, X., Zho, Z.L.X., and Lian, X.Y.J. (2011) Stereoselectivity of Michael addition of P(X)– *H*-type nucleophiles to cyclohexen-1-ylphosphine oxide: the case of base-selective transformation. *RSC Adv.*, **1** (4), 698–705.
68 Hammar, P., Vesely, J., Rios, R., Eriksson, L., and Cordova, A. (2008) Organocatalytic asymmetric hydrophosphination of α,β-unsaturated aldehydes: development, mechanism and DFT calculations. *Adv. Synth. Catal.*, **350** (11-12), 1875–1884.
69 Ibrahem, I., Rios, R., Vesely, J., Hammar, P., Eriksson, L., Himo, F., and Cordova, A. (2007) Enantioselective organocatalytic hydrophosphination of α,β-unsaturated aldehydes. *Angew. Chem. Int. Ed.*, **46** (24), 4507–4510.
70 Capuzzi, M., Perdicchia, D., and Jørgensen, K.A. (2008) Highly enantioselective approach to geminal bisphosphonates by organocatalyzed Michael-type addition of β–ketoesters. *Chem. Eur. J.*, **14** (1), 128–135.
71 Maerten, E., Cabrera, S., Kjrsgaard, A., and Jorgensen, K.A. (2007) Organocatalytic asymmetric direct phosphonylation of alpha,beta-unsaturated aldehydes: mechanism, scope, and application in synthesis. *J. Org. Chem.*, **72** (23), 8893–8903.
72 Sulzer-Mossé, S., Tissot, M., and Alexakis, A. (2007) First enantioselective organocatalytic conjugate addition of aldehydes to vinyl phosphonates. *Org. Lett.*, **9** (19), 3749–3752.
73 Sulzer-Moss, S., Alexakis, A., Mareda, J., Bollot, G., Bernardinelli, G., and Filinchuk, Y. (2009) Enantioselective organocatalytic conjugate addition of aldehydes to vinyl sulfones and vinyl phosphonates as challenging Michael acceptors. *Chem. Eur. J.*, **15** (13), 3204–3220.
74 Albrecht, Ł., Richter, B., Krawczyk, H., and Jorgensen, K.A. (2008) Enantioselective organocatalytic approach to α-methylene-δ-lactones and δ-lactams. *J. Org. Chem.*, **73** (21), 8337–8343.
75 Barros, M.T., Faísca, A.M., and Phillips, F. (2008) Enamine catalysis in the synthesis of chiral structural analogues of gem-bisphosphonates known to be biologically active. *Eur. J. Org. Chem.*, **2008** (15), 2525–2529.
76 Guang, J., Zhao, J.C.G., Guang, J., and Zhao, J.C.G. (2013) Organocatalyzed asymmetric Michael reaction of β–aryl–α–ketophosphonates and nitroalkenes. *Tetrahedron Lett.*, **54** (42), 5703–5706.

77 Samanta, S., Perera, S., and Zhao, C.G. (2010) Organocatalytic enantioselective synthesis of both diastereomers of hydroxyphosphinates. *J. Org. Chem.*, **75** (4), 1101–1106.
78 Dodda, R. and Zhao, C.G. (2006) Organocatalytic highly enantioselective synthesis of secondary α–hydroxyphosphonates. *Org. Lett.*, **8** (21), 4911–4914.
79 Perera, S., Naganaboina, V.K., Wang, L., Zhang, B., Guo, Q., Rout, L., and Zhao, C.G. (2011) Organocatalytic highly enantioselective synthesis of β–formyl-–hydroxyphosphonates. *Adv. Synth. Catal.*, **353** (10), 1729–1734.
80 Yao, Q. and Yuan, C. (2013) A synthetic study of chiral α–hydroxy-*H*-phosphinates based on proline catalysis. *Chem. Eur. J.*, **19** (20), 6080–6088.
81 Liu, T., Wang, Y., Wu, G., Song, H., Zhou, Z., and Tang, C. (2011) Organocatalyzed enantioselective Michael addition of 2-hydroxy-1,4-naphthoquinone to β,γ-unsaturated ketophosphonates. *J. Org. Chem.*, **76** (10), 4119–4124.
82 Smith, S.R., Leckie, S.M., Holmes, R., Douglas, J., Fallan, C., Shapland, P., Pryde, D., Slawin, A.M.Z., and Smith, A.D. (2014) α-ketophosphonates as ester surrogates: isothiourea-catalyzed asymmetric diester and lactone synthesis. *Org. Lett.*, **16** (9), 2506–2509.
83 Bhasker Gondi, V., Hagihara, K., and Rawal, V.H. (2009) Diastereoselective and enantioselective synthesis of tertiary α–hydroxy phosphonates through hydrogen-bond catalysis. *Angew. Chem. Int. Ed.*, **48** (43), 776–779.
84 Rassukana, Y.V., Yelenicha, I.P., Vlasenkoa, Y.G., and Onysko, P.P. (2014) Asymmetric synthesis of phosphonotrifluoroalanine derivatives via proline-catalyzed direct enantioselective C-C bond formation reactions of N-H trifluoroacetimidoyl phosphonate. *Tetrahedron: Asymmetry*, **25** (16–17), 1234–1238.
85 Dodda, R. and Zhao, C.G. (2007) Enantioselective synthesis of α-aminopropargylphosphonates. *Tetrahedron Lett.*, **48** (25), 4339–4342.
86 (a) Palacios, F., Aparicio, D., Ochoa de Retana, A.M., de los Santos, J.M., Gil, J.I., and Alonso, J.M. (2002) Asymmetric synthesis of 2 *H*-azirines derived from phosphine oxides using solid-supported amines. Ring opening of azirines with carboxylic acids. *J. Org. Chem.*, **67** (21), 7283–7288; (b) Palacios, F., Aparicio, D., de Retana, O.M.A., de los Santos, J.M., Gil, J.I., and Lopez de Munain, R. (2003) Asymmetric synthesis of 2 *H*-aziridine phosphonates, and α- or β-aminophosphonates from enantiomerically enriched 2 *H*-azirines. *Tetrahedron: Asymmetry*, **14** (6), 689–700.
87 Vicario, J., Ezpeleta, J.M., and Palacios, F. (2012) Asymmetric cyanation of α-ketiminophosphonates catalyzed by cinchona alkaloids: enantioselective synthesis of tetrasubstituted α-aminophosphonic acid derivatives from trisubstituted α-aminophosphonates. *Adv. Synth. Catal.*, **354** (14-15), 2641–2647.
88 Jiang, H., Paixão, M.W., Monge, D., and Jørgensen, K.A. (2010) Acyl phosphonates: good hydrogen bond acceptors and ester/amide equivalents in asymmetric organocatalysis. *J. Am. Chem. Soc.*, **132** (8), 2775–2783.
89 Kong, S., Fan, W., Wu, G., and Miao, Z. (2012) Enantioselective synthesis of tertiary α–hydroxy phosphonates catalyzed by carbohydrate/cinchona alkaloid thiourea organocatalysts. *Angew. Chem. Int. Ed.*, **51** (35), 8864–8867.
90 Chen, L., Wu, Z.-J., Zhang, M.-L., Yue, D.-F., Zhang, X.-M., Xu, X.-Y., and Yuan, W.-C. (2015) Organocatalytic asymmetric Michael/cyclization cascade reactions of 3-hydroxyoxindoles/3-aminooxindoles with α,β-unsaturated acyl phosphonates. *J. Org. Chem.*, **80** (11), 12668–12675.

91 Guang, J., Guo, Q., and Zhao, J.C.G. (2012) Acetylphosphonate as a surrogate of acetate or acetamide in organocatalyzed enantioselective Aldol reactions. *Org. Lett.*, **14** (12), 3174–3177.

92 Alvarez-Casao, Y., Marques-Lopez, E., and Herrera, R.P. (2011) Organocatalytic enantioselective Henry reactions. *Symmetry*, **3** (2), 220–245.

93 Mandal, T., Samanta, S., and Zhao, C. (2007) Organocatalytic highly enantioselective nitroaldol reaction of α-ketophosphonates and nitromethane. *Org. Lett.*, **9** (5), 943–945.

94 Albrecht, Ł., Albrecht, A., Krawczyk, H., and Jørgensen, K.A. (2010) Organocatalytic asymmetric synthesis of organophosphorus compounds. *Chem. Eur. J.*, **16** (1), 28–48.

95 Samanta, S. and Zhao, C.G. (2007) Organocatalyzed nitroaldol reaction of α-ketophosphonates and nitromethane revisited. *Arkivoc*, **XIII**, 218–226.

96 Vicario, J., Ortiz, P., Ezpeleta, J.M., and Palacios, F. (2015) Asymmetric synthesis of functionalized tetrasubstituted α-aminophosphonates through enantioselective Aza-Henry reaction of phosphorylated ketimines. *J. Org. Chem.*, **80** (1), 156–164.

97 Chen, X., Wang, J., Zhu, Y., Shang, D., Gao, B., Liu, X., Feng, X., Su, Z., and Hu, C. (2008) A secondary amine amide organocatalyst for the asymmetric nitroaldol reaction of α-ketophosphonates. *Chem. Eur. J.*, **14** (35), 10896–10899.

98 Zhang, Y., Liu, Y.K., Kang, T.R., Hu, Z.K., and Chen, Y.C. (2008) Organocatalytic enantioselective Mannich-type reaction of phosphorus ylides: synthesis of chiral N-boc-β-amino-α-methylene carboxylic esters. *J. Am. Chem. Soc.*, **130** (8), 2456–2457.

99 Hong, B.C., Jan, R.H., Tsai, C.W., Nimje, R.Y., Liao, J.H., and Lee, G.H. (2009) Organocatalytic enantioselective cascade Michael–Michael–Wittig reactions of phosphorus ylides: one-pot synthesis of the *all-cis* trisubstituted cyclohexenecarboxylates via the [1 + 2 + 3] annulation. *Org. Lett.*, **11** (22), 5246–5249.

100 Allu, S., Selvakumar, S., and Singh, V.K. (2010) Asymmetric organocatalytic Michael-type reaction of phosphorus ylides to nitroolefins: synthesis of γ-nitro-β-aryl-α-methylene carboxylic esters. *Tetrahedron Lett.*, **51** (2), 446–448.

101 Lin, A., Wang, J., Mao, H., Ge, H., Tan, R., Zhu, C., and Cheng, Y. (2011) Organocatalytic asymmetric Michael-type/Wittig reaction of phosphorus ylides: synthesis of chiral α-methylene-δ-ketoesters. *Org. Lett.*, **13** (16), 4176–4179.

102 Muci, A.R., Campos, K.R., and Evans, D.A. (1995) Enantioselective deprotonation as a vehicle for the asymmetric synthesis of C2-symmetric *P*-chiral diphosphines. *J. Am. Chem. Soc.*, **117** (35), 9075–9076.

103 Strohmann, C., Strohfeldt, K., Schildbach, D., McGrath, M.J., and O'Brien, P. (2004) Crystal structures of (+)-sparteine surrogate adducts of methyllithium and phenyllithium. *Organometallics*, **23** (23), 5389–5391.

104 Strohmann, C., Seibel, T., and Strohfeldt, K. (2003) Monomeric butyllithium compound [t-BuLi·(-)-Sparteine]: molecular structure of the first monomeric butyllithium compound. *Angew. Chem. Int. Ed.*, **42** (37), 4531–4533.

105 Genet, C., Canipa, S.J., O'Brien, P., and Taylor, S. (2006) Catalytic asymmetric synthesis of ferrocenes and *P*-stereogenic bisphosphines. *J. Am. Chem. Soc.*, **128** (29), 9336–9337.

106 Morisaki, Y., Imoto, H., Tsurui, K., and Chujo, Y. (2009) Practical synthesis of *P*-stereogenic diphosphacrowns. *Org. Lett.*, **11** (11), 2241–2244.

107 Morisaki, Y., Kato, R., and Chujo, Y. (2013) Synthesis of enantiopure *P*-stereogenic diphosphacrowns using *P*-stereogenic secondary phosphines. *J. Org. Chem.*, **78** (6), 2769–2774.

108 Dolhem, F., Johansson, M.J., Antonsson, T., and Kann, N. (2006) *P*-chirogenic α-carboxyphosphine boranes as effective pre-ligands in palladium-catalyzed asymmetric reactions. *Synlett*, (20), 3389–3394.

109 Johansson, M.J., Schwartz, L., Amedjkouh, M., and Kann, N. (2004) New chiral amine ligands in the desymmetrization of prochiral phosphine boranes. *Tetrahedron: Asymmetry*, **15** (22), 3531–3538.

110 Johansson, M.J., Schwartz, L.O., Amedjkouh, M., and Kann, N.C. (2004) Desymmetrization of prochiral phosphanes using derivatives of (-)-cytisine. *Eur. J. Org. Chem.*, **2004** (9), 1894–1896.

111 (a) Granander, J., Secci, F., Canipa, S.J., O'Brien, P., and Kelly, B. (2011) One-ligand catalytic asymmetric deprotonation of a phosphine borane: synthesis of *P*-stereogenic bisphosphine ligands. *J. Org. Chem.*, **76** (11), 4794–4799; (b) Johansson, M.J., Andersson, K.H.O., and Kann, N. (2008) Modular asymmetric synthesis of *P*-chirogenic -amino phosphine boranes. *J. Org. Chem.*, **73** (12), 4458–4463.

112 Imamoto, T., Nishimura, M., Koide, A., and Yoshida, K. (2007) *t*-Bu-QuinoxP* ligand: applications in asymmetric Pd-catalyzed allylic substitution and Ru-catalyzed hydrogenation. *J. Org. Chem.*, **72** (19), 7413–7416.

113 Morisaki, Y., Imoto, H., Ouchi, Y., Nagata, Y., and Chujo, Y. (2008) Stereospecific construction of a trans-1,4-diphosphacyclohexane skeleton. *Org. Lett.*, **10** (8), 1489–1492.

114 Imoto, H., Morisaki, Y., and Chujo, Y. (2010) Synthesis and coordination behaviors of *P*-stereogenic polymers. *J. Chem. Soc., Chem. Commun.*, (46), 7542–7544.

115 Ouchi, Y., Morisaki, Y., and Chujo, Y. (2006) Synthesis of photoresponsive polymers having *P*-chiral phosphine in the main chain. *Polym. Prepr.*, **47**, 708–709.

116 Ouchi, Y., Morisaki, Y., Ogoshi, T., and Chujo, Y. (2007) Synthesis of a stimuli-responsive *P*-chiral polymer with chiral phosphorus atoms and azobenzene moieties in the main chain. *Chem. Asian J.*, **2** (3), 397–402.

117 Oohara, N., Katagiri, K., and Imamoto, T. (2003) A novel *P*-chirogenic phosphine ligand, (S,S)-1,2-bis-[(ferrocenyl)methylphosphino]ethane: synthesis and use in rhodium-catalyzed asymmetric hydrogenation and palladium-catalyzed asymmetric allylic alkylation. *Tetrahedron: Asymmetry*, **14** (15), 2171–2175.

118 Hoge, G. (2003) Synthesis of both enantiomers of a *P*-chirogenic 1,2-bisphospholanoethane ligand via convergent routes and application to rhodium-catalyzed asymmetric hydrogenation of CI-1008 (Pregabalin). *J. Am. Chem. Soc.*, **125** (34), 10219–10227.

119 Maienza, F., Spindler, F., Thommen, M., Pugin, B., Malan, C., and Mezzetti, A. (2002) Exploring stereogenic phosphorus: synthetic strategies for diphosphines containing bulky, highly symmetric substituents. *J. Org. Chem.*, **67** (15), 5239–5249.

120 Miyazaki, T., Sugawara, M., Danjo, H., and Imamoto, T. (2004) Dihydroboronium derivatives of (S,S)-1,2-bis(t-butylmethylphosphino)ethane as convenient chiral ligand precursors. *Tetrahedron Lett.*, **45** (51), 9341–9344.

121 Gammon, J.J., Gessner, V.H., Barker, G.R., Granander, J., Whitwood, A.C., Strohmann, C., O'Brien, P., and Kelly, B. (2010) Synthesis of *P*-stereogenic compounds via kinetic deprotonation and dynamic thermodynamic resolution of phosphine sulfides: opposite sense of induction using (-)-Sparteine. *J. Am. Chem. Soc.*, **132** (39), 13922–13927.
122 Crepy, K.V.L. and Imamoto, T.P. (2003) Chirogenic phosphine ligands, in *Topics in Current Chemistry*, vol. **229** (ed. J.-P. Majoral), Springer-Verlag, Berlin, Heidelberg, pp. 1–41.
123 Wu, X., O'Brien, P., Ellwood, S., Secci, F., and Kelly, B. (2013) Synthesis of *P*-stereogenic phospholene boranes via asymmetric deprotonation and ring-closing metathesis. *Org. Lett.*, **15** (1), 192–195.
124 Tang, W., Wang, W., and Zhang, X. (2003) Phospholane–oxazoline ligands for Ir-catalyzed asymmetric hydrogenation. *Angew. Chem. Int. Ed.*, **42** (8), 943–946.
125 Liu D., Zhang X.Practical *P*-chiral phosphane ligand for Rh-catalyzed asymmetric hydrogenation. *Eur. J. Org. Chem.* 2005 **2005** (4), 646–649.
126 Imamoto, T., Crepy, K.V.L., and Katagiri, K. (2004) Optically active 1,10-di-tert-butyl-2,20-dibenzophosphetenyl: a highly strained *P*-stereogenic diphosphine ligand. *Tetrahedron: Asymmetry*, **15** (14), 2213–2218.
127 Gammon, J.J., O'Brien, P., and Kelly, B. (2009) Regioselective lithiation of silyl phosphine sulfides: asymmetric synthesis of *P*-stereogenic compounds. *Org. Lett.*, **11** (21), 5022–5025.
128 Wolfe, B. and Livinghouse, T. (1998) A direct synthesis of *P*-chiral phosphine-boranes via dynamic resolution of lithiated racemic tert-butylphenylphosphine-borane with (-)-sparteine. *J. Am. Chem. Soc.*, **120** (20), 5116–5117.
129 Ona-Burgos, P., Fernandez, I., Roces, L., Torre-Fernandez, L., Garcıa-Granda, S., and Lopez-Ortiz, F. (2008) Asymmetric deprotonation–substitution of N-pop-benzylamines using [RLi/(−)-sparteine]. Enantioselective sequential reactions and synthesis of N-heterocycles. *Org. Lett.*, **10** (15), 3195–3198.
130 Nielsen, M., Jacobsen, C.B., and Jørgensen, K.A. (2011) Asymmetric organocatalytic electrophilic phosphination. *Angew. Chem. Int. Ed.*, **50** (14), 3211–3214.
131 Weise, C.F., Lauridsen, V.H., Rambo, R.S., Iversen, E.H., Olsen, M.-L., and Jørgensen, K.A. (2014) Organocatalytic access to enantioenriched dihydropyran phosphonates via an inverse-electron-demand hetero-Diels–Alder reaction. *J. Org. Chem.*, **79** (8), 3537–3546.
132 Gicquel, M., Zhang, Y., Aillard, P., Retailleau, P., Voituriez, A., and Marinetti, A. (2015) Phosphahelicenes in asymmetric organocatalysis: [3+2] cyclizations of γ–substituted allenes and electron-poor olefins. *Angew. Chem. Int. Ed.*, **54** (18), 5470–5473.
133 Hayashi, M., Shiomi, N., Funahashi, Y., and Nakamura, S. (2012) Cinchona alkaloid amides/dialkylzinc catalyzed enantioselective desymmetrization of aziridines with phosphites. *J. Am. Chem. Soc.*, **134** (47), 19366–19369.
134 Phillips, A.M.F. and Barros, M.T. (2014) Enantioselective organocatalytic synthesis of α-cyclopropylphosphonates through a domino Michael addition/intramolecular alkylation reaction. *Eur. J. Org. Chem.*, **2014** (1), 152–163.
135 Kolodiazhnyi, O.I. (2015) in *Topics in Current Chemistry*, vol. **361** (ed. J.-L. Montchamp), Springer International Publishing, Switzerland, pp. 161–236.

6
Asymmetric Biocatalysis

6.1 Introduction

Enzyme catalysis (biocatalysis) has found increasing number of applications in organic chemistry including the synthesis of optically active compounds [1–3]. For example, the synthesis in the presence of lipases results in chiral compounds of different structures not only on the laboratory but also on the industrial scale [2]. A great advantage of biocatalysts is that they are available for almost all types of organic reactions. While 20–30 years ago, enzymes were used only in the synthesis of C-, H-, N-, and O-containing compounds, now they are used to obtain organoelement compounds as well, including organophosphorus compounds [3–7]. Research in the field of biocatalysis began with the synthesis of optically active hydroxyphosphonates which was followed by the synthesis of α- and β-aminophosphonates, phosphinous acid derivatives, tertiary phosphines, and phosphine oxides. Chiral organophosphorus compounds containing the phosphorus atom as stereogenic center are used in various fields of science and industry including biology, pharmacology, and asymmetric catalysis.

Currently, biocatalysis is one of the most convenient methods for the synthesis of chiral organophosphorus compounds. Research in this field progresses intensively and a large array of experimental data has been accumulated. Earlier, a number of reviews generalizing the results obtained in original studies were published. Among them, mention may be made of the reviews by Mikolajczyk and Kielbasin'ski [6] as well as by Kafarski and co-workers [7]. In 1988, Hammerschmidt [8] pioneered the systematic research on the biosynthesis of compounds with C–P bonds. At the same time, Natchev [9] reported the use of biocatalysis in the synthesis of some phosphorus-containing analogs of L-amino acids and peptides (in particular, the enzymatic synthesis of D- and L-phosphinothricin (2-amino-4-hydroxymethylphosphinylbutanoic acid, Pht) and a natural tripeptide antibiotic bialaphos (L-Pht-Ala-Ala)). Information on enzymatic and microbiological resolution of racemic phosphonic acids can be found in the patents [10, 11] issued in the early 1980s.

6.2 Enzymatic Synthesis of Organophosphorus Compounds

A major problem in carrying out enzymatic reactions involving organophosphorus compounds is poor solubility in water of most of the starting compounds. In addition,

Asymmetric Synthesis in Organophosphorus Chemistry: Synthetic Methods, Catalysis and Applications, First Edition. Oleg I. Kolodiazhnyi.

many organophosphorus compounds are unstable in aqueous solutions. Therefore, chiral organophosphorus compounds are usually synthesized in organic solvents. Enzymes can be introduced into organic media in the freeze-dried form, in the immobilized form on solid supports (ceramics or organic polymers), or by dispersion (via the formation of covalent enzyme–polymer complexes or microemulsions with surfactants). Many lipases retain their activity in organic solvents, which allows their use in the synthesis of chiral organophosphorus compounds.

6.2.1 Kinetic Resolution of Hydroxyphosphonates

Usually, the resolution of functionalized racemic phosphonates into enantiomers using biocatalysts is conducted under kinetic control. The kinetic resolution is possible if reactions of the (*R*)- and (*S*)-enantiomers with chiral reagents proceed at different rates [1, 2], for example, [12] (Scheme 6.1). Ideally, if only the (*R*)-enantiomer enters into the reaction ($k_S = 0$), the resulting mixture contains the unreacted starting compound (50%) and the product obtained from the (*R*)-enantiomer (50%). This method allows both optical isomers to be resolved with ease using a single enzyme. For example, hydrolysis of acylated racemic hydroxyphosphonate **1** in the presence of Chirazyme® P-2 protease in phosphate buffer (pH 7) mainly leads to hydroxyphosphonate (*R*)-**2**, while acyloxyphosphonate (*S*)-**1** remains unreacted. These compounds are then chromatographically separated and subsequent hydrolysis of the acyl group results in both optical isomers of the hydroxyphosphonate, namely, (*R*)-**2** and (*S*)-**2**, in enantiopure form [12].

To find out which of the two enantiomers of a racemate undergoes transesterification faster and to predict the absolute configurations of the products, the Kazlauskas rule is used [13]. This simple empiric model is an extended Prelog rule based on the assumption that enantioselectivity is proportional to the size difference between the large (L) and medium-size (M) substituents in the substrate [13, 14]. In α-hydroxyphosphonates $(RO)_2P(O)CH(OH)R'$, the phosphonic group is usually taken as the larger substituent, and R′ is the medium-size substituent (Figure 6.1).

In this case, the (*S*)-hydroxyphosphonate undergoes transesterification, whereas (*R*)-hydroxyphosphonate does not enter the reaction. Hydrolysis of the racemic α-acyloxyphosphonate results in (*S*)-hydroxyphosphonate while (*R*)-acyloxophosphonate remains in the mixture. For β-hydroxyphosphonates $(RO)_2P(O)CH_2CH(OH)R'$, it is taken that $L = (RO)_2P(O)CH_2$ and $M = R'$. Since in this case the order of priority of substituents at the stereogenic center changes, the (*R*)-hydroxyphosphonate undergoes transesterification, while the (*S*)-hydroxyphosphonate does not react. Correspondingly, hydrolysis of β-acyloxyphosphonates involves the (*R*)-enantiomer and gives the (*R*)-hydroxyphosphonate, whereas the (*S*)-acylphosphonate remains unchanged. This stereospecifity of the reactions of α- and β-hydroxyphosphonates is observed in most cases, although biocatalytic reactions involving hydroxyphosphonates sometimes proceed contrary to the Kazlauskas rule [13]. Recently, this rule has been widely used as

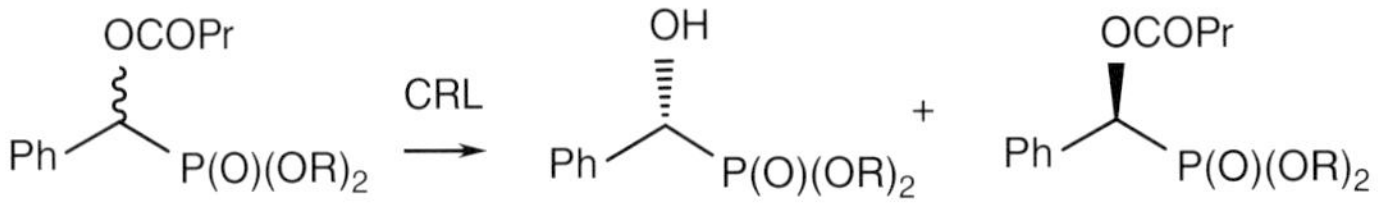

(*S*/*R*)-**1** (*R*)-**2** (*S*)-**1**

Scheme 6.1 Examples of kinetic resolution of hydroxyphosphonates.

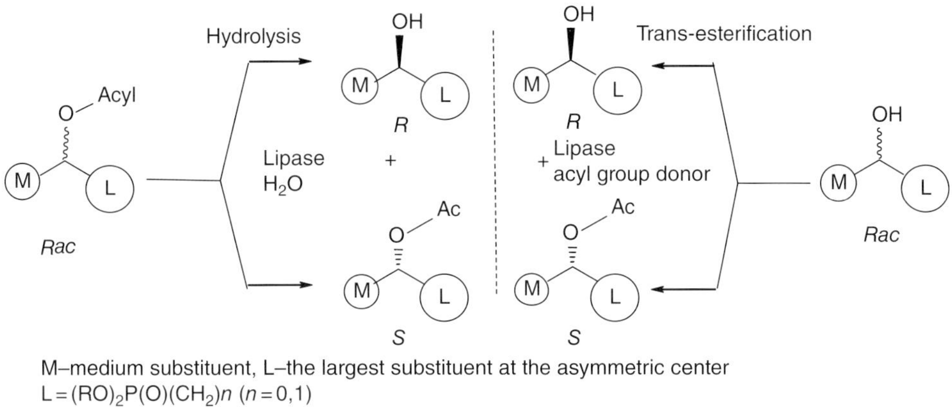

Figure 6.1 Symmetry and stereochemistry of enzymatic transesterification and hydrolysis processes.

a convenient empirical rule to predict the stereochemistry of the products of enzymatic resolution of secondary alcohols and amines.

6.2.2 Resolution of α-Hydroxyphosphonates by Biocatalytic Transesterification

Resolution of racemates using lipases is one of the most convenient methods for obtaining enantiopure compounds. In the case of α-hydroxyphosphonates, enzymatic *trans*-esterification in organic solvents is used as a rule [15–18]. To attain kinetic resolution, various acyl group donors are introduced into the reaction mixture. These are ethyl acetate, ethyl chloroacetate, ethyl benzoate, *p*-chlorophenyl acetate, vinyl acetate, or isopropenyl acetate [19]. Since enzymatic transesterification/hydrolysis reactions are reversible, and the same enzyme can catalyze both the forward and reverse reactions, vinyl acetate and isopropenyl acetate are considered as the best acylating agents. Vinyl acetate is transformed to vinyl alcohol and then to acetaldehyde due to keto–enol tautomerism. Isopropenyl acetate is transformed to acetone and the reaction becomes irreversible (Scheme 6.2).

The *Burkholderia cepacia* lipase (BCL) immobilized on diatomite is an effective biocatalyst for enantiomeric resolution of racemic α-hydroxyphosphonates [18, 19] (Scheme 6.3). In the presence of this lipase, vinyl acetate esterifies only the (*S*)-enantiomer of racemic α-hydroxyphosphonates **3a–c** to give a mixture containing 50% of α-acylphosphonate (*S*)-**4** and 50% of hydroxyphosphonate (*R*)-**3**. The latter can be separated by column chromatography. It was established that the esterification rate and optical purity of products depend little on the solvent (tetrahydrofuran (THF), toluene, and vinyl acetate), but they strongly depend on temperature and

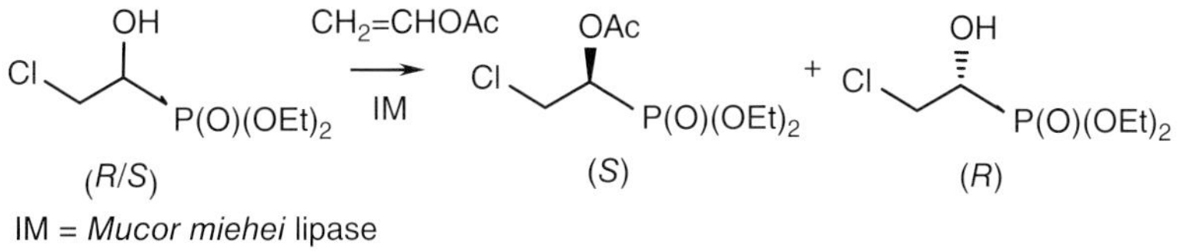

Scheme 6.2 Resolution of diethyl-2-chloro-1-hydroxymethylphosphonate by *Mucor miehei* lipase.

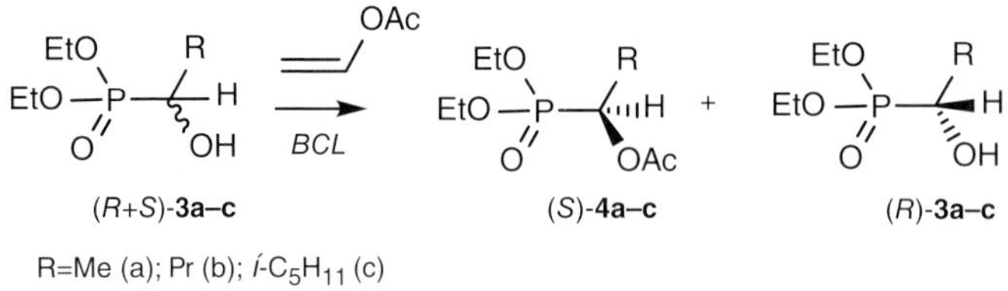

Scheme 6.3 Resolution of diethyl α-hydroxyphosphonate by *Burkholderia cepacia* lipase.

lipase excess. When temperature is raised to 40 or 60 °C, the reaction rate enhances approximately 1.5 and 2 times, respectively. The time it takes to reach 50% esterification of α-hydroxyphosphonates, that is, to the complete esterification of (*S*)-enantiomer, decreases as the biocatalyst amount increases.

Diethyl 1-hydroxyethylphosphonate (R)-**3**a (typical example). To a solution of 1.5 g (0.076 mol) of racemic (*S/R*)-hydroxyphosphonate in 3 ml of THF and 3 ml of vinyl acetate, 0.15 g of BCL was added. The mixture was stirred for 48 h at room temperature. The lipase was filtered off, the solution was concentrated, and the residue was subjected to chromatography on a silica gel to form two fractions, one of which (R_f 0.25, hexane–acetone 2 : 1) was optically pure alcohol (*R*)-**3a** (yield 48%, bp 85 °C (0.1 mm Hg), $[\alpha]_D^{20}$ −7.0 (c = 3, $CHCl_3$), ^{31}P NMR ($CDCl_3$), δ_P, ppm: 31.5) [2, 4]. The other fraction (R_f 0.55, hexane–acetone 2 : 1) was (*S*)-diethyl 1-acetyloxyethylphosphonate (*S*)-**4a** (yield 49%, $[\alpha]_D^{20}$ +25.0 (c = 2, $CHCl_3$), ^{31}P NMR ($CDCl_3$), δ_P, ppm: 25.8) [18].

The kinetic resolution of diastereomers in the presence of the lipases – CAL (immobilized lipase from *Candida antarctica*, Chirazyme L-2), and Amano AK (immobilized lipase from *Pseudomonas fluorescens*) was reported [20]. The racemic phosphinates **5a–c** containing two stereogenic centers were obtained by the reactions of ethylphenylphosphinate with aldehydes and separated by crystallization into the major (diastereomers (S_P,*S*)-**5** and (R_P,*R*)-**5**) and minor products (diastereomers (R_P,*S*)-**5** and (S_P,*R*)-**5**). Subsequent transesterification of the major product catalyzed by CAL and AK using vinyl acetate as the acyl group donor (Scheme 6.4) gave enantiopure acyl phosphinates (S_P,*S*)-**6a–c** (ee >98%) and hydroxyphosphinates (R_P,*R*)-**5a–c** at 50% conversion. The influence of the alkyl substituent Rin the biocatalyst on the acylation process was more pronounced for *C. antarctica* lipase B (CALB) compared to AK (Scheme 6.4). After protection of the OH group, ethyl (1-hydroxyethyl)phenylphophinate (R_P,*R*)-**5a** was transformed to P-chiral vinylphosphine oxide (R_P,*R*)-**7**, a convenient agent for the synthesis of other optically active phosphine oxides, by the reaction with vinylmagnesium bromide [20] (Scheme 6.5). In a similar way, Patel *et al.* [21] reported enantioselective acetylation of racemic (1-hydroxy-4-(3-phenoxyphenyl)butyl) phosphonate **8** by isopropenyl acetate in toluene in the presence of lipase from *Geotrichum candidum*. The resulting phosphonate (*S*)-**8** is an intermediate in the total synthesis of the squalene synthase inhibitor BMS-188494. The yield of the major product was 38% and its optical purity was 95% (Scheme 6.6). Transesterification of *cis*-1-diethylphosphonomethyl-2-hydroxymethylcyclohexane **10** with vinyl acetate catalyzed by the AK lipase afforded alcohol (+)-**15** and the corresponding acetate (+)-**11** in good yields and high enantiomeric excess. The reaction was solvent-free and

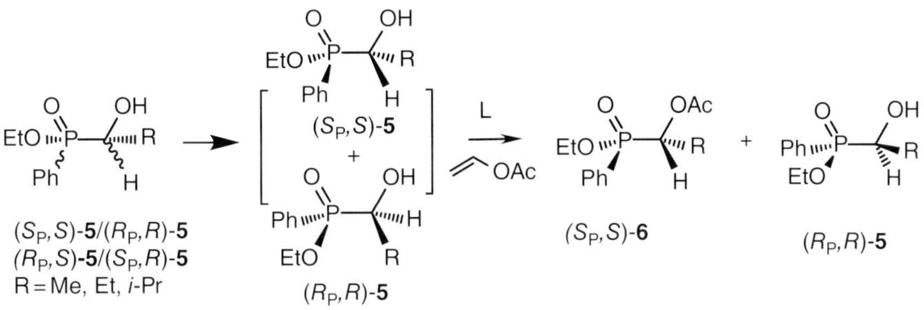

R	L = Lipase	Conversion (%)	ee,% acetate	ee,% alcohol
Me	CAL	50	98	98
Me	AK	50	98	98
Et	AK	50	98	98

Scheme 6.4 Kinetic resolution of ethyl (1-hydroxyalkyl) phenylphosphinate diastereomers.

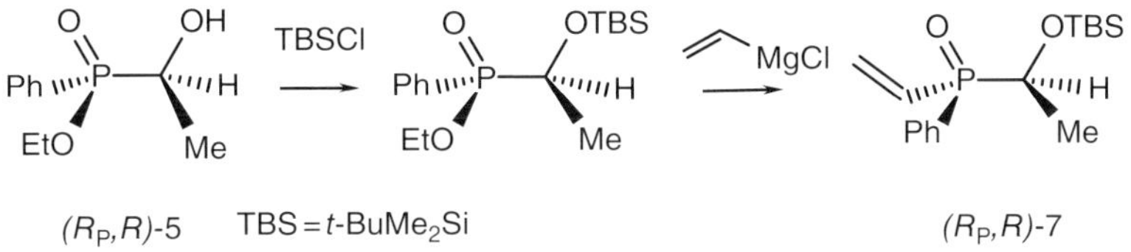

Scheme 6.5 Synthesis of P-chiral vinylphosphine oxide (R_p,R)-**7**.

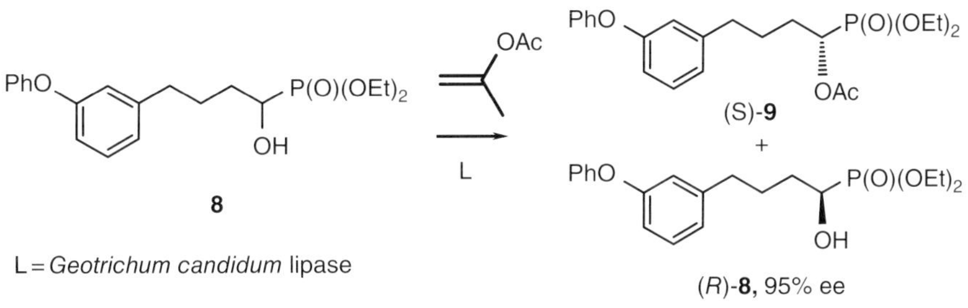

Scheme 6.6 Enantioselective acetylation of racemic hydroxyphosphonates **8**.

the enantioselectivity factor was 152. The alcohol (+)-**10** was then transformed to an optically active hydantoin, a starting compound for the synthesis of conformationally hindered (2*R*)-amino-5-phosphonopentanoate (AP 5) analogs. (Scheme 6.7) [22]. Racemic hydroxymethylphosphonates were resolved into enantiomers using lipase isolated from *P. fluorescens*. The products were used to obtain chiral phosphosulfonates possessing high herbicidal activity. Biological tests showed that the activity of (+)-phosphosulfonates is higher than that of the (−)-isomers and the racemate [23].

6.2.3 Resolution of α-Hydroxyphosphonates by Biocatalytic Hydrolysis

Enzymatic hydrolysis is a convenient method for resolution of chiral hydroxyphosphonates [24, 25]. Such reactions are usually carried out in two-phase systems in the

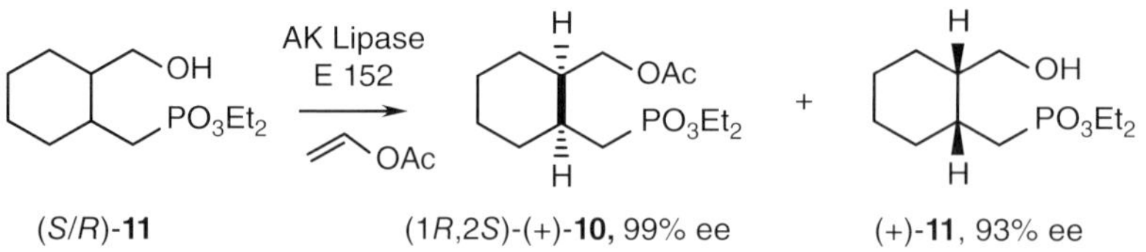

Scheme 6.7 Transesterification of (*S*/*R*)-**11** with vinyl acetate catalyzed by the AK lipase.

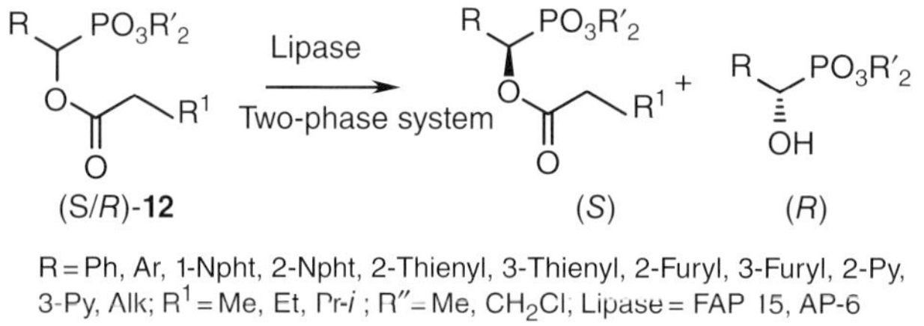

Scheme 6.8 Resolution of α-hydroxyphosphonates by enzymatic hydrolysis.

presence of a buffer maintaining pH 7 using a hexane-BuOMe mixture as the best solvent. Hydrophobic solvents decrease the activity of enzymes, but the addition of saturated aqueous solutions of salts ($MgCl_2$ or LiCl) considerably increases the enantioselectivity of the processes involving lipases (Scheme 6.8). For example, it was found that the hydrolysis of β-acyloxyphosphonates by lipases or pig liver esterase in the system *tert*-butyl methyl ether-hexane/enzyme/phosphate buffer (pH 7) at room temperature allowed to obtain enantiopure (*S*)- or (*R*)-hydroxyphosphonate [25, 26]. The pH value, which changed in the course of the reaction, was controlled by autotitrator. Lipases preferably hydrolyzed the (*S*)-enantiomers of the esters, while Chirazyme P-2 protease hydrolyzed the (*R*)-enantiomers [26]. The highest enantioselectivity was found for hydrolysis of (acetoxy)phenyl-methylphosphonates **12** by the lipase from *Rhizopus oryzae* (FAP 15) and *Aspergillus niger* (AP 6).

To increase the enantioselectivity of the reaction, and to decrease the time of reaction, enantioenriched mixtures can be used instead of racemates. Rowe and Spilling [27] employed this method to increase the optical purity of 1-hydroxyphosphonates containing unsaturated groups. The reactions of aldehydes with dialkyl phosphites catalyzed by chiral titanium alcoholates (Ti(OPr-*i*)$_4$ with a chiral diol as a ligand) resulted in enantioenriched α-hydroxyphosphonates (*R*)-**14** with 42–77% ee. The enantioenriched products **14** were then acetylated and the resulting acetates were subjected to lipase-catalyzed hydrolysis to give the alcohols (*R*)-**14** in good yields and with optical purity up to 99% ee. The hydrolysis of unreacted acetate was performed in *t*-BuOMe at room temperature and pH 7 to afford the hydroxyphosphonate (*S*)-**14** (Scheme 6.9 and Table 6.1).

Kafarski and co-workers [28, 29] resolved α-hydroxyphosphinates containing two asymmetric centers in the presence of lipases from various microorganisms including *Candida cylindracea, A. niger, Rhizopus niveus, Mucor javaniceus, Porcine pancreas, Pseudomonas cepacia,* and *Mucor circinelloides.* The kinetic resolution of ethyl (1-hydroxyethyl)phenylphosphinate **17** was achieved by enzymatic transesterification with vinyl butyrate or by hydrolysis of ethyl (1-terbutoxyphenyl) phenylphosphinate **16**

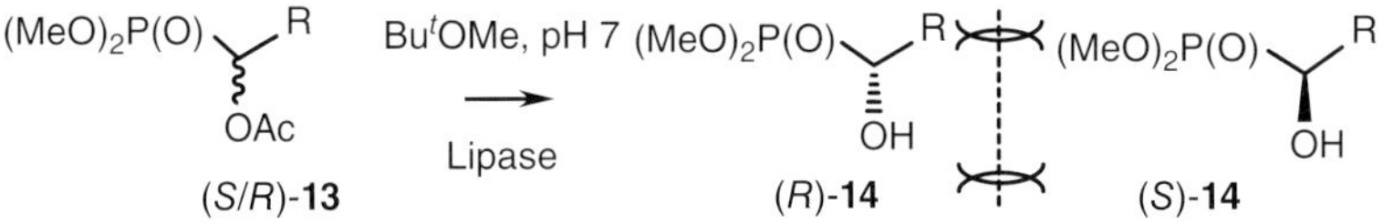

Scheme 6.9 Enzymatic hydrolysis of enantioenriched 1-acyloxyphosphonates.

Table 6.1 Enzymatic hydrolysis of enantioenriched 1-acyloxyphosphonates **13** (Scheme 6.9).

R	R′	Enzyme	Yield (%)	(*S*/*R*)-13		(*S*)-14*	
				ee (%)	Configuration	ee (%)	Configuration
Me	PhCH=CH	*Pseudomonas* sp.	68	73	*R*	99	*R*
Me	PhCH=CMe–	*Lipase* PSCII	75	77	*R*	99	*R*
Me	$C_5H_{11}C{\equiv}C$–	*Lipase* AY	59	49	*R*	92	*R*
Me	Ph	*Pseudomonas sp.*	72	70	*R*	99	*R*
Me	PhCH=CH	*Rhizopus arrizhus*	70	73	*S*	91	*R*
Me	MeCH=CH	F-API5	79	64	*S*	95	*R*
Me	$C_5H_{11}C{=}C$-	F-API5	74	69	*S*	95	*R*
Me	$C_5H_{11}C{\equiv}C$–	*Rhizopus arrizhus*	74	49	*S*	90	*R*
Me	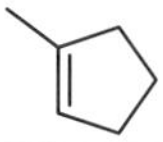	F-API5	72	42	*S*	79	*R*

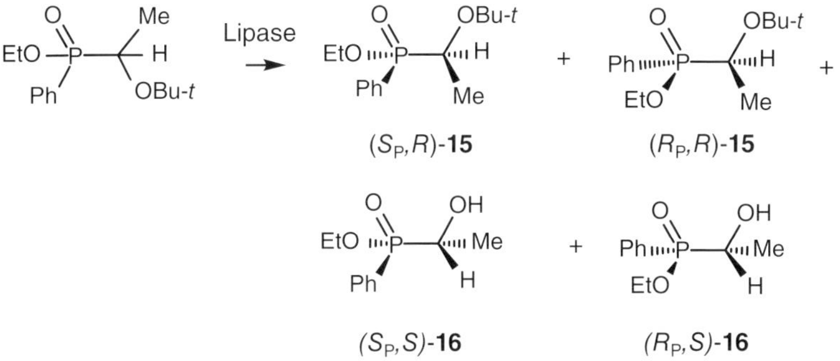

Scheme 6.10 Kinetic resolution of ethyl (1-hydroxyethyl)phenylphosphinate **15a**.

using lipases. Hydrolysis of compound **16** in the presence of porcine pancreatic lipase (PPL) resulted in two diastereomers **15** with ee greater than 98% at conversions varying from 10% to 49% depending on the reaction conditions (Scheme 6.10).

Optically active α-sulfonyl phosphonates **17** and the corresponding methyl sulfides **18** were synthesized from chiral (93–97% ee) α-hydroxyphosphonates **18** obtained by enzymatic resolution of compounds **17** [25]. Attempts to obtain chiral phosphonates bearing divalent sulfur atom at the α-carbon atom by enantioselective hydrolysis of corresponding racemic α-acetylthiophosphonates did not succeed. This is because hydrolysis in the presence of the AP 6 and Prozyme 6 lipases and the Chirazyme P-2 protease

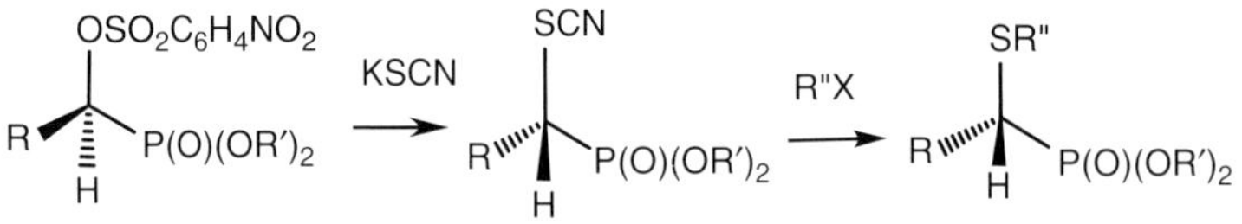

(*S*)-**17** (*S*)-**18**

R = Et, Ph; R′ = H, Me

Scheme 6.11 Synthesis of optically active α-sulfonyl phosphonates **17**.

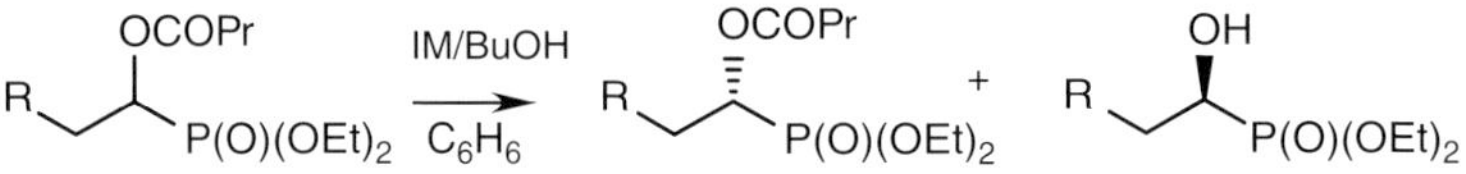

(*S/R*)-**19** (*R*)-**20** (*S*)-**19**

R = Me, Et, Pr-*i*, CH=CH$_2$

Scheme 6.12 Alcoholysis of butyroxy derivative **25** by *n*-butyl alcohol.

gave racemic products and the reactions in the presence of lipases isolated from *Candida rugosa, C. cylindracea,* and FAP 15 did not occur (Scheme 6.11).

Alcoholysis of butyroxy derivatives **19** by *n*-butyl alcohol in anhydrous benzene led to enantiopure hydroxyphosphonates (*S*)-**20** in 40% yield [19] (Scheme 6.12). The successful resolution of diisopropyl (2-azido-1-hydroxyethyl)phosphonates to enantiomers was attained using the SP 524 lipase (a genetically engineered hybrid of *Mucor* and *Aspergillus oryzae*) [23]. The racemic diisopropyl 2-azido-1-acetoxyethylphosphonate undergoes hydrolysis in the presence of SP 524 with the formation of α-hydroxyphosphonate (*S*)-**22** and ester (*R*)-**21**, which was then hydrolyzed to give stereoisomer (*R*)-**23** [23]. The phosphonates (*R*)-**21** and (*S*)-**22** were then transformed to α-phosphaisoserine (*R*)-**24** and α-phosphaserine (*S*)-**25** [30]. Yuan *et al.* [19] reported CALB to be efficient in resolving azide-containing racemic phosphonates into enantiomers [16] (Scheme 6.13).

Chiral α-hydroxy-*H*-phosphinates **26a–e** were obtained by enzymatic hydrolysis of the corresponding racemic acetates **25a–e** in the presence of lipase from *P. cepacia*. Enantiopure α-hydroxy-*H*-phosphinates (*R*,S_P)-**26a–e** were isolated from a mixture of four stereoisomeric α-acetoxy-*H*-phosphinates **25** [31]. It should be noted that hydroxyphosphonates with aromatic substituents are usually stable against lipase-catalyzed transesterification and hydrolysis, which is explained by both electronic and steric

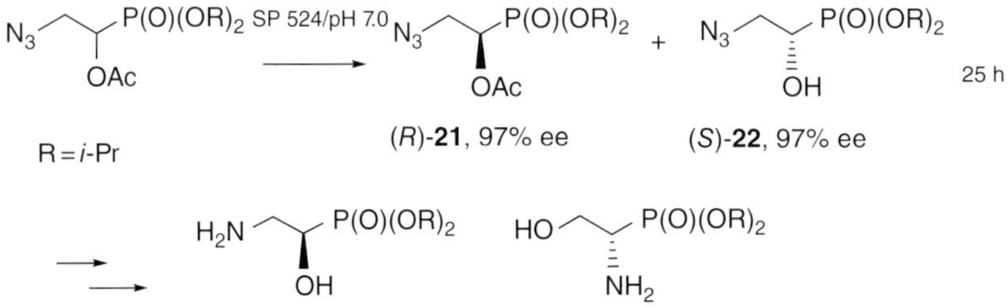

(*R*)-**23**, 91% ee (*S*)-**24**, 59% ee

Scheme 6.13 Resolution of diisopropyl (2-azido-1-hydroxyethyl)phosphonates to enantiomers.

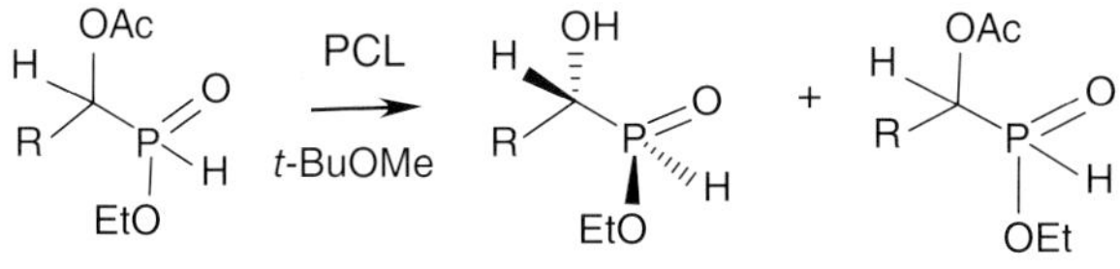

(S/R)-**25a–e** (R,S_P)-**26a–e** **27a–e**

R = n-C_5H_{11} (a), Bn (b), 4-XC_6H_4, where X = Me (c), H (d), Cl (e)
PCL = *Pseudomonas cepacia* lipase

Scheme 6.14 Resolution of diastereomeric hydroxyphosphonates by enzymatic hydrolysis.

factors. In this case, a possible reason is the substitution of the ethoxy group at the phosphorus atom for a smaller hydrogen atom. Only one of the four diastereomers **25c** present in the mixture, namely, the acetate (R,S_P)-**25c**, underwent hydrolysis under the action of PSC lipase in a hexane/*tert*-butyl methyl ether mixture in phosphate buffer (pH 7). The reaction led to the corresponding hydroxyphosphonate (R,S_P)-**27c** obtained in 18% yield and 99% ee. At the same time, the acetate (S,R_P)-**26c** was isolated in 33% yield (Scheme 6.14) [24].

6.2.4 Dynamic Kinetic Resolution of α-Hydroxyphosphonates

The ordinaire kinetic resolution of racemates allows to obtain each enantiomer in 50% yield. However, this limitation can be overcome using the dynamic kinetic resolution (DKR) technique [32–34]. In DKR, one deals with continuous isomerization of the substrate in the course of resolution, that is, the (R)- and (S)-enantiomers are in equilibrium. As a result, at $k_S = 0$, the starting (R)-enantiomer can be completely transformed into the (R)-product.

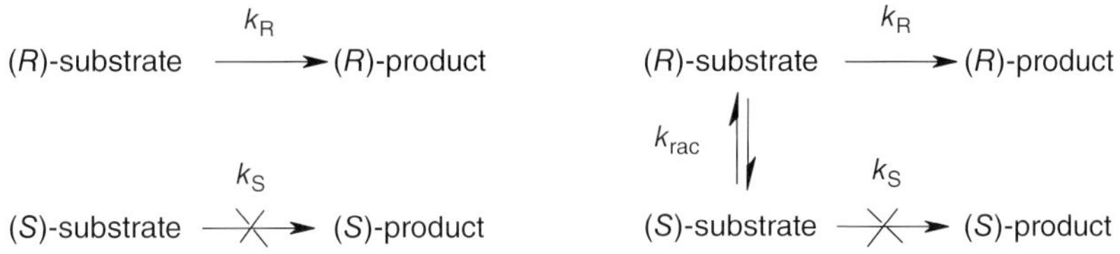

A combination of metal complex catalysis and biocatalysis was successfully used in the DKR of some hydroxyphosphonates. For instance, simultaneous action of an enzyme (lipase from *P. cepacia* or CALB (Novozym-435)) and a ruthenium complex **30** that catalyzes the isomerization of alcohols resulted in the transformation of the racemic hydroxyphosphonates **28** into enantiopure acetates **29** in high yields [34]. A series of racemic hydroxyphosphonates has been successfully transformed into enantiomerically pure acetates with 99% ee and 87% yields (Table 6.2).

6.2.5 Resolution of β- and ω-Hydroxyphosphonates

Similarly to α-hydroxyphosphonates, biocatalytic resolution of racemic phosphonates containing a hydroxyl group in β-, γ-, and δ-positions relative to the phosphorus atom (so-called ω-hydroxyphosphonates) is carried out in two ways, by transesterification of

Table 6.2 Dynamic kinetic resolution of α-hydroxyphosphonates.

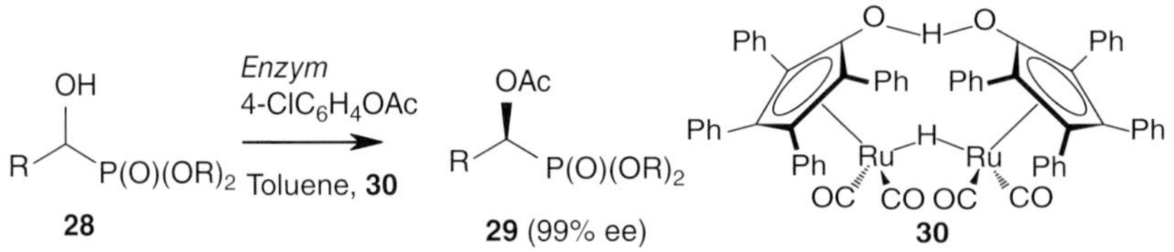

R	R′	Enzyme	*T* (°C)	Yield (%)	ee (%)
Me	Et	CALB	60	70	99
Me	Et	CALB	70	76	99
Me	Et	CALB	80	86	99
Me	Et	PS-C	60	69	99
Me	Et	PS-C	80	87	99
Me	Et	CALB	80	83	99
Et	Et	PS-C	80	85	99

Enzyme = PS-C (lipase of *Pseudomonas cepacia*) and CALB (Novozym-435).

carboxylic acid esters or by hydrolysis of the corresponding acetylated hhydroxyphosphonates [35–41]. Racemic β-hydroxyalkylphosphonates **31a–e** were acylated with vinyl acetate in the presence of lipases from *P. cepacia* and *P. fluorescens*, as well as the lipase Amano AH-S (LAH-S). The reactions were carried out under kinetic control. The unreacted substrates **31a–d** and the acetylated products **32a–e** were separated by column chromatography [35]. Successful resolution of β-hydroxyalkylphosphonates **31** (R^1 = Me, Et; R^2 = Me, Et, CH=CH_2, CH_2Cl) to (*R*)- and (*S*)-enantiomers by CALB-catalyzed transesterification with vinyl acetate was reported [17]. Compounds **32** with Me, Et, or vinyl group at the C_β-atom were acetylated quite smoothly with good enantioselectivity ($E > 100$). However, the compound **31e** (R^1 = CH_2Cl, R_2 = Et) was resolved with low enantioselectivity (50% ee, $E < 5$) (Scheme 6.15).

Attolini tested a number of enzymes (Liposyme, Amano AK, Amano PS, acylase from *Aspergillus melleus*, Amano AP 6, Amano AY, CALB, PPL, CRL (lipase

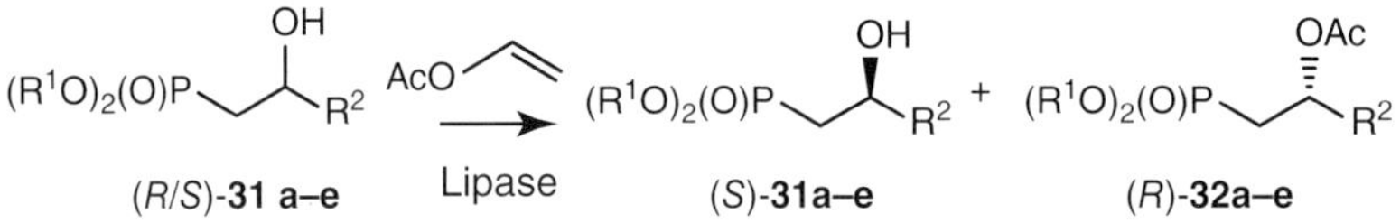

	R^1	R^2	Lipase	Alcohol ee %	Acetate, ee %
a	Et	Me	AK	93	90
s	Me	Bu	AK	61	92
c	Me	$(CH_2)_3Ph$	LAH-S	52	43
d	Et	1-Py	LPL	45	62
e	Et	CH_2Cl	CALB	50	

PCL = lipase *Pseudomonas cepacia*, AK = lipase *Pseudomonas fluorescens*, LAH-S = lipase *AH-S*

Scheme 6.15

from *C. rugosa*), Amano R10) for the ability to kinetically resolve racemic diethyl 3-hydroxybuten-1-ylphosphonate **33** [36]. The Liposyme lipase (Fluka) appeared to be the most efficient biocatalyst. In the presence of Liposyme, diethyl 3-acetoxybut-1-enylphophonate ((*R*)-**34**) was formed with 96% ee and the unreacted (*S*)-hydroxyphosphonate **33** was isolated with 99% ee. Similar results were obtained in the preparative-scale experiments with a few grams of the starting compounds. The yields were 46-48%. The Amano AK and Amano PS lipases also showed high efficiency in the kinetic resolution of racemic β-hydroxyphosphonate (in particular, an ee of 98% was achieved using the latter enzyme). The reaction was relatively fast and the enantioselectivity factor exceeded a value of 200. Good enantiopurity of the acylated product (98% ee) was also achieved in the presence of acylase from *A. melleus*, but the unreacted starting β-hydroxyphosphonate **33** was isolated only with 49% ee (Scheme 6.16).

Taking into account the interest in P-containing analogs of natural amino acids, such as carnitine and γ-amino-β-hydroxybutyric acid (GABOB), a number of methods for synthesis including enzymatic resolution procedures were reported [19, 37]. For instance, an efficient method was proposed for synthesis of enantiomerically pure phosphocarnitine starting from readily available 2-hydroxy-3-chloropropyl phosphonate **31e** [37]. Enantioenriched acetate (*S*)-(−)-**32e** with 88% ee obtained by enzymatic interesterification with the Amano AH-S lipase or the lipase from *P. fluorescens* was then subjected to additional enzymatic hydrolysis in the presence of the same lipase. This allowed one to substantially increase the purity of the hydroxyphosphonate. Hydrolysis of the acetate (*S*)-**32e** was carried out in diisopropyl ether saturated with a buffer (pH 7.2). The solution was stirred over a period of 25 days at 30 °C and the reaction was monitored by ^{31}P NMR spectroscopy. When the alcohol (*R*)-**31e** to acetate (*S*)-**32e** ratio reached 2.5 : 1, the reaction was stopped and the products were separated by column chromatography. The resulting β-hydroxy-phosphonate (S)-(−)-**31e** was obtained in 25% yield with 100% ee (Scheme 6.17).

Racemic 4-hydroxy-2-oxophosphonates **35** were resolved into enantiomers by CALB-catalyzed transesterification with vinyl acetate in benzene or by hydrolysis of the corresponding butyrates catalyzed by lipase from *C. rugosa* in diisopropyl ether containing a saturated magnesium chloride solution [42]. The reaction led to (*S*) (or (*R*))-*o*-hydroxyphosphonates **35** in 35–42% yields and acetates **36** in 48–51% yields. Reactions of compounds **36** with benzaldehyde in the presence of aqueous potassium carbonate gave chiral enones **37** (95–99% ee) (Scheme 6.18).

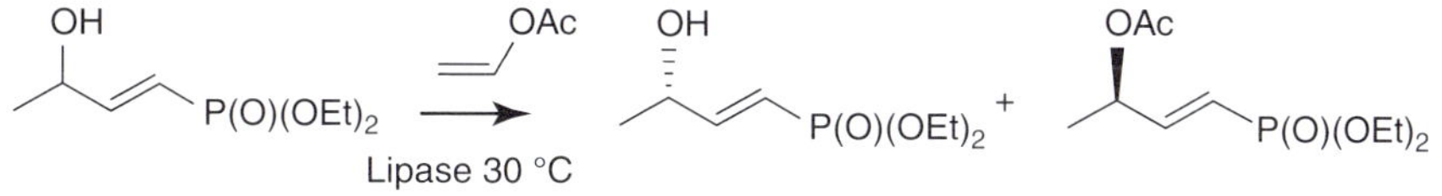

(*R*/*S*)-**33** (*S*)-**33** (*R*)-**34**

Lipase	Time (h)	Conversion (%)	ee %, (*R*)-**34**	ee %, (*S*)-**33**
Liposyme	21	51	99	96
Amano PS	17	45	81	98
PPL	49	19	22	99
CRL	10	13	8	54
Amano (*R*)-10	10	8	6	69

Scheme 6.16 Resolution of 3-hydroxybuten-1-ylphosphonate **33** by various lipases.

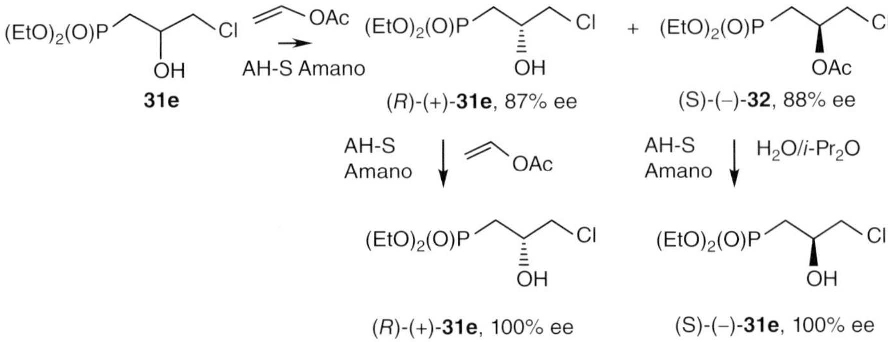

Scheme 6.17 Enzymatic resolution of 2-hydroxy-3-chloropropyl phosphonate.

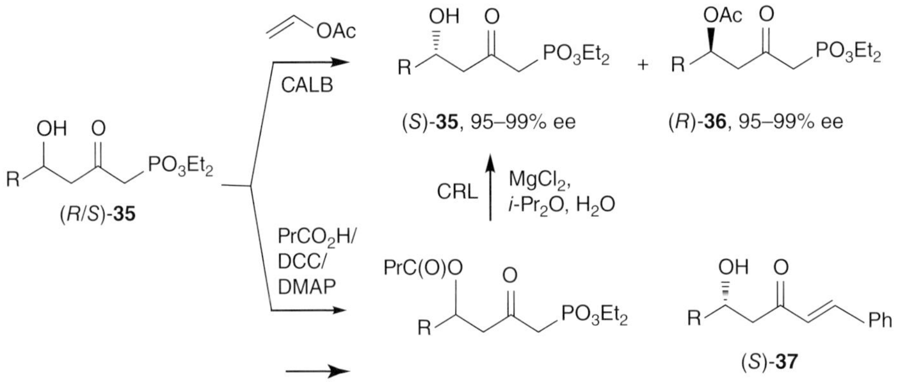

R	Lipase	Yield (%)		ee (%)	Configuration
		35	**36**	**37**	**37**
Me	CALB	40	48	99.1	*S*
Et	CALB	42	45	95	*S*
$CH{=}CH_2$	CALB	35	51	95	*R*
Ph	CRL	35	69	98.7	*R*
4-FC_6H_4	CRL	38	84	95.9	*R*
4-MeC_6H_4	CRL	37	81	100	*R*
4-$MeOC_6H_4$	CRL	32	92	96.8	*R*
4-ClC_6H_4	CRL	34	78	99.4	*R*
2-BrC_6H_4	CRL	36	86	98.0	*R*
2-ClC_6H_4	CRL	38	90	97.0	*R*
2-Furyl	CRL	31	95	85.9	*R*

Scheme 6.18 Resolution of racemic hydroxyphosphonates by CALB-catalyzed transesterification.

Racemic α-chloro-β-oxo-γ-(hydroxyalkyl)phosphonates **38** (R = Me, Et, CH_2=CH) were first resolved by CALB-catalyzed transesterification [42]. This resulted in optically active hydroxyketophosphonates (*S*)-**38** in 42–45% yields and acetylated products (*R*)-**39** in 45–50% yields. The acetates **39** were then additionally hydrolyzed to alcohols (*R*)-**38** in the presence of CALB (or CRL). Subsequent treatment of compounds **38** with benzaldehyde and potassium carbonate gave mixtures of geometric isomers of chiral α,β-unsaturated ketones, from which compounds **40** were isolated with enantiopurity >98% (Scheme 6.19).

Enantiopure cyclic diethyl (*S*)-3-hydroxycycloalk-1-enylphosphonates **41** were obtained by enzymatic resolution of the corresponding racemates [43]. The reaction also afforded the (*R*)-acetates **42a–c**. Among the lipases used (Amano AY, Amano PS, Amano AK, PPL), Amano AK and Amano PS showed the best results, that is, they allowed the enantiopurity of compounds (*S*)-**41a–c** to be as high as up to 99%. A somewhat different method was used to obtain a chiral six-membered hydroxyphosphonate (*S*)-**44**. Enzymatic resolution in the presence of the Amano AK lipase was carried out during the formation of bromocyclohexenol and resulted in the (*S*)-enantiomer **43** with 80% ee [43]. Subsequent palladium-catalyzed phosphonylation of **43** led to hydroxyphosphonate (*S*)-**44** with the same enantiopurity (Scheme 6.20).

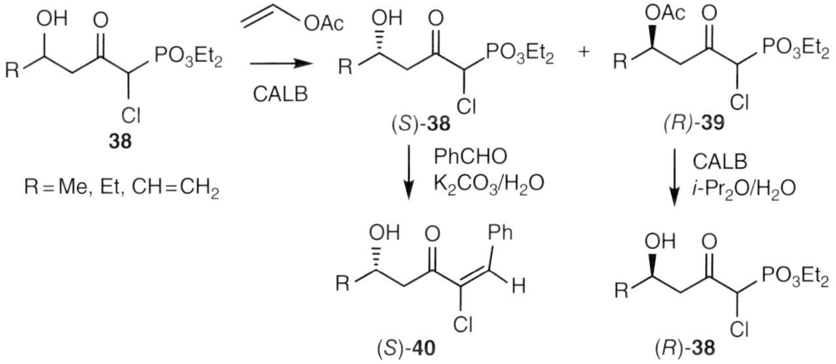

Scheme 6.19 Resolution of hydroxyalkyl)phosphonates **38** by transesterification.

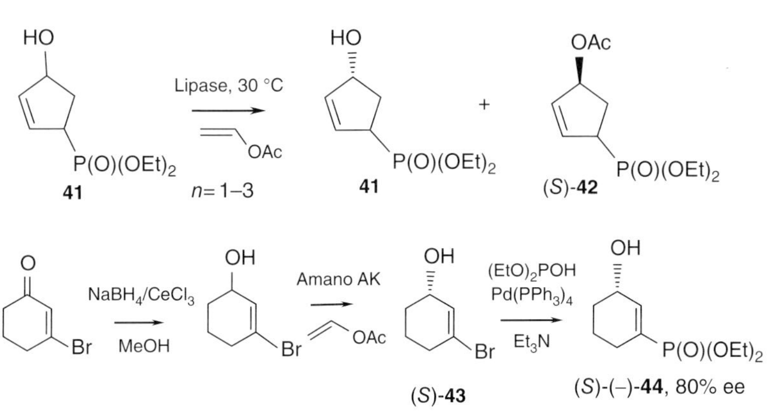

Scheme 6.20 Resolution of cyclic diethyl (*S*)-3-hydroxycycloalk-1-enylphosphonates **44** by biocatalytic trans esterification.

Enzymatic hydrolysis of β- and ω-hydroxyphosphonate esters is a convenient method for resolution of racemic mixtures of β- and ω-hydroxyphosphonates. In some cases, this method was used to obtain hydroxyphosphonate enantiomers that could not be resolved by enzymatic *trans*-esterification [44–47]. For instance, Hammerschmidt and co-workers [44] reported hydrolysis of some chloroacetoxyphosphonates by the lipase from *C. cylindracea* and by the protease subtilisin from *Bacillus subtilis* in a two-phase system and obtained (*S*)-hydroxyphosphonates with 51–92% ee. Hydrolysis of racemic chloroacetoxyphosphonates containing cyclic alkyl groups catalyzed by the AP 6 lipase from *A. niger* resulted in corresponding alcohols and (*S*)-enantiomers of the esters, obtained in high yields with good ee values [45–48]. Attolini *et al.* [36] reported enzymatic solvolysis of diethyl 3-hydroxybut-1-enylphosphonate **45** with isopropyl alcohol in diisopropyl ether, which acted as the acceptor of the acyl group. The acylated product **46** had an (*R*)-absolute configuration, while the unreacted hydroxyphosphonate had an (*S*)-absolute configuration. Yuan and co-workers proposed a convenient procedure for the synthesis of optically pure 2-hydroxy-2-aryl-ethylphosphonates. Enantioselective hydrolysis of the butyrates **45** was conducted in the presence of lipases in organic solvents. For instance, both compounds, (*S*)-**46** and (*R*)-**45**, were obtained in diisopropyl ether saturated with water in high yields with 95% ee. The CRL lipase showed the highest efficiency among the lipases studied. An alternative procedure involved CRL-catalyzed hydrolysis in diisopropyl ether saturated with 0.5% aqueous magnesium chloride (Scheme 6.21) [39].

A similar reaction of butyrates containing both aliphatic and aromatic substituents was studied [42]. The absolute configurations of the resulting hydroxyphosphonates were determined by X-ray diffraction analysis. The yields of products depended on substituents in the starting compounds and on the organic solvent used. Racemic α-chloro-β-oxo-δ-hydroxyalkylphosphonates **47** were resolved into enantiomers by CRL-catalyzed hydrolysis in diisopropyl ether saturated with aqueous $MgCl_2$ solution. The obtained alcohols **48** were used as chiral reagents in the Horner–Wadsworth–Emmons reaction which resulted in chiral α,β-unsaturated ketones [39] (Scheme 6.22). The hydroxyphosphonates bearing CF_3, CH_2Cl, CH_2N_3 groups resisted the enzymatic reaction owing to the strong electron-withdrawing effect of these groups [17]. However, Yuan and co-workers **67** succeeded in choosing suitable enzymes and developed methods for the resolution of racemic α- and β-hydroxyalkylphosphonates bearing the trifluoromethyl group into enantiomers [40].

The best results were obtained when alcoholysis catalyzed by *Mucor miehei* lipase or by CALB and hydrolysis catalyzed by the *C. rugosa* lipase was carried out in organic solvents. Esters of trifluoromethyl-containing β-hydroxy-phosphonates **49** were transformed to alcohols **50** under the action of *n*-butanol as a nucleophile in the presence of

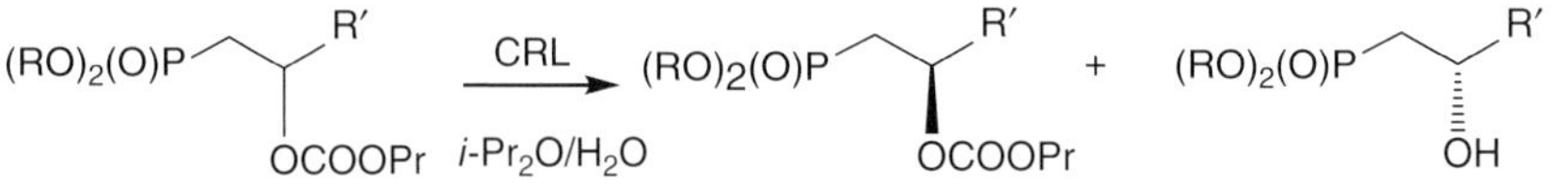

45 (*S*)-**46** (*R*)-**45**

R = Me, Et; Pr-*i*; R′ = XC_6H_4 (X = H, 2Cl, 4-F, 4-NO_2, 4-Et, 4-Br, 2-Br,2, 4-Cl_2, 3-Cl, 4-Me, 2CF_3, 2-Furyl, 2-Napht

Scheme 6.21 Enantioselective hydrolysis, catalyzed by CRL.

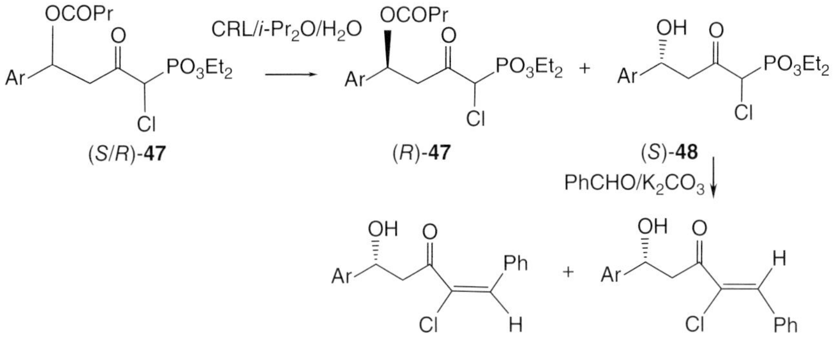

Scheme 6.22 Enantiomeric resolution of α-chloro-δ-hydroxy-δ-aryl-β-ketoalkylphosphonate.

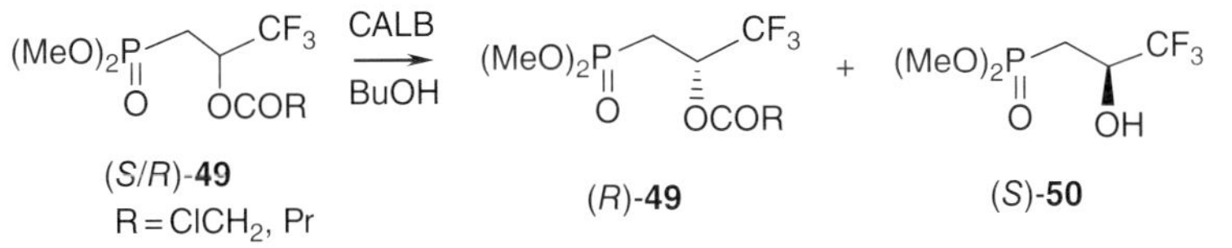

Scheme 6.23 Alcoholysis of trifluoromethyl-β-hydroxy-phosphonates catalyzed CALB.

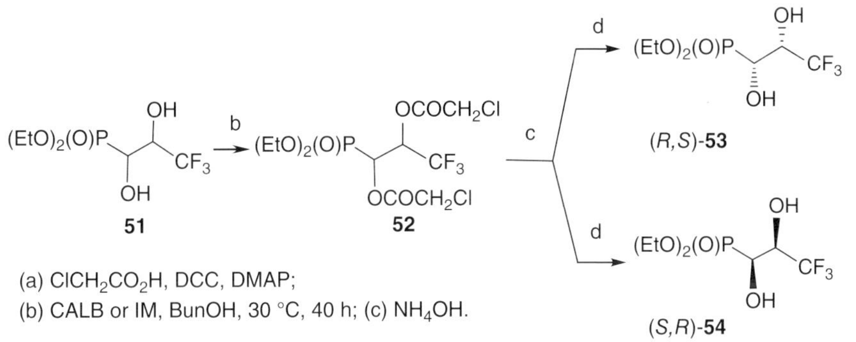

Scheme 6.24 Resolution of dihydroxypropylphosphonates **51**.

CALB in anhydrous benzene at 30 °C. As a 50% conversion was reached, the reaction was stopped and the mixture of chiral acyloxyalkylphosphonate **49** and alcohol **50** was separated chromatographically (Scheme 6.23).

A convenient and efficient method for obtaining chiral 2-trifluoromethyl-1,2-dihydroxypropylphosphonate **51** based on enzymatic catalysis was reported [41] (Scheme 6.24). The dihydroxylation of vinylphosphonate with potassium permanganate afforded the diol **51.** Enzymatic alcoholysis of racemic product **51** catalyzed by CALB or immobilized lipase IM led to optically active phosphonates (*R*,*S*)- and (*S*,*R*)-**52** in satisfactory yields with reasonable ee values (75–88% ee). These products were then hydrolyzed to optically active diols (*R*,*S*)-**53** and (*S*,*R*)-**54**. The absolute configurations of the obtained products follow the Kazlauskas rule if the diethylphosphonyl group is treated as the large substituent.

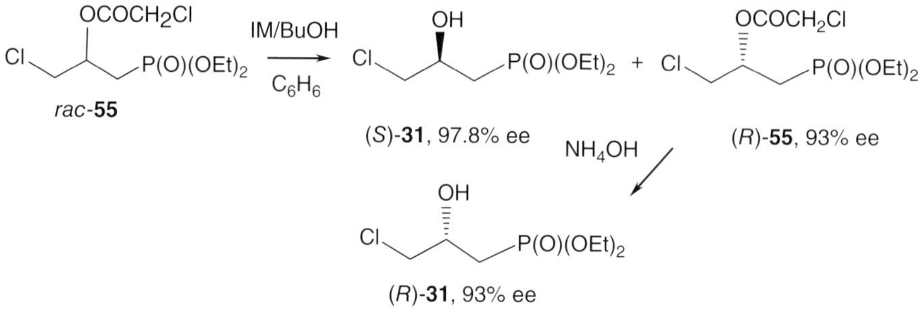

Scheme 6.25

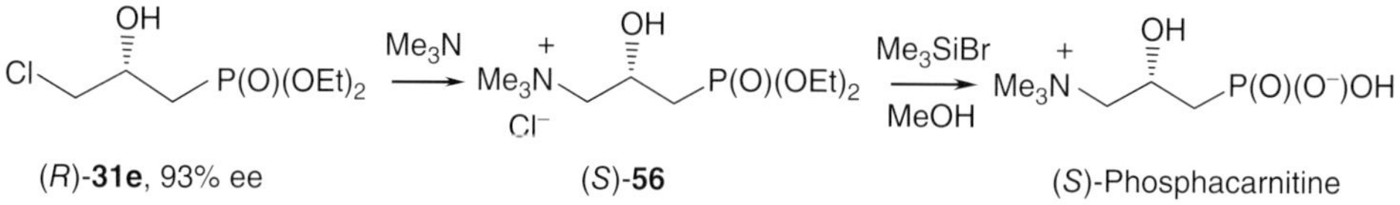

Scheme 6.26 Synthesis of (*S*)-P-Carnitine.

1-Chloro-2-hydroxypropanephosphonic acid was used as an important precursor of phosphocarnitine. CALB served as an effective biocatalyst in the resolution of 1- or 2-hydroxyalkanephosphonates [19]. Chromatographic purification afforded enantiopure hydroxyphosphonate (*S*)-**31e** and chloroacetate (*R*)-**55**, which was subsequently treated with ammonia (Scheme 6.25). Enantiomers (*S*)- and (*R*)-**31e** were used as the starting compounds in the synthesis of phosphocarnitine. The compound (*R*)-**31e** was converted to the trimethylammonium salt (*S*)-**58** which was treated with bromotrimethylsilane in methanol. Subsequent dehydrochlorination of the phosphocarnitine salt (*S*)-**56** resulted in the optically pure phosphorus analog of (*S*)-carnitine (Scheme 6.26). Enantiomers of ethyl ester of phospho-GABOB were prepared starting from diethyl 1-hydroxy-2-azidoethanephosphonate **57** [19] using biocatalytic hydrolysis by lipases [30]. The key step of the synthesis was enzymatic resolution of racemic acetate *rac*-**57** in the presence of lipases (CALB or IM) resulting in the (*R*)-enantiomer of ester **57** and the alcohol (*S*)-**58**, which were then separated by chromatography and converted into diethyl ester of (*S*)-phosphono-GABOB **59** (Scheme 6.27).

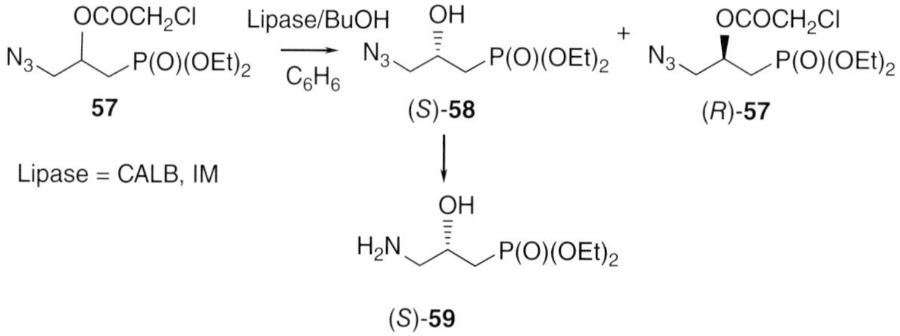

Scheme 6.27 Synthesis of enantiomers of phosphono-GABOB.

(S)-Diethyl 1-hydroxy-2-chloroethanephosphonate (*S*)-**31e**. Diethyl 1-butyryloxy-2-chloroethanephosphonate *rac*-**55** (5 mmol) was dissolved in absolute toluene (10 ml) and *n*-butanol (1.5 ml) and *M. miehei* lipase or CALB (400–500 mg) was added. The mixture was stirred at 30 °C with IM lipase or at 36 °C with CALB. The course of reaction was controlled by ^{31}P NMR (δ_P 21.0 (ester) and 26 (alcohol) ppm) or ^{1}H NMR (δ_H 5.1 (CHO(H) and 4.05 (CHO(Ac). The enzyme was filtered off when the ratio of these signals reached 1 : 1 and washed with acetone (3 ml). After evaporation of the solvent, the residue was separated by flash chromatography to furnish the title compound (S)-**31** (95 mg, 44%, ee > 95%) as a colorless oil $[\alpha]_D^{20}$ −15.5 (1,CH_3OH).

CALB-catalyzed alcoholysis of racemic chloroacetate **60** resulted in (*R,S*)-**61** and (*S,R*)-**60** with optical purity greater than 95% [19]. Satisfactory results were also obtained using the *M. miehei* lipase (IM). Hydrolysis of the optically pure chloroacetate (*R,S*)-**60** with ammonia in aqueous methanol was not accompanied by racemization and resulted in hydroxyphosphonate (*R,S*)-**60** in 85% yield. The latter was hydrolyzed to (1*R*,2*S*)-**59**, which undergoes cyclization to a (1*R*,2*S*)-fosfomycin sodium salt in alkaline medium (Scheme 6.28)

6.2.6 Dynamic Kinetic Resolution of β-Hydroxyphosphonates

DKR of racemic β-hydroxyphosphonates (R = Et (a), Me (f)) under the action of lipases was combined with isomerization of the alcohol catalyzed by a ruthenium complex **30** (Table 6.3) [34, 49]. To this end, racemic dialkyl 2-hydroxypropylphosphonates **60** were treated with 3 equivivalents of 4-chlorophenyl-acetate in toluene at 60 °C in the presence of different lipases. In all cases, the enantioselectivity of acylation was high enough, although the yields and enantiopurity of the products varied depending on the type of the lipase used. The CALB showed the highest activity, whereas other lipases (CRL, *P. pancreas, C. cylindracea* and *Aspergillus* sp.) were of low activity. DKR of racemic β-hydroxyphosphonates significantly differs from that of α-hydroxyphosphonates by longer resolution times (48 h vs 24 h, respectively), low yields of the products **61**, and the formation of the side ketophosphonates **62**. Bubbling hydrogen through the reaction mixture or the addition of 2,4-dimethylpentan-3-ol inhibited the formation of

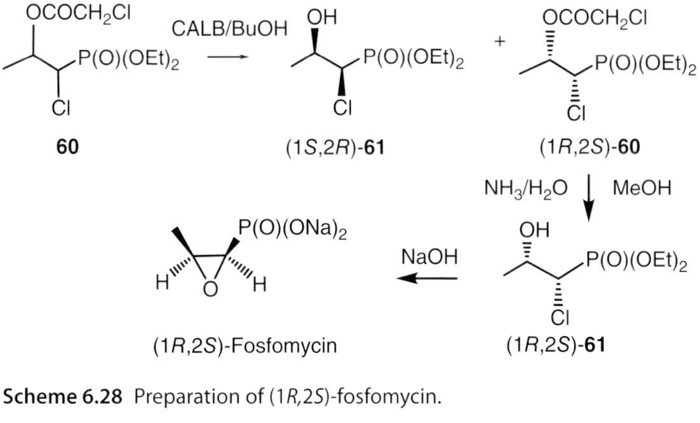

Scheme 6.28 Preparation of (1*R*,2*S*)-fosfomycin.

Table 6.3 Dynamic kinetic resolution of β-hydroxyphosphonates.

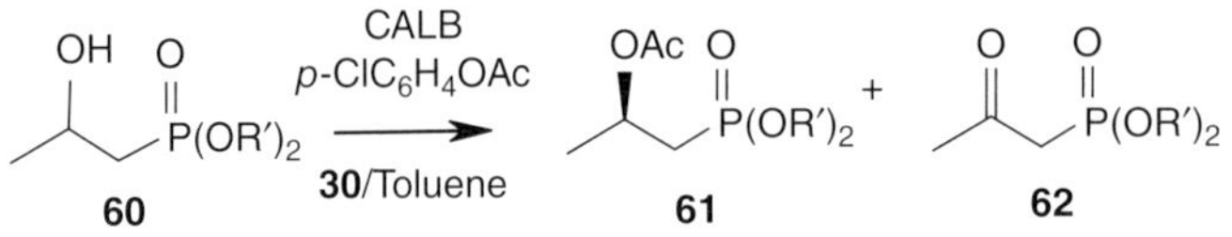

R	*T* (°C)	Yield (%) 61	ee (%)
Et	60	54	>99
Et	70	65	>99
Et	80	69	>99
Et	70	53	>99
Et	70	57	>99
Me	70	62	>99

ketophosphonates **62;** however, there was no decrease in the efficiency of DKR or the enantiopurity of the products **61** [34].

6.2.7 Resolution of Aminophosphonates

Aminophosphonic acids are of great interest as potential biologically active compounds. Therefore, attempts have been made to use biocatalysis for the synthesis of their enantiomers. Information on the use of chemoenzymatic methods for the synthesis of chiral aminophosphonic acids was first disclosed in patents issued to Hoechst Aktiengesellschaft in the 1980s [10]. Hammerschmidt and co-workers [8, 45, 47, 48] reported the chemoenzymatic synthesis of phosphonous analogs of phenylalanine, tyrosine, valine, β-aminoisocaproic acid, isoleucine, and α-amino-γ-methylthioacetic acid. The reaction sequence resulting in aminophosphonic acid (*R*)-**67** is shown below as an example. Enzymatic resolution of racemate **63** in the presence of *A. niger* lipase gave the optically active phosphonates (*R*)-**63** and (*S*)-**64**. The Mitsunobu reaction involving (*S*)-**64** proceeded via the formation of azide **65** and led to aminophosphonate (*R*)-**66**. Aminophosphonic acid (*R*)-**67** was isolated after subsequent acid hydrolysis of **66** [44–47] (Scheme 6.29).

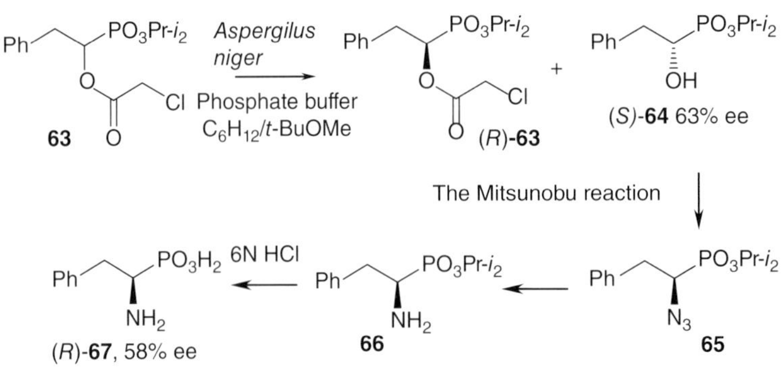

Scheme 6.29 Synthesis of aminophosphonates from hydroxyphosphonates.

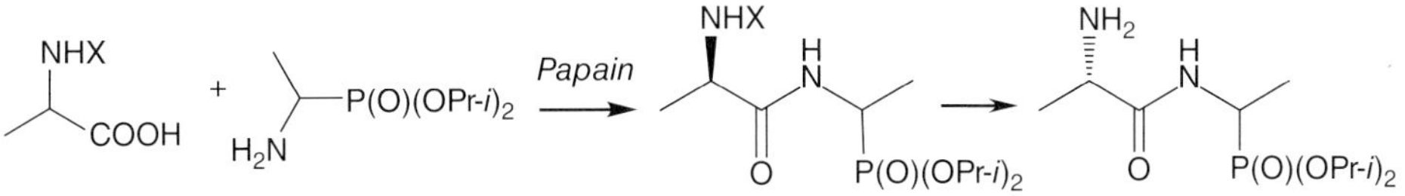

L-68 L/D- L,L-69 L,L-70

X = BnOCO

Scheme 6.30

The protease papaine (enzyme isolated from *Carica papaya* plant, which catalyzes hydrolysis of proteins and peptides) was used to resolve racemic aminophosphonic acid [40]. The reaction of *N*-benzyloxycarbonylalanine L-**68** with racemic aminophosphonate **67** was carried out in acetonitrile in the presence of papaine supported on solid polyamide at ambient temperature. This method was successfully used to obtain the optically active antibacterial agent alaphosphalin L,L-**70** (Scheme 6.30) [50–52].

Kukhar and co-workers [51] used *Penicillin acylase* (PA)-catalyzed hydrolysis for resolution of racemic *N*-phenyl-acetylaminophosphonic acids **71** into compounds D-**72** and L-**73** [50–52] (Scheme 6.31).

Khushi *et al.* [16] carried out one of the first enzymatic resolutions of aminophosphonates by hydrolysis of the corresponding esters of phosphono(amino)isocaproic acid by protease *Subtilisin Carlsberg* and *A. oryzae* lipase. The kinetically controlled reaction was carried out in aqueous solution (pH 7); the corresponding unreacted ester was isolated from the solution at pH 8.5 (Scheme 6.32).

Racemic β-aminoalkylphosphonates **74** were kinetically resolved into enantiomers by CALB-catalyzed *N*-acylation [53]. Ethyl acetate was used as the acylating agent because vinyl acetate, due to its high nucleophilicity, acylated the aminophosphonates even in the absence of CALB. As a result, acylation of diethyl 2-aminopropylphosphonate **74** (R^1 = Me, R^2 = Et) gave a mixture of corresponding enantioenriched aminophosphonate (*S*)-**74** (99.5% ee) and *N*-acylaminophosphonate (*R*)-**75** (78% ee), which was then separated by column chromatography. The efficiency of the process was affected by substituents at the phosphorus atom. It was found that the R group should be not too large for the CALB-catalyzed resolution to proceed successfully. For example, the enantioselectivity of resolution of diisopropyl 2-aminophosphonate **74** (R^1 = Me, R^2 = Pr-*i*)

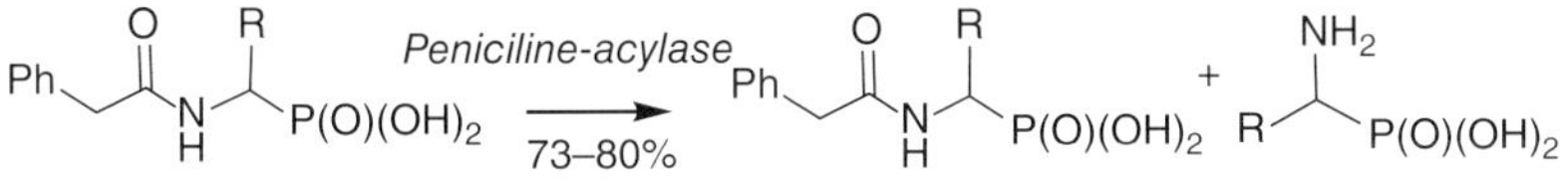

L/D-**71** R = Me, Ph D-**72** L-**73**

Scheme 6.31 Resolution of racemic *N*-phenyl-acetylaminophosphonic acids L/D-**71**.

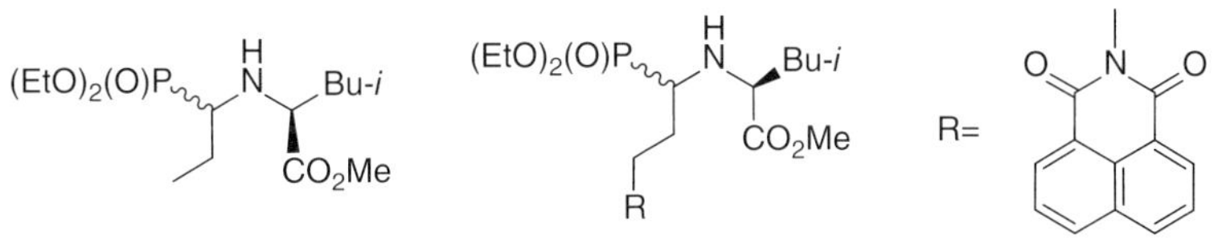

Scheme 6.32

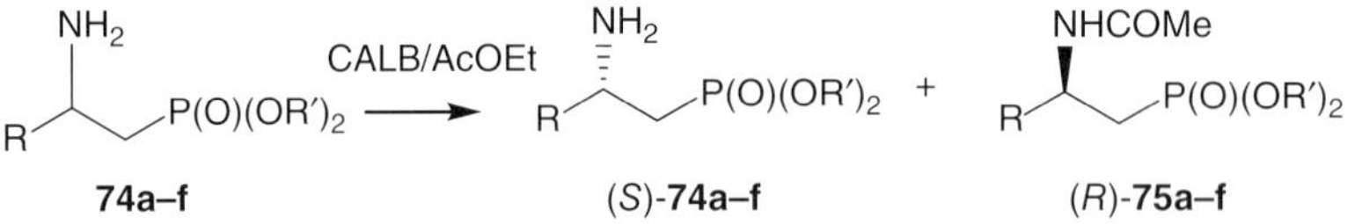

	R	R′	Time (h)	74 Yield (%)	74 ee (%)	75 Yield (%)	75 ee (%)
a	Me	Et	120	40	99.5	54	78
b	Me	Pr	120	41	100	53	76
c	Et	Pr-*i*	148	40	100	55	72
d	Et	Et	148	44	64	42	79
e	Et	Pr	—	41	56	40	74
f	Et	Pr-*i*	—	43	26	41	41

Scheme 6.33 Resolution of racemic α- and β-aminoalkylphosphonates.

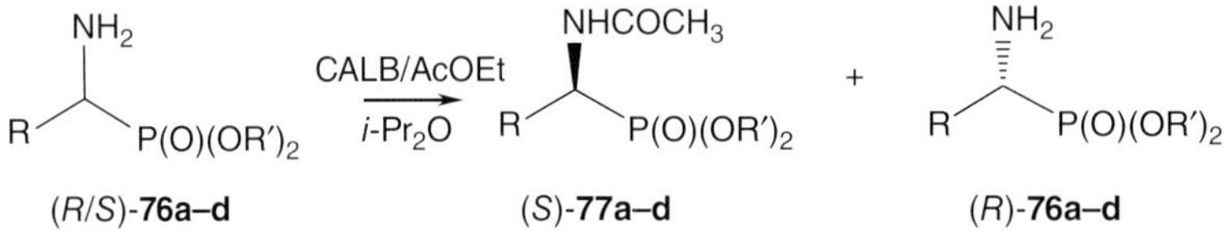

	R	R′	(S)-**77** Yield (%)	(S)-**77** ee (%)	(R)-**76** Yield (%)	(R)-**76** ee (%)
a	Me	Et	48	90	41	99.7
b	Me	Pr^i	42	90	42	90
c	Et	Me	43	98	44	96
d	Et	Et	10	100	73	18

Scheme 6.34 *N*-Acylation of α-aminophosphonates **76a–d** catalyzed by CALB.

under the same conditions was low (the ee values of the products were about 54% and 64%, respectively). However, the enantioselectivity was increased to $E > 70\%$ when the reaction of **74** (R^1 = Me, R^2 = Pr-*i*) was carried out in diisopropyl ether. The enantioselectivity of acylation also decreased when the vinyl group was used as the R substituent (Scheme 6.33). *N*-acylation of α-aminophosphonates **76a–d** with ethyl acetate in the presence of CALB was slow: a 50% conversion was achieved within 5 days, although the enantioselectivity in most cases was reasonably high (ee >90%). The reaction resulted in a mixture of compounds (*R*)-**87a–d** and (*S*)-**77a–d** (Scheme 6.34).

In some cases, a more reactive ethyl methoxyacetate was used instead of ethyl acetate. This improved, for example, the efficiency of the resolution of enantiomers of α-aminophosphonates **76d–f**. Then, the esters of aminophosphonic acids (*R*)-**76d–f** were separated from the acylated derivatives (*S*)-**78d–f** and transformed to compounds **79d**–f with protected amino group [53] (Scheme 6.35). The stereocourse of CALB-catalyzed acylation of all α-and β-aminoalkylphosphonates shown above follows the Kazlauskas rule (Scheme 6.35b).

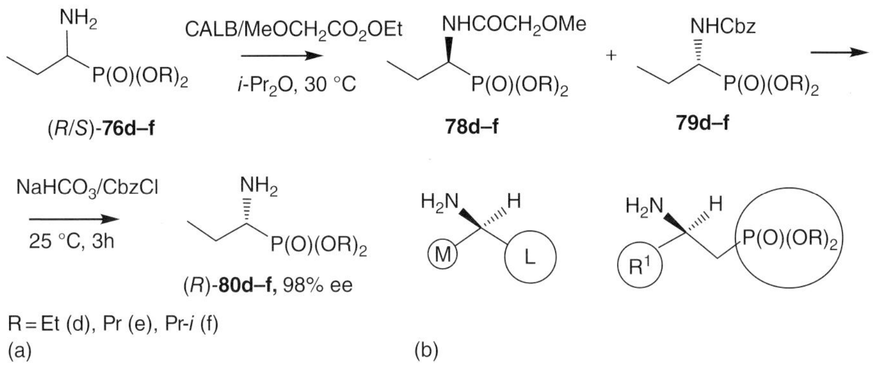

Scheme 6.35 (a) Resolution of α-aminophosphonates and (b) absolute configurations of (*S*)-stereoisomers of aminoalkylphosphonates resulting from CALB-catalyzed resolution.

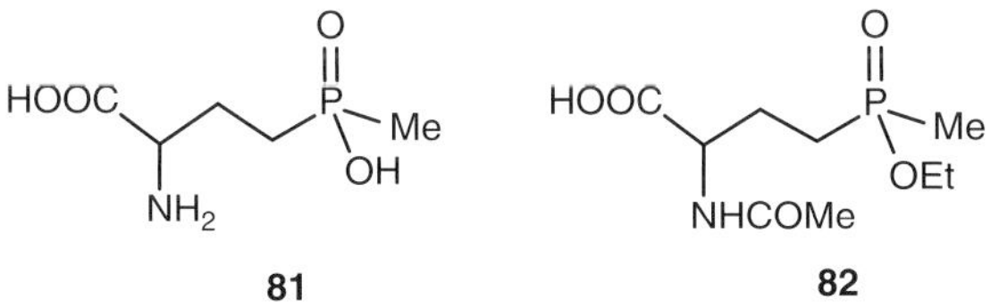

Scheme 6.36 Enzymatic synthesis of D- and L-phosphinothricin.

Natchev [54] reported enzymatic synthesis of D- and L-enantiomers of phosphinothricin (2-amino-4-hydroxymethylphosphinylbutanoic acid **81**) and its derivative **82**. L-Phosphinothricin is a known herbicide isolated from *Streptomyces hydroscopicus.* Successive enzymatic hydrolysis by phosphodiesterase, acylase, and glutaminase led to phosphinothricin **81.** The formation of the L-phosphinothricin derivative **82** was attained under the action of α-chymotrypsin (Scheme 6.36).

6.3 Biosynthesis of Compounds with C–P Bond

The discovery of natural phosphonates, such as 2-amino-ethylphosphonic acid, phosphinothricin, and fosfomycin, was followed by intensive research into the mechanism of C–P bond formation in biological systems [55–64]. It was established that the key step of phosphonate formation involves a rearrangement of phosphoenol pyruvate (PEP) to phosphopyruvate catalyzed by the enzymes responsible for the P–C bond formation. In particular, fosfomycin is formed as a result of biocatalyzed rearrangement of PEP **83** to phosphopyruvate **84** under the action of PEP phosphomutase (EC 5.4.2.9) isolated from *Tetrahymena pyriformis* and *Streptomyces hygroscopicus* [56, 57]. As a result of intramolecular rearrangement, PEP is transformed to phosphopyruvic acid which then undergoes decarboxylation with the formation of phosphonoacetaldehyde and, subsequently, of (1*R*,2*S*)-fosfomycin. (Scheme 6.37).

Using ^{18}O labels [61], it was proved that transfer of the phosphate group from oxygen to carbon atom in enol pyruvate **83** catalyzed by *T. pyriformis* is an intramolecular process proceeding with retention of the absolute configuration of the phosphorus

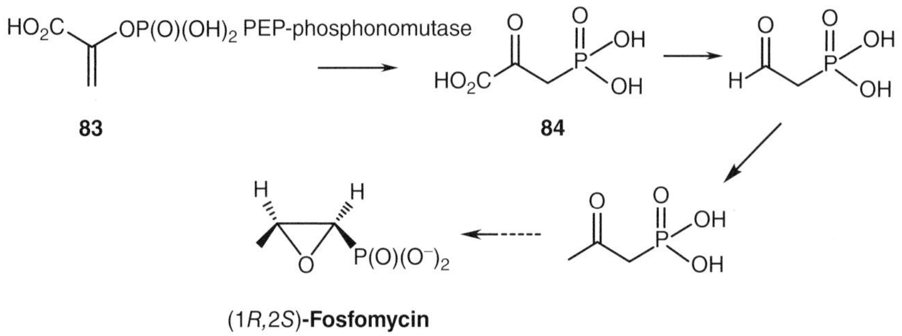

Scheme 6.37 Rearrangement of phosphoenol pyruvate (PEP) to phosphopyruvate and then to fosfomycin.

atom [59, 60]. The formation of the herbicide bialaphos (SF-1293) proceeds by a similar mechanism. This tripeptide antibiotic is produced by *Streptomyces veridochromogenes* bacteria and possesses herbicidal and fungicidal activity. Biocatalysis is carried out by CPEP-phosphonomutase (CPEP is carboxyphosphoenol pyruvate) isolated from infusoria *T. pyriformis.* This enzyme initiates the condensation of PEP with phosphonoformate and subsequent rearrangement of PEP to phosphonopyruvate with the formation of a P–C bond. The reaction catalyzed by CPEP-phosphonomutase also proceeded with retention of the absolute configuration of the phosphorus atom [65] (Scheme 6.38).

Biosynthesis of 2-amino-1-hydroxyethylphosphinic acid (OH-AEP) (*R*)-**85** was reported [9, 63]. Enantioselective hydroxylation of 2-aminoethylphosphinic acid (AEP) is carried out by the enzyme isolated from amoeba *Acanthamoeba castellanii* (Scheme 6.39) [63].

Hammerschmidt and co-workers [64, 66, 67] studied biocatalytic epoxidation of (*S*)-2-hydroxyalkylphosphonic acids **86** to epoxides **87** (fosfomycin analogs). The

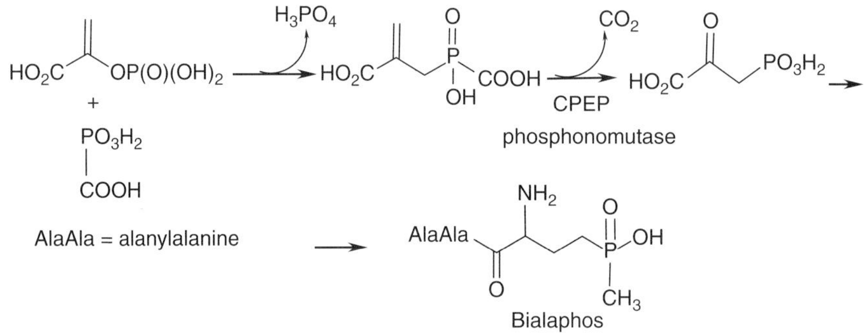

Scheme 6.38 Synthesis of the herbicide bialaphos.

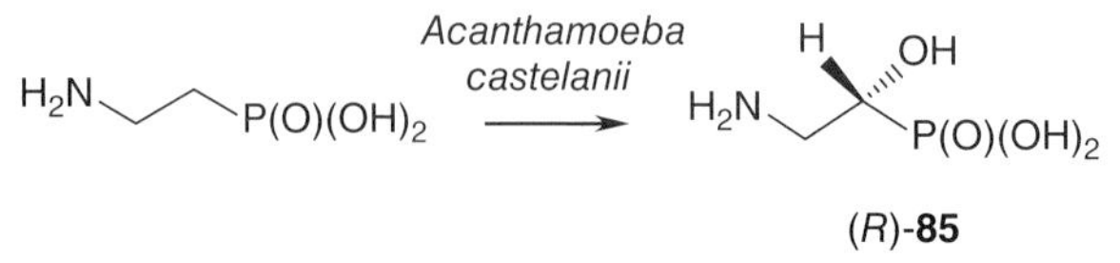

Scheme 6.39 Biosynthesis of 2-amino-1-hydroxyethylphosphinic acid.

reaction was initiated by the enzyme isolated from a *Streptomyces fradiae strain*. Replacement of hydrogen at the C^{α} atom of deuterated 2-hydroxypropylphosphonic acids catalyzed by this enzyme proceeded stereo specifically. It was found that the growth-supporting microenvironment of *S. fradiae* contains not only fosfomycin (*cis*-epoxide) but also 3% of its cometabolite (*trans*-epoxide). The formation of *cis*-epoxides involved inversion of the absolute configuration at the carbon atom, whereas *trans*-epoxides were formed with retained configuration. This fact was proved by transforming deuterated hydroxyphosphonates to epoxides. Deuterated (1*S*,2*S*)- and (1*R*,2*S*)-2-hydroxy-1-D-propylphosphonic acids were synthesized from (1*S*,2*S*)-2-benzyloxy-1-D-propanol obtained by catalytic reduction of the corresponding aldehyde in the presence of horse liver alcohol dehydrogenase (Scheme 6.40) [56].

A similar biocatalytic transformation of hydroxyphosphonate **88** to fosfomycin in the presence of (*S*)-2-hydroxypropylphosphonic acid epoxidase (Ps-HppE) isolated from the phytopathogenic bacteria *P. syringae* was reported [68, 69]. The purified Ps-HppE catalyzed the epoxidation of (*S*)-hydroxypropylphosphonic acid (*S*)-**88** with the formation of fosfomycin, while the oxidation of hydroxyphenylphosphonic acid (*R*)-**89** under the same conditions gave 2-oxopropylphosphonic acid **89** (Scheme 6.41). Acylation of hydroxyphosphonate **90** with vinyl acetate was carried out in isopropyl ether in the presence of lipase from *P. syringae*. The reaction was completed in 5 days at 30 °C. Chromatographic separation on silica gel followed by deacetylation afforded optically active hydroxyphosphonate (*R*)-**91** (46% yield), which was then transformed to a deuterated analog of (*S*,*R*)-fosfomycin. The isotope label experiment showed that the epoxide oxygen atom of (*S*,*R*)-fosfomycin is derived from the hydroxyl group of hydroxyphosphonate **91** [68] (Scheme 6.42).

6.4 Resolution of *P*-Chiral Phosphorus Compounds

The creation of a chiral center at the phosphorus atom by biocatalytic methods is a problem of great practical and theoretical importance. Efficient and general methods for the synthesis of chiral phosphine oxides and related compounds represents a permanent subject for research by organic chemists. Therefore, the application of biocatalytic

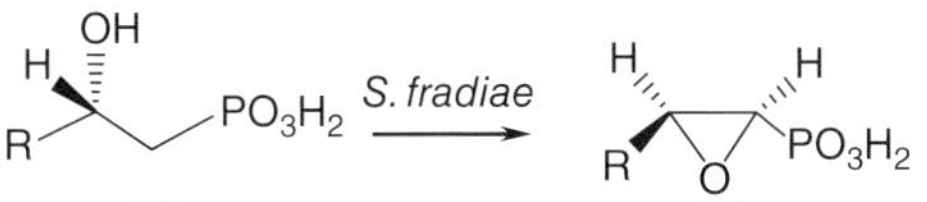

86

87

R = H, Et

Scheme 6.40 Biocatalytic epoxidation of (*S*)-2-hydroxyalkylphosphonic acids.

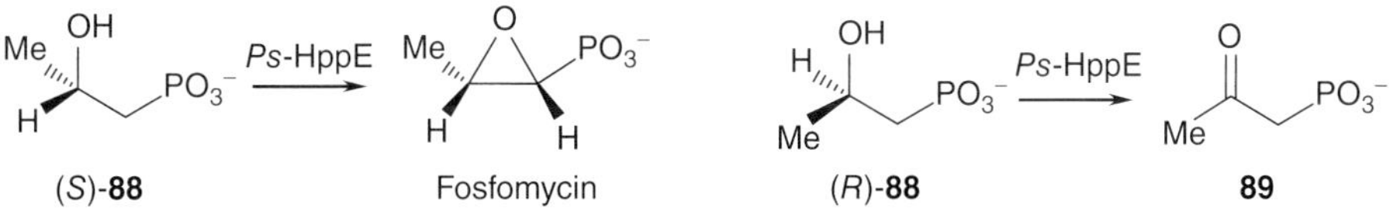

(*S*)-**88** Fosfomycin (*R*)-**88** **89**

Scheme 6.41 Biocatalytic synthesis of fosfomycin.

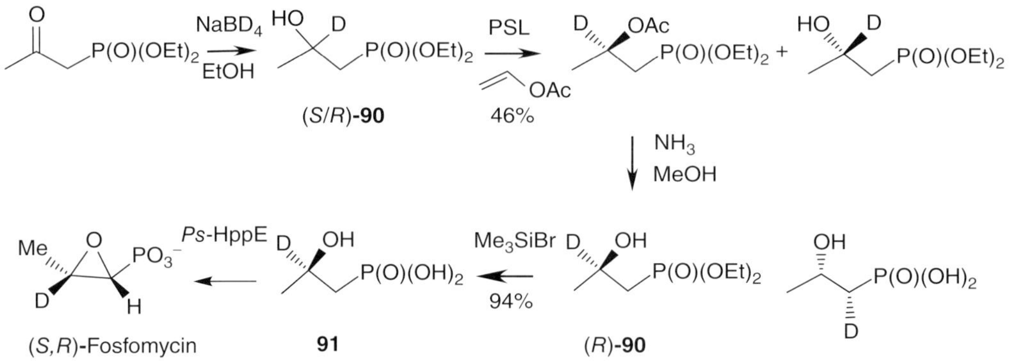

Scheme 6.42 Biocatalytic synthesis of deuterated Fosfomycin.

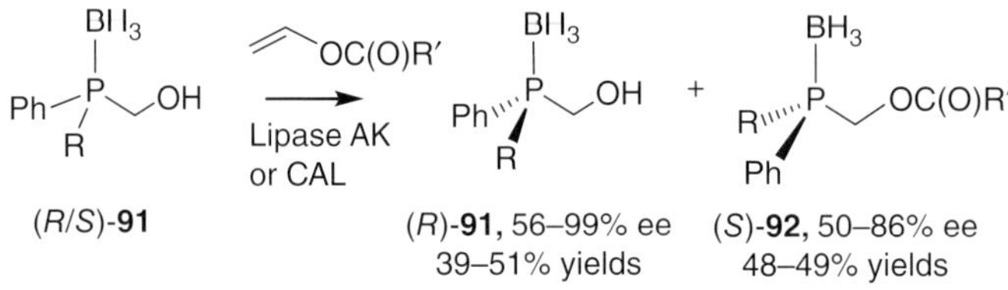

R=Et, Bu, c-C_6H_{11}, t-Bu, MeO, EtO, i-PrO, t-BuO; R′ = Me, Et, Pr. i-Pr;
Lipase = CALB, Amano AK, Amano PS, Amano AH, LPL

Scheme 6.43 Enzymatic resolution of racemic tertiary phosphines and phosphinites **91**.

methods for the preparation of optically active P-chiral compounds has attracted great attention and a number of successful syntheses have been described [20, 29, 70–76].

For example, Mikołajczyk and Kiełbasiński [70, 73] studied the acylation of phosphine–boranes **91** using CAL (Chirazyme®) lipase from *C. Antarctica* and Lipase AK from *P. fluorescens* (Scheme 6.43). The best enantioselectivity was attained in the lipase AK-catalyzed acylation of **91** in cyclohexane solution with vinyl butyrate as an acyl donor - 99% ee for unreacted hydroxyphosphinate **91** and 43% ee for the acylated product **92**. The *E*-values were on the level of 15. The enzymatic resolution of alcoxy (hydroxymethyl)phenyl-phosphine boranes (*R*/*S*)-**91** with vinyl acetate in the presence of CALB, Amano AK, Amano PS, Amano AH, and LPL was studied in various solvents. The best enantioselectivity was attained in cyclohexane (37% ee, conversion ~50%). Kiełbasiński [74] recently reported some additional data, including theoretical calculations and more accurate chemical correlation, which proved that the borane reduction of acyclic phosphine oxides proceeded with inversion of configuration at the phosphorus center. On this basis, the stereochemistry of the enzymatic reaction was ultimately determined.

Lipase-catalyzed acylation of ethyl (1-hydroxyalkyl)phenylphosphinates afforded a single diastereomer in high enantiomeric excess. The substituent effect of the alkyl group toward the acylation using CAL (Chirazyme®) was larger than that of an immobilized lipase AK from *P. fluorescens*. The kinetic separation of the major phosphinates (S_P,*S*)-**93** and (R_P,*R*)-**94** was carried out by CALB and lipase AK catalyzed acylation using vinyl acetate as an acyl donor. The influence of the alkyl substituent R in the biocatalyst on the acylation process was more pronounced for CALB than for AK (Scheme 6.44)

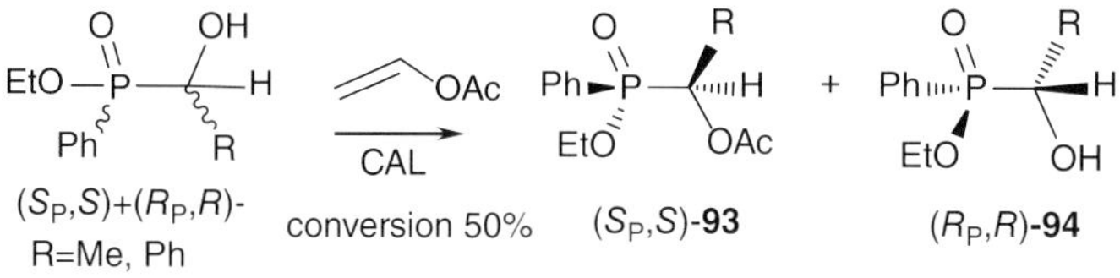

R	Lipase	ee (%) of **93**	ee (%) of **94**
Me	CAL	98	98
Et	CAL	98	28
Me	AK	98	98
Me	AK	98	98
Et	AK	98	98
Pr	AK	98	2

Scheme 6.44 Lipase-catalyzed acylation of ethyl (1-hydroxyalkyl)phenylphosphinates.

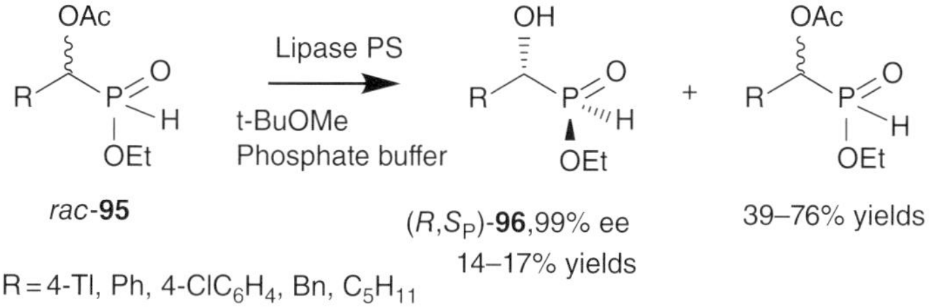

Scheme 6.45 Enzymatic synthesis of chiral α-hydroxy-*H*-phosphinates **95**.

[20, 29]. A synthesis of chiral α-hydroxy-*H*-phosphinates, bearing two asymmetric centers, was achieved via a lipase-catalyzed hydrolysis of acetate precursors. From a mixture of four kinds of stereoisomers (two enantiomers and two diastereomers) of α-acetoxy-*H*-phosphinates **95,** one isomer of α-hydroxy-*H*-phosphinates (R,S_P)-**96** was obtained with moderate yields and stereoselectivity up to 99% ee (Scheme 6.45) [31].

The kinetic resolution of 1,1-diethoxyethyl(hydroxymethyl)phosphinate *rac*-**97** possessing chirality at the phosphorus atom was achieved via a lipase-catalyzed acylation. The product **99** was transformed into the corresponding amine **100**, which is a useful precursor for the preparation of phosphinyl dipeptide isosteres. Acylation of **97** with vinyl acetate as an acyl donor in the presence of lipase Amano AK led to the formation of (*R*)-**98** (yield 59%, 88% ee) and (*S*)-**99** (yield 35%, 92% ee), which were separated by column chromatography. The enantiomeric purity of (R_P)-**98** was increased to 99% ee by an enzymatic double resolution under the same conditions (Scheme 6.46) [77].

Biocatalytic kinetic resolution of racemic hydroxymethylphosphinates **101** via their lipase-promoted acetylation in supercritical carbon dioxide as the reaction medium was investigated. The reaction was fastest when pressure was closer to the critical pressure: at 11 MPa, the reaction rate reached its maximum when the pressure was increased to 15 MPa. The optimal conditions were obtained at 13 MPa (yields ~50%, ~30% ee). The stereoselectivity of the reaction depended on solvent, substituents at phosphorus, and

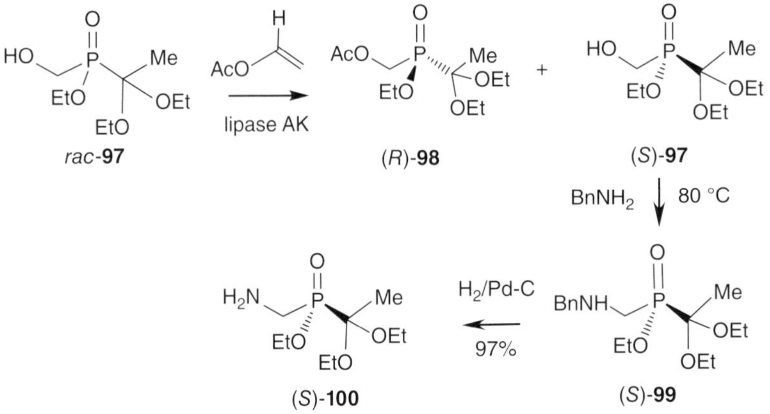

Scheme 6.46 The kinetic resolution of 1,1-diethoxyethyl(hydroxymethyl)phosphinate *rac*-**97**.

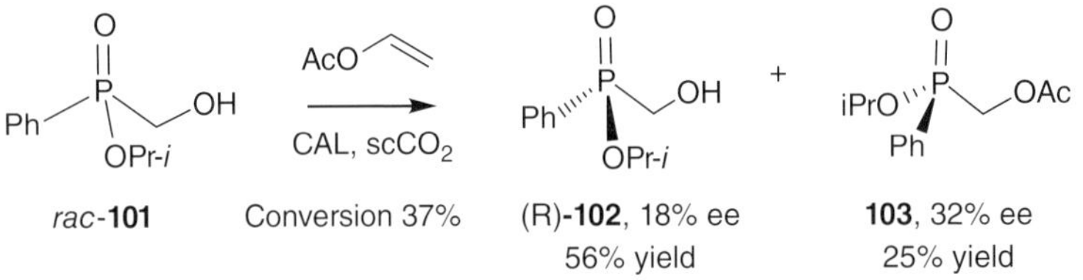

Scheme 6.47 Biocatalytic kinetic resolution of racemic hydroxymethylphosphinates **101**.

solubility of substrates in $scCO_2$. The best results were obtained with the *C. antarctica* lipase (Novozym 435) (Scheme 6.47) [78, 79].

The biocatalytic desymmetrization of various C_2-symmetric tertiary phosphine oxides was used for the preparation of P-chiral phosphines. Mikołajczyk *et al.* [65] studied the desymmetrization of bis-functional phosphinates and phosphine oxides. The hydrolysis of prochiral bis(methoxycarbonylmethyl)phenylphosphine oxide **104** was carried out in phosphate buffer in the presence of porcine liver esterase (PLE) to give a chiral monoacetate (*R*)-**105** in 92% yield and 72% ee. The chiral monoacetate (*R*)-**105** by decarboxylation was converted into chiral phosphine oxide (*R*)-**106** (Scheme 6.48).

The biocatalytic acetylation of prochiral bis(hydroxymethyl)phenylphosphine oxide **107** and the biocatalytic hydrolysis of prochiral bis(methoxycarbonylmethyl) phenylphosphine oxide **109** was subjected to hydrolysis in a phosphate buffer in the presence of several hydrolases (PLE, PFL, AHS, Amano-AK, and Amano-PS), of which only PLE proved to be efficient [80] (Scheme 6.49). The best results were attained with

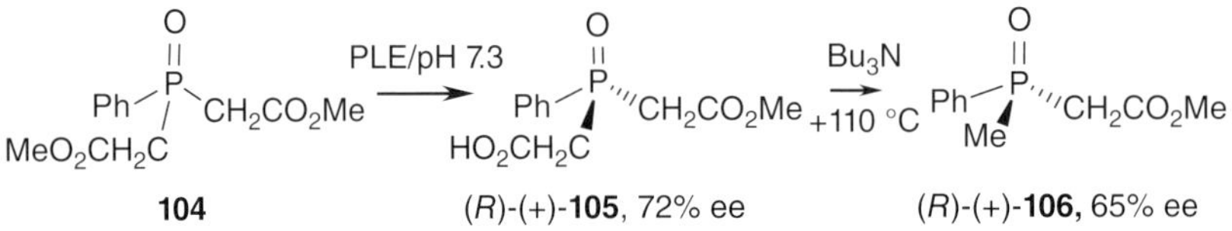

Scheme 6.48 The biocatalytic desymmetrization of bis-functional phosphinates.

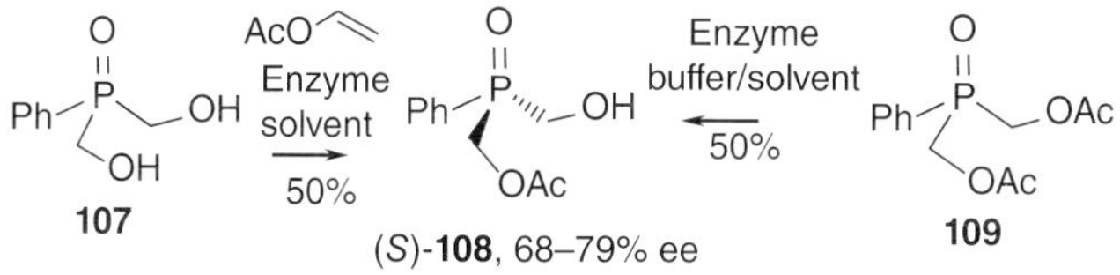

Scheme 6.49 Enzymatic preparation of tertiary (*R*)- or (*S*)-phosphine oxides.

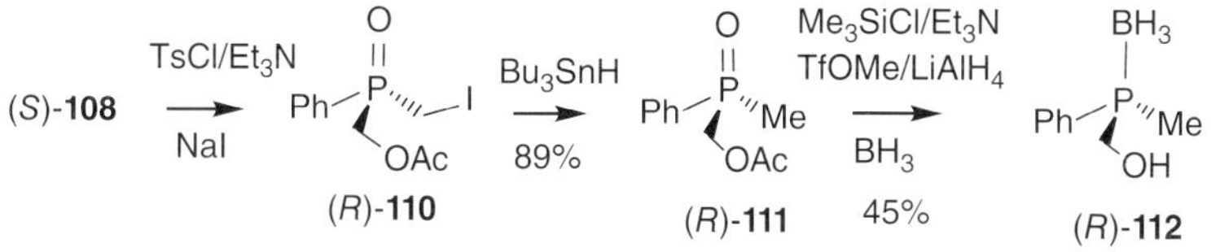

Scheme 6.50 Determination of absolute configuration of the (*S*)-**108** by chemical correlation.

P. fluorescens lipase (PFL) in chloroform that allowed to obtain the compound **109** in yield up to 76% and with ee up to 79%. Absolute configuration of the compound (*S*)-**108** was determined by means of chemical correlation to the earlier described compound (*R*)-**112** as shown in Scheme 6.50 [65].

Desymmetrization of the prochiral diol **113** was attained, using vinyl acetate as an acetylating agent and several lipases (CAL, AK, AH, PS, LPL, PFL), of which only PFL proved efficient. It was found that the use of various solvents led to opposite enantiomers of the product **114** and substantially influenced the stereoselectivity of the process. For example, the replacement of chloroform by isopropyl ether led to the formation of the optical antipode of **114** [73, 74]. Wiktelius *et al.* [75] reported that the *C. antarctica* lipase B (Novozym 435) afforded better results in the desymmetrization of prochiral phosphine-boranes than Amano ANL, Amano PS, PFL (Amano AK), and PPL (Fluka) lipases (Scheme 6.51).

Raushel *et al.* [81–86] have resolved racemic phosphinates **115,** bearing a phenol leaving-group at phosphorus, by means of fermentative hydrolysis catalyzed by phosphotriesterase (PTE) from *Pseudomonas diminuta.* The hydrolysis led to the formation of chiral phosphinates with optical purity greater than 99.8%, quantified by chiral electrophoresis. The catalyzed PTE hydrolysis of phosphinates (R_P/S_P)-**115** led to the formation of (S_P)-methylphenylphosphinate and unreacted phosphinate (R_P)-**115**. It was found that wild-type PTE preferably hydrolyzed (S_P)-phosphinates with formation of (R_P)-phosphinates **115**, and mutant PTE (TAGW) preferably hydrolyzed

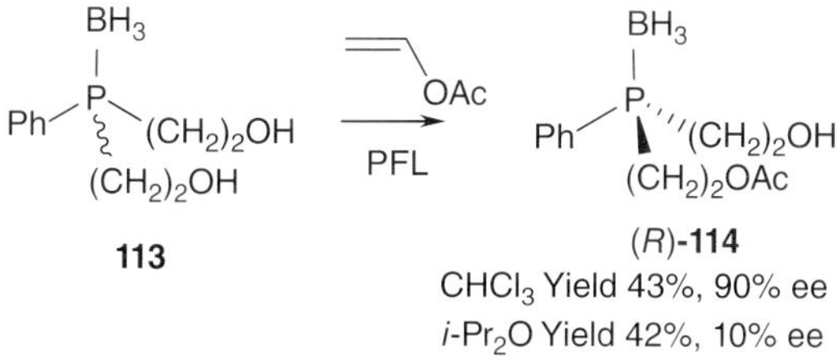

Scheme 6.51 Enzymatic desymmetrization of the prochiral diol **113**.

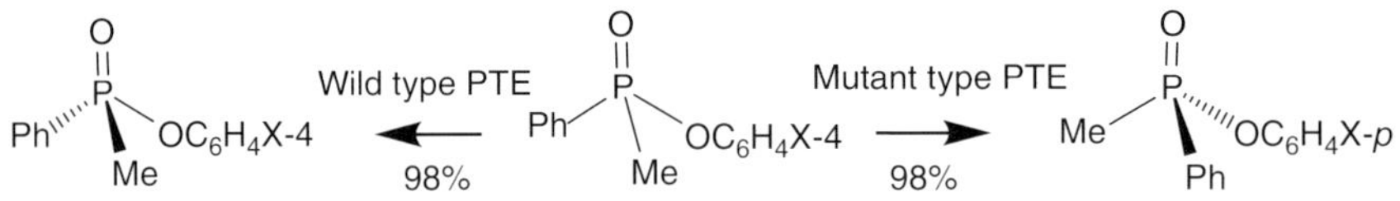

(R_P)-**115**, 99% ee
(R_P/S_P)-**115**
(S_P)-**115**, 99% ee
X = H, NO_2, CHO, CN, Ac, CO_2Me, Cl, F

Scheme 6.52 Hydrolysis of phenoxy (S_P)-phosphinates with wild or mutant PTE.

(R_P)-phosphinate with formation of (S_P)-phosphinates with the factor of enantioselectivity $E = 17$. The stereoselectivity of wild-type PTE was affected by the pK_a value of the phenol-leaving group. For the wild-type enzyme, the stereoselectivity has been enhanced in excess of three orders of magnitude catalyzed PTE hydrolysis of the most acidic phenolic substituents from an organophosphate triester. (R_P)-Stereoisomers **115** were purified by chromatography on silica gel and were obtained in 98% yield and with 99% ee (Scheme 6.52) [82–84].

It was found that wild-type PTE mainly hydrolyzes S_P-stereoisomers of phosphinates **115**, while mutant phosphotriesterase I106T/F132A/H254G/H257W (TAGW) mainly hydrolyzes (R_P)-phosphinates. In the latter case, the enantioselectivity factor for the formation of compounds (S_P)-**115** was equal to 17. The hydrolysis of phosphonate **116** (R = Me) by wild-type PTE' yielded mainly thioacids (*S*)-**117**, whereas the hydrolysis by mutant-type PTE led to the formation of (*R*)-thioacids **117**. In contrast, the hydrolysis of thioacid triesters **116** with either wild or mutant types of PTE in many cases yielded thioacids (*S*)-**117** with 99% ee. The chiral thiophosphates synthesized by these enzymatic methods were proposed as precursors for the synthesis of organic and organophosphorus compounds (Scheme 6.53) [85].

The biocatalytic oxidation of racemic *O,S*-dimethyl *O*-*p*-nitrophenyl phosphorodithioate (*S*/*R*)-**118** catalyzed by chloroperoxidase from *Caldariomyces fumago* led to the formation of the corresponding (−)-(*S*)-thiophosphate **119** and unoxidized substrate (+)-(*R*)-**118**. The thionoester (*S*/*R*)-**118** was subjected to oxidation with hydrogen peroxide in the presence of chloroperoxidase (CPO) in a mixture of citrate buffer, pH 5, and ethanol. The compounds were prepared with 99.6% and 97% ee, respectively. The thionation of the (−)-(*S*)-phosphate **119** with Lawesson's reagent gave (−)-(*S*)-phosphorodithioate **118** with full stereospecificity, while the oxidation of unreacted substrate (+)-(*R*)-**118** with iodoxybenzene resulted in the formation of (+)-(*R*)-**119** with 94.9% ee (Scheme 6.54) [87].

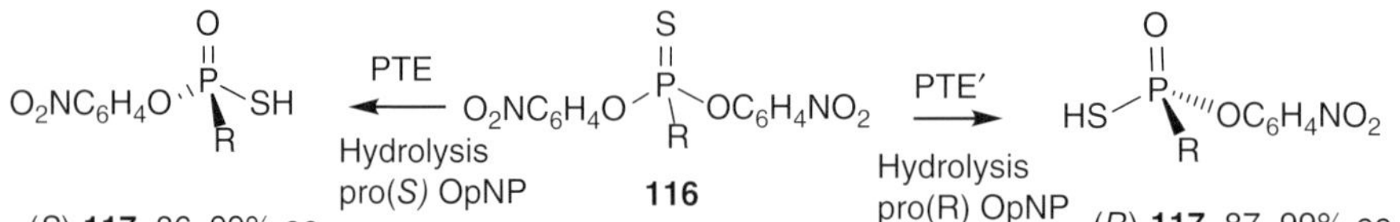

116

(*S*)-**117**, 86–99% ee
84–96% yield
R = Me, MeO, EtO, *i*-PrO

(*R*)-**117**, 87–99% ee
84–96% yield

Scheme 6.53 The hydrolysis of thioacid esters **116** by wild or mutant types of PTE.

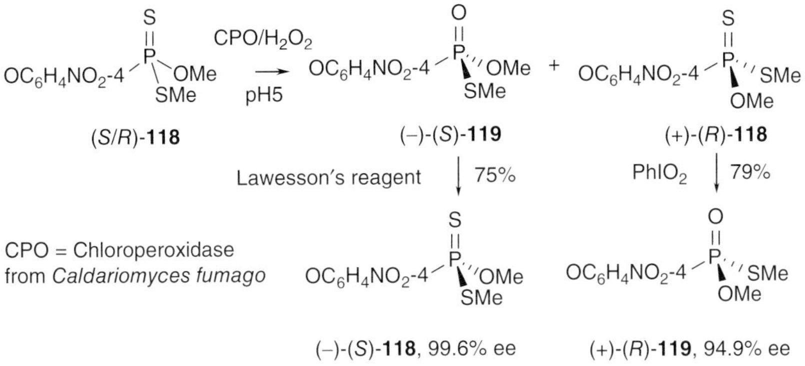

Scheme 6.54 The biocatalytic oxidation of racemic *O,S*-dimethyl *O*-*P*-nitrophenyl phosphorodithioate **118**.

Prochiral bis(cyanomethyl)phenylphosphine oxide **120** has been successfully transformed into the corresponding optically active monoamide **121** and monoacid **122** with enantiomeric excesses ranging from low (15%) to very high (up to 99%) using a broad spectrum of nitrile-hydrolyzing enzymes [88]. Enzymatic hydrolysis of prochiral bis(cyanomethyl) phenylphosphine oxide **121** was achieved using nitrile-converting enzymes under mild conditions (buffer solution of pH 7.2, 30 °C) with formation of cyanomethylphenyl-phosphinylacetamide **121** and cyanomethylphenylphosphinylacetic acid **122** in different proportions and enantioselectivities ranging from 15% to 99% ee. For example, the hydrolysis with nitrilase **106** led to the formation of products (*S*)-**121** and (*S*)-**122** in yields of 10.8% and 51.0% and with 99% and 70% ee (Scheme 6.55).

Hydrolysis of phosphonyl acetates **123a,b**, phosphoryl acetates **123c**, and phosphamide **123d** in the presence of PLE resulted in mixtures of P-chiral phosphorus-containing acetic acids **124a–d** and esters **123a–d** [89]. Enzymatic hydrolysis under kinetically controlled resolution was carried out until 50% conversion. PLE-catalyzed hydrolysis of phosphonyl acetates proceeded enantioselectively to give optically active products in good chemical yields, but the optical yields were relatively low. In all cases, the reaction mainly involved the (*R*)-enantiomers of acetates **123a–d** (Scheme 6.56).

Various organic media including ionic liquids and supercritical CO_2 were used for the resolution of racemic phosphinates [90]. Acetylation of racemic P-chiral hydroxymethyl-phosphinates and phosphine oxides in the presence of a lipase was carried out under kinetic resolution conditions in ionic liquids BMIM + X^- (BMIM is 1-*n*-butyl-3-methylimidazolium) [91]. Both lipases from *P. fluorescens*, namely, immobilized Amano AK and non-immobilized PFL, were six times more efficient in

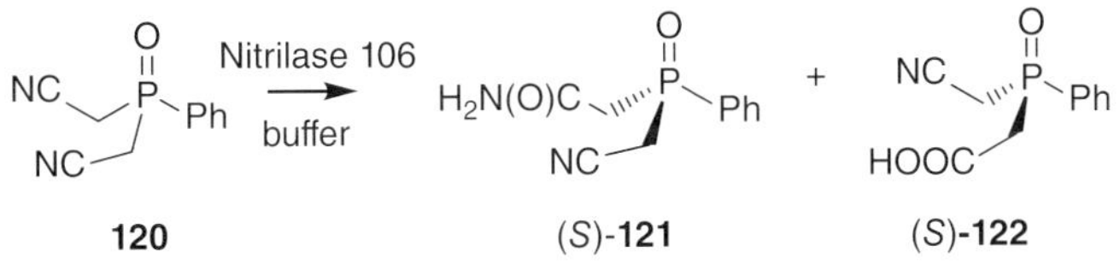

Scheme 6.55 Enzymatic hydrolysis of bis(cyanomethyl) phenylphosphine oxide **120**.

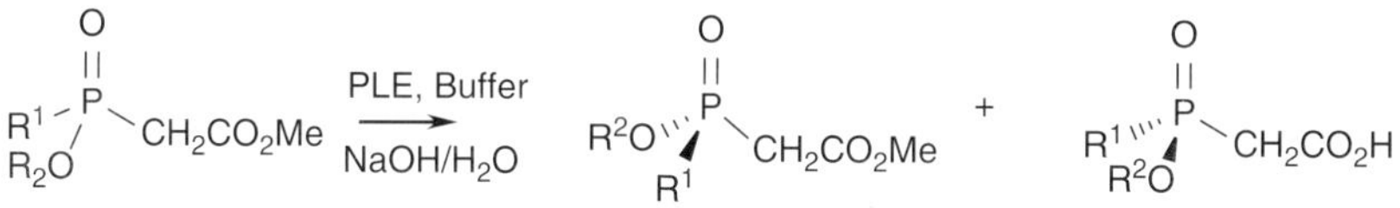

rac-**123a–d** (*S*)-**123a–d** (*R*)-**124a–d**

Compounds	R^1	R^2	(*S*)-**123** yield (%)	(*S*)-**123** ee (%)	(*R*)-**124** yield (%)	(*R*)-**124** ee (%)
a	Et	Me	50	38	34	24
b	Et	Et	40	—	60	—
c	PhO	Et	66	20	22	52
d	Et2N	Me	20	90	58	25

Scheme 6.56 PLE-catalyzed hydrolysis of phosphonyl acetates **123**.

BMIM + PF6 compared to organic solvents [92]. In contrast, reactions in BMIM + BF_4 showed almost no stereoselectivity. Racemic hydroxymethylphosphinates **128a–d** and hydroxymethylphosphine oxide **128e** were acetylated with vinyl acetate in the presence of lipases in ionic liquids. Of the two ionic liquids studied, $BMIM^+PF_6$ showed the best performance, which allowed one to increase the stereoselectivity of enzymatic resolution and to obtain compounds **128a–e** and **129a–e** with satisfactory optical purity. The *E*-values were three to six times higher when substrates containing bulky organic substituents were used (Scheme 6.57). The absolute configurations of compounds **124, 125** were determined by chemical correlations (conversion into tertiary phosphines: ethyl-methyl-phenylphosphine oxide **126**, and butyl-methyl-phenylphosphine oxide **127**) and by circular dichroism spectroscopy.

Kinetic resolution of racemic P-chiral hydroxymethyl-phosphinates **130a–d** by enzymatic transesterification in supercritical carbon dioxide (sc-CO_2) under different conditions was studied [78, 90]. It was found that the excess pressure affects the yields and enantiopurity of the products **130*a–d** and **131*a–d.** The reaction rate reached its maximum when the pressure approached a critical value of 11 MPa and decreased with further increase in the pressure up to 15 MPa. Optimal results were obtained at

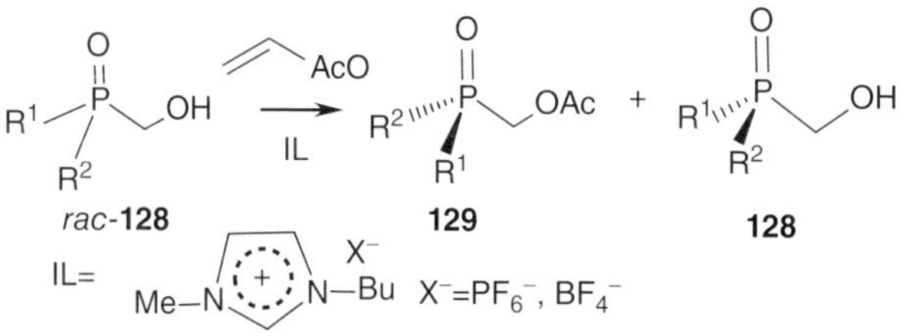

	R^1	R^2	Lipase	**128**, ee (%)	**129**, ee (%)
a	Ph	MeO	PF	89	89
b	Ph	EtO	AK	79	83
c	Ph	*i*-PrO	PF	96	80
d	Et	*i*-PrO	PF	95	50
e	Ph	*t*-Bu	AK	43	53

Scheme 6.57 Resolution of hydroxyphosphonates in ionic liquids.

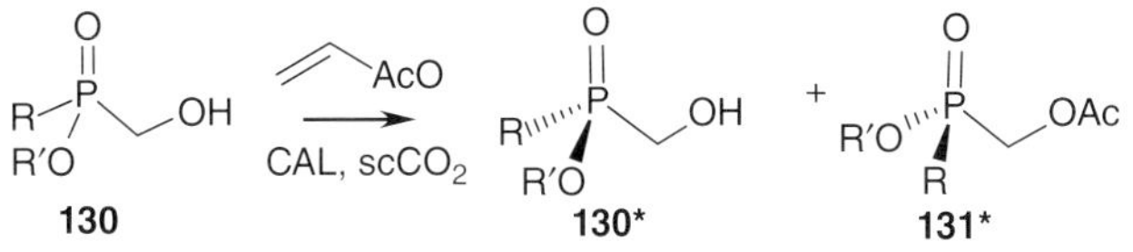

	R	R′	Conversion (%)	130*		131*	
				Yield (%)	ee (%)	Yield (%)	ee (%)
a	Ph	MeO	8	81	−1.1	5	7
b	Ph	EtO	93	6	−13.4	82	6
c	Ph	PriO	46	52	−6.2	46	27
d	Et	PriO	100	—	—	90	—

Scheme 6.58 Enzymatic resolution of racemic hydroxymethylphosphinates in supercritical CO_2.

a pressure of 13 MPa, namely, a 50% conversion was attained with ease and the enantioselectivity was highest. Thus, by varying the pressure, one can change the reactivity of the substrate and the stereoselectivity of the process. The stereoselectivity was also influenced by the solvent used, by the volume of substituents at the phosphorus atom and by the solubility of the substrates in sc-CO_2. The best results were obtained using lipase from *C. antarctica* (Novozym 435) (Scheme 6.58) [78]. Resolution of racemic hydroxymethylphosphinates **130a–c** by transesterification with vinyl acetate in the presence of lipase from *P. cepacia* (Amano PS) proceeded with moderate ee. In some cases, hydrolysis of the corresponding acetates **131a–c** was more efficient and led to an increase in ee. It was found that three successive hydrolyses of enantioenriched acetates **131a–c** catalyzed by PFL and Amano PS increased the enantiopurity of the products to more than 92% (Scheme 6.59) [93].

Among different classes of chiral phosphorus-containing compounds, chiral phosphine oxides are of particular importance. Therefore, the development of biocatalytic methods for the synthesis of phosphine oxides containing chiral phosphorus atoms is of

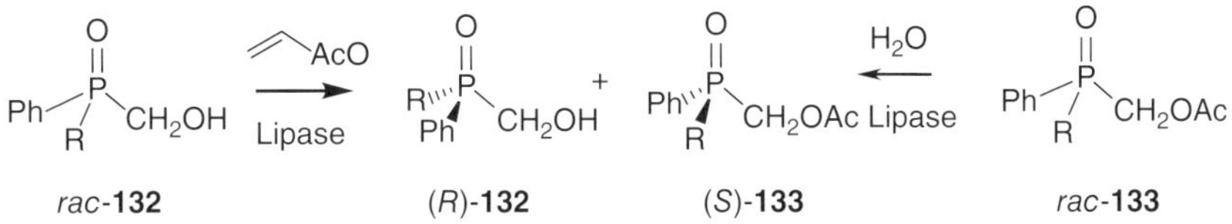

R	Lipase	132		133	
		Yield (%)	ee (%)	Yield (%)	ee (%)
MeO	PFL	44	80	39	89
MeO	AM	42	92	44	86
EtO	PFL	42	54	53	47
EtO	AM	30	54	68	21
i-PrO	PFL	52	36	45	24

Scheme 6.59 Resolution of racemic hydroxymethylphosphinates **132** by transesterification biocatalyzed by *Pseudomonas cepacia* lipase.

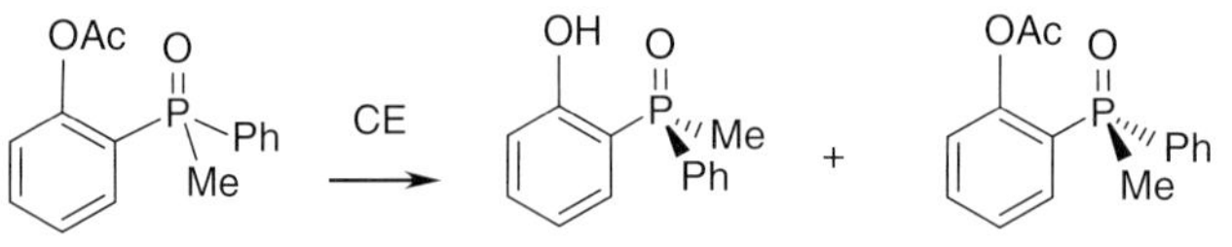

Scheme 6.60 Biocatalytic preparation of P-chirogenic phosphine oxides.

considerable interest. Serreqi and Kazlauskas [94] reported the preparation of P-chiral hydroxyarylphosphines and phosphine oxides by enzymatic acetylation. Among the commercially available enzymes tested, the best results were shown by cholinesterase (CE) and CRL. CE-catalyzed hydrolysis of a synthetic substrate (phosphine oxide acetate **134**) proceeded seven times faster than hydrolysis of natural cholesterol acetate under the same conditions. Similar results were obtained using CRL. The enantioselectivity of the resolution varied from low to high, but the optical purity of compound (*S*)-**135** attained a value of 95% ee (Scheme 6.60).

The same strategy was used to synthesize chiral tertiary naphthylphosphines **138**, the starting compounds for the synthesis of chiral Wittig reagents. Biocatalytic hydrolysis of *rac*-**136** catalyzed by CRL afforded a mixture of acetate (*S*)-**136** and phenol (*R*)-**137**. Subsequent methylation of (*S*)-**136** led to chiral methoxyphosphine oxide (99% ee) and the stereospecific reduction of this compound gave a chiral tertiary phosphine (*R*)-**138** (96% ee). Reactions using CRL and CE exhibited nearly the same enantioselectivity [94] (Schemes 6.61 and 6.62).

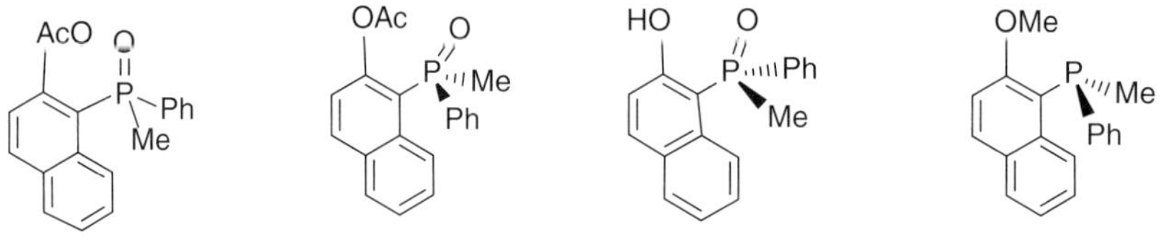

Scheme 6.61 The synthesis of P-chiral tertiary naphthylphosphines **138**.

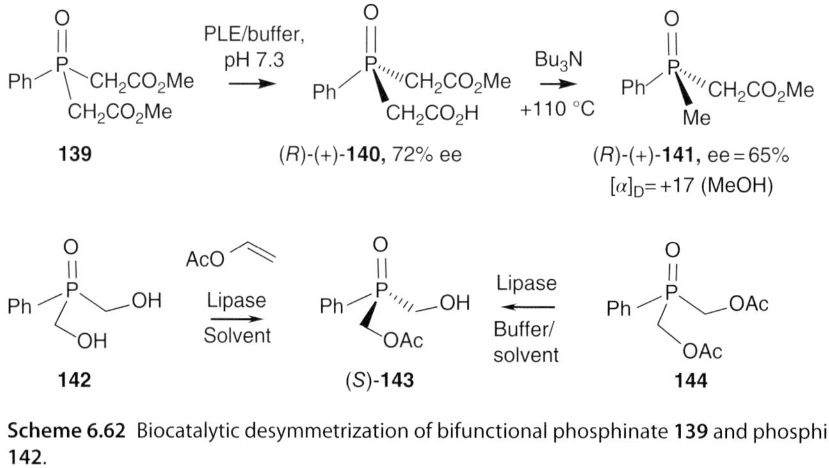

Scheme 6.62 Biocatalytic desymmetrization of bifunctional phosphinate **139** and phosphine oxide **142**.

Mikołajczyk and co-workers [65] studied biocatalytic desymmetrization of bifunctional phosphinates and phosphine oxides. Hydrolysis of bis(methoxycarbonylmethyl)-phenylphosphine oxide **138** was carried out in a phosphate buffer in the presence of pig liver esterase. Chiral monoacetate (*R*)-**139** was isolated by column chromatography in 92% yield and decarboxylated to give a chiral phosphine oxide **140**. The absolute configuration of compound **140** was known; this allowed one to determine the (*R*)-configuration of compound **139**. Biocatalytic acetylation of prochiral bis(hydroxymethyl)-phenylphosphine oxide **141** and biocatalytic hydrolysis of prochiral bis(acetoxymethyl)-phenylphosphine oxide **142** were carried out in the presence of various lipases including PFL, LAH-S, Amano AK, and Amano PS. Among them, the best results were obtained using PFL in chloroform. In this case, chiral compound (*S*)-**143** was formed in 50% yield with 79% ee. Replacement of the solvent (including the use of ionic liquid BMIM + PF_6) and the introduction of various additives had no effect on the selectivity. The absolute configuration of (*S*)-**143** was proved by chemical correlation, by conversion of (*S*)-**143** into tertiary phosphine borane (*R*)-**145** of known absolute configuration (Scheme 6.63). A similar technique was employed for desymmetrization of prochiral 2-(*o*-phosphono)alkylpropane-1,3-diols **146.** Acetylation of diol **146** catalyzed by lipase from *Pseudomonas* sp. led to chiral phosphonates **147** (93–98% ee), which were converted to protected chiral α-amino-*o*-phosphonic acids **148** [58] (Scheme 6.64).

In contrast to P-chiral hydroxymethylphosphine oxides, analogous P-borane complexes appeared to be poor substrates for lipase-catalyzed reactions [73, 95, 96]. The reactions proceeded slowly with low stereoselectivity, probably due to the size difference between the oxygen atom and BH_3 group and to different electronic effects of the P=O and P-BH_3 bonds. Mikolajczyk *et al.* reported biocatalytic resolution of alkoxy-(hydroxymethyl)phenylphosphineboranes **149** using transesterification with vinyl acetate. The reactions were carried out in diisopropyl ether or cyclohexane in the presence of lipases CAL, PSC, Amano AK, Amano PS, and Amano AH. The reaction in diisopropyl ether was slow (10–40 days to complete) and showed a low stereoselectivity (2–20% ee). The reaction in cyclohexane proceeded for 6.21 h, but the selectivity was also low (37% ee) (Scheme 6.65).

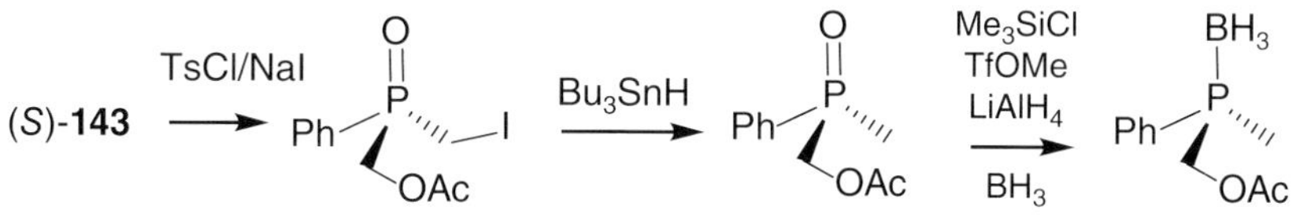

(*R*)-**145**

Scheme 6.63 Determination of absolute configuration of (*S*)-**143** by chemical correlation.

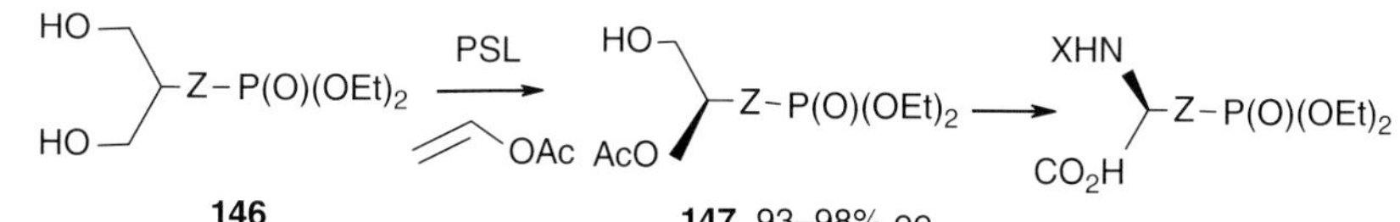

146 **147**, 93–98% ee **148**

Z = CH_2, CH_2CH_2, CH_2CF_3 X = Cbz, Boc

Scheme 6.64 Desymmetrization of prochiral 2-(*o*-phosphono)alkylpropane-1,3-diols **146**.

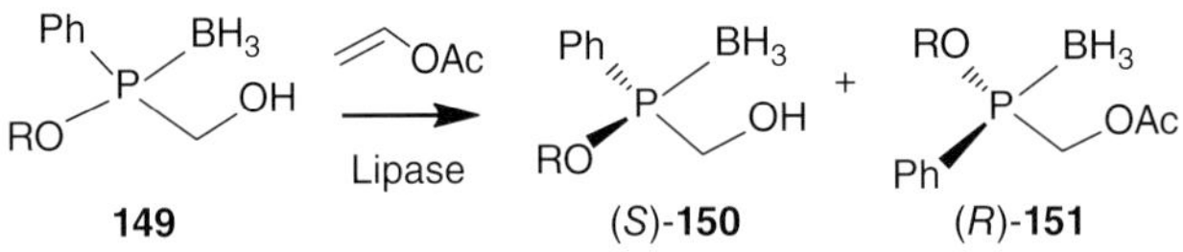

Scheme 6.65 Biocatalytic resolution of alkoxy-(hydroxymethyl)phenylphosphine boranes **149**.

The stereoselectivity of resolution increased in the case of a borane complex with 2-hydroxypropylphosphine **152** containing an asymmetric center in the side chain. A large number of enzymes was studied, namely, Amano R10, Amano AK, Amano AY, CALB, PPL, PFL, lipases from *M. miehei, Candida* sp., *A. niger, Candida lipolytica,* as well as acylase I and proteinase 6. The highest efficiency of resolution was achieved with CALB. For instance, the reaction of racemic (2-hydroxypropyl)diphenylphosphine-borane **152** with vinyl acetate in the presence of CALB gave the (*S*)-enantiomer of alcohol **153** (91% ee) and acetate **152**. The factor of enantioselectivity was 41, thus being reasonably high (Scheme 6.66) [95].

Desymmetrization of prochiral diol **154** was carried out using vinyl acetate as the acylating agent in the presence of various lipases including CAL, PFL, PSC, Amano AK, Amano AH, and Amano PS. Among them, only PLF appeared to be efficient. The nature of solvent had strong effect on the stereoselectivity of the process, namely, replacement of chloroform by diisopropyl ether resulted in an optical antipode of product **155.** This is probably due to different polarities of these solvents because the reaction in chloroform (solvent of lower polarity) showed a higher stereoselectivity (ee of the product was 90%) compared to the reaction in diisopropyl ether (10% ee) [73] (Scheme 6.67).

Wiktelius *et al.* [96] used various lipases (Amano A, Amano PS, CALB (Novozym 435), Amano AK, PPL (Fluka)) for desymmetrization of prochiral phosphine boranes **156** and **157** that gave enantiomers **158**. The best results were obtained using CALB, that is, compounds **158** formed with enantiopurity higher than 98% (Scheme 6.68).

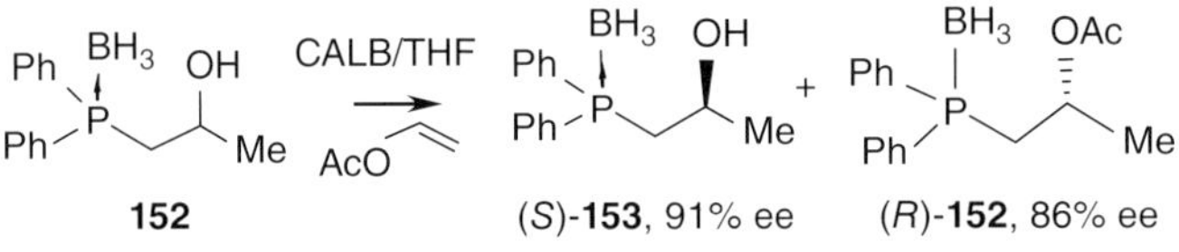

Scheme 6.66 Resolution of racemic **152** with vinyl acetate in the presence of CALB.

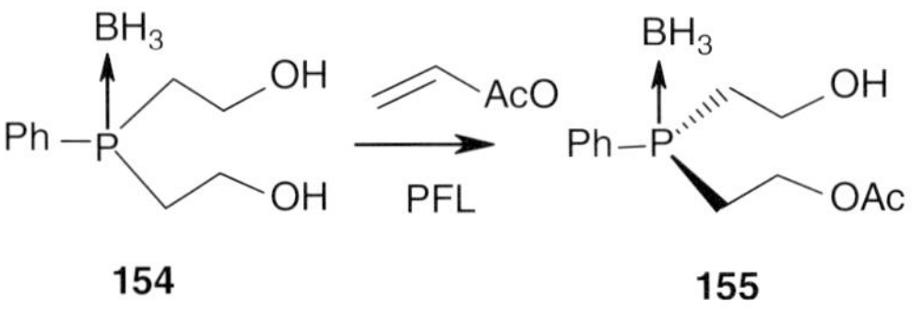

Scheme 6.67 Desymmetization of prochiral diol **154** by transesterification with vinyl acetate catalyzed by PFL.

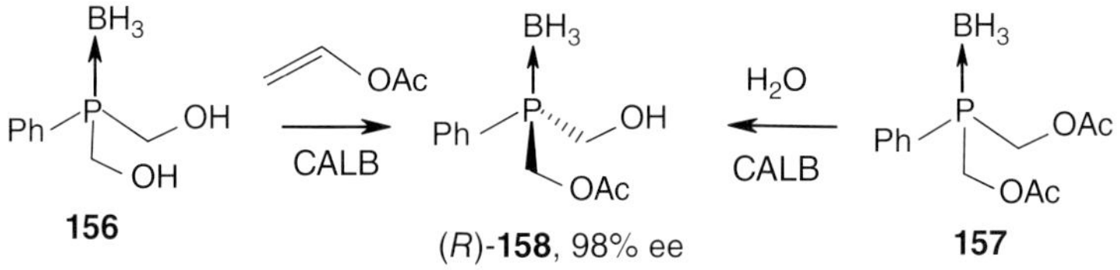

Scheme 6.68 Desymmetrization of prochiral phosphine boranes **156**, **157**.

Thus, lipases that are relatively cheap and experimentally convenient enzymes (especially when immobilized on solid supports) gradually replace microorganisms in everyday laboratory practice. Unfortunately, in the case of racemic organophosphorus compounds, lipases sometimes exhibit low activity and one should still use microorganisms in biocatalysis.

6.5 Microbiological Synthesis of Chiral Organophosphorus Compounds

Living microorganisms that excrete enzymes directly into the reaction medium can be used to carry out biocatalytic reactions in which racemic substrates are transformed to enantiopure compounds. [97–99]. Usually, microbiological catalysis is carried out in aqueous solutions [97], therefore, substrates to be used should be water-soluble. Recently, techniques were developed to allow one to carry out microbiological reactions in nonaqueous media [99]. This is essential for organophosphorus compounds that are often unstable or insoluble in water. Such techniques use two-phase systems, emulsions or anhydrous media. However, in all cases, proper conditions are required for living biocatalysts to be active and vital in organic media. In particular, freeze-drying, and immobilization increases the stability of biocatalysts against poisoning by toxic organic solvents. These techniques were employed for, for example, bioreduction of α-ketophosphonates, which are readily hydrolyzed in water with cleavage of the P–C-bond [100]. Biosynthesis of organophosphorus compounds is carried out using various yeasts, microscopic fungi and bacteria including *Acinetobacter baumanni, B. subtilis, Pseudomonas aeruginosa, P. fluorescens, Rhodococcus* sp., *Serratia liquefaciens* bacteria; *A. niger, Beauveria bassiana, Beauveria brongniartii, Cladosporium* sp. Op328, *Cunninghamella elegans, G. candidum, Penicillium citrinum, Penicillium oxalicum, Verticillium* sp., fungi, as well as *yeasts Saccharomyces cerevisiae, Rhodotorula rubra,* and *Rhodotorula glutinis.* Some of these organisms are sources of biocatalysts that are isolated and used separately. Freeze-dried microorganisms (this form is most convenient for practical use) are commercially available from chemical and biochemical supply companies (Sigma-Aldrich, Fluka, etc.).

6.5.1 Yeast-Catalyzed Synthesis

Baker's yeast (*S. cerevisiae*) is a readily available and versatile catalyst widely used in organic chemistry [98–107]. Ordinary yeast from a grocery store is often well suited for experiments. The best-studied process which is most often carried out in the presence of this biocatalyst is reduction of ketones including phosphorus-containing ones. For

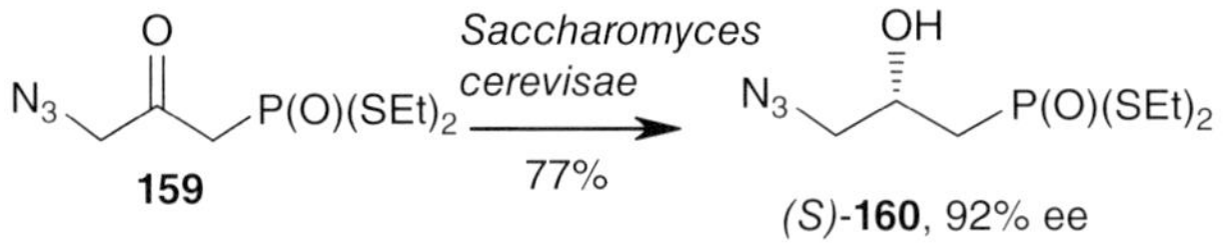

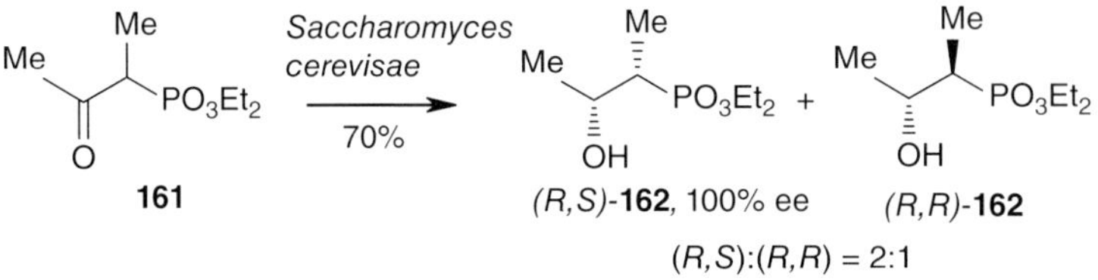

Scheme 6.69 Yeast-catalyzed asymmetric reduction of diethyl 2-oxoalkylphoshonates.

instance, yeast-catalyzed asymmetric reduction of diethyl 2-oxoalkylphoshonates **159** and **161**, respectively leads to 2-hydroxyalkylphosphonates **160** and **162** obtained in good yields with enantiopurity as high as about 97–100% [61–63] (Scheme 6.69). The reduction of phosphonate **161** afforded a mixture of diastereomers in a total yield of 70% and an (*R*,*S*):(*R*,*R*) ratio of 2 : 1.

A similar biocatalyzed reduction of diethyl 2-oxopropylphosphonate **163a** resulted in 2-hydroxyalkylphosphonate **31a** with ee 97% [98]. According to spectroscopic studies, this hydroxyphosphonate exists in a "frozen" conformation owing to the formation of an intramolecular hydrogen bond between the phosphonate oxygen atom and the hydroxyl hydrogen atom (Scheme 6.70).

Reduction of diethyl 2-hydroxy-2-phenylethylphosphonate by baker's yeast. Baker's yeast (50 g) was suspended in water (300 ml), then ketophosphonate (2 mmol) was added and the mixture was stirred for 5 days at 30 °C. The reaction mixture was centrifugated and then extracted with diethyl ether (3 × 30 ml) and with chloroform (2 × 30 ml). The combined extracts were dried over anhydrous Na_2SO_4 and the solvent was removed under reduced pressure. The residue was chromatographed on column with silica gel using ethyl acetate as eluent (colorless oil, 99% ee, 65% yield, $[\alpha]_D^{20}$ +3.0 ($c = 2$, MeOH), δ_P 31 ppm).

The efficiency of biocatalyzed reduction of β-ketophosphonates **161** and **163** using *S. cerevisiae* depends strongly on the nature of the substrate [95, 102]. For example, in the case of compound **161**, the yields decrease if the alkyl group in the immediate vicinity of the carbonyl group creates steric hindrance. Using anaerobic cultivated yeast, one can to some extent overcome this problem and increase the yield of products and the selectivity of reduction [103]. In particular, anaerobic preincubation allowed one to

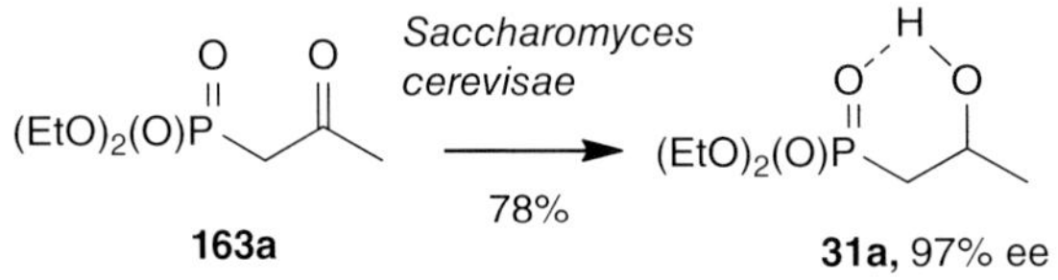

Scheme 6.70 Biocatalyzed reduction of diethyl 2-oxopropylphosphonate.

carry out the biocatalyzed reduction of diethyl-3-oxo-4-methylbutylphosphonate which could not be reduced under normal conditions and to obtain the corresponding hydroxyphosphonate with 85% ee [100, 103]. Microorganisms *R. rubra, R. glutinis, Cladosporium* sp., *Verticillium* sp., and *S. cerevisiae* were used for enantioselective reduction of diethyl-α-ketophosphonates **164** which were rapidly hydrolyzed in water with cleavage of the C–P bond. To suppress hydrolysis, the process was carried out under anhydrous conditions. The reaction catalyzed by freeze-dried cells immobilized on Celite R 630 resulted in α-hydroxyphosphonates [99, 100, 103]. The best results were obtained in the reduction of compounds **164** catalyzed by cells in anhydrous hexane and the products **28** were obtained with enantiopurity as high as 99% [99, 104, 105] (Scheme 6.71). Chiral (*R*)- and (*S*)-diethyl 2-hydroxypropylphosphonates of high enantiopurity were obtained by biocatalyzed reduction of ketophosphonate **163** in the presence of the following microorganisms: *R. rubra, R. glutinis, Rhodotorula gracilis,* and *S. cerevisiae*. Reduction in the presence of *S. cerevisiae* gave only (*S*)-enantiomers, while the reaction in the presence of *R. rubra* afforded (*R*)-hydroxyphosphonate **31**. The products were characterized by ee greater than 90% [107] (Scheme 6.72).

Asymmetric reduction of halogenated diethyl 2-ketoalkylphosphonates **163b,c–g** with dry yeast gave the corresponding diethyl 2-hydroxyphosphonates **31** in good yields with satisfactory enantiopurity (Table 6.4) [100, 108]. Reduction of ketophosphonates **163** was carried out at 30 °C under aerobic conditions. In the case of chemically nonreactive compounds, anaerobic reduction was used. The stereoselectivity of reduction depended on the nature of 3-substituted-2-oxopropylphosphonates. Electron-withdrawing groups in the side chain caused the optical yields of the products to decrease. For instance, the enantiomeric yields and purities of 1-bromo-2-hydroxypropylphosphonates **31** were higher than those of the corresponding chloro derivatives. Also, biocatalyzed reduction of ketophosphonates **163i,j** bearing CF_3 or C_3F_7 groups proceeded with low enantiomeric excesses [109]. Isopropyl groups at the phosphorus atom in compounds **163h,k** also hampered the process. 3-Substituted-2-hydroxyalkylphosphonates **31** thus obtained were used as chiral reagents for the synthesis of phosphorus-containing analogs of bioactive molecules such as P-GABOB and others [98, 109]. The addition of ethyl chloroacetate or methyl

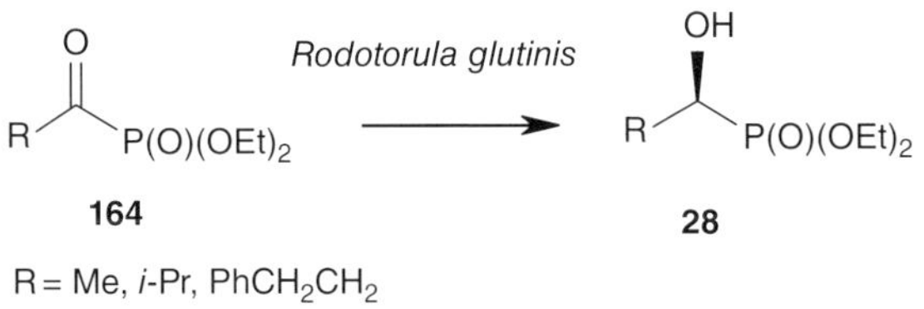

Scheme 6.71 Bioreduction of β-ketophosphonate by *Saccharomyces cerevisie.*

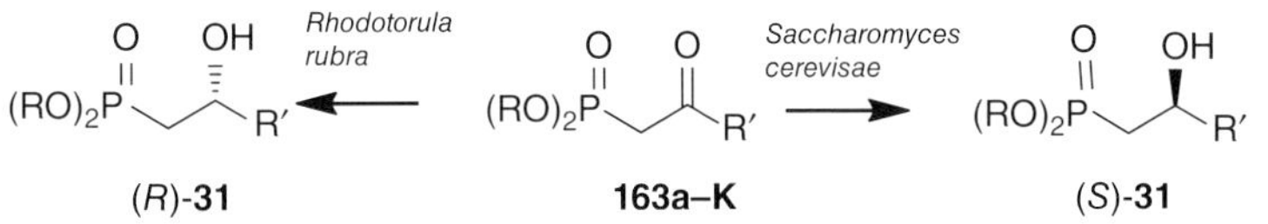

Scheme 6.72 Biocatalyzed reduction of ketophosphonate by *Rhodotorula rubra* and *Saccharomyces cerevisiae* (Table 6.4).

Table 6.4 Bioreduction of β-ketophosphonates **163** with dry yeast.

163	R	R′	Yield (%)	ee (%)	Configuration
a	Et	Me	70	95	*S*
b	Et	CH_2Cl	82	72	*R*
c	Me	CH_2Cl	74	70	*R*
d	Pr-i	CH_2Cl	57	13	—
e	Bu	CH_2Cl	88	70	*R*
f	Et	CH_2Br	35	83	*R*
g	Pr-*i*	CH_2Br	41	52	—
h	Bu	CH_2Br	55	87	*R*
i	Et	CF_3	86	52	—
j	Et	C_3F_7	55	20	—
k	Et	CH_2N_3	77	92	*S*

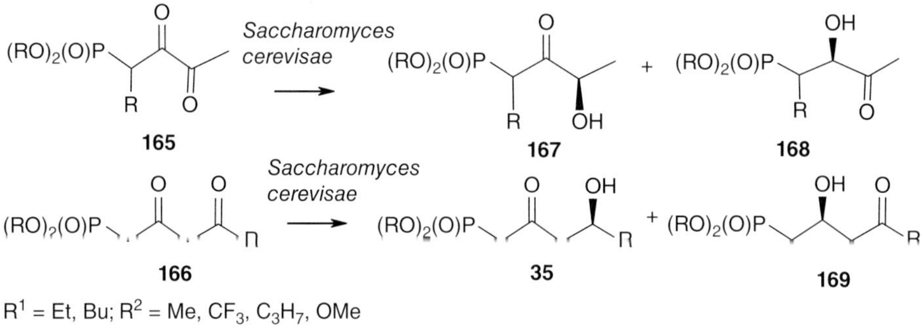

Scheme 6.73 Bioreduction of dioxoalkylphosphonates.

vinyl ketone to the baker's yeast reducing medium also led to changes in the absolute configuration of the product, namely, the (*R*)-stereoisomer of 2-hydroxyphosphonate formed. With no additives, (*S*)-2-hydroxyphosphonate formed with 99% ee. Bioreduction of dioxoalkylphosphonates **165** and **166** containing two C=O groups by baker's yeast gives mixtures of isomers of hydroxy(keto)alkylphosphonates **167**, **168** and **35**, **169**, respectively, in reasonable chemical yields, the optical yields ranging from moderate to good (from 80% ee to 92–94% ee). Some of these compounds were isolated as individual substances and used as chiral Horner–Wadsworth–Emmons reagents [102, 109] (Scheme 6.73).

Mikołajczyk *et al.* [37] developed a method for chemoenzymatic synthesis of P-carnitine enantiomers using microbiological reduction in the key step of the process (Scheme 6.74). P-carnitine was synthesized from diethyl 2-oxo-3-chloropropylphosphonate **163b** in three steps including reduction, additional enzymatic purification of enantioenriched phosphonates **31e** and **32e**, transformation of compound **31e** to phosphonic acid, and the final reaction with trimethylamine resulting in trimethylammonium salt (*R*)-**56**. This technique was used for the preparation of

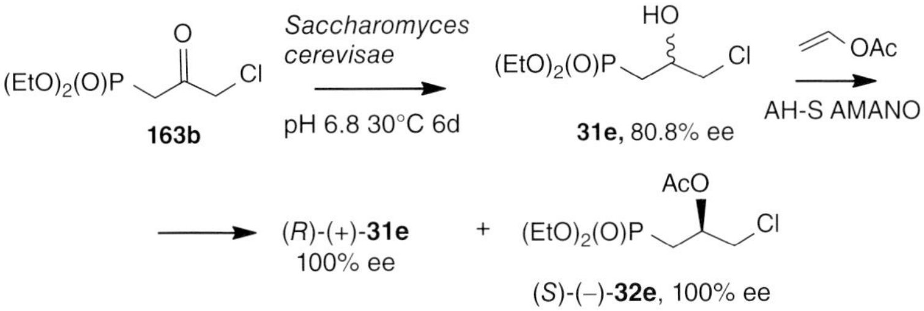

Scheme 6.74 Reduction of diethyl-2-ketoalkylphosphonates with yeast.

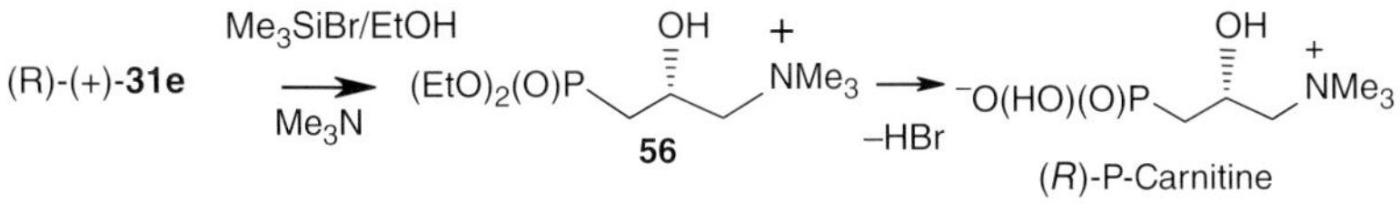

Scheme 6.75 Chemoenzymatic synthesis of P-carnitine.

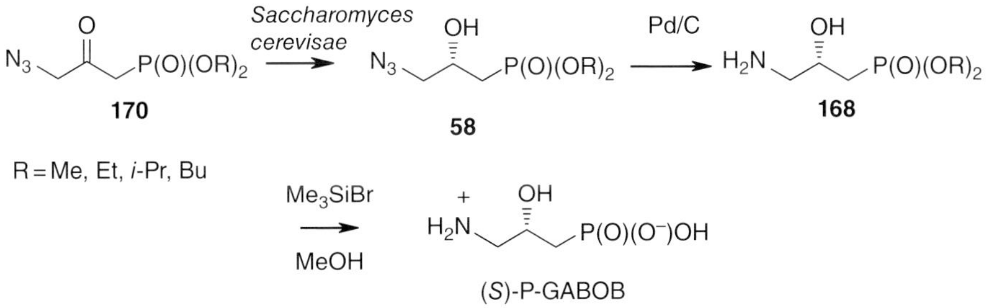

Scheme 6.76 Enantioselective synthesis of phosphono-GABOB.

enantiopure phosphonates (*R*)-(+)-**31e** and (*S*)-(−)-**32e**, which were then converted to P-carnitine (Scheme 6.75) and for the preparation of P-GABOB (Scheme 6.76). Subsequent reduction of azide (*S*)-**58** with hydrogen in the presence of palladium afforded 3-amino-2-hydroxypropylphosphonate (*S*)-**168**, which was treated with bromotrimethylsilane and methanol to form P-GABOB in high yield and with high ee.

Enantiopure hydroxy(chloro)alkylphosphonates **31e** and **59**, which were obtained by the microbiological method, were used in the synthesis of fosfomycin precursors. Treatment of chiral compounds (*R*)-**31e** and (*R,S*)-**59** with potassium carbonate in THF led to chiral epoxyphosphonates (*S*)-**171** and (*S*)-**172** with satisfactory enantiopurity (Scheme 6.77) [19, 103].

Attolini *et al.* [43, 110] carried out enantioselective reduction of cyclic dialkyl 3-oxoalk-1-enylphosphonates **173a–c** with various types of yeast, namely, fermented yeast, freeze-dried dry yeast, as well as acetone extracts from dry powdered yeast. Reduction of the six- and seven-membered enones **173b,c** with fermented yeast led to the corresponding phosphonates **44b,c** in satisfactory yields with enantiomeric excesses ranging from moderate to good. At the same time, reduction of five-membered enonephosphonate **173a** involved the C=C double bond and gave a cyclic ketone **174a** with low ee. Variation of the reaction conditions did not considerably increase the

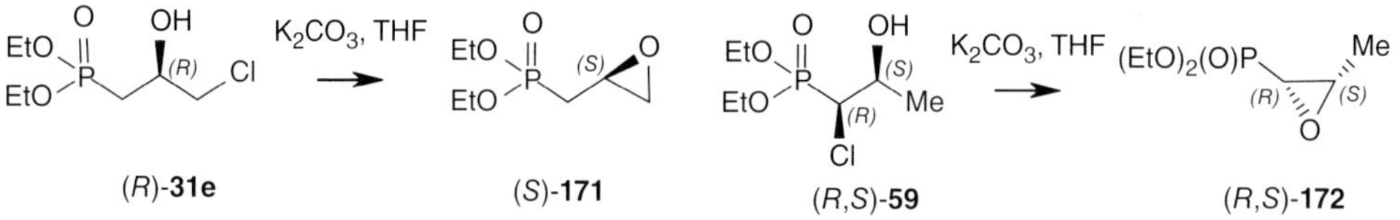

Scheme 6.77 Synthesis of fosfomycin precursors.

optical purity of reaction products. For the six-membered rings, it was found that bulky substituents at the phosphorus atom increase the enantioselectivity of the reduction to 95% ee. Reduction of six-membered enonphosphonates **173b** with acetone yeast extract increased the yields of the reduction products **44b**, although their optical purity remained almost the same. Reduction of cyclic enones with dry yeast in organic solvents did not increase the product yields and enantioselectivity of the reaction. As the volume of the alkyl substituent at the phosphorus atom in the six-membered compounds **173b** increased on going from Me to Et and *i*-Pr, the optical yields of compounds **44b** increased from 45% to 95%. The (*S*)-absolute configurations of the six-membered hydroxyphosphonates **44b** were determined by chemical correlations (Scheme 6.78).

The application of *Rhodospirillum toruloides* strain allowed resolving the chemically synthesized racemic mixtures of the following chiral aminophosphonic acids: 1-aminoethylphosphonic acid (1), 1-amino-1-iso-propyl-1-phosphonic acid (2), 1-amino-1-phenylmethylphosphonic acid (4), and 1-amino-2-phenylethylphosphonic acid (3). The applied protocols resulted in obtaining pure (*R*)-1-aminoethylphosphonic acid (100% of ee) and enantiomerically enriched mixtures of other phosphonates (73% ee of (*S*)-1-amino-1-phenylmethylphosphonic acid, 51% ee of (*R*)-1-amino-2-phenylethylphosphonic acid, and 40% ee of (*S*)-1-amino-2-methylpropylphosphonic acid) [111].

6.5.2 Synthesis Using Unicellular Fungi

G. candidum (milk mold) is a plant pathogenic fungus available as two different strains, IFO 4597 and IFO 5767, usually exhibit high stereoselectivity in the reduction of ketones of different structure. However, the reduction of 2-oxoalkylphosphonates with these fungi was efficient only for diethyl 2-oxopropylphosphonate **163** ($R^1 = Me$, $R^2 = Et$) which was converted to diethyl (+)-(*R*)-2-hydroxypropylphosphonate **31a** under the action of *G. candidum* with 98% ee in satisfactory yield. The ketophosphonates **163** containing larger substituents either were reduced in low yields or did not react at all (Scheme 6.79) [35].

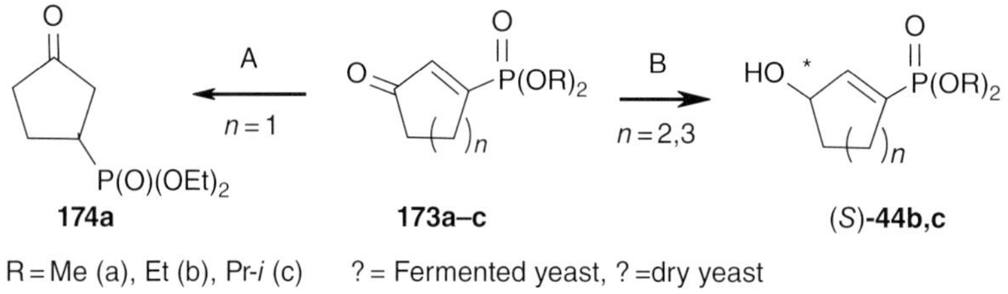

Scheme 6.78 Enantioselective reduction of cyclic dialkyl (3-keto-1-alkenyl) phosphonates.

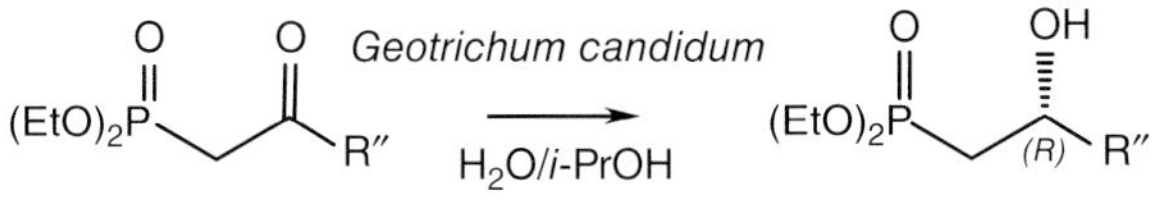

163

R′=Et, Me; R″=Me, Bu, Ph, $(CH_2)_3Ph$

(*R*)-**31a**, 50–90% ee

Scheme 6.79 Bioreduction of 2-oxoalkylphosphonates with fungi *Geotrichum candidum*.

Hydrolytic oxirane ring opening in substituted 1,2-epoxyphosphonates **175** was carried out under the action of microbial cell cultures [100, 112, 113]. In contrast to chemical hydrolysis, phosphorus-containing oxiranes mainly converted to *erythro*-1,2-dihydroxyphosphonates containing small amounts of *threo*-isomers under biocatalysis conditions [113]. For example, chemical hydrolysis of compound *trans*-**175** led to a mixture of *threo*- and *erythro*-stereoisomers **176** in 79% yield with 85 : 15 ratio. At the same time, microbiological hydrolysis by *B. bassiana* fungi gave these stereomers in a total yield of 59% and a 42 : 58 ratio [100, 113]. The *erythro*-isomer was obtained with 100% ee and the *threo*-isomer was obtained with 98% ee as (2*S*)-enantiomer. It was found that this type of reaction is typical of such microorganisms as *A. niger, C. elegans, B. bassiana, B. brongniartii, R. glutinis,* and *Rhodococcus* sp. (Scheme 6.80).

Biotransformation of diethyl 2-oxo-2-phenylethylphosphonate **162b** under the action of *B. bassiana* 271B, *R. rubra* 70403, *R. glutinis* 10134, *G. candidum* 6593, *P. oxalicum,* and *Cladosporium* sp. Op328 strains resulted in the optically active O-phosphorylated 2-hydroxy-2-phenylethylphosphonate **177** with 99% ee and a small amount of the corresponding hydroxyphosphonate **31**. [105, 112]. The biotransformation took 72 h to complete at ambient temperature with vigorous stirring using ethyl acetate as an additive [105]. A similar transformation catalyzed by *Cladosporium* sp. led solely to the (*S*)-stereoisomer of compound **31**. The reaction under the action of *B. bassiana* also gave the target compound (total yield was 85%) even with no additives. To increase the biocatalyst activity, fungi *G. candidum* were grown using chemical additives affecting their activity. For instance, the introduction of ethanol into the reaction mixture

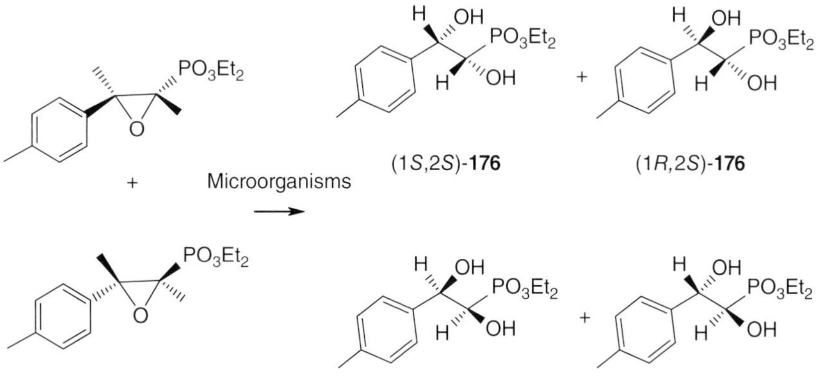

175

Scheme 6.80 Hydrolytic oxirane ring opening of substituted 1,2-epoxyphosphonates **175**.

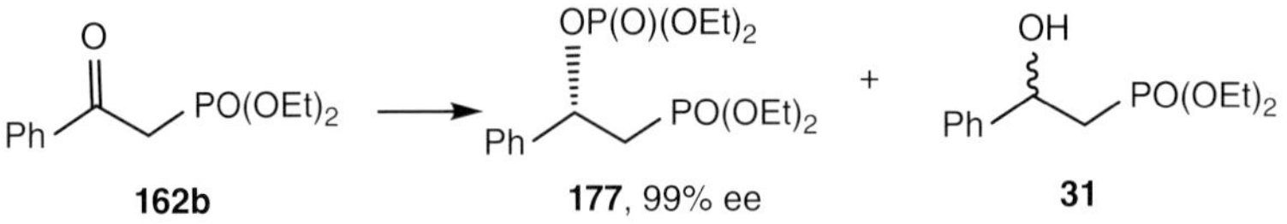

Scheme 6.81 Biotransformation of diethyl 2-oxo-2-phenylethylphosphonate under action of fungi.

increased the yield of compound **177**. Several different microorganisms, namely, *P. oxalicum, Aspergillus flavus, G. candidum,* and *Rhodotorula gracillis* were cultivated on dextrose potato agar and then on Chapeck growth medium that provided the conditions favorable for the growth of microorganisms (Scheme 6.81).

Racemic ethyl 1-hydroxybenzyl(phenyl)phosphinate **178** was used as substrate in the fungi-catalyzed process. The racemic mixture contained four stereoisomers, a levorotatory pair $(R_P,R)+(S_P,S)$ and a dextrorotatory pair $(R_P,S)+(S_P,R)$. Preincubation of the biological catalyst (pretreated fungi *G. candidum* with or without chemical additives) in the presence of oxalic acid to improve the efficiency of the biotransformation allowed one to resolve phosphinate **178** into pairs of diastereomers (R_P,R)-**178** and (S_P,R)-**178** with high ee values (>99%) [114] (Scheme 6.82).

Several fungal strains, *B. bassiana, Cuninghamella echinulata, Aspergillus fumigatus, Penicillium crustosum,* and *Cladosporium herbarum* were used as biocatalysts to resolve racemic mixtures of 1-aminoethanephosphonic acid **73** using L/D-amino acid oxidase activity [115]. The best result (42% ee of the (*R*)-isomer) was obtained with a strain of *C. echinulata*. Biotransformations were carried out in a phosphate buffer (pH 6.11). The application of *R. toruloides* strain allowed successful resolving of several racemic mixtures of aminophosphonic acids (Scheme 6.83). The applied protocols resulted in obtaining pure (*R*)-1-aminoethylphosphonic acid (100% ee) and enantiomerically enriched samples of several phosphonates **73** with 40–73% ee [116]

6.5.3 Synthesis Using Bacteria

Kafarski *et al.* [100, 108, 116] successfully used lipolytic bacteria to resolve racemic α-hydroxyalkylphosphonates **181** into enantiomers. For example, hydrolysis of racemic diethyl 1-butyryloxyalkylphosphonates **181** by *P. fluorescens* and *P. citrinum* bacteria resulted in chiral hydroxyphosphonates (*S*)-**182** with good enantioselectivity [104].

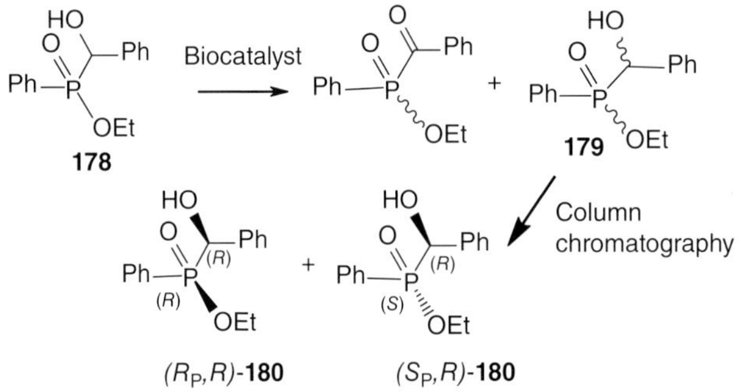

Scheme 6.82 Fungi-catalyzed resolution of ethyl 1-hydroxybenzyl(phenyl)phosphinate **178**.

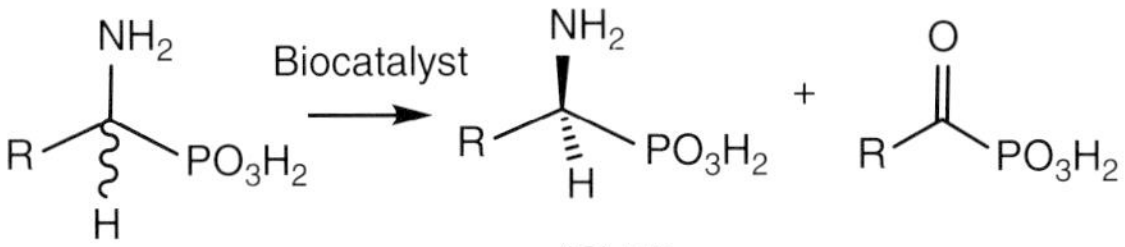

rac-**73** (*S*)-**73**

R = Me, Ph, *i*-Pr, $PhCH_2$

Scheme 6.83 Biocatalytic resolution of racemic aminoalkylphosphonic acid.

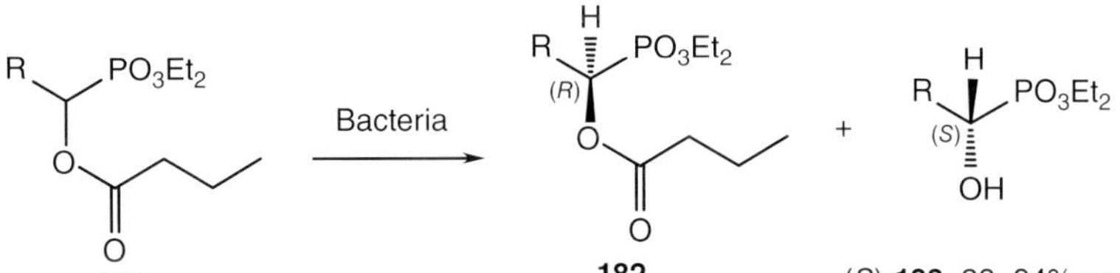

181 **182** (*S*)-**183**, 28–94% ee

Bacteria = *Pseudomonas fluorescens, Penicilium citrinum*
R = Et, Pr, Bu^i, Bu, Allyl, XC_6H_4 (X = H, 4-Cl, 4-MeO), $PhCH_2CH_2$, PhCH(Me)-, 2-Furyl

Scheme 6.84 Microbiological synthesis of chiral hydroxyphosphonates using bacteria.

P. citrinum bacteria hydrolyzed substrates containing aliphatic substituents in α-position, whereas *P. fluorescens* bacteria were more efficient in the hydrolysis of substrates with aromatic substituents. The degree of substrate conversion increased with time and this was accompanied by a decrease in the enantiopurity of the products (Scheme 6.84).

Bacteria appeared to be efficient in the resolution of diastereomeric mixtures of hydroxyphosphinates containing two asymmetric centers. For example, racemic phosphinate esters **184** (mixtures of four diastereomers) were at first hydrolyzed using four types of bacteria, namely, *B. subtilis, A. baumanni, S. liquefaciens,* and *P. aeruginosa* [108]. These bacteria predominantly hydrolyzed diastereomers having (*S*)-absolute configuration of the C^{α} atom. The best results (90% ee for products) were obtained with *B. subtilis.* Thus four diastereomers, namely, (S_P,*S*)-**185**, (R_P,*R*)-**186**, (R_P,*S*)-**185**, and (S_P,*R*)-**186**, gave two pairs of diastereomers, which were separated by column chromatography. Then, diastereoenriched acyloxyphosphinates **10** were hydrolyzed again with *B. subtilis.* The diastereomeric pairs thus obtained were purified by preparative high performance liquid chromatography (HPLC). As a result, some diastereomers, for example, hydroxyphosphonate ($R_{P,}$*S*)-**185** (R = Ph), were isolated in optically pure form. The absolute configuration of this compound was determined by X-ray analysis (Scheme 6.85) [108].

P. fluorescens bacteria and *P. citrinum* mold fungi were also used for kinetic resolution of 1-butoxyphosphonates to obtain optically active diethyl (*S*)-α-hydroxyalkylphosphonates in moderate to high yields with satisfactory optical purity [111, 117]. Using various strains of bacteria (*A. fumigatus, Proteus vulgaris*) and fungi *Actinomycetes*, *cis*-1,2-epoxypropylphosphonic acid was successfully resolved into (*S*)- and (*R*)-enantiomers (Scheme 6.84) [107, 108]. Recently Żymańczyk-Duda proposed Cyanobacteria as biocatalysts for the reduction of 2-oxoalkylphosphonates. Among the tested blue-green algae, only the applying of *Arthrospira maxima* and *Nodularia*

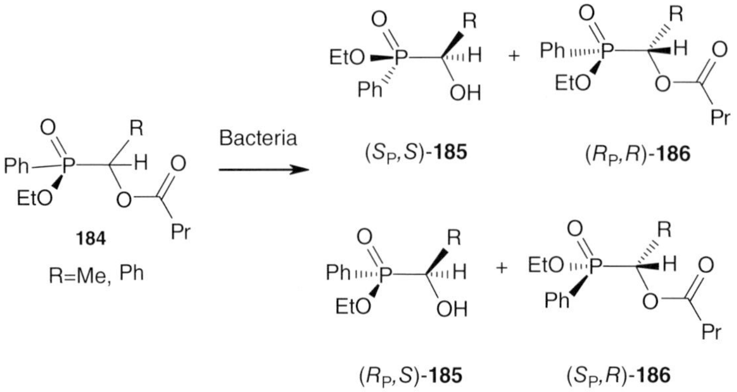

Scheme 6.85 Hydrolysis of phosphinate diastereomers by bacteria.

sphaerocarpa cultures in bioconversion of diethyl 2-oxopropylphosphonate allowed obtaining the corresponding diethyl 2-hydroxypropylphosphonate. The degree of conversion of the substrate was 26.4% and the optical purity of the product was over 99% [111].

6.6 Summary

We considered various procedures for enzymatic kinetic resolution and some other techniques that allow one to obtain individual enantiopure organophosphorus in high optical yield. It should be noted that, in spite of impressive achievements in the synthesis and research on the properties of chiral organophosphorus, some problems still persist. The search for enantioselective methods providing easy access to optically active phosphorus-containing acids, tertiary phosphines, and phosphine oxides remains topical. In this connection, elaboration of highly efficient methods of enzymatic and microbiological synthesis of chiral organophosphorus compounds is of particular importance. Currently, lipases are widely used to obtain chiral compounds for compounds chemistry. The possibility of carrying out enzymatic processes with high selectivity makes lipases very attractive catalysts, especially for transformations that cannot be performed under other conditions. We have demonstrated versatile applications of biocatalysts, for example, the possibility to obtain various biologically active organophosphorus compounds in organic solvents. Most biocatalytic reactions can be carried out in health- and environmentally safe conditions. The advantages of biocatalysis over chemical processes are due to its high stereoselectivity and applicability at ambient temperature and pressure. Enzymes can catalyze various types of chemical reactions, thus providing access to a broad spectrum of chiral substances. Therefore, combining biocatalysis and asymmetric synthesis will give an impetus to the design of novel synthetic strategies of practical and theoretical importance [118–120].

References

1 Faber, K., Fessner, W.-D., and Turner, N.J. (eds) (2015) *Biocatalysis in Organic Synthesis* (3 volumes), Thieme, Stuttgart.

2 Liese, A., Seelbach, K., and Wandrey, C. (eds) (2006) *Industrial Biotransformations*, 2nd edn, Wiley-VCH Verlag GmbH, Weinheim.
3 Gotor, V., Alfonso, I., and Garcıa-Urdiales, E. (eds) (2008) *Asymmetric Organic Synthesis with Enzymes*, Wiley-VCH Verlag GmbH, Weinheim, 325 pp.
4 Bornscheuer, U.T. and Kazlauskas, R.J. (2006) *Hydrolases in Organic Synthesis: Regio- or Stereoselective Biotransformations*, 2nd edn, Wiley-VCH Verlag GmbH, Weinheim.
5 Kolodiazhnyi, O.I. (2011) Enzymatic synthesis of organophosphorus compounds. *Russ. Chem. Rev.*, **80** (9), 883–910.
6 Kielbasinski, P. and Mikolajczyk, M. (2007) in *Future Directions in Biocatalysis* (ed. T. Matsuda), Elsevier, Amsterdam, p. 159.
7 Brzezinska-Rodak, M., Zymanczyk-Duda, E., Kafarski, P., and Lejczak, B. (2002) Application of fungi as biocatalysts for the reduction of diethyl 1-oxoalkylphosphonates in anhydrous hexane. *Biotechnol. Progr.*, **18** (6), 1287–1291.
8 Hammerschmidt, F. (1988) Biosynthese von Naturstoffen mit einer P-C-Bindung, I Einbau von ~D[6,6-D_2]-Glucose in (2-Aminoethy1)phosphonsaure in Tetrahymena thermophile. *Liebigs Ann. Chem.*, (6), 531–535.
9 Natchev, I.A. (1988) Organophosphorus analogues and derivatives of the natural L-amino carboxylic acids and peptides. I. Enzymatic synthesis of D-, DL-, and L-phosphinothricin and their cyclic analogues. *Bull. Chem. Soc. Jpn.*, **61** (11), 3699–3704.
10 Grabley, S. and Sauber, K. (1982) Process for the enzymatic preparation of L-2-amino-4-methylphosphinobutyric acid. US Patent 4389488, https://www.google.ch/patents/US4389488 (accessed 24 May 2016).
11 Goi, H., Miyado, S., Shomura, T., Suzuki, A., Niwa, T., and Yamada, Y. (1980) Process for the optical resolution of d,l-2-amino-4-methylphosphinobutyric acid. US Patent 4226941, Oct. 7, 1980, http://www.google.co.in/patents/US4226941 (accessed 24 May 2016).
12 Hammerschmidt, F. and Wuggenig, F. (1998) Enzymatic synthesis of organophosphorus compounds. *Phosphorus, Sulfur Silicon Relat. Elem.*, **141** (1), 231–238.
13 Kazlauskas, R.J., Weissfloch, A.N.E., Rappaport, A.T., and Cuccia, L.A. (1991) A rule to predict which enantiomer of a secondary alcohol reacts faster in reactions catalyzed by cholesterol esterase, lipase from *Pseudomonas cepacia*, and lipase from *Candida rugosa*. *J. Org. Chem.*, **56** (8), 2656–2665.
14 Whittall, J. and Sutton, P. (2010) *Practical Methods for Biocatalysis and Biotransformations*, John Wiley & Sons, Ltd, Chichester, 432 pp.
15 Otera, J. (1993) Transesterification. *Chem. Rev.*, **93** (4), 1449–1470.
16 Khushi, T., O'Toole, K.J., and Sime, J.T. (1993) Biotransformation of phosphonate esters. *Tetrahedron Lett.*, **34** (14), 2375–2378.
17 Zhang, Y., Yuan, C., and Li, Z. (2002) Kinetic resolution of hydroxyalkanephosphonates catalyzed by *Candida antarctica* lipase B in organic media. *Tetrahedron*, **58** (15), 2973–2978.
18 Kolodyazhna, O.O. and Kolodyazhnyi, O.I. (2010) Biocatalytic separation of α-hydroxyphosphonates with lipase of *Burkholderia cepacia*. *Russ. J. Gen. Chem.*, **80** (8), 1718–1719.
19 Wang, K., Zhang, Y., and Yuan, C. (2003) Enzymatic synthesis of phosphocarnitine, phosphogabob and fosfomycin. *Org. Biomol. Chem.*, **1** (20), 3564–3569.

20 Shioji, K., Tashiro, A., Shibata, S., and Okuma, K. (2003) Synthesis of bifunctional *P*-chiral hydroxyl phosphinates; lipase-catalyzed stereoselective acylation of ethy(hydroxyalkyl)-phenylphosphinates. *Tetrahedron Lett.*, **44** (5), 1103–1105.

21 Patel, R.N., Banerjee, A., and Szarka, L.J. (1997) Stereoselective acetylation of [1-(hydroxy)-4-(3-phenyl)butyl]phosphonic acid, diethyl ester. *Tetrahedron: Asymmetry*, **8** (7), 1055–1059.

22 Yokomatsu, T., Nakabayashi, N., Matsumoto, K., and Shibuya, S. (1995) Lipase-catalyzed kinetic resolution of cis-1-diethylphosphonomethyl-2-hydroxymethylcyclo-hexane. Application to enantioselective synthesis of 1-diethylphosphonomethyl-2-(5′-hydantoinyl)cyclohexane. *Tetrahedron: Asymmetry*, **6** (12), 3055–3062.

23 Spangler, L.A., Mikołajczyk, M., Burdge, E.L., Kiełbasinski, P., Smith, H.C., Łyzwa, P., Fisher, J.D., and Omelanczuk, J. (1999) Synthesis and biological activity of enantiomeric pairs of phosphosulfonate herbicides. *J. Agric. Food. Chem.*, **47** (1), 318–321.

24 Kolodiazhna, O.O., Kolodiazhna, A.O., and Kolodiazhnyi, O.I. (2013) Enzymatic preparation of (1S,2R)- and (1R,2S)-stereoisomers of 2-halocycloalkanols. *Tetrahedron: Asymmetry*, **24** (1), 37–42.

25 Drescher, M., Hammerschmidt, F., and Kiihlig, H. (1995) Enzymes in organic chemistry; part 3:1 enantioselective hydrolysis of 1-acyloxyalkylphosphonates by lipase from *Aspergillus niger* (Lipase AP 6). *Synthesis*, **10**, 1267–1272.

26 Hammerschmidt, F. and Wuggenig, F. (1999) Enzymes in organic chemistry. Part 9:[1]. Chemoenzymatic synthesis of phosphonic acid analogues of l-valine, l-leucine, l-isoleucine, l-methionine and l-α-aminobutyric acid of high enantiomeric excess. *Tetrahedron: Asymmetry*, **10** (9), 1709–1721.

27 Rowe, B.J. and Spilling, C.D. (2001) The synthesis of 1-hydroxy phosphonates of high enantiomeric excess using sequential asymmetric reactions: titanium alkoxide-catalyzed P-C bond formation and kinetic resolution. *Tetrahedron: Asymmetry*, **12** (12), 1701–1708.

28 Majewska, P., Doskocz, M., Lejczak, B., and Kafarski, P. (2009) Enzymatic resolution of α-hydroxyphosphinates with two stereogenic centres and determination of absolute configuration of stereoisomers obtained. *Tetrahedron: Asymmetry*, **20** (13), 1568–1574.

29 Majewska, P. and Kafarski, P. (2006) Simple and effective method for the deracemization of ethyl 1-hydroxyphosphinate using biocatalysts with lipolytic activity. *Tetrahedron: Asymmetry*, **17** (20), 2870–2875.

30 Hammerschmidt, F., Lindner, W., Wuggenig, F., and Zarbl, E. (2000) Enzymes in organic chemistry. Part 10:1 chemo-enzymatic synthesis of l-phosphaserine and l-phosphaisoserine and enantioseparation of amino-hydroxyethylphosphonic acids by non-aqueous capillary electrophoresis with quinine carbamate as chiral ion pair agent. *Tetrahedron: Asymmetry*, **11** (14), 2955–2964.

31 Yamagishi, T., Miyamae, T., Yokomatsu, T., and Shibuya, S. (2004) Lipase-catalyzed kinetic resolution of α-hydroxy-H-phosphinates. *Tetrahedron Lett.*, **45** (36), 6713–6716.

32 Pàmies, O. and Bäckvall, J.E. (2002) Enzymatic kinetic resolution and chemoenzymatic dynamic kinetic resolution of delta-hydroxy esters. An efficient route to chiral delta-lactones. *J. Org. Chem.*, **67** (4), 1261–1265.

33 Pamies, O. and Backvall, J.E. (2003) Combination of enzymes and metal catalysts. A powerful approach in asymmetric catalysis. *Chem. Rev.*, **103** (8), 3247–3262.
34 Pamies, O. and Backvall, J.E. (2003) An efficient route to chiral α- and β–hydroxyalkanephosphonates. *J. Org. Chem.*, **68** (12), 4815–4818.
35 Zurawinski, R., Nakamura, K., Drabowicz, J., Kiełbasinskia, P., and Mikołajczyk, M. (2001) Enzymatic desymmetrization of 2-amino-2-methyl-1,3-propanediol: asymmetric synthesis of (S)-N-Boc-N,O-isopropylidene-α-methylserinal and (4R)-methyl-4-[2-(thiophen-2-yl)ethyl]oxazolidin-2-one. *Tetrahedron: Asymmetry*, **12** (22), 3139–3142.
36 Attolini, M., Iacazio, G., Peiffer, G., and Maffei, M. (2002) Enzymatic resolution of diethyl (3-hydroxy-1-butenyl) phosphonate. *Tetrahedron Lett.*, **43** (47), 8547–8549.
37 Mikołajczyk, M., Łuczak, J., and Kiełbasinski, P. (2002) Chemoenzymatic synthesis of phosphocarnitine enantiomers. *J. Org. Chem.*, **67** (22), 7872–7875.
38 Zhang, Y., Li, Z., and Yuan, C. (2002) *Candida rugosa* lipase-catalyzed enantioselective hydrolysis in organic solvents. Convenient preparation of optically pure 2-hydroxy-2-arylethanephosphonates. *Tetrahedron Lett.*, **43** (17), 3247–3249.
39 Xu, C. and Yuan, C. (2004) Enzymatic synthesis of optically active α-chloro-δ-hydroxy-β-ketoalkanephosphonates and reactions thereof. *Tetrahedron*, **60** (12), 3883–3892.
40 Zhang, Y., Li, J., and Yuan, C. (2003) Enzymatic synthesis of optically active trifluoromethylated 1- and 2-hydroxyalkanephosphonates. *Tetrahedron*, **59** (4), 473–479.
41 Yuan, C., Li, J., and Zhang, W. (2006) A facial chemoenzymatic method for the preparation of chiral1,2-dihydroxy-3,3,3,-trifluoropropanephosphonates. *J. Fluorine Chem.*, **127** (1), 44–47.
42 Zhang, Y., Xu, C., Li, J., and Yuan, C. (2003) Enzymatic synthesis of optically active δ-hydroxy-β-ketoalkanephosphonates. *Tetrahedron: Asymmetry*, **14** (1), 63–70.
43 Attolini, M., Iacazio, G., Peiffer, G., Charmassona, Y., and Maffei, M. (2004) Enzymatic resolution of diethyl 3-hydroxycycloalkenyl phosphonates. *Tetrahedron: Asymmetry*, **15** (5), 827–830.
44 Woschek, A., Lindner, W., and Hammerschmidt, F. (2003) Enzymes in organic chemistry, 11. Hydrolase-catalyzed resolution of α- and β-hydroxyphosphonates and synthesis of chiral, non-racemic β-aminophosphonic acids. *Adv. Synth. Catal.*, **345** (12), 1287–1298.
45 Drescher, M. and Hammerschmidt, F. (1997) The cause of the rate acceleration by diethyl ether solutions of lithium perchlorate (LPDE) in organic reactions. Application to high pressure synthesis. *Tetrahedron*, **53** (13), 4627–4648.
46 Drescher, M., Li, Y.F., and Hammerschmidt, F. (1995) Enzymes in organic chemistry, part 2: lipase-catalysed hydrolysis of 1-acyloxy-2-arylethylphosphonates and synthesis of phosphonic acid analogues of L-phenylalanine and L-tyrosine. *Tetrahedron*, **51** (14), 4933–4946.
47 Wuggenig, F. and Hammerschmidt, F. (1999) Chemoenzymatic synthesis of α-aminophosphonic acids. *Phosphorus, Sulfur Silicon Relat. Elem.*, **147** (1), 439.
48 Hammerschmidt, F. and Wuggenig, F. (1998) Enzymes in organic chemistry VI [1]. Enantioselective hydrolysis of 1-chloroacetoxycycloalkylmethylphosphonates with lipase AP 6 from *Aspergillus niger* and chemoenzymatic synthesis of chiral, nonracemic 1-aminocyclohexylmethylphosphonic acids. *Monatsh. Chem.*, **129** (1), 423–436.

49 Huerta, F.F., Minidis, A.B.E., and Backvall, J.E. (2001) Racemisation in asymmetric synthesis. Dynamic kinetic resolution and related processes in enzyme and metal catalysis. *Chem. Soc. Rev.*, **30** (6), 321–331.
50 Solodenko, V.A. and Kukhar, V.P. (1989) Stereoselective papain-catalyzed synthesis of alafosfalin. *Tetrahedron Lett.*, **30** (49), 6917–6918.
51 Solodenko, V.A., Belik, M.Y., Galushko, S.V., Kukhar, V.P., Kozlova, E.V., Mironenko, D.A., and Svedas, V.K. (1993) Enzymatic preparation of both L- and D-enantiomers of phosphonic and phosphonous analogues of alanine using *Penicillin acylase. Tetrahedron: Asymmetry*, **4** (9), 1965–1968.
52 Solodenko, V.A., Kasheva, T.N., Kukhar, V.P., Kozlova, E.V., Mironenko, D.A., and Svedas, V.K. (1991) Preparation of optically active 1-aminoalkylphosphonic acids by stereoselective enzymatic hydrolysis of racemic N-acylated 1-aminoalkylphosphonic acids. *Tetrahedron*, **47** (24), 3989–3998.
53 Yuan, C., Xu, C., and Zhang, Y. (2003) Enzymatic synthesis of optically active 1- and 2-aminoalkanephosphonates. *Tetrahedron*, **59** (32), 6095–6102.
54 Natchev, I.A. (1989) Total synthesis and enzyme-substrate interaction of D-, DL-, and L-phosphinotricine, 'bialaphos'(SF-1293) and its cyclic analogues. *J. Chem. Soc., Perkin Trans. 1*, (1), 125–131.
55 Schweifer, A. and Hammerschmidt, F. (2008) On the conversion of structural analogues of (S)-2-hydroxypropylphosphonic acid to epoxides by the final enzyme of fosfomycin biosynthesis in S. fradiae. *Bioorg. Med. Chem. Lett.*, **18** (10), 3056–3059.
56 Hammerschmidt, F. (1993) Biosynthesis of natural products with a P-C bond: incorporation of d-[l-H1]glucose into 2-aminoethylphosphonic acid in tetmmena thermophila and of d-[l-Hl]glucose and l~methionine into fosfomycin in Sreptomyces fradiae. *Phosphorus, Sulfur Silicon Relat. Elem.*, **76** (1-4), 111–114.
57 McQueney, M.S., Lee, S., Swartz, W.H., Ammon, H.L., Mariano, S., and Dunaway-Mariano, D. (1991) Evidence for an intramolecular, stepwise reaction pathway for PEP phosphomutase catalyzed phosphorus-carbon bond formation. *J. Org. Chem.*, **56** (25), 7121–7130.
58 Yokomatsu, T., Sato, M., and Shibuya, S. (1996) Lipase-Catalyzed Enantioselective Acylation of Prochiral 2-(coPhosphono)alkyl-1,3-Propanediols: Application to the Enantioselective Synthesis of co-Phosphono-a-Amino Acids. *Tetrahedron: Asymmetry*, **7** (9), 2743–2754.
59 Hammerschmidt, F. (1994) Einbau von L-[Methyl-2H3]methionin und 2-[Hydroxy-^{18}O]hydroxyethylphosphonsaure in Fosfomycin in *Streptomyces fradiae*–ein ungewohnlicher Methyltransfer. *Angew. Chem.*, **106** (3), 334–335.
60 Hammerschmidt, F. (1991) Biosynthesis of natural products with a P–C bond. Part 8: on the origin of the oxirane oxygen atom of fosfomycin in *Streptomyces fradiae. J. Chem. Soc., Perkin Trans. 1*, (8), 1993–1996.
61 Freeman, S., Seidel, H.M., Schwalbe, C.H., and Knowles, J.R. (1989) Phosphonate biosynthesis: the sterochemical course of phosphoenolpyruvate phosphomutase. *J. Am. Chem. Soc.*, **111** (26), 9233–9234.
62 Hidaka, T. and Seto, H. (1989) Biosynthesis mechanism of carbon-phosphorus bond formation. Isolation of carboxyphosphonoenolpyruvate and its conversion to phosphinopyruvate. *J. Am. Chem. Soc.*, **111** (20), 8012–8013.

63 Hammerschmidt, F. (1988) Synthese der (R)- und (S)-(2-Amino[2-D1]ethyl) phosphonsaure und Hydroxylierung zu (2-Amino-1-hydroxyethyl) phosphonsaure in *Acanthamoeba castellanii* (Neff). *Liebigs Ann. Chem.*, (10), 961–964.

64 Hammerschmidt, F. (1991) Markierte Vertreter eines m6glichen Zwischenprodukts der Biosynthese yon Fosfomycin in *Streptomyces fradiae*: Darstellung yon (R,S)-(2-Hydroxypropyl)-, (R,S)-, (R)-,(S)-(2-Hydroxy-ll,l-2H_2]propyl)-und (R,S)-(2-[^{18}O] Hydroxypropyl)phosphonsaiure. *Monatsh. Chem.*, **122**, 389–398.

65 Kiełbasinski, P., Zurawinski, R., Albrycht, M., and Mikołajczyk, M. (2003) The first enzymatic desymmetrizations of prochiral phosphine oxides. *Tetrahedron: Asymmetry*, **14**, 3379–3384.

66 Woschek, A., Wuggenig, F., Peti, W., and Hammerschmidt, F. (2002) On the transformation of (S)-2-hydroxypropylphosphonic acid into fosfomycin in *Streptomyces fradiae*—A unique method of epoxide ring formation. *ChemBioChem*, **3** (9), 829–835.

67 Simov, B.P., Wuggenig, F., Lämmerhofer, M., Lindner, W., Zarbl, E., and Hammerschmidt, F. (2002) Indirect evidence for the biosynthesis of (1S,2S)-1,2-epoxypropylphosphonic acid as a co-metabolite of fosfomycin [(1R,2S)-1,2-epoxypropylphosphonic acid] by *Streptomyces fradiae. Eur. J. Org. Chem.*, **2002** (7), 1139–1142.

68 Munos, J.W., Moon, S.J., Mansoorabadi, S.O., Chang, W., Hong, L., Yan, F., Liu, A., and Liu, H.W. (2002) Indirect evidence for the biosynthesis of (1S,2S)-1,2-epoxypropylphosphonic acid as a co-metabolite of fosfomycin [(1R,2S)-1,2-epoxypropylphosphonic acid] by *Streptomyces fradiae. Eur. J. Org. Chem.*, **2002** (7), 1139–1142.

69 Yan, F., Moon, S.J.L., Zhao, Z., Lipscomb, J.D., Liu, A., and Liu, H.W. (2007) Determination of the substrate binding mode to the active site iron of (S)-2-hydroxypropylphosphonic acid epoxidase using ^{17}O-enriched substrates and substrate analogues. *Biochemistry*, **46** (44), 12628–12638.

70 Krasinski, G., Cypryk, M., Kwiatkowska, M., Mikołajczyk, M., and Kiełbasinski, P. (2012) Molecular modeling of the lipase-catalyzed hydrolysis of acetoxymethyl(i-propoxy)phenylphosphine oxide and its *P*-borane analogue. *J. Mol. Graphics Modell.*, **38**, 290–297.

71 Kiełbasiński, P. and Mikołajczyk, M. (2007) *Future Directions in Biocatalysis*, (ed. T. Matsuda) Chapter 9, pp. 159–203.

72 Shioji, K., Kurauchi, Y., and Okuma, K. (2003) Novel synthesis of P-chiral hydroxymethylphosphine–boranes through lipase-catalyzed optical resolution. *Bull. Chem. Soc. Jpn.*, **76** (4), 833–834.

73 Kiełbasinski, P., Albrycht, M., Zurawinski, R., and Mikołajczyk, M. (2006) Lipase-mediated kinetic resolution of racemic and desymmetrization of prochiral organophosphorus *P*-boranes. *J. Mol. Catal. B*, **39** (1-4), 45–49.

74 Kwiatkowska, M., Krasiński, G., Cypryk, M., Cierpiał, T., and Kiełbasiński, P. (2011) Lipase-mediated stereoselective transformations of chiral organophosphorus P-boranes revisited: revision of the absolute configuration of alkoxy(hydroxymethyl) phenylphosphine P-boranes. *Tetrahedron: Asymmetry*, **22** (14-15), 1581–1590.

75 Wiktelius, D., Johansson, M.J., Luthman, K., and Kann, N. (2005) A biocatalytic route to *P*-chirogenic compounds by lipase-catalyzed desymmetrization of a prochiral phosphine-borane. *Org. Lett.*, **7** (22), 4991–4994.

76 Willms, L., Fulling, G., and Keller R. (1998) Process for the enzymatic resolution of 2-amino-4-methyl-phosphinobutyric acid derivatives. US Patent 5879930 A, Feb. 9, 1998, http://www.google.com/patents/US5879930 (accessed 24 May 2016).
77 Yamagishi, T., Mori, J.-I., Haruki, T., and Yokomatsu, T. (2011) A chemo-enzymatic synthesis of optically active 1,1-diethoxyethyl(aminomethyl)phosphinates: useful chiral building blocks for phosphinyl dipeptide isosteres. *Tetrahedron: Asymmetry*, **22**, 1358–1363.
78 Albrycht, M., Kiełbasinski, P., Drabowicz, J., Mikołajczyk, M., Matsuda, T., Harada, T., and Nakamura, K. (2005) Supercritical carbon dioxide as a reaction medium for enzymatic kinetic resolution of *P*-chiral hydroxymethanephosphinates. *Tetrahedron: Asymmetry*, **16** (11), 2015–2018.
79 Shioji, K., Ueyama, T., Ueda, N., Mutoh, E., Kurisaki, T., Wakita, H., and Okuma, K. (2008) Evaluation of enantioselectivity in lipase-catalyzed acylation of hydroxyalkylphosphine oxides. *J. Mol. Catal. B: Enzym.*, **55** (3-4), 146–151.
80 Kiełbasiński, P., Żurawiński, R., Pietrusiewicz, K.M., Zabłocka, M., and Mikołajczyk, M. (1998) Synthesis of *P*-chiral, non-racemic phosphinylacetates via enzymatic resolution of racemates. *Pol. J. Chem.*, **72** (3), 564–572.
81 Li, Y., Aubert, S.D., Maes, E.G., and Raushel, F.M. (2004) Enzymatic resolution of chiral phosphinate esters. *J. Am. Chem. Soc.*, **126**, 8888–8889.
82 Tsai, P.-C., Bigley, A., Li, Y., Ghanem, E., Cadieux, C.L., Kasten, S.A., Reeves, T.E., Cerasoli, D.M., and Raushel, F.M. (2010) Stereoselective hydrolysis of organophosphate nerve agents by the bacterial phosphotriesterase. *Biochemistry*, **49**, 7978–7987.
83 Kim, J., Tsai, P.-C., Chen, S.-L., Himo, F., Almo, S.C., and Raushel, F.M. (2008) Structure of diethyl phosphate bound to the binuclear metal center of phosphotriesterase. *Biochemistry*, **47**, 9497–9504.
84 Li, Y., Aubert, S.D., and Raushel, F.M. (2003) Operational control of stereoselectivity during the enzymatic hydrolysis of racemic organophosphorus compounds. *J. Am. Chem. Soc.*, **125**, 7526–7527.
85 Li, W.S., Li, Y., Hill, C.M., Lum, K.T., and Raushel, F.M. (2002) Enzymatic synthesis of chiral organophosphothioates from prochiral precursors. *J. Am. Chem. Soc.*, **124**, 3498–3499.
86 Nowlan, C., Li, Y., Hermann, J.C., Evans, T., Carpenter, J., Ghanem, E., Shoichet, B.K., and Raushel, F.M. (2006) Resolution of chiral phosphate phosphonate and phosphinate esters by an enantioselective enzyme library. *J. Am. Chem. Soc.*, **128**, 15892–15902.
87 Mikołajczyk, M., Łuczak, J., Kiełbasinski, P., and Colonna, S. (2009) Biocatalytic oxidation of thiophosphoryl compounds: a new chemoenzymatic approach to enantiomeric insecticidal thionophosphates and their oxons. *Tetrahedron: Asymmetry*, **20**, 1948–1951.
88 Kiełbasiński, P., Rachwalski, M., Kwiatkowska, M., Mikołajczyk, M., Wieczorek, W.M., Szyrej, M., Sieroń, L., and Rutjes, F.P.J.T. (2007) Enzyme-promoted desymmetrisation of prochiral bis(cyanomethyl)phenylphosphine oxide. *Tetrahedron: Asymmetry*, **18**, 2108–2112.
89 Kielbasinski, P. and Mikolajczyk, M. (1996) Enzymatic synthesis of chiral, non-racemic phosphoryl compounds. *Phosphorus, Sulfur Silicon Relat. Elem.*, **109-110** (1-4), 497–500.

90 Matsuda, T., Harada, T., Nakamura, K., and Ikariya, T. (2005) Asymmetric synthesis using hydrolytic enzymes in supercritical carbon dioxide. *Tetrahedron: Asymmetry*, **16** (5), 909–915.

91 Baudequin, C., Bregeon, D., Levillain, J., Guillen, F., Plaquevent, J.-C., and Gaumont, A.-C. (2005) Chiral ionic liquids, a renewal for the chemistry of chiral solvents? Design, synthesis and applications for chiral recognition and asymmetric synthesis. *Tetrahedron: Asymmetry*, **16** (24), 3921–3945.

92 Kielbasinski, P., Albrycht, M., Luczak, J., and Mikolajczyk, M. (2002) Enzymatic reactions in ionic liquids: lipase-catalysed kinetic resolution of racemic, *P*-chiral hydroxymethanephosphinates and hydroxymethylphosphine oxides. *Tetrahedron: Asymmetry*, **13**, 735–738.

93 Kielbasinski, P., Omelanczuk, J., and Mikolajczyk, M. (1998) Lipase-promoted kinetic resolution of racemic, *P*-chiral hydroxymethylphosphonates and phosphinates. *Tetrahedron: Asymmetry*, **9** (18), 3283–3287.

94 Serreqi, N.A. and Kazlauskas, R.J. (1994) Kinetic resolution of phosphines and phosphine oxides with phosphorus stereocenters by hydrolases. *J. Org. Chem.*, **59** (25), 7609–7615.

95 Faure, B., Iacazio, G., and Maffei, M. (2003) First enzymatic resolution of a phosphane–borane complex. *J. Mol. Catal. B*, **26** (1-2), 29–33.

96 Wiktelius, D., Johansson, M.J., Luthman, K., and Kann, N. (2005) Gallium(III) chloride-catalyzed double insertion of isocyanides into epoxides. *Org. Lett.*, **7** (26), 4991–4993.

97 Borges, K.B., Borges, W.S., Durán-Patrón, R., Pupo, M.T., Bonato, P.S., and Collado, I.G. (2009) Stereoselective biotransformations using fungi as biocatalysts. *Tetrahedron: Asymmetry*, **20** (4), 385–397.

98 Żymańczyk-Duda, E., Lejczak, B., Kafarski, P., Grimaud, J., and Fischer, P. (1995) Enantioselective reduction of diethyl 2-oxoalkylphosphonates by baker's yeast. *Tetrahedron*, **51** (43), 11809–11814.

99 Zymanczyk-Duda, E. (2008) Biocatalyzed reactions in optically active phosphonate synthesis. *Phosphorus, Sulfur Silicon Relat. Elem.*, **183** (2-3), 369–382.

100 Kafarski, P. and Lejczak, B. (2004) Application of bacteria and fungi as biocatalysts for the preparation of optically active hydroxyphosphonates. *J. Mol. Catal. B*, **29** (1-6), 99–104.

101 Yuan, C.Y., Wang, K., and Li, Z.Y. (2001) Enantioselective reduction of 2-keto-3-haloalkane phosphonates by Baker's yeast. *Heteroat. Chem*, **12** (6), 551–556.

102 Wang, K., Li, J.-F., Yuan, C.Y., and Li, Z.Y. (2002) Regio- and stereoselective bioreduction of diketo-n-butylphosphonate by Baker"s yeast. *Chin. J. Chem.*, **20** (11), 1379–1387.

103 Zymanczyk-Duda, E., Brzezinska-Rodak, M., and Lejczak, B. (2000) Reductive biotransformation of diethyl β-, γ- and δ-oxoalkylphosphonates by cells of baker's yeast. *Enzyme Microb. Technol.*, **26** (2-4), 265–270.

104 Ziora, Z., Maly, A., Lejczak, B., Kafarski, P., Holband, J., and Wojcik, G. (2000) Reactions of N-phthalylamino acid chlorides with trialkyl phosphites. *Heteroat. Chem.*, **11**, 232–239.

105 Zymanczyk-Duda, E., Brzezinska-Rodak, M., and Lejczak, B. (2004) Stereochemical control of asymmetric hydrogen transfer employing five different kinds of fungi in anhydrous hexane. *Enzyme Microb. Technol.*, **34** (6), 578–582.

106 Żymańczyk-Duda, E. and Klimek-Ochab, M. (2012) Stereoselective biotransformations as an effective tool for the synthesis of chiral compounds with P-C bond – Scope and limitations of the methods. *Curr. Org. Chem.*, **16**, 1408.

107 Zymanczyk-Duda, E., Klimek-Ochab, M., Kafarski, P., and Lejczak, B. (2005) Stereochemical control of biocatalytic asymmetric reduction of diethyl 2-oxopropylphosphonate employing yeasts. *J. Organomet. Chem.*, **690** (10), 2593–2596.

108 Majewska, P., Kafarski, P., Lejczak, B., Bryndal, I., and Lis, T. (2006) An approach to the synthesis and assignment of the absolute configuration of all enantiomers of ethyl hydroxy(phenyl) methane(*P*-phenyl)phosphinate. *Tetrahedron: Asymmetry*, **17** (18), 2697–2701.

109 Wang, K., Li, Z., and Yuan, C. (2002) Enantioselective reduction of 2-ketoalkanephosphonate by Baker's yeast. *Phosphorus, Sulfur Silicon Relat. Elem.*, **177** (6), 1797–1800.

110 Attolini, M., Bouguir, F., Iacazio, G., Peiffer, G., and Maffei, M. (2001) Enantioselective synthesis of cyclic dialkyl (3-hydroxy-1-alkenyl) phosphonates by Baker's yeast-mediated reduction of the corresponding enones. *Tetrahedron*, **57** (3), 537–543.

111 Górak, M. and Żymańczyk-Duda, E. (2015) Application of cyanobacteria for chiral phosphonate synthesis. *Green Chem.*, **17**, 4570–4578.

112 Pedragosa-Moreau, S., Archelas, A., and Furstoss, R. (1996) Microbiological transformation 32: use of epoxide hydrolase mediated biohydrolysis as a way to enantiopure epoxides and vicinal diols: application to substituted styrene oxide derivatives. *Tetrahedron*, **52** (13), 4593–4606.

113 Majewska, P. (2015) Biotransformations of 2-hydroxy-2-(ethoxyphenylphosphinyl)acetic acid and the determination of the absolute configuration of all isomers. *Bioorg. Chem.*, **61** (1), 28–35.

114 Klimek-Ochab, M., Zymanczyk-Duda, E., Brzezinska-Rodak, M., Majewska, P., and Lejczak, B. (2008) Effective fungal catalyzed synthesis of *P*-chiral organophosphorus compounds. *Tetrahedron: Asymmetry*, **19** (4), 450–453.

115 Brzezińska-Rodak, M., Klimek-Ochab, M., Żymańczyk-Duda, E., and Kafarski, P. (2011) Biocatalytic resolution of enantiomeric mixtures of 1-aminoethanephosphonic acid. *Molecules*, **16**, 5896–5904.

116 Żymańczyk-Duda, E., Brzezińska-Rodak, M., Kozyra, K., and Klimek-Ochab, M. (2015) Fungal platform for direct chiral phosphonic building blocks production. Closer look on conversion pathway. *Appl. Biochem. Biotechnol.*, **175**, 1403–1411.

117 Skwarczynski, M., Lejczak, B., and Kafarski, P. (1999) Enantioselective hydrolysis of 1-butyryloxyalkylphosphonates by lipolytic microorganisms: *pseudomonas fluorescens* and *Penicillium citrinum*. *Chirality*, **11** (2), 109–114.

118 Then, J., Aretz, W., and Sauber, K. (1996) Process for the preparation of L-phosphinothricin using transaminases of different specificities in a linked process. US Patent 5587319, https://www.google.ch/patents/US5587319 (accessed 24 May 2016).

119 Zimmermann, G., Maier, J., and Gloger, M. (1989) Process for the preparation of stereoisomers of 1-aminoalkylphosphonic and phosphinic acids.

US Patent 4859602, http://www.google.ch/patents/US4859602 (accessed 24 May 2016).

120 Lejczak, B., Haliniarz, E., Skwarczynski, M., and Kafarski, P. (1996) The use of lypolitic microorganisms *Pseudomonas fluorescens* and *Penicillium citrinum* for the preparation of optically active 1-hydroxyalkylphosphonates. *Phosphorus, Sulfur Silicon Relat. Elem.*, **111** (1-4), 86–87.

Index

Asymmetric Synthesis in Organophosphorus Chemistry: Synthetic Methods, Catalysis and Applications, First Edition. Oleg I. Kolodiazhnyi.

b

i

j

k

l

m

q

r